RICHARD J. LARSEN · MORRIS L. MARX

Vanderbilt University   University of Oklahoma

An Introduction to Mathematical Statistics and its Applications

Prentice-Hall, Inc., Englewood Cliffs, New Jersey   07632

Library of Congress Cataloging in Publication Data

LARSEN, RICHARD J
   An introduction to mathematical statistics and
its applications.

   Bibliography: p.
   Includes index.
   1. Mathematical statistics.   I. Marx, Morris L.,
joint author.   II. Title.
QA276.L314   519.5   80-14807
ISBN   0-13-487744-6

To Our Parents

Printed in the United States of America

10   9   8   7   6   5   4   3   2

Editorial/production supervision by Linda Mihatov and Kathleen M. Lafferty
Interior design by Jayne Conte and Linda Mihatov
Cover design by Jayne Conte
Manufacturing buyers: Edmund W. Leone and John Hall

PRENTICE-HALL INTERNATIONAL, INC., *London*
PRENTICE-HALL OF AUSTRALIA PTY. LIMITED, *Sydney*
PRENTICE-HALL OF CANADA, LTD., *Toronto*
PRENTICE-HALL OF INDIA PRIVATE LIMITED, *New Delhi*
PRENTICE-HALL OF JAPAN, INC., *Tokyo*
PRENTICE-HALL OF SOUTHEAST ASIA PTE. LTD., *Singapore*
WHITEHALL BOOKS LIMITED, *Wellington, New Zealand*

CONTENTS

CHAPTER

THREE
Random Variables

CHAPTER

FOUR
Special Distributions

CHAPTER

FIVE
Estimation

SIX
Hypothesis Testing

SEVEN
The Normal Distribution

EIGHT
Two-Sample Problems

CHAPTER

NINE
Goodness-of-Fit Tests

CHAPTER

TEN
Regression

CHAPTER

ELEVEN
The Analysis of Variance

CHAPTER

TWELVE
Randomized Block Designs

CHAPTER

THIRTEEN
Nonparametric Statistics

Appendix A1

Answers to Selected Questions and Review Exercises A28

Bibliography A44

Index A57

This book gives an introduction to the mathematical theory of statistics and to the application of that theory to the "real world." The recommended mathematical prerequisite is three semesters of calculus, although a strong two-semester course would suffice for much of what is covered. The intended audiences are sophomore, junior, and senior math majors, as well as upper level or graduate students in related disciplines. By a judicious choice of topics, the book can be adapted to courses lasting one semester, two quarters, or two semesters.

In writing this text, it was our intention to select and organize material in such a way that the book could be used successfully by two basically different groups: those wanting a terminal course and those seeking a background of the subject as preparation for a more advanced treatment later on. For both groups we felt justified in sacrificing certain specifics for the sake of a well-motivated overview. To this end we emphasize throughout the book the interrelationship between probability theory, mathematical statistics, and data analysis. We believe that for all three to be represented is vitally important, particularly so for those students who take only one statistics course during their college careers.

The goal of integrating data analysis into an introduction to mathematical statistics—an objective that may seem somewhat unreasonable at first glance—has, in fact, not forced us to depart radically from the standard treatments of the latter. For the most part, the chief differences are in emphasis and placement rather than in the omission of familiar topics or the incorporation of new ones. As an example, the introduction of probability models as being potentially useful abstractions of empirical observations comes very early in our presentation, as do the basic notions of estimation and significance testing. (All three of these have already been described in the context of several different case studies by the time the reader finishes page 12.) By bringing these critically important ideas in early, even in a very informal way, we find it much easier to motivate them mathematically when they reappear later in more formal settings.

We have interspersed throughout the book a fair amount of historical information, our thought being that much of this material can be unified and, indeed, illuminated, by putting it in this perspective. This approach is particularly apparent in Chapter 4, which profiles a number of probability models that are especially important, application-wise, and in Chapter 10, where the principles of correlation and regression are taken up.

With the exception of Chapter 4, our treatment of basic probability and random variable theory is fairly standard. In beginning the study of statistics proper, though, we do make a few changes. First, no examples involving the normal distribution are used in Chapter 5, which deals with estimation. We elected instead to collect all the usual proofs that $\bar{X}$ is efficient, sufficient, and so on, and put them in Chapter 7, which is devoted solely to the normal distribution and its central role in the one-sample problem. That way the beginning of Chapter 7 serves as a welcome review of Chapter 5 and, more importantly, the all-important normal theory results are highlighted and do not get lost in a maze of Poisson, binomial, and gamma examples. Part of Chapter 6, dealing with hypothesis testing, also represents a slight departure from "standard" presentations. Specifically, we emphasize the generalized likelihood ratio criterion and relegate the Neyman-Pearson lemma to an Appendix. Our justification for this comes from our classroom experience: students at this level have a terribly difficult time understanding the mechanics of the Neyman-Pearson lemma, particularly when it is used to extend a best test to one that is uniformly most powerful. Actually, as a "tool" the generalized likelihood ratio criterion is more important anyway—very few real-life hypothesis testing procedures appeal to the Neyman-Pearson lemma for their derivation.

The rest of the book is perhaps a little more conventional in outlook. In Chapter 7 and Chapter 8 are assembled many of the well-known results concerning the normal distribution and the various one-sample and two-sample inference procedures that are associated with it. Chapter 9 discusses goodness-of-fit tests and tests for independence. Chapter 10, as already noted, describes the regression and correlation models. Next comes a development of the analysis of variance—first, for the completely randomly design (Chapter 11) and then for the randomized block design (Chapter 12). In these latter two chapters considerable attention is given to testing subhypotheses and to the multiple comparison problem. We have tried hard to keep students from thinking that a single overall F test would ever be an adequately thorough statistical analysis for a complicated experiment. Finally, in Chapter 13 we take a look at some of the more common nonparametric techniques.

In courses lasting less than two semesters, not everything in the text can be covered. Guidelines for deciding what should be kept and what should be omitted are difficult to pontificate because of their obvious dependence on the backgrounds of the students and the goals of the course. Perhaps our experience at Vanderbilt, though, where the book is used for a one-semester course taught primarily to sopho-more and junior math majors, will provide some helpful insights. We read Chapter 1, do essentially all of Chapter 2, and cover most of Chapter 3, leaving out the material on bivariate distributions. Then we work through all of Chapter 4 and most of Chapter 5, the only omission from the latter being the material on suffi-ciency. We do the first three sections of Chapter 6 and, if time permits, make a quick pass at the formulation of a generalized likelihood ratio test. By this time the semester has only a few weeks remaining and we spend that period skimming over the results of Chapter 7, Chapter 8, and sometimes Chapter 9. At this point our course takes a more applied bent and we focus on methodological problems—constructing confidence intervals, doing t tests, analyzing binomial data, and so on.

We have tried hard to make this book as "teachable" as possible. Motivation, in particular, has been given high priority. Chapter introductions describe at some length the nature of the material to come, how it relates to what has already been covered, and to what use it will ultimately be put. An extensive set of homework exercises is provided, ranging in difficulty from routine numerical applications to theoretical results not taken up in the text. Solutions or hints to more than five hundred of these exercises can be found in the Answer to Selected Exercises. Also, we have included a very detailed and extensively cross-referenced index. The binomial distribution, for example, appears as a general heading and is then divided into some twenty sublistings. Of most help to the teacher, though, in presenting this material will be the examples. The Bibliography lists over two hundred references, most of which are the original sources for the many case studies and examples used throughout the book. With these as a backdrop, it is easy for the student to see the relevance and applicability of statistics.

We would like to acknowledge the following for use of the chapter opening photographs and epigraphs: Chapter 1 from *Natural Inheritance* by Francis Galton (1908); Chapter 2 courtesy of the History of Science Collections, University of Oklahoma; Chapters 3, 4, and 6 from *Studies in the History of Statistical Method* by Helen M. Walker (Williams & Wilkins, Baltimore, 1929); Chapters 5, 11, and 12 from *R. A. Fisher: The Life of a Scientist* by Joan Fisher Box (John Wiley & Sons, New York, 1978); Chapter 8 from the Annals of Eugenics, Vol. 9, 1939; Chapter 9 from *Karl Pearson: An Appreciation of Some Aspects of His Life and Work* by Egon Sharpe Pearson (Cambridge University Press, Cambridge, 1938), Chapter 10 from *The Life, Letters and Labours of Francis Galton, Vol. IIIA* (Cambridge University Press, Cambridge, 1930); and Chapter 13 from "Effect of Non-normality on the Power Function of the Sign Test" by Jean D. Gibbons (*Journal of the American Statistical Association, Vol 59*, 1964). We wish to thank Marcia Goodman of the History of Science Collections, University of Oklahoma Libraries for her assistance in obtaining the photographs.

We would like to express our indebtedness to the editors at J. R. Geigy, Biometrika, and McGraw-Hill for letting us use the tables that appear in the Appendix and to all the many researchers whose data we have used for examples. We also wish to thank the numerous persons and organizations for permission to reprint tables and figures. Acknowledgements for these permissions are incorporated in the bibliography. For their valuable detailed reviews, criticisms, and suggestions we thank the reviewers: Jay Devore, California Polytechnic State University; Fred Fischer, State University of New York at Oswego; Michael Fligner, Ohio State University; Marian MacDonald; Jerry E. Mann, Virginia Polytechnic Institute and State University; Elizabeth Papousek, Fisk University; and James Stapleton, Michigan State University. Finally, we would like to thank Kathleen Lafferty and Linda Mihatov for their invaluable assistance in editing the manuscript. Their counsel on matters of form and content was sincerely appreciated.

RJL
Nashville, Tennessee

MLM
Norman, Oklahoma

CHAPTER ONE
Introduction

FRANCIS GALTON

Some people hate the very name of statistics, but I find them full of beauty and interest. Whenever they are not brutalized, but delicately handled by the higher methods, and are warily interpreted, their power of dealing with complicated phenomena is extraordinary. They are the only tools by which an opening can be cut through the formidable thicket of difficulties that bars the path of those who pursue the Science of man.

1.1 A BRIEF HISTORY

The astragalus, a cube-shaped bone in the heel of dogs, sheep, and horses, has been found far more often than other types of bones in a number of archeological excavations. Hypotheses of ancient beasts with many legs, or of wild pagan foot feasts, come quickly to mind, but the correct, albeit less dramatic, explanation can be read off Egyptian tomb inscriptions: the bones were used as primitive dice—for gambling! While this implies that some knowledge of probability, however rudimentary, was every bit as old as human socialization itself, it is nevertheless true that the formal mathematization of chance events—what this book is all about—is a much more recent development. We will begin our story in seventeenth-century France.

There are many early writings on the art of gambling, some of them even mathematical, but the date when the so-called calculus of probability was first applied to games of chance is 1654. In that year began a correspondence between the renowned mathematicians Blaise Pascal and Pierre de Fermat. The French nobleman, Antoine Gombaud, Chevalier de Mere, brought to Pascal's attention two old problems, one concerning odds in dice games and the other relating to the equitable division of stakes in case a game is interrupted before completion and one competitor holds an advantage. Pascal was intrigued by the problems and communicated his solutions to Fermat, thus prompting an exchange of letters that some consider to be the genesis of mathematical probability.

A year later a young Dutchman, Christiaan Huygens (best known for his work in optics), visited Paris and was told of the work of Pascal and Fermat. Not able to meet with Pascal, who had withdrawn to religious and philosophical contemplation, Huygens elected to pursue the subject on his own. Two years later, in 1657, his efforts produced the first published treatise on probability, *De Ratiociniis in Aleae Ludo* (*Calculations in Games of Chance*).

The newly conceived science began to receive serious consideration in the years after 1657. The birth process can be thought of as completed with the posthumous publication of Jacob Bernoulli's *Ars Conjectandi* (*The Art of Conjecture*) in 1713. Among the contents of *Ars Conjectandi* was the first published theorem on the limiting behavior of probability in repetitions of a simple chance experiment. Variations on this idea are still topics of research today.

Let us leave probability for a moment and turn to the other subject of our inquiry, *statistics*, a word denoting collections of data as well as the methodology for drawing inferences from data. To be sure, lists of data are as old as the art of writing, but the first comprehensive attack on the numerical interpretation of biological and social phenomena is due to John Graunt. In 1662 Graunt published *Natural and Political Observations Made upon the Bills of Mortality*, a book based on birth and death records gathered in plague-stricken England. Shortly thereafter, Sir William Petty began developing what became known as Political Arithmetic, "the art of reasoning by figures upon things relating to government." Finally, by

1700, the work of Edmund Halley (of Halley's Comet fame) on mortality tables and life expectancy had established statistical inference as a science both viable and eminently worthwhile.

By the time of the publication of *Ars Conjectandi*, the disciplines of probability and statistics were both well recognized but for the most part separate. Over the next century and a half, though, they were amalgamated into a single theory of mathematical statistics. A number of the important contributions to that union are discussed in later chapters, so we will limit ourselves here to just a few brief comments.

The scholarship of the eighteenth century in statistics and probability culminated in the work of Pierre-Simon Laplace. In 1812 his influential book, *Theorie Analytique des Probabilities*, presented all his previous contributions, unified the basic research in the field, and discussed a number of applications of probability, particularly to the theory of observational errors. By the end of the first quarter of the nineteenth century it was well established that the theory of probability could be profitably applied to empirical data in the physical sciences.

About this same time, Adolphe Quetelet, a most remarkable Belgian academician, began to demonstrate the use of probability models in describing social and biological phenomena. Quetelet had studied briefly with Laplace, and the latter's influence on the Belgian was unmistakable. Quetelet traveled widely throughout Europe in the ensuing years, spreading with fervor the statistical "gospel."

By the 1870s "modern" mathematical statistics was poised and ready for its debut. Among its leaders at this point was Sir Francis Galton, the great British scientist. Galton's story and the rapid evolution of the subject near the turn of the century will be told later. We look now at four examples that raise some of the sorts of questions we will eventually be considering.

1.2 SOME EXAMPLES

Do stock and commodity markets rise and fall randomly? Is there a common element in the aesthetic standards of the ancient Greeks and the Shoshoni Indians? Can external forces, such as phases of the moon, affect admissions to mental hospitals? What kind of relationship exists between exposure to radiation and cancer mortality?

Contentwise, these questions are quite diverse, but they have some important traits in common. They are all, for example, difficult or impossible to study in a laboratory, and none of them admits any self-evident axioms from which one may reason deductively. Indeed, such questions are usually attacked by collecting data, making guesses about the processes generating those data, and then testing those guesses. As it turns out, this approach leads to still another similarity—the element of chance or uncertainty that affects each data point recorded.

The goal of this text is to develop the mathematical tools and concepts necessary for incorporating the chance element into the methodology for describing or making predictions about real-world phenomena. The examples that follow are cases in point.

CASE STUDY

1.1

Each evening the radio and TV reporters offer an often bewildering array of averages, indices, and the like that presumably indicate the state of the stock market. But do they? Are these numbers conveying any really useful information? Some financial analysts would say "No," arguing that speculative markets tend to rise and fall randomly, much as though some hidden roulette wheel were spinning out the figures. In this example we attempt to examine quantitatively the reasonableness of this "random-movement" hypothesis.

We begin by formulating a theoretical construct, or *model*, that should describe the behavior of the market, *if the* (*random*) *hypothesis were true*. To this end, we translate the term "random movement" into two assumptions:

(a) The chances of the market's rising or falling on a given day are unaffected by its actions on any previous days.
(b) The market is equally likely to go up or down.

Measuring the day-to-day randomness, or lack of randomness, in the market's movements can be accomplished by looking at the lengths of *runs*. We define a *run of length n* (actually, these are runs *up* of length *n*) to be *n* consecutive days of market rises, with falls occurring on the day immediately before the first rise and immediately after the *n*th rise. If the actual behavior of the market's run lengths is markedly different than what assumptions (a) and (b) would predict, we can safely reject the random-movement hypothesis as being unrealistic. In this case, what our model predicts in terms of run length is not too difficult to figure out, even without the benefit of any formal training in probability.

Suppose a fall has occurred, followed by a rise. For a run of length one, the market must next fall. By the two assumptions, this happens half the time, so the probability of a run of length one is $\frac{1}{2}$; our notation for this will be $P(1) = \frac{1}{2}$. The other half the time the market rises, giving the sequence (fall, rise, rise). A run of length *two* occurs if there is now a fall. Again, this happens half the time, making its probability half the half giving the (fall, rise, rise) sequence. Thus, the probability of a run of length two is $\frac{1}{2} \cdot \frac{1}{2} = \frac{1}{4} = P(2)$, and the overall pattern should be clear: the probability of a run of length n is $P(n) = (\frac{1}{2})^n$. Furthermore, if a total of T runs are observed, it seems reasonable to expect $T \cdot (\frac{1}{2})^n$ of them to be of length n.

Table 1.1 gives the distribution of runs observed in weekly cash prices at the Chicago Wheat Market between 1883 and 1934 (1). The third column of the table gives the corresponding expected numbers, as calculated from the expression $T \cdot (\frac{1}{2})^n$.

TABLE 1.1   Runs in the cash prices
at the Chicago Wheat Market

Run Length, n	Observed	Expected $[= 586 \cdot P(n)]$
1	280	293.00
2	147	146.50
3	86	73.25
4	38	36.62
5	15	18.31
6	13	9.16
7+	7	9.16
	586	586.00

Here the agreement between the actual and the predicted run frequencies seems good enough to lend some credence to assumptions (a) and (b). Before jumping to any rash conclusions, though, we should apply the same two assumptions to a different set of data to see whether a similar level of agreement can be achieved a second time. This is done in Table 1.2, where the first three columns list the observed and expected run lengths in Standard and Poor's Composite Stock Price Index during the period from January 1918 to March 1956.

TABLE 1.2   Runs in Standard and Poor's Composite Stock Price Index

Run Length, n	Observed	Expected ($p = \frac{1}{2}$)	Expected ($p = 0.617$)
1	31	46.00	35.27
2	21	23.00	21.75
3	16	11.50	13.41
4	10	5.75	8.27
5	5	2.87	5.10
6+	9	2.88	8.20
	92	92.00	92.00

This time there is substantial disagreement between columns 2 and 3, a turn of events reflecting unfavorably on the model. We now have two choices: abandon the model or modify it so it fits the data better. One way to do the latter would be to allow for the possibility that the market's rising or falling on any given day might not be a 50--50 proposition. This suggests that assumption (b) should be replaced with

    (b′) The likelihood of a rise in the market is some number p, where $0 \leq p \leq 1$.

The quantity p is known as a *parameter* of the model. Of course, (b) is a special case of (b′), where $p = \frac{1}{2}$.

Invoking the new assumptions, we can recalculate the probabilities of the various run lengths. For example, following a (fall, rise) sequence, a fall would be expected $100(1 - p)\%$ of the time, so $P(1) = 1 - p$. A fraction p of the time there is another rise. Of this fraction, there is a likelihood $1 - p$ of a fall. Hence, the probability of the sequence (fall, rise, rise, fall)—that is, a run of length 2—is $P(2) = p(1 - p)$. In general, $P(n) = p^{n-1}(1 - p)$.

Two questions now arise. Which of the two would be more important for further study depends on the needs and interests of the model maker.

    1. Is the initial assumption that $p = \frac{1}{2}$ justified?
    2. Given the observed data, what is the best choice (or *estimate*) for p?

To answer question 1 we need to decide whether the discrepancies between the observed and expected run lengths are small enough to be attributed to chance or large enough to render the model invalid. One way to answer question 2 is to seek out that value of p that best "explains" the observations, in terms of maximizing their probability. In the Standard and Poor example, this type of estimate turns out to be $p = 0.617$. The corresponding expected values, based on $P(n) = p^{n-1}(1 - p) = (0.383)(0.617)^{n-1}$, are given in column 4 of Table 1.2. Note the dramatic improvement in the agreement between the observed values and the new expected values. Based on what we have seen thus far, we would have to conclude that neither set of data offers any serious refutation of the hypothesis that speculative markets rise and fall randomly.

Comment. The two questions raised at the conclusion of Case Study 1.1 touch on the dual themes of statistical inference: hypothesis testing and estimation. While it is necessary to be vague now about the mathematical techniques these two will require, it should be clear from the example that the notions of probability and expected value will play prominent roles. ∎

Among the useful criteria for classifying experiments into statistical "types" is the nature of the set of possible outcomes. Case Study 1.1, for example, described an experiment with a *finite* number of outcomes. A simple extension of that idea leads to experiments with outcomes that are *countably infinite*. Such an outcome set would have been appropriate had we idealized the stock-market model to allow for runs of *any* length. In either case, though, probabilities are calculated as *sums*. The probability that a run will be of either length 1 or length 2, for instance, is

$$P(1) + P(2) = p^0(1 - p)^1 + p^1(1 - p)^1 = 1 - p^2.$$

There are many circumstances, however, where the possible outcomes are so numerous and so close together that it becomes convenient to define the outcome set as an interval of real numbers (that is, as an *uncountably infinite* set). Probabilities are then represented as integrals, the continuous analogs of sums. The next case study is based on data that would be considered "continuous."

CASE STUDY

1.2

Not all rectangles are created equal. Since antiquity, societies have expressed aesthetic preferences for rectangles having certain width (w) to length (l) ratios. Plato, for example, wrote that rectangles whose sides were in a $1 : \sqrt{3}$ ratio were especially pleasing. (These are the rectangles formed from the two halves of an equilateral triangle.)

Another "standard" calls for the width-to-length ratio to be equal to the ratio of the length to the sum of the width and the length. That is,

$$\frac{w}{l} = \frac{l}{w + l} \qquad (1.1)$$

Equation 1.1 implies that the width is $\frac{1}{2}(\sqrt{5} - 1)$, or approximately 0.618, times as long as the length. The Greeks called this the golden rectangle and used it often in their architecture (see Figure 1.1). Many other cultures were similarly inclined. The Egyptians, for example, built their pyramids out of stones whose faces were golden rectangles. Today, in our society, the golden rectangle remains an architectural and artistic standard, and even items such as drivers' licenses, business cards, and picture frames often have w/l ratios close to 0.618.

The study described here is an example from a field known as experimental aesthetics. The data are width-to-length ratios of beaded rectangles used by the Shoshoni Indians to decorate their leather goods. The question at issue is whether the golden rectangle can be considered an aesthetic standard for the Shoshonis, just as it was for the Greeks and the Egyptians.

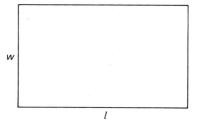

Figure 1.1   A golden rectangle: $w/l \doteq 0.618$.

It must be realized, of course, that even if the 0.618 ratio is adopted by a society as an aesthetic standard, there might still be considerable variability in the width-to-length ratios of individual rectangles. What should be close to 0.618, though, is the *average w/l* ratio. This suggests that the hypothesis (H) to be tested might be written

> **H:** *The "true" average width-to-length ratio characteristic of Shoshoni artistic work is 0.618.*

Table 1.3 presents the *w/l* ratios for 20 rectangles found on Shoshoni handicraft (39).

TABLE 1.3   Width-to-length ratios of Shoshoni rectangles

0.693	0.749	0.654	0.670
0.662	0.672	0.615	0.606
0.690	0.628	0.668	0.611
0.606	0.609	0.601	0.553
0.570	0.844	0.576	0.933

The average of these observed ratios is 0.661. The decision to be made, then, is whether H is true and the difference between 0.618 and 0.661 is solely the result of chance fluctuations, or whether H is false and the two societies do have different aesthetic standards. Ultimately, the choice will depend on how likely it is to obtain a *sample* average of 0.661 if, in fact, the *true* average is 0.618. Computing that likelihood, however, is not a simple task, so our final decision will have to be deferred. We return to the problem in Chapter 7.

Case Study 1.2 asked whether a given set of observations (coming from a single population) conformed to a theoretically determined standard ($w/l = 0.618$). Another common type of question asks if *several* populations differ among themselves in some particular aspect. For the latter situation, the data will consist of several sets of observations, one set from each of the populations being compared. The next case study is a typical "k-sample" problem.

CASE STUDY

1.3

In folklore, the full moon is often portrayed as something sinister, a kind of evil force possessing the power to control our behavior. Over the centuries, many prominent writers and philosophers have shared this belief (121). Milton, in *Paradise Lost*, refers to

> Demoniac frenzy, moping melancholy
> And moon-struck madness.

And Othello, after the murder of Desdemona, laments

> It is the very error of the moon,
> She comes more near the earth than she was wont
> And makes men mad.

On a more scholarly level, Sir William Blackstone, the renowned eighteenth-century English barrister, defined a "lunatic" as

> one who hath . . . lost the use of his reason and who hath lucid intervals, sometimes enjoying his senses and sometimes not, and that frequently depending upon changes of the moon.

The possibility of lunar phases' influencing human affairs is a theory not without supporters among the scientific community. Studies by reputable medical researchers have attempted to link the "Transylvania effect," as it has come to be known, with higher suicide rates, pyromania, and even epilepsy. In this example, we look at still another context in which this phenomenon might be expected to occur. Table 1.4 shows the admission rates to the emergency room of a Virginia mental health clinic *before*, *during*, and *after* the 12 full moons from August 1971 to July 1972 (12). Notice that for this particular set of data, the average admission rate *is* higher during the full moon than during the rest of the month.

TABLE 1.4   Admission rates (patients/day)

Month	Before Full Moon	During Full Moon	After Full Moon
Aug	6.4	5.0	5.8
Sep	7.1	13.0	9.2
Oct	6.5	14.0	7.9
Nov	8.6	12.0	7.7
Dec	8.1	6.0	11.0
Jan	10.4	9.0	12.9
Feb	11.5	13.0	13.5
Mar	13.8	16.0	13.1
Apr	15.4	25.0	15.8
May	15.7	13.0	13.3
Jun	11.7	14.0	12.8
Jul	15.8	20.0	14.5
Averages	10.9	13.3	11.4

For reasons that we will discuss in Chapter 6, hypothesis tests are always set up so that what is being tested is the absence or negation of any differences from population to population. Following that principle here leads us to state the hypothesis as

H: On the average, there is no difference in mental-hospital admission rates before, during, and after the full moon.

The decision to reject H will be made only if it can be demonstrated that averages as different as 10.9, 13.3, and 11.4 are extremely unlikely to arise by chance alone. An analysis done in Chapter 12 suggests that these averages *are* considerably different, implying that the data could be used as evidence in support of the existence of a Transylvania effect.

Many scientific laws are, from a mathematical viewpoint, discoveries of functional relationships between variable quantities. A familiar example is the formula $s(t) = 16t^2$, where $s(t)$ is the distance in feet an object initially at rest falls in t seconds (neglecting air resistance). The final case study in this section applies an empirical curve-fitting procedure to the problem of estimating the relationship between radiation exposure and cancer mortality rates.

CASE STUDY

1.4

The oil embargo of 1973 raised some very serious questions about energy policies in the United States. One of the most controversial is whether nuclear reactors should assume a more central role in the production of electric power. Those in favor point to their efficiency and to the availability of nuclear material; those against warn of nuclear "incidents" and emphasize the health hazards posed by low-level radiation.

Since nuclear power is relatively new, there is not an abundance of past experience to draw on. One notable exception, though, is a government reactor that has been in continuous operation for 30 years. What happened there is what environmentalists fear will be a recurrent problem if nuclear reactors are proliferated.

Since World War II, plutonium for use in atomic weapons has been produced at an AEC facility in Hanford, Washington. One of the major safety problems encountered there has been the storage of radioactive wastes. Over the years, significant quantities of these substances—including strontium 90 and cesium 137—have leaked from their open-pit storage areas into the nearby Columbia River, which flows through parts of Oregon and eventually empties into the Pacific Ocean.

To measure the health consequences of this contamination, an index of exposure was calculated for each of the nine Oregon counties having frontage on either the Columbia River or the Pacific Ocean. This particular index was based on several factors, including the county's stream distance from Hanford and the average distance of its population from any water frontage. As a covariate, the cancer mortality rate was determined for each of these same counties.

Table 1.5 shows the index of exposure and the cancer mortality rate (deaths per 100,000) for the nine Oregon counties affected (41). Higher index values represent higher levels of contamination.

A simple graph of the data (see Figure 1.2) suggests that the cancer mortality rate (y) and the index of exposure (x) vary *linearly*—that is, $y = a + bx$. Finding the numerical values for a and b that position the line so it best fits the data is an important problem in an area

TABLE 1.5   Radioactive contamination
and cancer mortality in Oregon

County	Index of Exposure	Cancer Mortality per 100,000
Umatilla	2.49	147.1
Morrow	2.57	130.1
Gilliam	3.41	129.9
Sherman	1.25	113.5
Wasco	1.62	137.5
Hood River	3.83	162.3
Portland	11.64	207.5
Columbia	6.41	177.9
Clatsop	8.34	210.3

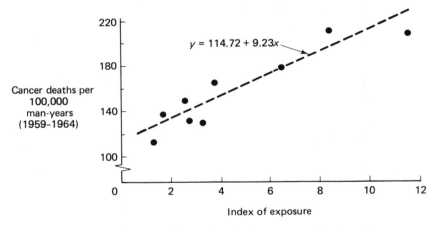

Figure 1.2   Cancer mortality and radiation exposure in nine Oregon counties.

of statistics known as regression analysis. Here, the optimal line, based on methods derived in Chapter 10, has the equation $y = 114.72 + 9.23x$.

1.3 A CHAPTER SUMMARY

The concepts of probability lie at the very heart of all statistical problems, the case studies of Section 1.2 being typical examples. Acknowledging that fact, the next two chapters take a close look at some of those concepts. Chapter 2 states the axioms of probability and investigates their consequences. It also covers the basic skills for algebraically manipulating probabilities and gives an introduction to combinatorics, the mathematics of counting. Chapter 3 reformulates much of the material in Chapter 2 in terms of *random variables*, the latter being a concept of great convenience in applying probability to statistics. Over the years, particular measures of probability have emerged as being especially useful: the most prominent of these are profiled in Chapter 4.

Our study of statistics proper begins with Chapter 5, which is a first look at the theory of parameter estimation. Chapter 6 introduces the notion of hypothesis testing, a procedure that, in one form or another, commands a major share of the remainder of the book. From a conceptual standpoint, these are very important chapters: any application of statistical methodology will involve either parameter estimation or hypothesis testing, or both.

Among the probability functions featured in Chapter 4, the *normal distribution*—more familiarly known as the bell-shaped curve—is sufficiently important to merit even further scrutiny. Chapter 7 derives in some detail many of the properties and applications of the normal distribution as well as those of several related probability functions. Much of the theory that supports the methodology appearing in Chapters 8 through 12 comes from Chapter 7.

Chapters 8, 11, and 12 continue the work of Chapter 7, but the emphasis is now on the comparison of several populations, as typified by Case Study 1.3. In Case Study 1.1 we discussed, in an offhand manner, the extent to which a set of observed values and expected values agreed. Chapter 9 offers a formal approach to that same idea. Linear relationships, such as the radiation exposure index–cancer mortality dependence described in Case Study 1.4, are taken up in Chapter 10.

The final chapter is an introduction to nonparametric statistics. The objective there is to develop procedures for answering some of the same sorts of questions raised in Chapters 7, 8, 11, and 12, but with fewer initial assumptions.

As a general format, each chapter contains numerous examples and case studies, the latter being actual experimental data taken from recently published journals and periodicals. It is hoped that these examples will make it abundantly clear from the very beginning that while the general orientation of this text is theoretical, the consequences of that theory are never too far from having direct applications in the "real world."

PIERRE DE FERMAT

BLAISE PASCAL

CHAPTER TWO
Probability

PIERRE DE FERMAT (1601–1665)

One of the most influential of seventeenth-century mathematicians, Fermat earned his living as a lawyer and administrator in Toulouse. He shares with Descartes credit for the invention of analytic geometry, but his most important work may have been in number theory. Fermat did not write for publication, preferring instead to send letters and papers to friends. His correspondence with Pascal was the starting point for the development of a mathematical theory of probability.

BLAISE PASCAL (1623–1662)

Pascal was the son of a nobleman. A prodigy of sorts, he had already published a treatise on conic sections by the age of 16. He also invented one of the early calculating machines to help his father with accounting work. Pascal's contributions to probability were stimulated by his correspondence, in 1654, with Fermat. Later that year he retired to a life of religious meditation.

2.1 INTRODUCTION

The evolution of probability as both a mathematical discipline and a tool of applied science was commented on briefly in Chapter 1. Among the inevitable consequences of probability's widespread adoption by the scientific community was the demand for a rigorous and orderly treatment of its logical foundations, one unfettered by the constraints imposed by specific subjects. In 1933 the great Russian mathematician, A. Kolmogorov, responded to the call for "axiomatization" and published *Grundbegriffe der Wahrscheinlichkeitsrechnung* (*Foundations of the Theory of Probability*).

Grundbegriffe was a landmark piece of work. By extending the approach taken by Richard von Mises, a German mathematician who had a similar objective in mind, Kolmogorov succeeded in reducing the whole of probability to just four axioms, each one simple in statement and intuitively appealing. Yet, like Hilbert's axioms for Euclidean geometry, they were sufficiently abstract to allow theorems to be proved independently of whatever meaning was given to the underlying concepts.

Central to the Kolmogorov-von Mises development of probability is the notion of an *experiment*—a procedure that (1) can be repeated, theoretically, an infinite number of times and (2) has a well-defined set of possible outcomes. Thus, rolling a pair of dice qualifies as an experiment; so does measuring a hypertensive's blood pressure or determining the lead content of moon rocks.

Corresponding to each of these possible outcomes, or set of outcomes, is a *probability*, a number between 0 and 1 that can be thought of as the long-range relative frequency with which that outcome (or set of outcomes) occurs. For example, if an experiment is defined to be the act of drawing a card from a standard poker deck, a face card (Jack, Queen, or King) should appear, *on the average*, $\frac{3}{13}$ of the time. Thus, we will say that the probability of drawing a face card is $\frac{3}{13}$.

The purpose of Chapter 2 is to examine in some depth the Kolmogorov-von Mises assumptions and their consequences. We will do this by looking separately at each component of their model: first, the set of outcomes associated with the experiment; next, the manner in which probabilities are assigned to outcomes; and, finally, the axioms themselves. The remainder of the chapter builds on these fundamentals to derive the basic rules for combining and manipulating probabilities. Out of all this will come some very useful and very powerful problem-solving techniques.

2.2 THE SAMPLE SPACE

An immediate consequence of the Kolmogorov-von Mises concept of an "experiment" is the notion of a *sample space*. Denoted S, the sample space is the set of all possible outcomes the experiment might lead to. For example, consider the simple experiment of flipping a coin three times. What comes to mind as possible outcomes are the eight "triples" listed below:

$$S = \{HHH, HHT, HTH, THH, HTT, THT, TTH, TTT\}.$$

It should be noted that S defined in this way is only an idealization of what might actually happen: there is no provision, for instance, for a coin standing on its edge or rolling away. However, to omit such trivialities from S greatly simplifies the mathematical model, yet does not diminish its usefulness.

It will often be the case that the focus of a problem is not any single member of S but, rather, a collection of members that share some particular characteristic. Such an aggregation is called an *event* and is necessarily a subset of S. (Each sample outcome, of course, qualifies as an event, as does S itself.) We say that an event *occurs* if the result of the experiment is one of the event's sample outcomes. In the coin-tossing example, if E were defined to be the event "majority of heads," then E would be a subset consisting of four outcomes:

$$E = \{HHH, HHT, HTH, THH\}.$$

The next five examples illustrate some simple applications of these definitions.

EXAMPLE 2.1.   Suppose the throw of two dice, one red and one green, is defined to be an experiment, with the data recorded being the numbers on the upturned faces. This makes the sample space a set of 36 ordered pairs, the first number of each pair being, say, the number showing on the red die. Specifically,

$$S = \{(1, 1), (1, 2), \ldots, (1, 6), (2, 1), \ldots, (2, 6), \ldots, (6, 1), \ldots, (6, 6)\}.$$

If this "experiment" were being conducted in a smoke-filled back room—amid cries such as "baby needs a new pair of shoes"—the event, E, that a "7 is thrown,"

$$E = \{(1, 6), (6, 1), (4, 3), (3, 4), (2, 5), (5, 2)\},$$

would take on special significance. In the game of craps, if E occurs on a player's *first* roll, he wins; if it occurs on any subsequent roll, he loses (see Question 2.2.1).

EXAMPLE 2.2.   Suppose a coin is flipped until the first tail comes up. That first "T" might appear on the first toss, or on the second toss, or on the third toss, and so on. Of course, it might *never* appear. Thus, S contains infinitely many outcomes,

$$S = \{T, HT, HHT, HHHT, HHHHT, \ldots\}.$$

EXAMPLE 2.3.   A local TV station advertises two newscasting positions; both have identical job descriptions, background requirements, and salary schedules. If three women $\{W_1, W_2, W_3\}$ and two men $\{M_1, M_2\}$ apply, the "experiment" of hiring two of them generates a sample space of ten outcomes:

$$S = \{\{W_1, W_2\}, \{W_1, W_3\}, \{W_1, M_1\}, \{W_1, M_2\}, \{W_2, W_3\}, \{W_2, M_1\},$$
$$\{W_2, M_2\}, \{W_3, M_1\}, \{W_3, M_2\}, \{M_1, M_2\}\}.$$

Notice that here, by contrast with Example 2.1, the order within an outcome is irrelevant—hiring W_1 and M_2 is the same as hiring M_2 and W_1. How would S be changed if the two positions being filled were different?

EXAMPLE 2.4.   Some years ago, a study was made of the weekly number of major labor strikes initiated in the United Kingdom (91). The period covered spanned the

12 years from 1948 to 1959. Although the weekly figures, which we will see in Example 2.13, were all quite low, we will find it convenient to let S be the countably infinite set of all nonnegative integers:

$$S = \{0, 1, 2, \ldots\}.$$

EXAMPLE 2.5. In the home, the amount of radiation emitted by a color television set does not pose a health problem of any consequence, but the same may not be true in department and appliance stores, where as many as 15 or 20 sets are turned on at the same time, and in a relatively confined area. Suppose a radiological physicist, at the request of the State Health Department, were to measure the amount of radiation emitted on the sales floors of three appliance stores. Even though the meter reading will necessarily be quantized, the number of possible radiation levels is so large that for all practical purposes the range for each measurement can be thought of as the entire set of nonnegative real numbers:

$$S = \{(X_1, X_2, X_3): 0 \leq X_i < \infty, i = 1, 2, 3\}.$$

The National Council on Radiation Protection has set the safety limit in these situations at 0.5 milliroentgens per hour (87). Thus, the event E that Stores 1 and 2 were in compliance with the NCRP limit, but Store 3 was not, would be denoted

$$E = \{(X_1, X_2, X_3): 0 \leq X_1 \leq 0.5, 0 \leq X_2 \leq 0.5, X_3 > 0.5\}.$$

Comment. Depending on the nature of S, there may be restrictions as to what qualifies as an event. If the sample space is finite or countably infinite (as was the case in Examples 2.1–2.4), there are no restrictions and *any* subset of S can be an event. When S is uncountably infinite—for example, an interval of real numbers—events must be Borel sets [named for the French mathematician Emile Borel (1871–1956)]. In practice, though, this is no restriction at all, because every event we will be interested in will be a Borel set. For a theoretical treatment of these ideas, see (4). As an operating rule, any set over which it is possible to integrate a real-valued function is a legitimate event. ∎

QUESTION 2.2.1   In the game of craps, the person rolling the dice (the *shooter*) wins outright if his first toss is a 7 or an 11. If his first toss is a 2, 3, or 12, he loses outright. If his first roll is something else, say, a 9, that number becomes his "point" and he keeps rolling the dice until he either rolls another 9, in which case he wins, or a 7, in which case he loses. Characterize the sample outcomes contained in the event "Shooter wins with a point of 9."

QUESTION 2.2.2   A woman has her purse snatched by two teenagers. She is subsequently shown a police lineup consisting of six suspects, including the two guilty ones. What is the sample space associated with the experiment "Woman picks two suspects out of lineup"? Which outcomes are in the event, E, that she makes at least one incorrect identification?

QUESTION 2.2.3   Recently a study was undertaken to characterize the sleeping patterns of the nine-banded armadillo (*Dasypus novemcinctus*). One of the variables considered was "hours slept per day." Listed below are the results found for six different armadillos (184).

Armadillo	Hours Slept Per Day
CB	17.4
JE	18.5
CM	20.0
FR	16.1
SS	15.3
TR	17.2

Describe the sample space for this experiment.

It is often necessary to work with events that are made up in various ways of other events. Consider one of the rules given in Question 2.2.1 for the game of craps: the shooter wins on the first toss if either a 7 or an 11 turns up. In the familiar terminology of Boolean algebra, the event "7 or 11" is the *union* of the events "7" and "11." Some of the other notions carrying over from set theory that we will find especially useful are reviewed in the next several definitions and examples.

DEFINITION 2.1   Let A and B be any two events defined over the same sample space S. Then:
1. The *intersection* of A and B, written $A \cap B$, is the event whose outcomes belong to both A and B.
2. The *union* of A and B, written $A \cup B$, is the event whose outcomes belong to either A or B or both.

EXAMPLE 2.6.   Let A be the event that occurs if and only if an ace is drawn from a poker deck:

$A = \{$ace of hearts, ace of diamonds, ace of clubs, ace of spades$\}$.

Let B be the event that occurs if a heart is drawn:

$B = \{2$ of hearts, 3 of hearts, . . . , king of hearts, ace of hearts$\}$.

Then

$A \cap B = \{$ace of hearts$\}$

and

$A \cup B = \{2$ of hearts, . . . , ace of hearts, ace of diamonds, ace of clubs, ace of spades$\}$.

What would S be?

DEFINITION 2.2   Events A and B defined over the same sample space are said to be *mutually exclusive* if they have no outcomes in common—that is, if $A \cap B = \varnothing$, the null set.

EXAMPLE 2.7.   Let A be the set of all possible coin-tossing sequences ending in a tail. Let B be the set of all possible coin-tossing sequences ending in a head. Clearly,

$$A \cap B = \varnothing,$$

since a coin-tossing sequence cannot simultaneously end with both a head and a tail. What does $A \cup B$ equal?

DEFINITION 2.3 Let A be any event defined on a sample space S. The *complement* of A, written A^c, is the event consisting of all the outcomes in S except those that are in A.

EXAMPLE 2.8.   For the two events defined in Example 2.6, A^c is the set of 48 cards, none of which is an ace:

$A^c = \{2$ of hearts, $\ldots$, king of hearts, $\ldots$, 2 of spades, $\ldots$, king of spades$\}$.

B^c is the set whose 39 members are either diamonds, clubs, or spades.

Venn diagrams provide an easy and often very useful method for picturing the sorts of relationships that can exist among events. Figure 2.1 shows Venn diagrams illustrating an intersection, a union, a complement, and two events that are mutually exclusive. In each case, the interior of a set is intended to represent the outcomes it includes.

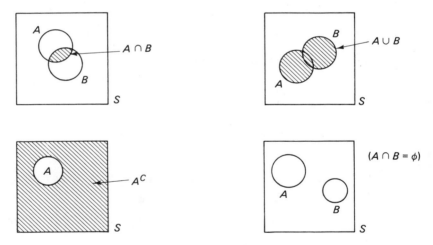

Figure 2.1   Venn diagrams.

Comment. We leave it to the reader to convince himself that if A and B are any two events defined over S, the event that

1. *exactly one* (of the two) occurs is $(A \cap B^c) \cup (B \cap A^c)$;
2. *at most one* (of the two) occurs is $(A \cap B)^c$.

How would the event "At least one occurs" be written?   ∎

QUESTION 2.2.4   Let A and B be two events defined over the same sample space S. Use a Venn diagram to suggest a formula for $A \cup B$.

QUESTION 2.2.5   During Orientation Week, "Star Wars" was shown twice at State University. Among the entering group of 200 freshmen, 85 went to see it the first time, 69 the second time, while 73 failed to see it either time. How many saw it twice?

QUESTION 2.2.6   Find $A \cup B$ and $A \cap B$ if
$$A = \{(x, y): 0 < x < 3, 0 < y < 3\}$$
and
$$B = \{(x, y): 2 < x < 4, 2 < y < 4\}.$$
Find C^c if
$$C = \{(x, y): x^2 + y^2 \leq 1\}.$$

2.3 THE PROBABILITY FUNCTION

Recall from Section 2.1 that the probability of any event E will be thought of as the (long-term) relative frequency with which that event occurs. This suggests that a probability *function* be set up—one that is real-valued and has for its domain the collection of admissible events defined over the sample space S. We will denote such a function P and write $P(E)$ to represent the probability of the event E. (Keep in mind that any *single* outcome of an experiment—say, s—also qualifies as an event, as does the entire sample space itself.)

It would be a simple matter to write down a rather lengthy list of properties that P, being a relative frequency, should possess; Kolmogorov's axioms, however, referred to earlier, make it possible to shrink that list to either three or four, depending on whether S is finite or infinite. Then from these few properties all the others can be derived.

Listed below are the necessary and sufficient axioms for P when the sample space is finite.

AXIOM 1   Let A be any event defined over S. Then $P(A) \geq 0$.
AXIOM 2 $P(S) = 1$.
AXIOM 3   Let A and B be any two mutually exclusive events defined over S. Then
$$P(A \cup B) = P(A) + P(B).$$

When S is infinite, a fourth axiom is needed. It appears in *Grundbegriffe* in the following form:

AXIOM 4   If $A_1 \supset A_2 \supset A_3 \supset \cdots$ and $\bigcap_{n=1}^{\infty} A_n = \varnothing$, then $\lim_{n \to \infty} P(A_n) = 0$.

Comment. Axiom 4 can be replaced by the infinite analog of Axiom 3:

AXIOM 4' Let $A_1, A_2, \ldots$ be events defined over S. If $A_i \cap A_j = \varnothing$ for each $i \neq j$, then

$$P(\bigcup_{i=1}^{\infty} A_i) = \sum_{i=1}^{\infty} P(A_i).$$

Concluding this section are six elementary properties of the probability function. Note that they all agree with how we would expect a relative frequency function to behave. Applications of these results will be deferred to the next two sections.

THEOREM 2.1 $P(A^c) = 1 - P(A)$.

PROOF   By Axiom 2 and Definition 2.3,
$$P(S) = 1 = P(A \cup A^c).$$
But A and A^c are mutually exclusive, so
$$P(A \cup A^c) = P(A) + P(A^c),$$
and the result follows.

THEOREM 2.2 $P(\varnothing) = 0$.

PROOF   Since $\varnothing = S^c, P(\varnothing) = P(S^c) = 1 - P(S) = 0$.

THEOREM 2.3 If $A \subset B$, then $P(A) \leq P(B)$.

PROOF   Note that the event B may be written in the form
$$B = A \cup (B \cap A^c),$$
where A and $(B \cap A^c)$ are mutually exclusive. Therefore,
$$P(B) = P(A) + P(B \cap A^c),$$
which implies that $P(B) \geq P(A)$, since $P(B \cap A^c) \geq 0$.

THEOREM 2.4 For any event $A, P(A) \leq 1$.

PROOF   The proof follows immediately from Theorem 2.3, because $A \subset S$ and $P(S) = 1$.

THEOREM 2.5 Let $A_1, A_2, \ldots, A_n$ be events defined over S. If $A_i \cap A_j = \varnothing$ for $i \neq j$, then

$$P(\bigcup_{i=1}^{n} A_i) = \sum_{i=1}^{n} P(A_i).$$

PROOF   The proof is a straightforward induction argument with Axiom 3 as the starting point.

> **THEOREM 2.6** $P(A \cup B) = P(A) + P(B) - P(A \cap B)$.

PROOF   The Venn diagram for $A \cup B$ certainly suggests that the statement of the theorem is true (recall Question 2.2.4). More formally, though, we have from Axiom 3 that

$$P(A) = P(A \cap B^c) + P(A \cap B)$$

and

$$P(B) = P(B \cap A^c) + P(A \cap B).$$

Adding these two equations gives

$$P(A) + P(B) = [P(A \cap B^c) + P(B \cap A^c) + P(A \cap B)] + P(A \cap B).$$

But by Theorem 2.5, the sum in the brackets is $P(A \cup B)$. Subtract $P(A \cap B)$ from both sides to get the statement of the theorem. (A generalization of this result, giving a formula for the probability of the union of n events, is proved in Appendix 2.1.)

The probability functions that we will deal with are broadly categorized into two groups: (1) those associated with a sample space that is either finite or countably infinite and (2) those associated with a sample space having outcomes that are *not* countably infinite. The first are referred to as *discrete*, the second as *continuous*. In the next two sections we will describe these two types of probability functions and, in the process, show how Theorems 2.1–2.6 are put into practice.

2.4 DISCRETE PROBABILITY FUNCTIONS

Suppose the sample space for a given experiment is either finite or countably infinite. Then any P such that

1. $0 \le P(s)$ for each s in S
2. $\sum\limits_{\text{all } s \in S} P(s) = 1$

is said to be a *discrete probability function*. Of course, it follows that the probability of any event A is the sum of the probabilities associated with A's outcomes:

$$P(A) = \sum\limits_{\text{all } s \in A} P(s).$$

A very important special case of a discrete probability function (when S is finite) is the *equally-likely model*: if S contains n outcomes, this model assigns $P(s) = 1/n$, for all s. Thus, if some event A is composed of m outcomes, $P(A) = m/n$.

EXAMPLE 2.9.   Consider again the situation described in Example 2.1. If the two dice were balanced, the 36 outcomes in S would each have probability $\frac{1}{36}$. This would make the probability of the event "7 is thrown" $\frac{6}{36}$, since there are six ways to roll a 7. Similarly, if the TV station in Example 2.3 did hire in a completely nondiscriminatory fashion—and if the five applicants were all equally qualified—each of the pairs listed in S would have a 1-out-of-10 chance of being selected.

EXAMPLE 2.10.   Suppose a rather limited lottery is set up that contains only 12 tickets, numbered 1 through 12. One ticket is to be drawn. Suppose A is defined to be the event "number on ticket drawn is divisible by 2" and B the event "number on ticket drawn is divisible by 3." What is $P(A \cup B)$? Assuming no one cheats, this is clearly a situation where the equally-likely model is appropriate. By simple enumeration, the ticket numbers that are divisible by 2 or 3 (or both) are 2, 3, 4, 6, 8, 9, 10, and 12. That is,

$$A \cup B = \{2, 3, 4, 6, 8, 9, 10, 12\}$$

in which case $P(A \cup B) = \frac{8}{12}$.

Note that Theorem 2.6 gives the same answer for the probability of the union, but in a more formal way. Specifically (see Figure 2.2),

$$A = \{2, 4, 6, 8, 10, 12\},$$
$$B = \{3, 6, 9, 12\},$$

and

$$A \cap B = \{6, 12\},$$

so that

$$P(A \cup B) = P(A) + P(B) - P(A \cap B)$$
$$= \tfrac{6}{12} + \tfrac{4}{12} - \tfrac{2}{12} = \tfrac{8}{12}.$$

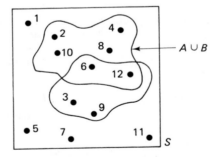

Figure 2.2   Lottery example.

QUESTION 2.4.1
If three fair dice are tossed—one red, one white, and one green—what is the probability that the sum of the faces showing will be less than or equal to 5? Greater than or equal to 5?

We will look at many more examples of the equally-likely model later in this chapter. The remainder of this section, though, is devoted to the "general" discrete model, where $P(s_i)$ is not necessarily equal to $P(s_j)$ for $i \neq j$.

EXAMPLE 2.11.   To encipher, or "code," a message, each letter of the original text is replaced by some other letter, according to a prescribed plan. Thus, if the substitution scheme called for the new letters to be shifted one place to the right of the old, the word "DOG" would become "CNF." To *break* a cipher, it is necessary to know the letter frequency—that is, the probability function—of the language being "coded." For English, this gives rise to a sample space of 26 outcomes, with the probabilities at issue being $P(A)$, $P(B)$, ..., $P(Z)$. It would be foolish to postulate an equally-likely model here because we know from experience that E's, for example, occur much more often in English text than, say, X's. The question of how much more often was investigated by Dewey (36), who made a tally of some 438,023 letters and got the relative frequencies shown in Table 2.1.

TABLE 2.1   Relative frequencies of letters in English text

Letter	Relative Frequency	Letter	Relative Frequency
E	0.1268	F	0.0256
T	0.0978	M	0.0244
A	0.0788	W	0.0214
O	0.0776	Y	0.0202
I	0.0707	G	0.0187
N	0.0706	P	0.0186
S	0.0634	B	0.0156
R	0.0594	V	0.0102
H	0.0573	K	0.0060
L	0.0394	X	0.0016
D	0.0389	J	0.0010
U	0.0280	Q	0.0009
C	0.0268	Z	0.0006

Now we can quantify precisely the relative likelihoods of certain letters' appearing in English text—E's, for example, are almost 80 times as common as X's: $P(E)/P(X)$ = 0.1268/0.0016 = 79.3.

Suppose the event "vowel is chosen" were defined, where Y was considered a vowel. Then

$$P(\text{vowel is chosen}) = P(A) + P(E) + P(I) + P(O) + P(U) + P(Y)$$
$$= 0.0788 + 0.1268 + 0.0707 + 0.0776 + 0.0280 + 0.0202$$
$$= 0.4021.$$

A similar approach could be used to determine the probability that a consonant is chosen—that is, adding up the relative frequencies of the 20 consonants—but Theorem 2.1 suggests a much easier solution. Since the events "vowel is chosen" and "consonant is chosen" are complementary,

$$P(\text{consonant is chosen}) = 1 - P(\text{vowel is chosen})$$
$$= 1 - 0.4021$$
$$= 0.5979.$$

QUESTION 2.4.2   Use Table 2.1 to translate the following quotation:

HVOCZHVODXDVIN VMZ GDFZ AMZIXCHZI;
RCVOZQZM TJP NVT OJ OCZH,
OCZT OMVINGVOZ DIOJ OCZDM JRI GVIBPVBZ
VIY AJMOCRDOC DO DN NJHZOCDIB ZIODMZGT YDAAZMZIO.

BJZOCZ

EXAMPLE 2.12.   Suppose the coin in Example 2.2 is balanced so that heads and tails are equally likely. If $P(k)$ denotes the probability that the first tail appears on the kth toss, it is clear that $P(1) = \frac{1}{2}$. Half the time a coin that came up heads on the first toss will show tails on the second toss, so $P(2) = \frac{1}{4}$. In general, $P(k) = (\frac{1}{2})^k$ (recall the similar derivation of run length in Case Study 1.1). As a partial check that our reasoning is correct, we note that $P(k)$ *does* sum to 1:

$$\sum_{k=1}^{\infty} \left(\frac{1}{2}\right)^k = \sum_{k=0}^{\infty} \left(\frac{1}{2}\right)^k - \left(\frac{1}{2}\right)^0 = \frac{1}{1 - \frac{1}{2}} - 1 = 1.$$

Let E be the event "First tail occurs on an odd-numbered toss." Then

$$P(E) = \sum_{i=0}^{\infty} P(2i + 1) = \sum_{i=0}^{\infty} \left(\frac{1}{2}\right)^{2i+1} = \frac{1}{2} \sum_{i=0}^{\infty} \left(\frac{1}{2}\right)^{2i}$$

$$= \frac{1}{2} \sum_{i=0}^{\infty} \left(\frac{1}{4}\right)^i = \left(\frac{1}{2}\right)\left(\frac{1}{1 - \frac{1}{4}}\right) = \frac{2}{3}.$$

Comment. Because of its similarity to the series of the same name, the P just defined is often referred to as the *geometric* probability function. This important family of probability models will be described more fully in Chapter 4. ∎

QUESTION 2.4.3   A die is loaded in such a way that the probability of any particular face's showing is directly proportional to the number on that face. What is the probability that an even number appears?

EXAMPLE 2.13.   The first three columns of Table 2.2 summarize the strike data referred to in Example 2.4.

TABLE 2.2 Strike data

Weekly Number of New Strikes, k	Frequency	Relative Frequency	$P(k)$
0	252	0.403	0.407
1	229	0.366	0.366
2	109	0.174	0.165
3	28	0.045	0.045
4+	8	0.012	0.017
	626	1.00	

For reasons that will be discussed in Chapter 4, the behavior of k can be described very well by a probability function of the form

$$P(k) = \frac{e^{-0.90}(0.90)^k}{k!}, \qquad k = 0, 1, 2, \ldots. \tag{2.1}$$

The last column of Table 2.2 gives the "predicted" probabilities for k based on Equation 2.1. Notice how good the agreement is. This particular model is a special case of what is known as the *Poisson* probability function.

2.5 CONTINUOUS PROBABILITY FUNCTIONS

We will call S a *continuous* sample space if it contains an interval of real numbers, either bounded or unbounded. Associated with such an S will be a *continuous probability function, f*: if A is any event defined over S, f is a real-valued function having the property that

$$P(A) = \int_A f(x)\,dx.$$

Analogous to the conditions imposed on P, f must satisfy the following:

1. $0 \leq f(x)$ for each x in S;

2. $\int_S f(x)\,dx = 1.$

EXAMPLE 2.14.   Suppose an enemy aircraft flies directly over the Alaskan pipeline and fires a single air-to-surface missile. If the missile hits anywhere within 20 feet of the pipeline, major structural damage will be incurred and oil flow will be disrupted. Suppose that with a certain sighting device, the probability function describing the missile's point of impact is given by

$$f(x) = \begin{cases} \dfrac{60 + x}{3600} & \text{for } -60 < x < 0 \\[2mm] \dfrac{60 - x}{3600} & \text{for } 0 \leq x < 60 \\[2mm] 0 & \text{elsewhere,} \end{cases}$$

where x is the perpendicular distance from the pipeline to the point of impact.

Let A be the event "Flow is disrupted." Then the probability of A is the area under $f(x)$ above the interval $(-20, +20)$:

$$\begin{aligned} P(A) &= \int_{-20}^{20} f(x)\,dx = \int_{-20}^{0} \frac{60 + x}{3600}\,dx + \int_{0}^{20} \frac{60 - x}{3600}\,dx \\[2mm] &= \left[\frac{60x}{3600} + \frac{x^2}{7200}\right]_{-20}^{0} + \left[\frac{60x}{3600} - \frac{x^2}{7200}\right]_{0}^{20} \\[2mm] &= \frac{1200}{3600} - \frac{400}{7200} + \frac{1200}{3600} - \frac{400}{7200} \\[2mm] &= 0.55. \end{aligned}$$

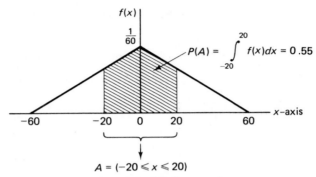

Figure 2.3   Missile example.

Figure 2.3 is a graph of $f(x)$ and shows the area representing $P(A)$.

QUESTION 2.5.1   Verify that the missile impact function is a true probability function.

QUESTION 2.5.2   Suppose that absolutely no damage would result if the missile failed to land within 45 feet of the pipeline. What is the probability of that happening?

EXAMPLE 2.15.   The lifetime of a piece of equipment (or a person, for that matter) is a variable whose sample space is clearly continuous. Specifically, if x denotes the equipment's life, we would write $S = (0 \le x < \infty)$. Many empirical studies have shown that electrical equipment lifetimes often have a probability function of the form

$$f(x) = \begin{cases} (1/\lambda)e^{-x/\lambda} & \text{for } x > 0 \\ 0 & \text{elsewhere.} \end{cases} \qquad (2.2)$$

The factor λ is assigned a numerical value that depends on the life expectancy of the equipment. For example, many 60-watt light bulbs are advertised as having an average life of 1000 hours; we will see in Chapter 5 that this means λ should be set equal to 1000.

Davis (35) applied the exponential model of Equation 2.2 to the lifetimes of V805 transmitter tubes, equipment that was once standard in aircraft radar systems. The results for some 903 tubes are summarized in Table 2.3.

For these particular tubes, the life expectancy is 179 hours, so the probability function being proposed as a model for the length of a V805's life is

$$f(x) = \begin{cases} (\tfrac{1}{179})e^{-x/179} & \text{for } x > 0 \\ 0 & \text{elsewhere.} \end{cases}$$

Suppose the event A were defined to be "a tube fails in 40 hours or less." The corresponding probability would be the integral of $f(x)$ over the interval $(0, 40)$:

$$P(A) = \int_0^{40} (\tfrac{1}{179})e^{-x/179}\, dx$$

$$= \int_0^{40/179} e^{-u}\, du = 1 - e^{-40/179}$$

$$= 0.201.$$

TABLE 2.3 Lifetimes of 903 V805 transmitter tubes

Lifetime (hours)	Number of Tubes	Relative Frequency	Probability from Model
0– 40	166	0.184	0.201
40– 80	151	0.167	0.160
80–120	132	0.146	0.128
120–160	98	0.108	0.103
160–200	73	0.081	0.082
200–240	45	0.050	0.066
240–280	53	0.059	0.052
280–320	40	0.044	0.042
320–360	23	0.025	0.033
360–400	26	0.029	0.027
400–440	24	0.026	0.021
440–480	9	0.010	0.017
480–520	9	0.010	0.014
520–560	8	0.009	0.011
560–600	9	0.010	0.009
600–700	17	0.019	0.015
700–800	8	0.009	0.008
800–up	12	0.013	0.011

That is, the probability function "predicts" that 20% of the tubes will fail in less than 40 hours. What actually happened was that $\frac{166}{903}$, or 18%, failed that soon. The last two columns of Table 2.3 show that the agreement between the empirical and postulated probability functions is quite good.

QUESTION 2.5.3   Compute the *predicted* probability for the "800-up" class.

QUESTION 2.5.4   According to the model, about how many of the 903 tubes should have "survived" to within 50 hours (plus or minus) of the 179-hour average?

The simplest f that can be defined over a continuous (and bounded) S is the *uniform*, or *rectangular*, probability function. Such a function is constant wherever it has nonzero probability:

$$f(x) = \begin{cases} \dfrac{1}{b-a} & \text{for } x \text{ in the interval } (a, b) \\ 0 & \text{elsewhere.} \end{cases}$$

(See Figure 2.4.) Thus, the probability of x's lying in any subinterval in (a, b) of

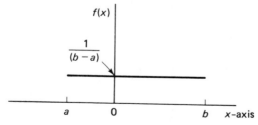

Figure 2.4 The uniform probability function.

length t is the same, regardless of where in (a, b) the interval is located. (The uniform model, of course, is the continuous analog of the equally-likely model described in Section 2.4.) Parzen (124) gives the following example of a uniform probability function that points out a rather curious property of the real numbers.

EXAMPLE 2.16.   Suppose x is a number randomly selected from the interval $(0, 1)$. That is, $f(x) = 1$ for $0 < x < 1$, and zero elsewhere. What is the probability that the second digit in $\sqrt{x}$ is equal to k, $k = 0, 1, 2, \ldots, 9$? We will show that even though x is uniform, the second digit in $\sqrt{x}$ is not.

Let A_k denote the set of numbers in $(0, 1)$ whose square roots have a second digit equal to k. We are looking for $P(A_k)$. Since

$$\sqrt{x} = 0.a_1 a_2 a_3 \ldots$$

it follows that

$$10\sqrt{x} = a_1.a_2 a_3 \ldots$$

and, for some m, $m = 0, 1, \ldots, 9$,

$$m + \frac{k}{10} \leq 10\sqrt{x} \leq m + \frac{k+1}{10}.$$

These inequalities could be written equivalently as

$$\frac{1}{100}\left(m + \frac{k}{10}\right)^2 \leq x \leq \frac{1}{100}\left(m + \frac{k+1}{10}\right)^2$$

from which the *length* of the interval is seen to be $(1/10,000)(20m + 2k + 1)$:

$$\frac{1}{100}\left(m + \frac{k+1}{10}\right)^2 - \frac{1}{100}\left(m + \frac{k}{10}\right)^2 = \frac{1}{10,000}(20m + 2k + 1).$$

However, x was presumed uniform over $(0, 1)$, so the probability of A_k is simply the sum of the above lengths for the ten possible values of m. That is,

$$P(A_k) = \sum_{m=0}^{9} \frac{1}{10,000}(20m + 2k + 1)$$
$$= 0.091 + 0.002k, \qquad\qquad (2.3)$$

which is *not* a uniform probability—the likelihood that the square root of the second digit is k increases as k increases. Table 2.4 lists the A_k probabilities derived in Equation 2.3.

TABLE 2.4   Probability that second
digit in $\sqrt{x}$ is k

k	Probability
0	0.091
1	0.093
2	0.095
3	0.097
4	0.099
5	0.101
6	0.103
7	0.105
8	0.107
9	0.109

2.6 CONDITIONAL PROBABILITY

In this section we will see that the probability of the event A may have to be recomputed if we know for certain that some other event, say, B, has already occurred. That probabilities *should* change in the light of additional information is certainly not unreasonable and is a concept easily demonstrated. Consider a fair die being tossed with A defined as the event "The number 2 appears." Clearly, $P(A) = \frac{1}{6}$. But suppose the die has already been tossed—by someone who refuses to tell us whether or not A occurred but does enlighten us to the point of confirming that the event B, "An even number appears," *did* occur. What are the chances of A now? The answer is obvious: there are three equally likely even numbers making up the event B, one of them satisfies the event A, so the "updated" probability is $\frac{1}{3}$.

Notice that the effect of additional information, such as the knowledge that B has occurred, is to revise—indeed, to *shrink*—the original sample space S to a new sample space S'. Here, the original S contained six outcomes, the conditional sample space, three. (See Figure 2.5.)

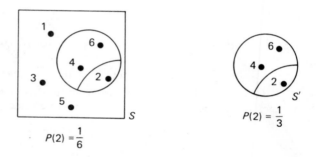

Figure 2.5   A conditional sample space.

The symbol $P(A \mid B)$, read "the probability of A given B," will be used to denote a conditional probability. It will prove convenient to have a formula for $P(A \mid B)$ that can be evaluated in terms of the original S, rather than the revised S'. To motivate that formula, we need to consider a slightly more general version of the dice problem. Suppose S is a finite sample space with n outcomes and P is the equally-likely probability function. Suppose A and B are two events containing a and b outcomes, respectively, and let c denote the number of outcomes in the intersection of A and B. Then, from what we have just seen, the conditional probability of A given B is the ratio of c to b. But

$$\frac{c}{b} = \frac{c/n}{b/n},$$

a form that suggests the next definition.

DEFINITION 2.4   Let B be an event such that $P(B) > 0$. Then the conditional probability of the event A, given that B has already occurred, is the ratio

$$P(A \mid B) = \frac{P(A \cap B)}{P(B)}.$$

EXAMPLE 2.17.   Consider families with two children and assume that the four possible "outcomes"—(younger is a boy, older is a boy), (younger is a boy, older is a girl), (younger is a girl, older is a boy), and (younger is a girl, older is a girl)—are equally likely. What is the probability that both children are boys *given that at least one is a boy*? As a first thought, it may seem that the answer should be $\frac{1}{2}$, but it isn't. The event "at least one is a boy" reduces the original sample space from four outcomes to three: $S' = \{(B, G), (G, B), (B, B)\}$. Only one of these, though, is a two-boy family, so the correct probability is $\frac{1}{3}$. Reassuringly, Definition 2.4 gives the same answer. If A is the event "both children are boys" and B, "at least one child is a boy,"

$$P(A \mid B) = \frac{P(A \cap B)}{P(B)} = \frac{P(A)}{P(B)}$$

$$= \frac{\frac{1}{4}}{\frac{3}{4}} = \frac{1}{3}.$$

QUESTION 2.6.1   Suppose we ignored the *age* of children in two-child families and distinguished *three* family types, rather than four: $\{B, B\}$, $\{G, B\}$, and $\{G, G\}$. Would the conditional probability of both children being boys given that at least one is a boy be different than the $\frac{1}{3}$ that was calculated in Example 2.17? Explain.

Definition 2.4 can be rewritten to give a very useful formula for the probability of the intersection of two events. If $P(A \mid B) = P(A \cap B)/P(B)$, then

$$P(A \cap B) = P(A \mid B)P(B). \tag{2.4}$$

A similar formulation holds for intersections of higher order. For example, the probability of the intersection of events A, B, and C can be written as the product of three probabilities—two conditional and one unconditional:

$$P(A \cap B \cap C) = P(A \cap (B \cap C)) = P(A \mid B \cap C)P(B \cap C)$$

$$= P(A \mid B \cap C)P(B \mid C)P(C).$$

EXAMPLE 2.18.   Suppose an urn (there seem to be three main types of urns: burial urns, Grecian urns, and probability urns) contains $r = 3$ red chips and $w = 6$ white chips. A chip is drawn at random. If it is red, it and $k = 2$ additional red chips are put back into the urn. If it is white, the chip is simply returned to the urn. Suppose, now, a second chip is drawn. What is the probability of both selections being red? If we let R_1 be the event "Red chip is selected on first draw" and R_2, "Red chip is selected on second draw," it should be clear that

$$P(R_1) = \tfrac{3}{9} \quad \text{and} \quad P(R_2 \mid R_1) = \tfrac{5}{11}.$$

Substituting these probabilities into Equation 2.4 gives

$$P(R_1 \cap R_2) = P(\text{both chips are red}) = P(R_1)P(R_2 \mid R_1)$$

$$= (\tfrac{3}{9})(\tfrac{5}{11}) = \tfrac{15}{99}.$$

Comment. The urn problem just described can serve as a probability model for the spread of contagious diseases. Think of the red chips as, say, measles cases. The constants r and w characterize the initial status of the population at risk, while k reflects the strength of the contagion. ▌

EXAMPLE 2.19. The possibility of importing liquefied natural gas (LNG) from Algeria has been suggested as one way of easing the United States' energy crunch. Complicating matters, though, is the fact that LNG is highly volatile and poses an enormous safety hazard. Any major spill occurring near a U.S. port could result in a fire of catastrophic proportions. The question, therefore, of the *likelihood* of a spill becomes critical input for the policymakers who will have to decide whether or not to go ahead with the proposal.

Two numbers need to be taken into account: (1) the probability that a tanker will have an accident near a port and 2) the probability that a major spill will develop *given* that an accident has happened. Although no significant spills of LNG have yet occurred anywhere in the world, these probabilities can be approximated from records kept on similar tankers transporting less dangerous cargo. On the basis of such data, it has been estimated (42) that the probability is 8/50,000 that an LNG tanker will have an accident on any one trip. Similarly, given that an accident *has* occurred, it is estimated that only three times out of 15,000 will the damage be sufficiently severe and located in those parts of the ship that a major spill would be the consequence. Thus, the single-trip probability for a major LNG disaster computes to 3.2×10^{-8}:

$$P(\text{accident occurs and spill develops}) = P(\text{spill} \mid \text{accident})P(\text{accident})$$

$$= \left(\frac{3}{15,000}\right)\left(\frac{8}{50,000}\right)$$

$$= 0.000000032.$$

QUESTION 2.6.2   A man has n keys on a chain, only one of which opens the door to his apartment. Having celebrated a little too much one evening, he comes home and finds himself unable to distinguish one key from another. He proceeds to choose a key at random and try it. If it fails to open the door he discards it and chooses at random one of the remaining $n - 1$ keys, and so on. What is the probability he opens the door with the ith key tried?

Equation 2.4 can be extended in still another way. Suppose S is a sample space partitioned by a set of mutually exclusive and exhaustive events, $A_1, A_2, \ldots, A_n$—that is, every sample outcome in S belongs to one and only one A_i. Let B be some other event defined over S (Figure 2.6). Theorem 2.7 offers a formula for $P(B)$ in terms of the n conditional probabilities, $P(B \mid A_i)$. As we will soon see, this is a fundamentally important result and finds numerous applications.

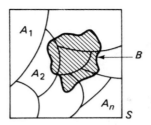

Figure 2.6 Partitioned sample space.

THEOREM 2.7 Let $\{A_i\}_{i=1}^n$ be a set of events defined over S such that $S = \bigcup_{i=1}^n A_i$ and $A_i \cap A_j = \varnothing$, for $i \neq j$. Also, let $P(A_i) > 0$, for $i = 1, 2, \ldots, n$. Then, for any event B,

$$P(B) = \sum_{i=1}^n P(B \mid A_i)P(A_i).$$

PROOF

Because of the two conditions imposed on the A_i's,

$$P(B) = P(B \cap S)$$

$$= P\left(B \cap \left(\bigcup_{i=1}^n A_i\right)\right)$$

$$= P\left(\bigcup_{i=1}^n (B \cap A_i)\right) \qquad \text{(see Review Exercise 15)}$$

$$= \sum_{i=1}^n P(B \cap A_i).$$

Now, apply Equation 2.4 to each of the intersection probabilities to get the result:

$$P(B) = \sum_{i=1}^n P(B \mid A_i)P(A_i).$$

EXAMPLE 2.20. A toy manufacturer buys ball bearings from three different suppliers —50% of his total order comes from Supplier 1, 30% from Supplier 2, and the rest from Supplier 3. Past experience has shown that the quality control standards of the three suppliers are not all the same. Of the ball bearings produced by Supplier 1, 2% are defective; Suppliers 2 and 3 produce defective bearings 3% and 4% of the time, respectively. What the toy manufacturer wants to know is the proportion of ball bearings in his inventory that are defective.

Let A_i be the event "Bearing came from Supplier i," $i = 1, 2, 3$. Let B be the event "Bearing in toy manufacturer's inventory is defective." Then

$$P(A_1) = 0.5, \quad P(A_2) = 0.3, \quad P(A_3) = 0.2$$

and

$$P(B \mid A_1) = 0.02, \quad P(B \mid A_2) = 0.03, \quad P(B \mid A_3) = 0.04.$$

Combining these probabilities according to Theorem 2.7 gives

$$P(B) = (0.02)(0.5) + (0.03)(0.3) + (0.04)(0.2)$$

$$= 0.027,$$

meaning that 2.7% of the manufacturer's ball bearing stock is defective.

EXAMPLE 2.21.   Consider the following con game. Three cards are put into a hat. Card 1 is white on both sides, card 2 is red on both sides, and card 3 is red on one side and white on the other. A card is drawn from the hat and placed on a table, without the underside being seen. Suppose the side showing is red. The con man argues that what has been drawn is clearly not the white-white card so the underside is equally likely to be red or white. He offers to bet on "red" at even money. A bit of empirical observation, though, will reveal that the con man wins such a bet $\frac{2}{3}$ of the time, rather than $\frac{1}{2}$ of the time. Why?

A correct analysis of this game recognizes that a conditional probability is involved. Let A_i be the event of drawing card i, $i = 1, 2, 3$. Then $P(A_1) = P(A_2) = P(A_3) = \frac{1}{3}$. Let D be the event "Red side is down" and U, the event "Red side is up." The bet as proposed is on the occurrence of D given the occurrence of U. With the problem reduced to these terms, the solution is a simple application of Definition 2.4 and Theorem 2.7:

$$P(D|U) = \frac{P(D \cap U)}{P(U)} = \frac{P(D \cap U)}{P(U|A_1)P(A_1) + P(U|A_2)P(A_2) + P(U|A_3)P(A_3)}$$

$$= \frac{\frac{1}{3}}{0(\frac{1}{3}) + 1(\frac{1}{3}) + \frac{1}{2}(\frac{1}{3})}$$

$$= \frac{2}{3}.$$

(Note the similarity between this problem and Example 2.17.)

QUESTION 2.6.3   Urn I contains two red chips and four white chips; urn II contains three red and one white. A chip is drawn at random from urn I and transferred to urn II. Then a chip is drawn from urn II. What is the probability the chip drawn from urn II is red?

Described in the next example is an unusual application of Theorem 2.7 to the art of survey sampling. It is no secret that opinion polls, although widely used, are treacherous to interpret: subtle nuances in the way a question is worded or the context in which it is asked can profoundly influence the way a respondent is likely to answer. This situation, bad under the best of conditions, obviously degenerates when the question itself is of a sufficiently "sensitive" nature that respondents feel pressured to answer in a certain way. Surveys addressing topics such as drug use or sexual preferences clearly fall into this category. Example 2.22 shows that by using what is called a randomized response survey, it is possible to ask personal questions and still convince the respondents that their anonymity is completely guaranteed (so they have nothing to fear by being truthful).

EXAMPLE 2.22.   A sociologist is interested in estimating the percentage of college students who smoke marihuana regularly. One hundred students are randomly selected to make up her sample. The question she wants to ask is, "Do you smoke marihuana at least once a week?" but to allay any fears they might have about their confidentiality being breached, she does it indirectly. First, each student is asked to pick a random number between 00 and 99 (without revealing its identity). Anyone whose number is between, say, 00 and 69 is then asked to answer the marihuana question. The other students are instructed to answer a trivial, nonsensitive question such as "Does your student ID end with an odd digit?" Each person writes his answer on a slip of paper, without ever indicating which question he is responding to. Suppose it turns out that 44% of all the responses were "Yes." What fraction would we estimate was answering "Yes" to the marihuana question?

The answer is 41% and it comes from a direct application of Theorem 2.7. Note that

$$P(\text{"Yes" answer}) = P(\text{"Yes"}\,|\,\text{sensitive question})P(\text{sensitive question})$$
$$+ P(\text{"Yes"}\,|\,\text{nonsensitive question})P(\text{nonsensitive question})$$

Substituting for the four probabilities we know gives

$$0.44 = P(\text{"Yes"}\,|\,\text{sensitive question})(0.7) + (0.5)(0.3),$$

making it a simple matter to solve for the fifth:

$$P(\text{"Yes"}\,|\,\text{sensitive question}) = P(\text{smoke marihuana at least once a week})$$
$$= \frac{0.44 - (0.5)(0.3)}{0.7}$$
$$= 0.41.$$

Strange as it may sound, what we have done here is analyze a set of answers without knowing what the questions were (20)!

QUESTION 2.6.4   If men constitute 47% of the population and tell the truth 78% of the time, while women tell the truth 63% of the time, what is the probability that a person selected at random will answer a question truthfully?

The last theorem in this section has an interesting history. Its first explicit statement, coming in 1812, was due to Laplace, but its name derives from the Reverend Thomas Bayes, whose 1763 paper (published posthumously) had already outlined the result. On one level, the theorem is a relatively minor extension of the definition of conditional probability. When viewed from a loftier perspective, though, it takes on some rather profound philosophical implications. These implications, in fact, have precipitated a schism among practicing statisticians: "Bayesians" analyze data one way—"non-Bayesians," another.

It is not our purpose to contrast these two schools of inference. The basic tenets of Bayesian inference are set forth in any number of texts [see, for example, (129) and (98)]. Our use of Bayes' theorem will be restricted to its original role as a formula for evaluating certain conditional probabilities.

THEOREM 2.8 (Bayes) Let $\{A_i\}_{i=1}^n$ be a set of n events, each with positive probability, that partition S in such a way that $\bigcup_{i=1}^n A_i = S$ and $A_i \cap A_j = \varnothing$ for $i \neq j$. For any event B, where $P(B) > 0$, and any j between 1 and n,

$$P(A_j \mid B) = \frac{P(B \mid A_j)P(A_j)}{\sum_{i=1}^n P(B \mid A_i)P(A_i)}.$$

PROOF

From Definition 2.4,

$$P(A_j \mid B) = \frac{P(A_j \cap B)}{P(B)} = \frac{P(B \mid A_j)P(A_j)}{P(B)}.$$

But Theorem 2.7 allows the denominator to be written as

$$\sum_{i=1}^n P(B \mid A_i)P(A_i),$$

and the result follows.

Bayes' theorem has been applied with considerable success to the problem of diagnosing medical conditions—that is, estimating the probability that a patient has a certain disease given that a particular diagnostic procedure says he does. The next example is one such application, and it makes a significant point: that when the disease being looked for is very rare, the number of incorrect diagnoses can be alarmingly high.

EXAMPLE 2.23. Consider the problem of "screening" for cervical cancer (143). Let C be the event of a woman's having the disease and B, the event of a positive biopsy—that is, B occurs when the diagnostic procedure indicates that she *does* have cervical cancer. We will assume that $P(C) = 0.0001$, $P(B \mid C) = 0.90$ (the test correctly identifies 90% of all the women who do have the disease), and $P(B \mid C^c) = 0.001$ (the test gives one false positive, on the average, out of every 1000 patients). The question is, what is $P(C \mid B)$, the probability that a woman actually has cervical cancer given that her biopsy says she does?

Appealing to Bayes' theorem, we see that that probability is surprisingly low:

$$P(C \mid B) = \frac{P(B \mid C)P(C)}{P(B \mid C)P(C) + P(B \mid C^c)P(C^c)}$$

$$= \frac{(0.9)(0.0001)}{(0.9)(0.0001) + (0.001)(0.9999)}$$

$$= 0.08.$$

That is, only 8% of those women identified as having the disease actually do! Table 2.5 shows the strong dependence of $P(C \mid B)$ on $P(C)$ and $P(B \mid C^c)$. In light of these figures, one must question the practicality of large-scale screening programs directed at diseases with low prevalence.

TABLE 2.5 Screening example

$P(C)$	$P(B\mid C^c)$	$P(C\mid B)$
0.0001	0.001	0.08
	0.0001	0.47
0.001	0.001	0.47
	0.0001	0.90
0.01	0.001	0.90
	0.0001	0.99

QUESTION 2.6.5   A family has two dogs (Rex and Rover) and a little boy (Russ). None of them is particularly fond of the mailman. Given that they are outside, Rex and Rover have a 30% and a 40% chance, respectively, of biting the mailman. Russ, if he is outside, has a 15% chance of doing the same thing. Suppose that one and only one of the three is outside when the mailman comes. If Rex is outside 50% of the time, Rover 20% of the time, and Russ 30% of the time, what is the probability the mailman will be bitten? If the mailman *is* bitten, what are the chances that Russ did it?

2.7 INDEPENDENCE

The preceding section dealt with reevaluating the probability of a given event in light of additional information that some other event already occurred. Sometimes though, the probability of the given event remains unchanged, regardless of the outcome of the other event—that is, $P(A\mid B) = P(A) = P(A\mid B^c)$. Events sharing this property are said to be *independent*. Definition 2.5 gives a necessary and sufficient condition for two events to be independent.

DEFINITION 2.5   Two events A and B are said to be *independent* if $P(A \cap B) = P(A)P(B)$.

Comment. The fact that the probability of the intersection of two independent events is equal to the product of their individual probabilities follows immediately from our "first" definition of independence, that $P(A\mid B) = P(A)$. Recall that the definition of conditional probability holds true for *any* two events A and B [provided $P(B) > 0$]:

$$P(A\mid B) = \frac{P(A \cap B)}{P(B)}.$$

But $P(A\mid B)$ can equal $P(A)$ only if $P(A \cap B)$ factors into $P(A)$ times $P(B)$. ∎

EXAMPLE 2.24.   Let K be the event of drawing a king from a standard poker deck and D the event of drawing a diamond. Then, by Definition 2.5, K and D are independent because the probability of their intersection—drawing a king of diamonds—is equal to $P(K)P(D)$:

$$P(K \cap D) = \tfrac{1}{52} = \tfrac{1}{13} \cdot \tfrac{1}{4} = P(K)P(D).$$

EXAMPLE 2.25.   Suppose A and B are independent events. Does it follow that A^C and B^C are also independent? That is, does $P(A \cap B) = P(A)P(B)$ guarantee that $P(A^C \cap B^C) = P(A^C)P(B^C)$? The answer is "Yes": the proof is accomplished by equating two different expressions for $P(A^C \cup B^C)$. First, by Theorem 2.6,

$$P(A^C \cup B^C) = P(A^C) + P(B^C) - P(A^C \cap B^C).$$

But the union of two complements is also the complement of their intersection. (Readers familiar with set theory will recognize that statement as one of de Morgan's laws.) Therefore,

$$P(A^C \cup B^C) = 1 - P(A \cap B).$$

Therefore,

$$1 - P(A \cap B) = 1 - P(A) + 1 - P(B) - P(A^C \cap B^C),$$

in which case

$$
\begin{aligned}
P(A^C \cap B^C) &= 1 - P(A) + 1 - P(B) - [1 - P(A)P(B)] \\
&= [1 - P(A)][1 - P(B)] \\
&= P(A^C)P(B^C).
\end{aligned}
$$

QUESTION 2.7.1   If A and B are independent, show that A and B^C are independent.

It is not obvious how to extend Definition 2.5 to, say, three events. To call A, B, and C independent, should we require that the probability of the three-way intersection factors into the product of the three original probabilities,

$$P(A \cap B \cap C) = P(A)P(B)P(C), \tag{2.5}$$

or should we impose the definition we already have on the three *pairs* of events:

$$
\begin{aligned}
P(A \cap B) &= P(A)P(B), \\
P(B \cap C) &= P(B)P(C), \\
P(A \cap C) &= P(A)P(C)?
\end{aligned}
\tag{2.6}
$$

As the next two examples show, neither condition by itself is sufficient. If three events satisfy Equations 2.5 *and* 2.6, we will call them independent, but Equation 2.5 does not imply Equation 2.6, nor does Equation 2.6 imply Equation 2.5.

EXAMPLE 2.26.   Suppose two fair dice (one red and one green) are thrown, and events A, B, and C are defined as follows:

A: a 1 or a 2 shows on the red die,
B: a 3, 4, or 5 shows on the green die,
C: the dice total is 4, 11, or 12.

It can be readily verified, then, that $P(A) = \tfrac{1}{3}$, $P(B) = \tfrac{1}{2}$, $P(C) = \tfrac{1}{6}$, $P(A \cap B) = \tfrac{1}{6}$, $P(A \cap C) = \tfrac{1}{18}$, $P(B \cap C) = \tfrac{1}{18}$, and $P(A \cap B \cap C) = \tfrac{1}{36}$. What follows is that Equation 2.5 is satisfied,

$$P(A \cap B \cap C) = \tfrac{1}{36} = P(A)P(B)P(C) = (\tfrac{1}{3})(\tfrac{1}{2})(\tfrac{1}{6}),$$

but Equation 2.6 is not,

$$P(B \cap C) = \tfrac{1}{18} \neq P(B)P(C) = (\tfrac{1}{2})(\tfrac{1}{6}) = \tfrac{1}{12}.$$

EXAMPLE 2.27.   A roulette wheel has 36 numbers, colored red or black according to the pattern indicated in Figure 2.7. Let R be the event "Red number appears," E

1	2	3	4	5	6	7	8	9	10	11	12	13	14	15	16	17	18
R	R	R	R	R	B	B	B	B	R	R	R	R	B	B	B	B	B
36	35	34	33	32	31	30	29	28	27	26	25	24	23	22	21	20	19

Figure 2.7   Roulette-wheel pattern.

the event "Even number appears," and T the event "Total is ≤ 18." Then $P(R) = P(E) = P(T) = \tfrac{1}{2}$, $P(R \cap E) = \tfrac{1}{4}$, $P(E \cap T) = \tfrac{1}{4}$, and $P(R \cap T) = \tfrac{1}{4}$. It should be clear that R, E, and T are all "pairwise" independent (Equation 2.6 holds), and yet the probability of the three-way intersection does not factor:

$$P(R \cap E \cap T) = \tfrac{4}{36} = \tfrac{1}{9} \neq P(R)P(E)P(T) = (\tfrac{1}{2})^3 = \tfrac{1}{8}.$$

The upshot of Examples 2.26 and 2.27 is that for n events to be independent, the probabilities of *all* possible intersections must factor into the product of the probabilities of the component events. Definition 2.6 gives the formal statement.

DEFINITION 2.6   Events $A_1, A_2, \ldots, A_n$ are said to be *independent* if for every set of indices $i_1, i_2, \ldots, i_k$ between 1 and n,

$$P(A_{i_1} \cap A_{i_2} \cap \cdots \cap A_{i_k}) = P(A_{i_1})P(A_{i_2}) \cdots P(A_{i_k}).$$

EXAMPLE 2.28.   Suppose a fair coin is flipped three times. Let H_1 be the event of a head on the first flip, T_2 a tail on the second flip, and H_3 a head on the third flip. Then

$$P(H_1) = P(T_2) = P(H_3) = \tfrac{1}{2}.$$

Also,

$$P(H_1 \cap T_2) = P(\text{HTH, HTT}) = \tfrac{2}{8} = \tfrac{1}{4} = P(H_1)P(T_2).$$

Similarly,

$$P(H_1 \cap H_3) = \tfrac{1}{4} = P(H_1)P(H_3)$$

and

$$P(T_2 \cap H_3) = \tfrac{1}{4} = P(T_2)P(H_3).$$

Finally,

$$P(H_1 \cap T_2 \cap H_3) = P(\text{HTH}) = \tfrac{1}{8} = P(H_1)P(T_2)P(H_3).$$

By Definition 2.6, then, events H_1, T_2, and H_3 are independent.

EXAMPLE 2.29.   A legal controversy recently arose in the California Supreme Court over the appropriateness of Definition 2.6 as a measure of culpability (43). In 1964, a woman shopping in Los Angeles had her purse snatched by a young, blond female wearing a pony tail. The suspect fled on foot but was seen shortly thereafter getting into a yellow automobile driven by a black male who had a mustache and a beard.

A police investigation subsequently turned up a suspect, one Janet Collins, who was blond, wore a pony tail, and associated with a black male who drove a yellow car and had a mustache. An arrest was made.

Not having any tangible evidence, and no reliable witnesses, the prosecutor sought to build his case on the "unlikelihood" of Ms. Collins and her companion sharing these characteristics and not being the guilty parties. First, the bits of evidence that were available were assigned probabilities. For example, it was estimated that the probability of a female wearing a pony tail in Los Angeles was $\frac{1}{10}$. Table 2.6 lists the probabilities quoted for the six "facts" agreed on by the victim and the eyewitnesses.

TABLE 2.6 Probabilities in purse-snatching case

Characteristic	Probability
Yellow automobile	$\frac{1}{10}$
Man with mustache	$\frac{1}{4}$
Woman with pony tail	$\frac{1}{10}$
Woman with blond hair	$\frac{1}{3}$
Black man with beard	$\frac{1}{10}$
Interracial couple in car	$\frac{1}{1000}$

The prosecutor multiplied these six numbers together and claimed that the product, $(\frac{1}{10})(\frac{1}{4}) \cdots (\frac{1}{1000})$, or 1 in 12 million, was the probability of the intersection—that is, the probability of a random couple's fitting this description. This figure is so small, he contended, it can only be concluded that the defendants were guilty. The jury agreed, and handed down a verdict of second-degree robbery. The Supreme Court of California, however, *disagreed*. Ruling on an appeal filed by the defendants, the higher court reversed the decision, claiming that the probability argument was incorrect and misleading.

Comment. We will look at these same data from another angle in Chapter 3—and come away with an entirely different interpretation. ∎

QUESTION 2.7.2   If you were counsel for the defense, how would you counter the prosecutor's probability arguments?

2.8 REPEATED INDEPENDENT TRIALS

It is not unusual for an experiment to be made up of a series of identical, repeated "subexperiments," each performed under essentially the same conditions. What we want to consider in this section is the relationship between the sample spaces of the individual subexperiments and the sample space of the overall experiment. The two are not the same.

Recall the game of craps, whose rules were given in Question 2.2.1. The one or more rolls of two dice that comprise each game qualify as subexperiments. For the shooter to win with, say, a point of 9 on the fourth roll, he must roll a 9 on the first throw, something other than a 9 or a 7 on the second and third throws, and a 9 on the fourth. If E_1, E_2, E_3, and E_4 denote those particular outcomes,

$$P(\text{shooter wins on fourth roll with a point of 9}) = P(E_1 \cap E_2 \cap E_3 \cap E_4).$$

On intuitive grounds, we would like to argue that because the rolls of the dice are independent, the probability of the intersection factors according to Equation 2.7:

$$P(E_1 \cap E_2 \cap E_3 \cap E_4) = P(E_1)P(E_2)P(E_3)P(E_4). \tag{2.7}$$

In point of fact, Equation 2.7 *is* correct, but it does need some justification. Definition 2.6, which talks about the probability of the intersection of independent events, does not apply here because the sample space for the intersection is different than the sample spaces for the individual E_i's. To shore up the mathematics, we must appeal to the notion of a *Cartesian product*.

Phrasing the problem more generally, suppose an experiment is repeated n times under "identical" conditions. Let $E_1, E_2, \ldots, E_n$ denote events defined on the first, second, $\ldots$, and nth trials. Let $S_k = \{s_i^{(k)}\}_i$ denote the sample space for the kth trial, $k = 1, 2, \ldots, n$. (For simplicity, we are assuming the S_k's are discrete.) The Cartesian product, then, of $S_1, S_2, \ldots, S_k$ is defined to be the set of n-tuples,

$$S = \{(s_{i_1}^{(1)}, s_{i_2}^{(2)}, \ldots, s_{i_n}^{(n)})\}_{i_1, i_2, \ldots, i_n}.$$

(In mathematics texts, this would often be written $S = S_1 \otimes S_2 \otimes \cdots \otimes S_n$.)

Set up in this way, S serves as a common denominator for events related to individual trials as well as to groups of trials. For example, if E_j is just a single outcome in S_j—$E_j = \{s_{j^*}^{(j)}\}$—then, relative to S, E_j can be written

$$E_j = \{(s_{i_1}^{(1)}, \ldots, s_{i_{j-1}}^{(j-1)}, s_{j^*}^{(j)}, s_{i_{j+1}}^{(j+1)}, \ldots, s_{i_n}^{(n)})\}_{i_1, \ldots, i_{j-1}, i_{j+1}, \ldots, i_n}.$$

At the other extreme, we can use S to express the intersection of the n (single-outcome) E_j's:

$$E_1 \cap E_2 \cap \cdots \cap E_n = \{(s_{1^*}^{(1)}, s_{2^*}^{(2)}, \ldots, s_{n^*}^{(n)})\}.$$

QUESTION 2.8.1   Let $S_1 = (2, 4, 5)$ and $S_2 = (2, 4, 5)$. Write down the members of the Cartesian product of S_1 and S_2. What sample outcomes (in $S = S_1 \otimes S_2$) belong to the event "First digit is even"?

Now we can define precisely the notion of repeated independent trials.

DEFINITION 2.7 Let $E_1, E_2, \ldots, E_n$ denote the outcomes of a series of n trials. If, for all j, the probability of any given outcome on the jth trial does not depend on the outcomes of the preceding $j - 1$ trials, the trials are said to be *independent*.

Comment. The analog of Definition 2.5 holds for repeated independent trials:

$$P(E_1 \cap E_2 \cap \cdots \cap E_n) = P(E_1)P(E_2) \cdots P(E_n). \quad \blacksquare$$

EXAMPLE 2.30.   The great French mathematician, Pierre Simon Laplace (1749–1827), proposed an independent-trials problem that led to some rather curious interpretations. Suppose we have a coin whose probability of coming up heads is unknown but is equally likely to be $1/N, 2/N, \ldots$, or N/N. That is, if $p = P$ (heads),

$$P\left(p = \frac{i}{N}\right) = \frac{1}{N}, \qquad i = 1, 2, \ldots, N.$$

The question is this: if the first n tosses of the coin turn up heads, what is the probability that the $(n + 1)$st toss will also be heads?

Casting this into a workable framework requires that we define $(N + 2)$ events:

1. Let B be the event that the first n tosses come up heads.
2. Let A be the event that the $(n + 1)$st toss comes up heads.
3. Let C_i be the event that $p = i/N, i = 1, 2, \ldots, N$.

The objective, of course, is to find $P(A \mid B)$. As a starting point, note that

$$B = (B \cap C_1) \cup (B \cap C_2) \cup \cdots \cup (B \cap C_N)$$

and, because the $(B \cap C_i)$'s are mutually exclusive,

$$P(B) = \sum_{i=1}^{N} P(B \cap C_i) = \sum_{i=1}^{N} P(B \mid C_i)P(C_i).$$

But the coin tosses qualify as independent events, so

$$P(B \mid C_i) = P((\text{1st coin is heads} \cap \cdots \cap n\text{th coin is heads}) \mid C_i)$$
$$= \left(\frac{i}{N}\right)^n.$$

Therefore,

$$P(B) = \frac{1}{N} \sum_{i=1}^{N} \left(\frac{i}{N}\right)^n.$$

In a similar way, the intersection $A \cap B$ can be written

$$A \cap B = [(A \cap B) \cap C_1] \cup [(A \cap B) \cap C_2] \cup \cdots \cup [(A \cap B) \cap C_N],$$

in which case

$$P(A \cap B) = \sum_{i=1}^{N} P(A \cap B \cap C_i) = \sum_{i=1}^{N} P(A \cap B \mid C_i)P(C_i)$$
$$= \frac{1}{N} \sum_{i=1}^{N} \left(\frac{i}{N}\right)^{n+1}.$$

From Definition 2.4, then,

$$P(A \mid B) = \frac{P(A \cap B)}{P(B)} = \frac{(1/N) \sum_{i=1}^{N} (i/N)^{n+1}}{(1/N) \sum_{i=1}^{N} (i/N)^n}. \qquad (2.8)$$

A more convenient expression for the conditional probability, though, can be gotten by treating the numerator and denominator of Equation 2.8 as Riemann sums:

$$\frac{1}{N} \sum_{i=1}^{N} \left(\frac{i}{N}\right)^{n+1} \doteq \int_0^1 x^{n+1} \, dx = \frac{1}{n+2},$$

$$\frac{1}{N} \sum_{i=1}^{N} \left(\frac{i}{N}\right)^n \doteq \int_0^1 x^n \, dx = \frac{1}{n+1}.$$

Then

$$P(A \mid B) \doteq \frac{1/(n+2)}{1/(n+1)} = \frac{n+1}{n+2}. \tag{2.9}$$

Equation 2.9 is known as *Laplace's rule of succession*. Given the presumed equally-likely structure, it says that if some particular event has occurred in n consecutive trials, the probability is $(n+1)/(n+2)$ that it will occur on the very next trial. Laplace suggested that Equation 2.9 would be one way to estimate the probability of the sun's rising tomorrow. Assume history goes back 5000 years (or 1,826,213 days). Since the sun rose on each of those days, the probability of its making an appearance tomorrow is 1,826,214/1,826,215.

QUESTION 2.8.2   Use Laplace's rule of succession to get a probability for the sun's *not* rising at least once during the next 5000 years.

EXAMPLE 2.31.   In the game of craps, the person rolling the dice is called the *shooter*. The game is played at even money, but we will see in this example that the shooter actually has a slightly less than 50–50 chance of winning.

There are two basic ways the shooter can win (other than by cheating): (1) by throwing either a 7 or an 11 on his first roll (this is called a *natural*) or (2) by throwing either a 4, 5, 6, 8, 9, or 10 on his first roll and then throwing that number again *before* he rolls a 7 (this is called making his *point*). Let A_1 be the event the shooter throws a natural, and let A_4, A_5, A_6, A_8, A_9, and A_{10} be the events that the shooter eventually wins when his point is a 4, 5, 6, 8, 9, or 10, respectively. The A_i's are mutually exclusive, so

$$P(\text{shooter wins}) = P(A_1) + P(A_4) + P(A_5) + P(A_6) + P(A_8) + P(A_9) + P(A_{10}).$$

The first of these probabilities is gotten immediately:

$$P(A_1) = P(7 \text{ or } 11) = P(7) + P(11) = \frac{6}{36} + \frac{2}{36} = \frac{8}{36}.$$

To determine the remaining $P(A_i)$'s, we need to think of the game as a series of repeated, independent trials. For example, the shooter will win with a point of 4 if he rolls a 4 on the first throw and a 4 on the second *or* a 4 on the first, something other than a 4 or a 7 on the second, and a 4 on the third *or* a 4 on the first, something other than a 4 or a 7 on the second and third, and a 4 on the fourth, and so on. Appealing again to the fact that these possibilities are all mutually exclusive, we can write

$$P(A_4) = P(4 \text{ on 1st} \cap 4 \text{ on 2nd}) + P(4 \text{ on 1st} \cap \{4, 7\}^C \text{ on 2nd} \cap 4 \text{ on 3rd})$$
$$+ P(4 \text{ on 1st} \cap \{4, 7\}^C \text{ on 2nd} \cap \{4, 7\}^C \text{ on 3rd} \cap 4 \text{ on 4th}) + \cdots.$$

By inspection, $P(4) = \frac{3}{36}$ and $P(\{4, 7\}^C) = \frac{27}{36}$. Then, because each roll is an independent trial,

$$P(A_4) = \left(\frac{3}{36}\right)\left(\frac{3}{36}\right) + \left(\frac{3}{36}\right)\left(\frac{27}{36}\right)\left(\frac{3}{36}\right) + \left(\frac{3}{36}\right)\left(\frac{27}{36}\right)\left(\frac{27}{36}\right)\left(\frac{3}{36}\right) + \cdots$$
$$= \left(\frac{3}{36}\right)^2 \sum_{k=0}^{\infty} \left(\frac{27}{36}\right)^k = \left(\frac{3}{36}\right)^2 \left(\frac{1}{1 - \frac{27}{36}}\right)$$
$$= \frac{1}{36}.$$

The other $P(A_i)$'s are calculated similarly. Of course, because of symmetry we need only determine $P(A_4)$, $P(A_5)$, and $P(A_6)$: since the probability of throwing a 4 is the same as the probability of throwing a 10, $P(A_4) = P(A_{10})$—also, $P(A_5) = P(A_9)$ and $P(A_6) = P(A_8)$. Table 2.7 summarizes these computations.

TABLE 2.7   Probabilities for the shooter

Winning Event, A_i	$P(A_i)$
A_1	$\frac{8}{36}$
A_4	$\frac{1}{36}$
A_5	$\frac{16}{360}$
A_6	$\frac{25}{396}$
A_8	$\frac{25}{396}$
A_9	$\frac{16}{360}$
A_{10}	$\frac{1}{36}$

Adding the entries in the second column of Table 2.7 gives the probability in question:

$$P(\text{shooter wins}) = \tfrac{8}{36} + \tfrac{1}{36} + \cdots + \tfrac{1}{36}$$
$$= 0.493.$$

This confirms what was claimed at the outset, that the odds are *against* the shooter—although not by much, and as games of chance go, craps is relatively "fair."

QUESTION 2.8.3   Suppose a dice game consists of two players alternately rolling two fair dice. Player A rolls first. The winner is the person who throws the first "easy" 8 (a 2 and a 6 or a 5 and a 3). What is the probability that Player A wins?

An important special case of repeated independent trials, one of great utility in spite of its apparent triviality, occurs when S_j reduces to just two outcomes. For example, the n trials might be n epileptics taking the same medication, with the outcome for each being whether or not his seizures are prevented. Or, the n trials might be n hurricanes seeded with silver iodide, and the outcomes whether or not the hurricane's wind speed is diminished. On a very simple level, the n trials might he flips of a coin, with heads or tails being the possible outcomes at each trial.

For the sake of generality, we will denote the sample space for the jth two-outcome trial as $S_j = \{s, f\}$, where long tradition requires that the s outcome be referred to as a "success" and the f as a "failure." The probabilities associated with s and f will be denoted p and q—that is, $P(s) = p$ and $P(f) = q = 1 - p$. (The fact that the conditions under which the trials are being conducted are presumed identical precludes our having to subscript p and q.) Any set of repeated independent trials having this sort of dichotomous structure are called *Bernoulli trials*.

Comment. James Bernoulli (1654–1705) was one of a family of famous Swiss mathematicians and scientists. In 1713 his *Ars Conjectandi (The Art of Conjecture)* was posthumously published, a book that became a major force in elevating probability to the status of respectable mathematics. Included in *Ars Conjectandi* was a discussion of the properties of the repeated trials that now bear Bernoulli's name. ∎

With Bernoulli trials, the particular *order* in which a series of successes and failures occurred is usually not the information of primary interest. Frequently, all we really want to know is *how many* successes occurred. Thus, in the clinical trial example cited earlier, which particular epileptics achieved seizure control would be of less interest (to the person evaluating the effectiveness of the medication) than the total number that did. Put more formally, we will be concerned with a function, Y, defined on $S = S_1 \otimes S_2 \otimes \cdots \otimes S_n$ whose range is the set of integers from 0 to n. Specifically,

$$Y(E_1 \cap E_2 \cap \cdots \cap E_n) = \text{total number of successes in } n \text{ trials.}$$

The next example is a case in point.

EXAMPLE 2.32. Among the games that interested the Chevalier de Mere (recall Section 1.1) was one that involved the appearance, or lack of appearance, of a 6 in four rolls of a die. Put into the context of Bernoulli trials, the probability structure of the game is easily discerned. If "6 occurs" is called a success, each S_j has $p = P(6) = \frac{1}{6}$ and $q = P(\text{non-}6) = \frac{5}{6}$, for $j = 1, 2, 3, 4$. Table 2.8 lists the 16 four-tuples in the Cartesian product of the S_j's. The corresponding probabilities, calculated from the Comment following Definition 2.7, are given in the second column.

TABLE 2.8   A probability distribution for rolling four dice

Outcome in S	Probability
(s, s, s, s)	$(\frac{1}{6})(\frac{1}{6})(\frac{1}{6})(\frac{1}{6}) = (\frac{1}{6})^4$
(s, s, s, f)	$(\frac{1}{6})(\frac{1}{6})(\frac{1}{6})(\frac{5}{6}) = (\frac{1}{6})^3(\frac{5}{6})$
(s, s, f, s)	$(\frac{1}{6})(\frac{1}{6})(\frac{5}{6})(\frac{1}{6}) = (\frac{1}{6})^3(\frac{5}{6})$
(s, f, s, s)	$= (\frac{1}{6})^3(\frac{5}{6})$
(f, s, s, s)	$= (\frac{1}{6})^3(\frac{5}{6})$
(s, s, f, f)	$(\frac{1}{6})(\frac{1}{6})(\frac{5}{6})(\frac{5}{6}) = (\frac{1}{6})^2(\frac{5}{6})^2$
(s, f, s, f)	$= (\frac{1}{6})^2(\frac{5}{6})^2$
(f, s, s, f)	$= (\frac{1}{6})^2(\frac{5}{6})^2$
(s, f, f, s)	$= (\frac{1}{6})^2(\frac{5}{6})^2$
(f, s, f, s)	$= (\frac{1}{6})^2(\frac{5}{6})^2$
(f, f, s, s)	$= (\frac{1}{6})^2(\frac{5}{6})^2$
(s, f, f, f)	$(\frac{1}{6})(\frac{5}{6})(\frac{5}{6})(\frac{5}{6}) = (\frac{1}{6})(\frac{5}{6})^3$
(f, s, f, f)	$= (\frac{1}{6})(\frac{5}{6})^3$
(f, f, s, f)	$= (\frac{1}{6})(\frac{5}{6})^3$
(f, f, f, s)	$= (\frac{1}{6})(\frac{5}{6})^3$
(f, f, f, f)	$(\frac{5}{6})(\frac{5}{6})(\frac{5}{6})(\frac{5}{6}) = (\frac{5}{6})^4$

Suppose E were defined as the event "three successes occur." Then

$$E = \{(s, s, s, f), (s, s, f, s), (s, f, s, s), (f, s, s, s)\}.$$

In functional notation, E occurs if and only if $Y = 3$. The probability of E, or the probability that $Y = 3$, follows immediately from Table 2.8:

$$P(E) = P(Y = 3) = (\tfrac{1}{6})^3(\tfrac{5}{6}) + (\tfrac{1}{6})^3(\tfrac{5}{6}) + (\tfrac{1}{6})^3(\tfrac{5}{6}) + (\tfrac{1}{6})^3(\tfrac{5}{6})$$

$$= 4(\tfrac{1}{6})^3(\tfrac{5}{6}).$$

Similarly, $P(Y = 0) = (\tfrac{5}{6})^4$, $P(Y = 1) = 4(\tfrac{1}{6})(\tfrac{5}{6})^3$, $P(Y = 2) = 6(\tfrac{1}{6})^2(\tfrac{5}{6})^2$, and $P(Y = 4) = (\tfrac{1}{6})^4$.

The bet, for even money, that at least one 6 would appear in four tosses is a money-maker, because the probability of winning is 0.52:

$$P(\text{at least one 6}) = 1 - P(\text{no 6's})$$

$$= 1 - P(Y = 0) = 1 - (\tfrac{5}{6})^4$$

$$= 0.52.$$

Comment. One of the problems communicated to Pascal by de Mere was to determine the smallest number, n, of throws of two dice so that

$$P(\text{at least one double 6 in } n \text{ throws}) \geq 0.5.$$

There was an old gambling rule proposed by Cardano a century earlier that de Mere had used to obtain a solution. The rule stated that if the probability of success in one trial is $p = 1/N$, and if n is, as before, the least number of trials making the probability of one or more successes at least 0.5, then n/N is a constant. The rule is false, although it is approximately true for large N. Applying it here, de Mere got the proportion $n/36 = \tfrac{4}{6}$, or $n = 24$. This is the wrong answer: n should be 25. Pascal gave de Mere the correct solution and recorded the nobleman's reaction in a letter to Fermat (122): "This was a great scandal which made him proclaim loudly that the theorems were not constant and Arithmetic belied herself. But you can easily see the reason for this result by the principles you possess." ∎

QUESTION 2.8.4   Compute the correct odds in de Mere's second game: find the probability of rolling at least one double 6 in 24 throws of two dice.

Emerging clearly from Example 2.32, and Table 2.8, is the fact that any two sets of trials having the same number of successes—say, k—have the same probability, $p^k q^{n-k}$. It follows that to compute the probability that k successes occur (in n trials), we need simply multiply $p^k q^{n-k}$ times the number of ways to "arrange" k successes and $n - k$ failures. Finding that number will be one of the objectives of the next section, where we take a brief excursion into the realm of combinatorial mathematics.

QUESTION 2.8.5   Suppose n people draw a single card from n poker decks. Let A be the event "At least one ace is drawn in n trials." Find the smallest n so that $P(A) \geq 0.8$.

2.9 COMBINATORICS

Combinatorics is a time-honored branch of mathematics concerned with counting, arranging, and ordering. While blessed with a wealth of early contributors [there are references to combinatorial problems in the Old Testament (135)], its emergence as a separate discipline is often credited to the German mathematician and philosopher, Gottfried Wilhelm Leibniz (1646–1716), whose 1666 treatise, *Dissertatio de arte combinatoria*, was perhaps the first monograph written on the subject (104).

Applications of combinatorics are rich in both diversity and number. Users range from the molecular biologist trying to determine how many ways genes can be positioned along a chromosome, to the ecologist using a capture-recapture technique to estimate the prevalence of an endangered species (see Example 2.43), to a psychologist modeling the way we learn, to a weekend poker player wondering whether he should draw to a straight or to a flush. Priorities dictate that we limit our brief treatment of combinatorics to those results that will reappear later in the book; readers interested in a more thorough coverage are referred to (194) and (45).

As a starting point, we consider the problem of selecting k objects from among a set of n and arranging them in order. Such an arrangement is called a *permutation of length k* (or a *permutation of n objects taken k at a time*). A license plate, for example, is a permutation, often of length 6, of the 36 objects (0, 1, . . . , 9, A, B, . . . , Z). Here order is obviously important, since ML-1711 is a different plate than ML-1171. Depending on the situation, it may or may not be permissible to choose the same object more than once. Thus, we speak of permutations *with repetition* and permutations *without repetition*. In the case of license plates, both the letters and the numbers can be selected *with* repetition.

THEOREM 2.9

(a) There are $n(n-1)(n-2) \cdots (n-k+1) = n!/(n-k)!$ permutations (without repetition) of n objects taken k at a time.

(b) There are n^k permutations (with repetition) of n objects taken k at a time.

PROOF

Consider part (a). The first object to be chosen can be any one of the original n. The second can be any except the first. Thus, the first two can be selected in $n(n-1)$ ways. Then, for any choice of the first two, there are $n-2$ possibilities left for the third, so the first *three* can be chosen (or arranged) in $n(n-1)(n-2)$ ways. Continuing in this fashion, we see that the kth selection can be any of the remaining $n-(k-1)=n-k+1$ objects, and the conclusion follows.

The proof for part (b) is similar. With no restrictions being imposed on the number of times any of the objects can be selected, the first position can be filled in *n* ways, the second position in *n* ways, . . . , and the *k*th position in *n* ways. The total number of arrangements, then, is n^k.

EXAMPLE 2.33.   Recently an experiment was set up (133) to see whether language skills could be taught to anthropoid apes. Among the subjects was an African-born, five-year-old chimpanzee named Sarah. For one of the tests designed to measure comprehension, Sarah was asked to "read" a sentence and respond appropriately. She was to answer by first selecting a set of four plastic symbols (from among a set of eight) and then arranging the four in descending order on a board. For each sentence there was only one correct answer.

As a prerequisite for analyzing the results, investigators needed to find the probability that Sarah would be able to form the correct response just by chance. To do that, note that the number of ways to select and order the symbols is the same as the number of permutations (without repetition) of eight objects taken four at a time, which by Theorem 2.9 is $8!/(8 - 4)!$, or 1680. Since random selection means that each permutation is equally likely, the desired probability of an "accidental" lucky guess is 1/1680, or 0.0006. With this latter figure being so small, it follows that a correct answer should be viewed as something much more than just plain luck. (Chimpanzee lovers and Sarah's many friends will be pleased to know that she was correct 75% of the time.)

EXAMPLE 2.34.   Suppose the format for license plates in a certain state is two letters followed by four numbers.

(a) How many different plates can be made?

Since repetition is not prohibited, part (b) of Theorem 2.9 gives the answer as $26^2 \cdot 10^4$, or 6,760,000.

(b) How many different plates are there if the letters can be repeated but no two numbers can be the same?

Here both parts of Theorem 2.9 come into play, and the number is reduced to $26^2 \cdot 10 \cdot 9 \cdot 8 \cdot 7$, or 3,407,040.

(c) How many different plates can be made if repetition is allowed except that no plate can have four zeros?

For the number part of the plate, there would be 10^4 possibilities if no restrictions were imposed. Eliminating 0000 leaves $10^4 - 1$, or 9999 admissible numerical sequences. Thus, the total number of plates is $26^2 \cdot 9999$, or 6,759,324.

(d) What is the probability that a plate has no repeated digits?

Assuming an equally-likely model, the probability of no repeated digits is the quotient of the answer to part (b) divided by the answer to part (a): $3,407,040/6,760,000 = 0.504$.

QUESTION 2.9.1   An octave contains 12 distinct notes (on a piano, five black keys and seven white keys). How many different eight-note melodies within a single octave can be written using the white keys only?

QUESTION 2.9.2   If *n* people, among whom are A and B, are seated in a row, what is the probability that there will be exactly *r* people between them? Assume all seating arrangements are equally likely.

The next example is a very famous one in probability, because its answer goes contrary to our intuition.

EXAMPLE 2.35. Suppose a group of k randomly selected individuals is assembled. What is the probability that at least two of them will have the same birthday? Most people would guess that for k relatively small—say, less than 50—a match would be very unlikely. It can be easily shown, though, that the odds of at least one match are better than 50–50 if as few as 23 people are present. And when the group does number 50, there is a 97% chance of at least one match!

The solution to the birthday problem is similar to the solution to part (d) of Example 2.34. Omitting leap year, we see that there are 365 possible birthdays for each of the k people. If we imagine the people to be ordered, there are 365^k corresponding permutations of their k birthdays, repetitions allowed. If repetitions are not allowed, the number of permutations of length k reduces to $365!/(365 - k)!$. Assume each person has an equal chance of being born on any particular day. Then

$$P(\text{all } k \text{ birthdays are different}) = \frac{(365)!/(365 - k)!}{(365)^k}$$

and

$$P(\text{at least two have the same birthday}) = 1 - P(\text{all birthdays are different})$$

$$= 1 - \frac{(365)!/(365 - k)!}{(365)^k} = P_k.$$

Table 2.9 gives the value of P_k for $k = 15, 22, 23, 40, 50,$ and 70.

TABLE 2.9   Birthday matchings

k	$P_k = P$ (at least one match)
15	0.253
22	0.475
23	0.507
40	0.891
50	0.970
70	0.999

Comment. To facilitate the computation of the P_k's, it was assumed that the equally-likely model would describe the distribution of birthdays in the general population. That, of course, is not entirely true, since births are more common during the summer than during the winter. It has been shown, though, that any such nonuniformity serves only to *increase* the value of P_k (117). Thus, with 23 people, the smallest possible probability of at least one match is 0.507, and that occurs when each birthday is equally likely. ∎

Comment. Presidential biographies offer one opportunity to "confirm" the unexpectedly large values Table 2.9 gives for P_k. And they do. Among the $k = 38$ presidents before Carter, two did have the same birthday—Harding and Polk were both born on November 2. More surprising, though, are the death dates of the presidents, where there were *four* matches: Adams, Jefferson, and Monroe all died on July 4 and Fillmore and Taft both died on March 8.   ∎

QUESTION 2.9.3   An apartment building has eight floors. If seven people get on the elevator on the first floor, what is the probability they all want to get off on different floors? on the same floor? that six want to get off on one floor and one on another floor? Assume all the passengers are equally likely to get off on any of the seven floors.

Sometimes in choosing k objects from a set of n it makes no sense to order what is finally selected. Thus, a poker player is dealt $k = 5$ cards from a deck of $n = 52$, but whether he receives the ace of hearts, the king of clubs, the 10 of diamonds, the 6 of clubs, and the 3 of clubs (in that order), or first the king of clubs, then the ace of hearts, the 10 of diamonds, the 6 of clubs, and the 3 of clubs— or any of the other $5! - 2 = 118$ ways to permute five objects—is irrelevant: the hand is still the same. We will call an *unordered* selection a *combination of n objects taken k at a time*.

> **THEOREM 2.10**   The number of combinations of n objects taken k at a time is equal to
>
> $$\frac{n!}{k!(n-k)!},$$
>
> which is denoted by the symbol $\binom{n}{k}$.

PROOF   Let $\binom{n}{k}$ denote the (unknown) number of ways to select k objects out of n, without ordering. If each of these combinations were to be multiplied by $k!$, the number of ways to order k objects, the product, $\binom{n}{k} k!$, would clearly be the number of *permutations* of n objects taken k at a time (without repetition). But, from Theorem 2.9, this latter number is $n!/(n-k)!$. Solving the equation

$$\frac{n!}{(n-k)!} = \binom{n}{k} k!$$

gives the result.

EXAMPLE 2.36.   How many straight lines can be drawn between five points (A, B, C, D, and E), no three of which are collinear?

By inspecting Figure 2.8, we see the number is 10, but the answer can also be

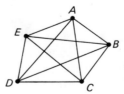

Figure 2.8 A representation of lines as combinations.

gotten by a simple application of Theorem 2.10. Each line corresponds to a combination of length $k = 2$ chosen from among the available $n = 5$ points. (Combinations rather than permutations are appropriate here, because the lines are not thought of as being directed.) Thus, the number of possibilities is

$$\binom{5}{2} = \frac{5!}{2!\,3!} = 10.$$

EXAMPLE 2.37.   Suppose the license-plate format of Example 2.34 were modified so that a tag had to include two letters and four numbers but they could be in any order. Now how many plates could be made?

Here, as in many combinatorial problems, the solution involves both permutations and combinations. First, consider positioning the two letters and the four numbers. This can obviously be accomplished in $\binom{6}{2}$, or 15, ways, since any set (combination) of positions for the two letters automatically positions the four numbers. Then, once a particular positioning has been decided on, the letters and numbers can be selected and permuted in $(26)^2(10)^4$ ways. It follows that the total number of plates is $(15)(26)^2(10)^4$, or 101,400,000.

QUESTION 2.9.4   Four men and four women are to be seated in a row of chairs numbered 1 through 8.

(a) How many total arrangements are possible?

(b) How many arrangements are possible if the men are required to sit in alternate chairs?

(c) How many arrangements are possible if the four men are considered indistinguishable and the four women are considered indistinguishable?

(d) How many arrangements are possible if the four men are considered indistinguishable but the four women are considered distinguishable?

One of the more instructive—and to some, one of the more useful—applications of combinatorics is the calculation of probabilities associated with various poker hands. It will be assumed in what follows that five cards are dealt from a poker deck and that no other cards are showing, although some may already have been dealt. The sample space is the set of all $\binom{52}{5} = \frac{52!}{5!\,47!} = 2{,}598{,}960$ different hands, each having probability $1/2{,}598{,}960$. Example 2.38 derives the probabilities of being dealt a full house, one pair, two pairs, a straight, and a flush. Probabilities for other kinds of hands are gotten in much the same way.

EXAMPLE 2.38

(a) Full house. A full house consists of three cards of one denomination and two of another. Denominations for the three-of-a-kind can be chosen in $\binom{13}{1}$ ways. Then, given that a denomination has been chosen, the three requisite suits can be selected in $\binom{4}{3}$ ways. Applying the same reasoning to the pair gives $\binom{12}{1}$ available denominations, each having $\binom{4}{2}$ possible choices of suits. Thus,

$$P(\text{full house}) = \frac{\binom{13}{1}\binom{4}{3}\binom{12}{1}\binom{4}{2}}{\binom{52}{5}} = 0.00144.$$

(b) One pair. To qualify as a one-pair hand, the five cards must include two cards of the same denomination and three "single" cards—cards whose denominations match neither the pair nor each other. For the pair, there are $\binom{13}{1}$ possible denominations and, once selected, $\binom{4}{2}$ possible suits. Denominations for the three single cards can be chosen $\binom{12}{3}$ ways (see Question 2.9.5), and each card can have any of $\binom{4}{1}$ suits. Multiplying all these factors together and dividing by $\binom{52}{5}$ gives a probability of 0.42:

$$P(\text{one pair}) = \frac{\binom{13}{1}\binom{4}{2}\binom{12}{3}\binom{4}{1}\binom{4}{1}\binom{4}{1}}{\binom{52}{5}} = 0.42.$$

(c) Two pairs. A typical two-pair hand is, say, a king of diamonds and a king of clubs, a 7 of hearts and a 7 of clubs, and a 5 of diamonds—that is, the two pairs must not have the same denomination (if they did, the hand would be labeled "four-of-a-kind") and the denomination of the fifth card must not match that of either of the pairs (otherwise, the hand would be a full house). For the two pairs there are $\binom{13}{2}$ combinations of denominations and $\binom{4}{2}\binom{4}{2}$ suits; for the fifth card, $\binom{11}{1}$ denominations and $\binom{4}{1}$ suits. Thus,

$$P(\text{two pairs}) = \frac{\binom{13}{2}\binom{4}{2}\binom{4}{2}\binom{11}{1}\binom{4}{1}}{\binom{52}{5}} = 0.048.$$

(d) Straight. A straight is five cards having consecutive denominations—for example, 4 of diamonds, 5 of hearts, 6 of hearts, 7 of clubs, and 8 of diamonds. An ace may be counted "high" or "low": that means (10, jack, queen, king, and ace) is a straight and so is (ace, 2, 3, 4, and 5). Altogether, there are 10 sets of consecutive denominations of length 5: (ace, 2, 3, 4, 5), (2, 3, 4, 5, 6), . . . , (10, jack, queen, king, ace). No restrictions are put on the suits, so each card can be either a diamond, heart, club, or spade. The probability of a straight, then, computes to 0.00394:

$$P(\text{straight}) = \frac{10\binom{4}{1}^5}{\binom{52}{5}} = 0.00394.$$

What we have just calculated is not entirely correct. While it *is* true that the proportion of hands that qualify as a straight is 0.00394, included in that number are hands that are straights in the same suit. These are called straight flushes and constitute a hand "type" themselves. Obviously, a straight flush, being rarer, would beat a straight. To calculate the probability, then, that a straight would be "called" a straight, we need to subtract from the numerator all the hands that are straight flushes. The reader is referred to Question 2.9.6.

(e) Flush. A flush is five cards in the same suit, denominations being of no importance. Thus, for a flush in, say, hearts, five cards must be selected from among the 13 denominations. This can be done in $\binom{13}{5}$ ways. A similar number of hands qualify as flushes in each of the other three suits, implying that

$$P(\text{flush}) = \frac{4\binom{13}{5}}{\binom{52}{5}} = 0.0020.$$

QUESTION 2.9.5   For one-pair hands, why is the number of denominations for the three single cards equal to $\binom{12}{3}$ rather than $\binom{12}{1}\binom{11}{1}\binom{10}{1}$?

QUESTION 2.9.6   What is the probability of being dealt a *straight flush*? Note: A straight flush is five cards in the same suit having denominations that are consecutive. What is the probability that a poker player is dealt a hand that he *calls* a straight—i.e., one that qualifies as a straight but is not a straight flush? What is the probability a player is dealt a hand that he *calls* a straight flush?

QUESTION 2.9.7   What is the probability of four-of-a-kind?

QUESTION 2.9.8   A poker player is dealt a 3 of diamonds, an 8 of clubs, a 9 of clubs, a 10 of spades, and an ace of hearts. If he discards the 3 and the ace, what are the chances he ends up with a straight?

The numbers $\binom{n}{k}$ are called *binomial coefficients* in deference to their role in the expansion of $(a + b)^n$. For n an integer, the binomial theorem says

$$(a + b)^n = \sum_{k=0}^{n} \binom{n}{k} a^k b^{n-k}.$$

The definition of $\binom{n}{k}$ can be extended, though, to arbitrary real numbers, n—and, by so doing, we can obtain a generalization of the binomial theorem. The proof will be left as an exercise.

DEFINITION 2.8   For any real number x and positive integer k, let
$$\binom{x}{k} = \frac{x(x - 1) \cdots (x - k + 1)}{k!}.$$
Also, let $\binom{x}{0} = 1$.

> **THEOREM 2.11 (Newton's binomial formula)**   For real numbers t and x,
> $$(1 + t)^x = \sum_{k=0}^{\infty} \binom{x}{k} t^k.$$

PROOF   (See Question 2.9.9.)

We shall see in Chapter 4 that these generalized binomial coefficients admit a probabilistic interpretation. For the present, though, we restrict our attention to the more familiar case where n is an integer.

QUESTION 2.9.9   Prove Newton's binomial formula using Taylor's theorem.

QUESTION 2.9.10   For integer n, verify two different ways that

$$\sum_{k=0}^{n} \binom{n}{k} = 2^n.$$

Hint: Consider the total number of ways to form groups—of any size—using the original n objects.

2.10 THE BINOMIAL AND HYPERGEOMETRIC DISTRIBUTIONS

We conclude this chapter by examining two very useful probability models, the *binomial* and the *hypergeometric*. Both have functional forms that draw on the combinatorial ideas of the previous section.

First, the binomial. Recall from Section 2.8 that the most relevant information to be gleaned from a series of n Bernoulli trials is often not which trials ended in success and which in failure but, rather, *how many* ended in success. Let Y denote that number. What we would like to find is the probability that Y takes on the value k, where k is some integer between 0 and n, inclusive. The derivation is really very straightforward: $P(Y = k)$ will equal the number of ways to position k successes (and $n - k$ failures) in n trials times the probability associated with any particular ordering. The latter figure, we have already seen, is $p^k q^{n-k}$; the former, by virtue of Theorem 2.10, is $\binom{n}{k}$. (Table 2.8 illustrates this argument for $n = 4$.)

> **THEOREM 2.12**   Let Y denote the number of successes in n Bernoulli trials, where the probability of a success at any particular trial is p (and the probability of a failure is $q = 1 - p$). Then Y is said to have a *binomial distribution* and
> $$P(Y = k) = \binom{n}{k} p^k q^{n-k}, \qquad k = 0, 1, \ldots, n.$$

EXAMPLE 2.39.   In a nuclear reactor, the fission process is controlled by inserting into the radioactive core a number of special rods whose purpose is to absorb the neutrons emitted by the critical mass. The effect is to slow down the nuclear chain reaction. When functioning properly, these rods serve as the first-line defense against a disastrous core meltdown.

Suppose that a particular reactor has ten of these control rods (in "real life" there would be more than 100), each operating independently, and each having a 0.80 probability of being properly inserted in the event of an "incident." Furthermore, suppose that a meltdown will be prevented if at least half the rods perform satisfactorily. What is the probability that, upon demand, the system will fail?

If Y denotes the number of control rods that function as they should, a system failure occurs if $Y \leq 4$. By Theorem 2.12, the probability of this happening is 0.007:

$$P(\text{system will fail}) = P(Y \leq 4) = \sum_{k=0}^{4} \binom{10}{k}(0.80)^k(0.20)^{10-k}$$

$$= \binom{10}{0}(0.80)^0(0.20)^{10} + \cdots + \binom{10}{4}(0.80)^4(0.20)^6$$

$$= 0.000 + 0.000 + 0.000 + 0.001 + 0.006$$

$$= 0.007.$$

QUESTION 2.10.1   Assume that births can be treated as Bernoulli trials, where the probability of a boy is $\frac{1}{2}$. Let Y be the number of boys in a six-child family. Find the probability distribution for Y—that is, calculate $P(Y = k)$ for $k = 0, 1, \ldots, 6$.

EXAMPLE 2.40.   When p is a rational number, the binomial probability function can be thought of in terms of an urn model. Suppose an urn contains N chips, of which r are red and w are white. For each of n "trials," a chip is to be drawn at random, its color noted, and then returned to the urn. Let Y denote the number of red chips eventually drawn. Since $p = r/N$,

$$P(Y = k) = \binom{n}{k}\left(\frac{r}{N}\right)^k\left(\frac{w}{N}\right)^{n-k}, \qquad k = 0, 1, \ldots, n.$$

QUESTION 2.10.2   As k goes from 0 to n, binomial probabilities first increase and then decrease. Under what conditions will the maximum value for $\binom{n}{k}p^k q^{n-k}$ be *shared* by two values of k?

The urn model just described called for the n samples to be drawn *with replacement*. What we want to consider next is the slightly modified problem where the sampling is done *without replacement*. Theorem 2.13 shows the effect of such a change on the probability function for Y.

THEOREM 2.13 Suppose an urn contains r red chips and w white chips ($r + w = N$). If n chips are drawn at random, without replacement, and Y denotes the total number of red chips selected, then Y is said to have a *hypergeometric distribution* and

$$P(Y = k) = \frac{\binom{r}{k}\binom{w}{n-k}}{\binom{N}{n}}, \qquad k = 0, 1, \ldots, \min(r, n).$$

PROOF   Consider the chips to be distinguishable. By Theorem 2.10, the total number of ways to select a sample of size n is $\binom{N}{n}$. By the same reasoning, there are $\binom{r}{k}$ ways to select a sample of k red chips, and, for each of those, $\binom{w}{n-k}$ ways to select enough white chips $(n-k)$ to fill out the sample. Since each selection is presumed equally likely, it follows that the probability of drawing k red chips will be the product $\binom{r}{k}\binom{w}{n-k}$ divided by the total number of samples, $\binom{N}{n}$.

Comment. The name "hypergeometric" derives from a series introduced by the Swiss mathematician and physicist, Leonhard Euler, in 1769:

$$1 + \frac{ab}{c}x + \frac{a(a+1)b(b+1)}{2!\,c(c+1)}x^2 + \frac{a(a+1)(a+2)b(b+1)(b+2)}{3!\,c(c+1)(c+2)}x^3 + \cdots.$$

This is an expansion of considerable flexibility: given appropriate values for a, b, and c, it reduces to many of the standard infinite series used in analysis. In particular, if a is set equal to 1, and b and c are set equal to each other, it reduces to the familiar *geometric* series,

$$1 + x + x^2 + x^3 + \cdots,$$

hence the name *hypergeometric*. The relationship of the probability function of Theorem 2.13 to Euler's series becomes apparent if we set $a = -n$, $b = -r$, $c = w - n + 1$, and multiply the series by $\binom{w}{n}\big/\binom{N}{n}$. Then the coefficient of x^k will be

$$\frac{\binom{r}{k}\binom{w}{n-k}}{\binom{N}{n}},$$

the value the theorem gives for $P(Y = k)$. ∎

EXAMPLE 2.41.   Urn I contains five red chips and four white chips; urn II contains four red and five white. Two chips are to be transferred from urn I to urn II. Then a single chip is to be drawn from urn II. What is the probability that the chip drawn from the second urn will be white? (See Figure 2.9.)

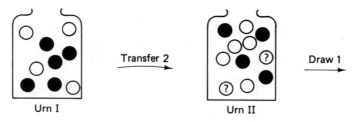

Figure 2.9   An application of the hypergeometric distribution.

Let W be the event "White chip is drawn from urn II." Let A_i, $i = 0, 1, 2$, denote the event "i white chips are transferred from urn I to urn II." Then, by Theorem 2.7,

$$P(W) = P(W|A_0)P(A_0) + P(W|A_1)P(A_1) + P(W|A_2)P(A_2).$$

Note that $P(W|A_i) = (5 + i)/11$ and that $P(A_i)$ is gotten directly from Theorem 2.13. Therefore,

$$P(W) = \left(\frac{5}{11}\right)\frac{\binom{4}{0}\binom{5}{2}}{\binom{9}{2}} + \left(\frac{6}{11}\right)\frac{\binom{4}{1}\binom{5}{1}}{\binom{9}{2}} + \left(\frac{7}{11}\right)\frac{\binom{4}{2}\binom{5}{0}}{\binom{9}{2}}$$

$$= \left(\frac{5}{11}\right)\left(\frac{10}{36}\right) + \left(\frac{6}{11}\right)\left(\frac{20}{36}\right) + \left(\frac{7}{11}\right)\left(\frac{6}{36}\right)$$

$$= \frac{53}{99}.$$

QUESTION 2.10.3   An urn contains eight chips, numbered 1 through 8. Four are selected, without replacement. Let X represent the number of the second smallest chip drawn. Find $P(X = 3)$.

EXAMPLE 2.42.   The hypergeometric distribution finds frequent application in the field of *acceptance sampling*. Consider the plight of the businessman who subcontracts the manufacture of some component that eventually becomes part of the product that he sells. Upon receiving a shipment of these components, he would like some assurance that they meet his specifications and requirements. He could, of course, inspect each and every item; but this would be costly, time-consuming, and, in some cases, impossible (what if the component were a flashbulb?). A more reasonable approach is to take a random sample from the shipment, inspect each member of the sample, and then accept the shipment (as being of sufficiently high quality) if the number of defectives found is fewer than some specified number.

As a simple example of this technique, suppose the shipment contains 100 items, out of which he selects a sample of size $n = 2$. Let Y be the number of defectives found in the sample: if $Y \geq 1$, he will return the shipment. Now suppose, in fact, the shipment is 10% defective. What the hypergeometric shows is that he will still accept the shipment 81% of the time:

$$P(\text{accepts shipment}) = P(Y = 0) = \frac{\binom{90}{2}\binom{10}{0}}{\binom{100}{2}}$$

$$= \frac{(4005)(1)}{4950} = 0.81.$$

If the shipment were 20% defective,

$$P(\text{accepts shipment}) = \frac{\binom{80}{2}\binom{20}{0}}{\binom{100}{2}}$$

$$= 0.64.$$

Figure 2.10 is a graph plotting the manufacturer's acceptance probability as a function of the shipment percent defective. Graphs of this sort are referred to as

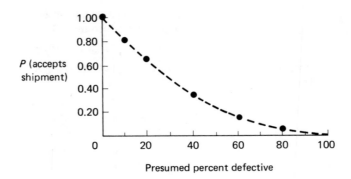

Figure 2.10   Operating characteristic curve.

operating characteristic curves: they show the sensitivity of the sampling plan in detecting lapses in shipment quality.

Among the information that Figure 2.10 imparts is the realization that the suggested sampling plan is really not very good. If the shipment were 60% defective, for example—certainly an intolerable state of affairs for the businessman—the plan would still recommend acceptance 16% of the time. To get a "steeper" operating characteristic curve, it would be necessary to increase the sample size (see Question 2.10.4).

QUESTION 2.10.4   Suppose five items are to be sampled from the shipment (of 100), with acceptance requiring that *Y* be less than or equal to 1. Construct the corresponding operating characteristic curve.

EXAMPLE 2.43.   Estimating the size of wildlife populations is a problem often confronting ecologists. Among the procedures for getting such estimates is the *capture-recapture* method, which is based on the hypergeometric distribution. In one recent study where this technique was used (23), the objective was to estimate the number of largemouth bass in Dryden Lake, a small body of water in central New York State. First, a total of $T = 213$ largemouth bass were caught, tagged, and released. Then, after sufficient time had been allowed for the tagged fish to mix with the untagged fish, a second sample of $k = 104$ bass were taken. Among these 104 were $r = 13$ that had been previously tagged. Given this information, how many largemouth bass should we estimate to be in Dryden Lake?

Let N denote the unknown number of largemouth bass in Dryden Lake and $\hat{N}$, our estimate of that number. If $\hat{N}$ were set equal to, say, 500, the probability of recapturing only 13 out of 213 tagged fish (with a sample of 104) would be

$$\frac{\binom{213}{13}\binom{287}{91}}{\binom{500}{104}}.$$

But this is a very small probability, less than 10^{-10}, and does not lend much credence to the speculation that N equals 500. At the other extreme, had we estimated N to be very large—say, 5000—the corresponding probability would be

$$\frac{\binom{213}{13}\binom{4787}{91}}{\binom{5000}{104}},$$

also a miniscule figure.

This should suggest that as an estimation procedure it would not be unreasonable to select as $\hat{N}$ the value that maximizes the likelihood of what has occurred. That is, given T, r, and k, maximize

$$\frac{\binom{T}{r}\binom{N-T}{k-r}}{\binom{N}{k}}$$

as a function of N. To do this, we consider the ratio of the probabilities for two successive values of N:

$$\frac{\binom{T}{r}\binom{N-T}{k-r}}{\binom{N}{k}}\Bigg|\frac{\binom{T}{r}\binom{N-1-T}{k-r}}{\binom{N-1}{k}} = \frac{(N-T)(N-k)}{N(N-T-k-r)}.$$

Note that this ratio is larger than 1 (i.e., the probabilities are increasing with N) if and only if

$$(N-T)(N-k) > N(N-T-k-r)$$

or, equivalently, if and only if

$$\frac{kT}{r} > N.$$

Therefore, we will set

$$\hat{N} = \frac{kT}{r}.$$

(From this follows the intuitively appealing result that the *population* ratio, $\hat{N}/T$, is equal to the *sample* ratio, k/r.) For these data, $\hat{N} = (213)(104)/13$, or 1704.

Comment. We will examine the "method of maximum likelihood" in more detail in Chapter 5. The technique is an extremely important one in applied statistics. ∎

QUESTION 2.10.5   Show that as $N \longrightarrow \infty$ such that $r/N \longrightarrow p$, sampling without replacement is equivalent to sampling with replacement. That is, show that hypergeometric probabilities converge to a corresponding binomial probability.

APPENDIX 2.1 THE MATCHING PROBLEM

Theorem 2.6 gave a formula for the probability of the union of two events:

$$P(A \cup B) = P(A) + P(B) - P(A \cap B). \tag{2.10}$$

A Venn diagram for *three* events, A, B, and C, as shown in Figure 2.11, would suggest that the probability of the three-way union can be expressed as

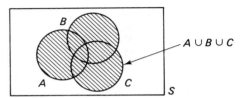

Figure 2.11 Venn diagram for $A \cup B \cup C$.

$$P(A \cup B \cup C) = P(A) + P(B) + P(C) - P(A \cap B) - P(A \cap C)$$
$$- P(B \cap C) + P(A \cap B \cap C).$$

In this appendix we first give a formal proof of the general n-event extension of Equation 2.10 and then apply the result to the so-called "matching problem," which dates back to the early eighteenth century.

THEOREM 2.14 For any n events $A_1, A_2, \ldots, A_n$ defined on a sample space S,

$$P\left(\bigcup_{i=1}^{n} A_i\right) = \sum_{i=1}^{n} P(A_i) - \sum_{i<j} P(A_i \cap A_j) + \sum_{i<j<k} P(A_i \cap A_j \cap A_k)$$
$$- \sum_{i<j<k<l} P(A_i \cap A_j \cap A_k \cap A_l) + \cdots$$
$$+ (-1)^{n+1} \cdot P(A_1 \cap A_2 \cap \cdots \cap A_n).$$

PROOF   We can partition the union of the A_i's into a collection of disjoint subsets whose outcomes belong to a certain set of A_i's and to no others. The result will be proved if it can be shown that the probability of each of these subsets is counted exactly once by the right-hand side of the statement of the theorem. (Draw a Venn diagram to convince yourself that that argument makes sense.)

Consider an event B whose outcomes belong to each of the events $A_1, A_2, \ldots,$ A_k but to none of the other events $A_{k+1}, \ldots, A_n$. That is,

$$B = A_1 \cap \cdots \cap A_k \cap A_{k+1}^C \cap \cdots \cap A_n^C.$$

Since B is a subset of k of the A_i's, $P(B)$ will be counted exactly k times in the expression $\sum_{i=1}^{n} P(A_i)$, exactly $\binom{k}{2}$ times in $\sum_{i<j} P(A_i \cap A_j)$, $\binom{k}{3}$ times in $\sum_{i<j<k}$ $P(A_i \cap A_j \cap A_k)$, and so on. Thus, it appears a total of

$$C = \binom{k}{1} - \binom{k}{2} + \binom{k}{3} - \binom{k}{4} + \cdots + (-1)^{k+1}\binom{k}{k}$$

times. But $C = 1$ because

$$(1 - 1)^k = \sum_{j=0}^{k} \binom{k}{j}(-1)^j(1)^{k-j} = \binom{k}{0} - \binom{k}{1} + \binom{k}{2} - \binom{k}{3} + \cdots = 0,$$

and this implies that

$$\binom{k}{0} = 1 = \binom{k}{1} - \binom{k}{2} + \binom{k}{3} - \cdots.$$

Since k was arbitrary, it follows that every point in the union is counted exactly once by the formula given in the theorem.

Comment. A statement analogous to Theorem 2.14 can be made for the probability of the union of an *infinite* number of events defined on a sample space S. The proof (by induction) is left to the interested reader. ■

The "matching problem" (or *probleme des rencontres*) has a long and rich history, and some of its variations and extensions are still topics of research today. The first general solution was given in 1718 by DeMoivre in his *Doctrine of Chances*, although several years earlier Montmort and Bernoulli had worked out solutions for certain special cases (124). In recent years the problem has found applications in a number of different fields, psychology and genetics being just two.

DeMoivre's statement of the problem is interesting:

> Any number of letters a, b, c, d, e, f, etc., all of them different, being taken promiscuously as it happens; to find the Probability that some of them shall be found in their places according to the rank they obtain in the alphabet and that others of them should at the same time be displaced.

One way to pose the problem in a more contemporary framework would be the following:

> Suppose n married couples sign up for a dance class. If the instructor pairs the n men with the n women *at random*, what is the probability that exactly r men will be dancing with their wives?

Actually, we will look at a somewhat simpler problem: *what is the probability that at least one husband dances with his wife?* The solution is based on a direct application of Theorem 2.14.

Let A_i be the event that occurs if husband i dances with his wife, $i = 1, 2, \ldots, n$. Note that

$$P(\text{at least one husband dances with his wife}) = P\left(\bigcup_{i=1}^{n} A_i\right).$$

Since each husband has a $1/n$ chance of being assigned to his wife,

$$\sum_{i=1}^{n} P(A_i) = n\left(\frac{1}{n}\right) = 1.$$

Similarly, the probability that husbands 1 and 2 are both dancing with their wives is $(1/n)\,[1/(n-1)]$. The same is true for any two husbands, not just 1 and 2; therefore,

$$\sum_{i<j} P(A_i \cap A_j) = \binom{n}{2} \frac{1}{n(n-1)} = \frac{1}{2!}.$$

Similarly,

$$\sum_{i<j<k} P(A_i \cap A_j \cap A_k) = \binom{n}{3} \frac{1}{n(n-1)(n-2)} = \frac{1}{3!},$$

and so on. Finally, the probability that *all* husbands are assigned to their wives is

$$P(A_1 \cap A_2 \cap \cdots \cap A_n) = \frac{1}{n!}.$$

It follows from Theorem 2.14, then, that

$$P\left(\bigcup_{i=1}^{n} A_i\right) = P(\text{at least one husband dances with his wife})$$

$$= 1 - \frac{1}{2!} + \frac{1}{3!} - \frac{1}{4!} + \cdots + (-1)^{n+1} \frac{1}{n!}.$$

Table 2.10 shows the probability of the union for various values of *n*. Note that for large *n* the probability converges (and quite rapidly) to $1 - e^{-1}$:

TABLE 2.10   Probabilities for
pairing example

n	$P(\bigcup_{i=1}^{n} A_i)$
2	0.5000
3	0.6667
4	0.6250
5	0.6333
6	0.6319
∞	0.632

$$\lim_{n\to\infty} P\left(\bigcup_{i=1}^{n} A_i\right) = \lim_{n\to\infty} \left(1 - \frac{1}{2!} + \frac{1}{3!} - \frac{1}{4!} + \cdots + (-1)^{n+1} \frac{1}{n!}\right)$$

$$= 1 - e^{-1}$$

$$= 0.632.$$

REVIEW EXERCISES FOR CHAPTER 2

1. A certain fraternity man rates each of his dates as having been either a "success" or a "failure." (His criteria are beyond the scope of this text.) Suppose he has three dates lined up for December. Write out an appropriate sample space. What elementary outcomes are in the event, "Second success occurs on third date?" What outcomes are in the event, "First success never occurs?"

2. An ecologist has trapped a male bear, a female bear, and a cub. She intends to fit each of them with a radio-transmitting collar so they can be kept under electronic surveillance after they are released. To allow each bear to be distinguishable from the other two, each collar will transmit on a different frequency. Write out a sample space describing the ecologist's options in assigning the three collars to the three bears.

3. Urn I contains three red chips and two white chips. Urn II contains two red chips and one white chip. Two chips are to be transferred from urn I to urn II *sequentially*. Then two chips are to be drawn from urn II *simultaneously*. Write out the appropriate sample space. Assume all the red chips are alike and all the white chips are alike. Let A be the event, "A red chip and a white chip are transferred and a red chip and a white chip are drawn." Let B be the event, "First chip transferred is red." Find (a) $A \cap B$ and (b) $A \cup B$.

4. Engine blocks coming off an assembly line are numbered serially. During one particular work shift, the six blocks produced are numbered 17850 through 17855. An inspector selects two of the blocks at random to examine in detail. Write out an appropriate sample space.

5. A probability-minded despot offers a convicted murderer a final chance to win his release. The prisoner is given 20 chips, 10 white and 10 black. All 20 are to be placed into two urns, according to any allocation scheme the prisoner wishes, with the one proviso that each urn must contain at least one chip. The executioner will then pick one of the two urns at random and from that urn, pick one chip at random. If the chip selected is white, the prisoner will be set free; if it is black, the less said the better. Write out a sample space describing the prisoner's possible allocation schemes. Write out a second sample space describing the executioner's role and its consequences. Assuming the prisoner wants to live, how should he allocate the chips?

6. Let $E_i = \{x : 0 \le x < 1/i\}$, $i = 1, 2, \ldots, k$. Describe the sets

   (a) $\displaystyle\bigcup_{i=1}^{k} E_i.$ (b) $\displaystyle\bigcap_{i=1}^{k} E_i.$

7. A standard deck of cards is dealt to the four players, North, South, East, and West. For each integer j between 0 and 13, define N_j to be the event that North has at least j spades. Define similarly the events S_j, E_j, and W_j. Which of the following are true?
   (a) $S_6 \cap E_8 = \varnothing.$ (b) $S_0 = S$ (the sample space).
   (c) $S_3 \cap N_3 \cap E_3 \cap W_3 = \varnothing.$ (d) $S_4 \cup N_4 \cup E_4 \cup W_4 = S.$
   (e) $S_{13}^c = N_1 \cup E_1 \cup W_1.$

8. Using the notation of Review Exercise 7, describe the following events:
   (a) North has exactly four spades.
   (b) North has four spades or less.
   (c) Everyone has the same number of spades.
   (d) North has at least four spades, but South has no more than two.
   (e) North and South have at least seven spades between them.

9. A certain backwoods religious cult requires all its members to demonstrate their faith by handling rattlesnakes. A total of six snakes are put into an open pit—two are poisonous, the other four have been defanged. Church rules require that a person reach in at random and grab one snake with each hand. Write out an appropriate sample space. Assume all six snakes are distinguishable, but consider the events "Snake A in left hand and snake B in right hand" and "Snake A in right hand and snake B in left hand" to be the same.

10. An electronic system has four components, divided into two pairs. The two components of a pair are wired in parallel while the pairs themselves are in series. Let A_{ij}

denote the event "The ith component in the jth pair fails," $i = 1, 2; j = 1, 2$. Let A be the event, "System fails." Write an expression for A in terms of the A_{ij}'s.

11. Let A and B be any two events. Use Venn diagrams to show that (a) the complement of the intersection is the union of the complements:

$$(A \cap B)^C = A^C \cup B^C$$

and (b) the complement of the union is the intersection of the complements:

$$(A \cup B)^C = A^C \cap B^C.$$

Prove the same results *formally*. (See Appendix 2.1.)

12. Let A be the set of x's for which $x^2 + 2x = 8$ and let B be the set of x's for which $x^2 + x = 6$. Find $A \cap B$ and $A \cup B$.

13. Consider the general quadratic equation,

$$x^2 + 2bx + c = 0.$$

Let A be the event "Equation has complex roots." Characterize the event A in terms of a set of (b, c) values.

14. (a) Let A, B, and C be any three events, not necessarily disjoint. Use Venn diagrams to suggest a formula for $P(A \cup B \cup C)$.
    (b) Use your answer to part (a) to suggest a formula for $P(A \cup B \cup C \cup D)$.

15. Prove that the following two distributive laws hold for unions and intersections:
    (a) $A \cap (B \cup C) = (A \cap B) \cup (A \cap C)$.
    (b) $A \cup (B \cap C) = (A \cup B) \cap (A \cup C)$.

16. A high school senior applies for admission to College A and to College B. She estimates the probability of acceptance at A to be 0.7, of acceptance at B to be 0.4, and of the rejection of at least one of her applications to be 0.6. What is the probability that she will be admitted to at least one of the colleges?

17. Let A and B be any two events defined on S. Suppose $P(A) = 0.4, P(B) = 0.5$, and $P(A \cap B) = 0.1$. Find the probability that A or B but not both occur.

18. Doomsday Airlines has an old propeller plane that will fly if either or both of its engines function. Suppose P(port engine fails) $= 0.10$, P(starboard engine fails) $= 0.15$, and P(both engines fail) $= 0.015$. What is the probability the plane will safely complete its next flight? Do the problem two different ways.

19. Consolidated Industries has come under considerable pressure to eliminate its discriminatory hiring practices. Company officials have agreed that during the next five years, 60% of their new employees will be females and 30% will be black. One out of four new employees, though, will be white males. What percentage of black females are they committed to hiring?

20. An urn contains six chips, numbered 1 through 6. Three chips are drawn out at random. What is the probability that the largest chip in the sample is a 5?

21. Suppose two fair dice are rolled. What is the probability that the number showing on one will be exactly twice the number appearing on the other?

22. For any element s in the set $S = \{s : s = 2, 3, 4, \ldots\}$, let $P(s) = c(\frac{2}{3})^s$. Determine the c that makes $P(s)$ a probability function.

23. (a) Let $f(x) = k(x - 60)^2$, $60 \leq x \leq 80$. Find the value of k that makes $f(x)$ a probability function.

    (b) Suppose $f(x)$ is the probability function describing the likelihood that a motorist traveling x miles per hour will be caught in a police radar trap. What is the probability that a motorist traveling in excess of 75 mph will be caught?

24. In the kinetic theory of gases, the distance, X, that a molecule travels before colliding with another molecule is described probabilistically by the function

$$f(x) = \left(\frac{1}{\lambda}\right)e^{-x/\lambda}, \qquad x > 0.$$

The average distance between collisions is called the *mean free path*, μ, and is defined by the expression

$$\mu = \int_0^\infty x \cdot f(x)\, dx.$$

What is the probability that the distance a molecule travels between consecutive collisions is less than half the mean free path?

25. It can be shown that

$$f(x) = \frac{1}{(\alpha - 1)!\, \beta^\alpha} x^{\alpha-1} e^{-x/\beta}, \qquad x > 0, \quad \beta > 0, \alpha \text{ a positive integer}$$

is a probability function (see Chapter 4). Assuming that to be true, evaluate

$$\int_0^\infty x^5 e^{-3x}\, dx.$$

26. Assume the reaction time of motorists over the age of 70 to a certain visual stimulus is described by a continuous probability function of the form,

$$f(x) = xe^{-x}, \qquad x > 0,$$

where the units of x are seconds. Let A be the event "Motorist requires longer than 1.5 seconds to respond." Find $P(A)$.

27. The time it takes a commuter to travel from home to the nearest train station is uniformly distributed between 15 and 20 minutes. His train leaves at exactly 7:30 A.M. Find the probability he catches the train if he leaves home at 7:12.

28. One choice for the probability function describing the life of an ordinary 60-watt bulb is

$$f(x) = \tfrac{1}{1000}e^{-(1/1000)x}, \qquad x > 0.$$

    (a) What is the probability that such a bulb will burn more than 1000 hours?

    (b) Find the bulb's *median* life—that is, find the number m such that the probability of a lifetime less than m equals the probability of a lifetime greater than m.

29. A batch of small-caliber ammunition is accepted as satisfactory if one shell, selected at random from the batch, is fired and lands within 2 feet of the center of a target. Assume that for a given batch, the probability function describing r, the distance from the target center to a shell's point of impact, is

$$f(r) = \frac{2re^{-r^2}}{1 - e^{-9}}, \qquad 0 < r < 3.$$

Find the probability the batch will be accepted.

30. One card is drawn from a standard deck of cards.
   (a)   Find the probability the card is a king given that the card is a club.
   (b)   Find the probability the card is a club given that the card is a king.
   (c)   What is the probability of the card's being a 10 given that it is not a face card?

31. The Starship *Enterprise* is planning a surprise attack against the Klingons at 0100 hours. The possibility of a meteor shower, however, is causing Captain Kirk and Mr. Spock to reassess their strategy. According to Spock's calculations, the probability of encountering a heavy meteor shower at 0100 hours is 0.6, a light meteor shower, 0.3, and no meteor shower, 0.1. Captain Kirk feels that the probability of the attack's being a success is 0.8 if there is a heavy meteor shower, 0.7 if there is a light meteor shower, and 0.2 if there is no meteor shower. Spock claims the attack will be a tactical misadventure if its probability of success is not at least 0.7306. Should they attack?

32. The governor of a certain state has decided to come out strongly for prison reform and is preparing a new early release program. Its guidelines are fairly simple: if a prisoner is related to a member of the governor's staff, he has a 90% chance of being pardoned; if he is not a relative, his chances for release are 0.01. Suppose that 40% of all inmates are related to someone on the governor's staff. What is the probability that a prisoner selected at random will be released? Suppose we know that a prisoner was released. What is the probability that that prisoner was, in fact, related to someone in the administration?

33. Two myopic deer hunters fire rifles simultaneously at a nearby rooster. The probability of hunter A's shot killing the rooster is 0.2; hunter B's, 0.3. Suppose the rooster is hit and killed by only one bullet. What is the probability that hunter B fired the fatal shot?

34. Two sections are being taught of a certain statistics course. From what she has heard about the two instructors listed, Jasmine estimates that her chances of passing the course are 0.85 if she gets instructor 1 and 0.60 if she gets instructor 2. Which section she gets put into is determined by the registrar. Suppose that her chances of being assigned to instructor 1 are 0.40. Fifteen weeks later we learn that Jasmine did, indeed, pass the course. What is the probability she was enrolled in instructor 1's section?

35. Let S be a sample space and P a probability function defined on the events of S. Suppose H is a subset of S with $P(H) > 0$. Define a probability function P_H on S by $P_H(A) = P(A|H)$, for any $A \subset S$. Show that P_H satisfies Axioms 1, 2, and 3 of Section 2.3.

36. In a certain congressional race, the Republican candidate (R) is running unopposed. There are three Democrats (D_1, D_2, D_3) seeking the nomination. Assume these latter three have probabilities 0.35, 0.40, and 0.25, respectively, of being nominated. Furthermore, it is thought that R's chances of winning the general election over D_1, D_2, and D_3 are 0.40, 0.35, and 0.60, respectively. What is the overall probability that R will win the election?

37. During a power blackout, 100 persons are arrested on suspicion of looting. Each is given a polygraph test. From past experience it is known that the polygraph is 90% reliable when administered to a guilty suspect and 98% reliable when given to someone who is innocent. Suppose that of the 100 persons taken into custody, only 12 were actually involved in any looting. What is the probability that a given suspect is innocent given that the polygraph says he is guilty? ⁊⧸

38. What is the maximum release probability for the execution lottery described in Review Exercise 5?

39. One urn contains three red chips and one white chip. A second urn contains two red chips and two white chips. One chip is drawn from each urn and placed in the other urn. Then a chip is drawn from the first urn. What is the probability that that chip is red?

40. Recently the U.S. Senate Committee on Labor and Public Welfare was investigating the feasibility of a national screening program to detect child abuse. A team of consultants estimated the following probabilities: (1) approximately 1 child in 100 is abused, (2) a physician can detect an abused child about 90% of the time, and (3) a national screening program would incorrectly label about 3% of the nonabused children as abused. What is the probability a child is actually abused given the screening program diagnoses him as such? How does this probability change if the incidence of abuse is 1 in 1000? one in 50?

41. Three prisoners, A, B, and C, all with identical records, are up for parole and the Board decides to release two of them, without saying which two. Clearly, $P(A$ is released$) = \frac{2}{3}$. It so happens, though, that A has befriended one of the jailers who knows the board's decision. Curious about his fate, A considers asking the jailer to tell him which prisoner, other than himself, will be released (he deems it too bold to ask about his own fate directly). After thinking about it, however, he decides not to pursue the matter because it seems that by asking the jailer for information, he will have diminished his chances for release. For example, if the jailer says, "B will be the other prisoner released," then the Board will release either A and B or B and C, suggesting that A's chances have now fallen to $\frac{1}{2}$. Is A's reasoning correct?

42. Show that the following argument is incorrect. A study has shown that 7 out of 10 people, in calling a coin toss, will call heads. But heads occurs only 5 times out of 10. Thus, it is to your advantage to let the other person call the toss.

43. The 330 voters in a certain precinct in a large U.S. city have the following age and race composition:

	Under 30 (A)	30 + (A^C)
Black (B)	50	60
White (B^C)	100	120

Are A and B independent? Explain.

44. Mark is not a terribly bright student. His chances of passing chemistry are 0.35, his chances of passing mathematics are 0.40, and his chances of passing both are 0.12. Are the events "Mark passes chemistry" and "Mark passes mathematics" independent? What are his chances of failing both subjects?

45. Suppose two events A and B, each with nonzero probability, are mutually exclusive. Are A and B dependent or independent? Explain.

46. An insurance company has three clients whose estimated probabilities of living to the year 1985 are 0.7, 0.9, and 0.3, respectively. What is the probability that by 1985 the company will have had to pay death benefits to exactly one of the three policy-holders? Assume the fates of the three clients are independent.

47. Show that $P(A \cap B \cap C) = P(A|B \cap C)P(B|C)P(C)$.

48. There are four spiders living in four student post-office boxes. One box belongs to a freshman, one to a sophomore, one to a junior, and one to a senior. If undisturbed by any incoming mail, each spider will spin a web in the box each day. Suppose the probability that the freshman gets mail on any given day is 0.7; the sophomore, 0.5; the junior, 0.4; and the senior, 0.3. Assume the mail delivered to each student is independent of what is delivered to any other student. What is the probability that on any given day exactly three of the spiders will have a chance to spin a web?

49. In a roll of a pair of dice (one red and one green), let A be the event that the red die shows a 3, 4, or 5, let B be the event the green die shows a 1 or a 2, and let C be the event that the dice total is 7. Show that A, B, and C are independent.

50. In a roll of a pair of dice (one red and one green), let A be the event of an odd number on the red die, B, an odd number on the green die, and C, an odd sum. Show that any pair of these events is independent but that A, B, and C are not mutually independent.

51. Diane and Lew are health physicists employed by the Department of Public Health. One of their duties is being "on call" during nonworking hours to handle any nuclear-related incidents that might endanger the public safety. Each carries a beeper that can be activated by personnel at Civil Defense Headquarters. When a call is received, they are to investigate the incident immediately and take whatever action is warranted. Lew is a conscientious worker and is within earshot of his beeper 80% of the time. Diane is somewhat less reliable and would be capable of responding to a beeper alert only 40% of the time. If Lew and Diane respond independently, what is the probability that at least one of them could be contacted in the event of a nuclear emergency? Suppose Mike, who has a 60% chance of hearing his beeper go off, is added to the emergency team. How much would his presence increase the team's response probability?

52. Define a probability function P' by

$$P'(A) = \frac{P(A \cap B)}{P(B)}.$$

Show that P' satisfies Axioms 1–3 of Section 2.3.

53. Suppose A is the set of points either on the circumference or in the interior of a circle of radius 1. Let B be the set, $B = \{x: 0 \leq x \leq 1\}$. Geometrically, what is the Cartesian cross product, $A \otimes B$?

54. A, B, and C are to fight a three-cornered pistol duel. All combatants know that A's chance of hitting his target is 0.3, C's is 0.5, and B never misses. They are to fire at their choice of target in succession, and cyclically, in the order A, B, C until only one man is left unhit. (If a combatant is hit he no longer participates, either as a shooter or as a target.) Show that A's optimal strategy (assuming he wants to live) is to deliberately fire his first shot into the ground (115).

55. The quality control policy at a certain auto manufacturer requires that one out of every n cars coming off the assembly line be thoroughly road tested. Unknown to the inspectors, the assembly line personnel include in each group of n cars one out-and-out "lemon." Assume all the other cars in each group of n could pass the road inspection. Suppose n such groups are inspected. As n gets large, what is the limit of the probability that none of the lemons will ever be detected by the inspectors?

56. The astragalus bones mentioned in Chapter 1 have four sides, whose probabilities of occurrence are as follows: broad convex side, 0.39; opposite broad side, 0.37; other two sides, 0.12 each. The Greeks threw four astragali; the best throw was a *Venus*, which occurred when all four sides were different. What is the probability of throwing a Venus?

57. A shifty stockbroker elects to invest all of his client's money in Ne'er-do-well, Inc., currently selling at $6 a share. The closing price of Ne'er-do-well fluctuates randomly from day to day, going up 1 point with probability $\frac{1}{2}$ and down 1 point with probability $\frac{1}{2}$. If it ever drops to $5 a share, the client will go bankrupt. What is the probability of that happening?

58. Suppose n people each draw a single card from one of n poker decks. Let A be the event that at least one face card is drawn. Find the smallest n so that $P(A) \geq 0.95$.

59. Suppose that since the early 1950s some 10,000 independent UFO sightings have been reported to civil authorities. If the probability of any particular sighting's being genuine is on the order of 1 in a 100,000, what is the probability that at least one sighting was genuine?

60. Players A, B, and C toss a fair coin in order. The first to throw a head wins. What are their respective chances of winning?

61. You are playing for the Monopoly Championship of the World. Your opponent is on GO. It is your turn and you have enough money to put a house on either Oriental Avenue, Vermont Avenue, or Connecticut Avenue. The three properties are 6, 8, and 9 spaces away from GO, respectively. Where should you put the house?

62. The British Museum has a crystal die made in classical times. The die was rolled 204 times with the following results:

Face	1	2	3	4	5	6
Frequency	30	38	31	34	34	37

Using these data to estimate the necessary probabilities, compute the likelihood of rolling a seven in two throws.

63. A fair die is rolled until a 6 shows. What is the probability it will take k rolls for that to happen? What is the probability of the first 6 appearing on an even-numbered roll?

64. Suppose three points are chosen at random on a circle—that is, the points are chosen independently from the uniform probability function over the interval $(0, 2\pi)$. What is the probability of the three points all lying in the same semicircle?

65. Suppose four people (A, B, C, and D) each toss a fair die in order (A first, then B, and so on) until the "6" face appears for the first time. What is the probability that C is the one who rolls the first 6?

66. A certain kind of word puzzle found in many newspapers has 20 sentences, each requiring that a word be filled in. For each sentence two choices are provided, only one being correct. How many entries must a person submit before he can be certain that one of his solutions is entirely correct?

67. In the Czechoslovakian game of "Sportka," a player selects six numbers from 0, 1, 2, . . . , 49. Once a week six numbers from this set are drawn and a player wins if he gets three, four, five, or six correct. What are the probabilities of each of these wins?

68. A chemist wishes to observe the effect of temperature, pressure, and the amount of catalyst on the yield of a particular chemical in a certain reaction. If the experimenter chooses to include two different temperatures, three pressures, and two catalysts, how many experiments must be conducted in order to run each temperature-pressure-catalyst combination exactly once?

69. The Alpha Beta Zeta sorority is trying to fill a pledge class of nine new members during Fall Rush. Among the 25 available candidates, 15 have been judged marginally acceptable and ten highly desirable. How many ways can the pledge class be chosen to give a two-to-one ratio of highly desirable to marginally acceptable candidates?

70. A history examination consists of ten questions. Students are to omit two of the first four and two of the last six. How many different sets of questions can a student elect to answer?

71. If four Americans, three Frenchmen, and three Englishmen are to be seated in a row at a U.N. banquet, how many arrangements are possible if the Americans are to occupy the first four seats, the Frenchmen the next three, and the Englishmen the last three? How many arrangements are possible if members of the same nationality must sit next to one another? How many arrangements are possible if no constraints are placed on where anyone can sit?

72. Suppose each of ten sticks is broken into a long part and a short part. The 20 parts are arranged into ten pairs and glued back together, so that again there are ten sticks. What is the probability that each long part will be paired with a short part? (Note: This problem is a model for the effects of radiation on a living cell. Each chromosome, as a result of being struck by ionizing radiation, breaks into two parts, one part containing the centromere. The cell will die unless the part containing the centromere recombines with one not containing a centromere.)

73. A certain cabaret singer always opens his act by telling four jokes. His current engagement is scheduled to run for four months. If he gives one performance a night and never wants to repeat the same set of jokes on any two nights, what is the minimum number of jokes he must have in his repertoire?

74. What is the probability that the four aces in a well-shuffled poker deck will all be adjacent? Hint: As a "model," consider the aces to be joined together and shuffled as a single card.

75. The crew of *Apollo 17* consisted of two pilots and one geologist. Suppose that NASA had actually trained a total of nine pilots and four geologists. How many possible *Apollo 17* crews could have been formed? (a) Assume the two pilot positions have identical duties. (b) Assume the two pilot positions are really a pilot and a copilot.

76. Suppose there are 3 million registered passenger cars in Tennessee, with license plates numbered consecutively from 1. If a car is selected at random, what is the probability that the first digit of the license plate will be 1?

77. Thirteen tombstones in a country churchyard are arranged in three rows, four in the first row, five in the second, and four in the third. Suppose that two women are buried in each row. Assuming each arrangement to be equally likely, what is the probability that in each row the women occupy the two leftmost positions? Is the assumption of equally likely arrangements a reasonable one here?

78. Ten basketball players meet in the school gym for a pickup game. How many ways can they be split up into two teams of five each?

79. A national cosmetic manufacturer has a line of eye shadows that comes in seven different colors. If any number of them can be mixed to create new shades, how many different colors are possible?

80. How many poker hands will have exactly three cards less than or equal to 10 in either clubs or spades (or both) and two cards higher than 10 in either hearts or diamonds (or both)?

81. A Scrabble set consists of 54 consonants and 44 vowels. What is the probability that your initial draw (of seven letters) will be all consonants? six consonants and one vowel? five consonants and two vowels?

82. Suppose 20 economists, including Milton Friedman and John Kenneth Galbraith, line up for a picture. What is the probability that there are exactly three economists standing between Friedman and Galbraith? For philosophical reasons, assume that Friedman never stands to the left of Galbraith.

83. How many five-card poker hands will have two jacks and three numerical cards whose sum is 28?

84. Samuel Pepys was the greatest diarist of the English language. He was also a friend of Sir Isaac Newton and, in 1693, sought the latter's advice in a matter related to gambling. Phrased in modern terminology, Pepys' question can be stated as follows: Is it more likely to get at least one 6 when six dice are rolled, at least two 6's when 12 dice are rolled, or at least three 6's when 18 dice are rolled? After considerable correspondence [see (155)], Newton was finally able to convince a skeptical Pepys that the former has the greatest likelihood. Compute these three probabilities.

85. If five cards are dealt off the top of a poker deck, we know that P(full house) = 0.0014 (recall Example 2.38). Suppose, though, one card is dealt off the deck face down and *then* a five-card hand is dealt. What is the probability that this "second" five-card hand will be a full house?

86. A bleary-eyed freshman awakens one morning and pulls two socks at random out of a drawer that contains ten black, six brown, and two blue socks, all randomly "arranged." What is the probability that the two he draws are a matched pair?

87. Use a combinatorial argument to show that

$$\binom{n+1}{k} = \binom{n}{k} + \binom{n}{k-1}.$$

Hint: Consider a set of $n+1$ distinguishable elements, $a_1, a_2, \ldots, a_{n+1}$, and the number of ways any particular one, say, a_1, can appear in subsets of size k.

88. Urn I contains four white chips and five red chips. Urn II contains five white chips and four red chips. Two chips are transferred from urn I to urn II. Then a single chip is drawn from urn II. What is the probability that the chip drawn from urn II will be white?

89. A coke hand in bridge is one containing no card higher than a 9 (aces are high). Find its probability.

90. Six dice are rolled at one time. What is the probability that each of the six faces appears?

91. Each day a stock price moves up one point or down one point with probabilities $\frac{1}{4}$ and $\frac{3}{4}$, respectively. What is the probability that after four days the stock will have returned to its original price? Assume the daily price fluctuations are independent events.

92. In five-card poker, do four aces beat a royal flush? Answer the question by calculating the two probabilities involved. (A royal flush is a 10, J, Q, K, A of the same suit.)

93. Six politicans meet at a party. How many handshakes are exchanged if each politician shakes hands with every other politician once and only once?

94. A pinochle deck has 48 cards, two of each of six denominations (9, J, Q, K, 10, A), and the usual four suits. Among the many hands that count for meld is a *roundhouse*, which occurs when a player has a king and a queen of each suit. In a hand of 12 cards, what is the probability of getting a "bare" roundhouse (a king and queen of each suit and no other kings or queens)?

95. A certain chain of fast-food restaurants offers its customers a choice of eight "extras" to put on their hamburgers (pickles, onions, and so on). How many different hamburgers can be made? Assume the order in which the extras are added is irrelevant.

96. License plates in a certain state are restricted to six characters—either two letters and four numbers or one letter and five numbers. If two letters are used, they must not be adjacent. Also, the letters must be different. In single-letter plates, the letter can appear anywhere, but the numbers must all be different. How many different plates can be made?

97. A five-card hand is dealt from a standard poker deck. In the game of Night Whammy, a blue turtle is defined to be a hand consisting of two face cards (J, Q, or K) and three numerical cards (2 through 10) of different but *even* denominations. What is the probability of a blue turtle?

98. Given a set of n numbers, we define their *range* to be the difference between the largest and the smallest. Suppose four chips are drawn (without replacement) from an urn containing ten chips numbered 1 through 10. What is the probability that the range of the sample drawn will be 6?

99. A burglar is casing a floor of rooms in your dormitory. He narrows his attention to ten equally vulnerable rooms, one of which is yours and another your best friend's. If the burglar intends to hit four rooms during one night, what is the probability that both you and your best friend will be robbed?

100. A game is played with a 45-card deck composed of nine denominations (6, 7, 8, 9, 10, J, Q, K, A) and five suits (hearts, diamonds, spades, clubs, and fishes). A hand consists of four cards. Find the probability of a *glorp* (two cards less than or equal to 10 and two cards higher than 10).

101. Show that

(a) $\sum_{k=0}^{n} \binom{n}{k} = 2^n.$ (b) $\sum_{k=1}^{n} k \binom{n}{k} = n2^{n-1}.$

102. A certain freshman has been having a little trouble getting dates for Saturday night. On the average, only one out of every ten girls he asks will go out with him (and no one has ever gone out with him twice). Suppose he asks the next six girls he sees for a date. What is the probability that at least one says "Yes"? What is the probability that the first two say "Yes" and the last four say "No"? What is the probability that exactly two say "Yes"? How many girls does he have to ask in order to be at least 90% certain of having a date Saturday night? Assume that each girl is an independent Bernoulli trial and that he asks all n before getting any replies.

103. Consider again the probability function $f(r)$ defined in Review Exercise 29. With what probability will the batch of ammunition be accepted if the criterion for acceptance is that fewer than two out of a sample of five shots fall more than two feet from the center of the target?

104. Four army officers who contracted malaria during the Second World War are being treated with a new quinine derivative. Suppose the drug is 75% successful. What is the probability that an odd number of the officers will be cured?

105. In the spring of 1978 Pete Rose of the Cincinnati Reds set a National League record by hitting safely in 44 consecutive games. Assume that Rose is a .300 hitter and that he comes to bat (officially) four times each game. If each at-bat is assumed to be an independent and identically distributed Bernoulli trial, what is the probability of such a streak? Include in your answer the probability of his not getting a hit in the game immediately before, and immediately after, the 44 games.

106. When a certain manufacturing process is "in control," the probability of a defective fuse is $\frac{1}{100}$. At periodic intervals a box of six fuses is inspected. If two or more of the six are defective, production is halted while trouble-shooters look for an assignable cause. What is the probability that production will be halted, even though the process is functioning as it should?

107. In Review Exercise 100, a card hand known as a *glorp* is defined. The probability of being dealt such a hand is 0.38. Furthermore, the rules of the game state that if a player is dealt exactly two glorps in three consecutive hands, he is entitled to jump up and yell "Snard," at which time 50 points will be added to his score. What is the probability that a player will be dealt four glorps in six hands but only be entitled to yell "Snard" once?

108. Experience has shown that only $\frac{1}{3}$ of all patients having a particular disease will recover if given the standard treatment. A new drug is to be tested on a group of 12 volunteers. If the FDA requires that at least seven of these patients should recover before it will license the new drug, what is the probability the drug will be discredited even if it increases the individual recovery rate to $\frac{1}{2}$?

109. (a)   Let

$$f(x) = \begin{cases} x/k & \text{for } 0 < x < 2 \\ 0 & \text{elsewhere.} \end{cases}$$

Find the value of k that makes $f(x)$ a probability function.

(b)   Suppose the outcome of an experiment is described by the $f(x)$ given in part (a). If the experiment is repeated (independently) six times, what is the probability that exactly four of the outcomes will be in the interval $(0, 1)$?

110. Let d and e be any real numbers and let k be a positive integer. Prove that

$$\sum_{j=0}^{k} \binom{d}{j}\binom{e}{k-j} = \binom{d+e}{k}.$$

Hint: Compare the coefficients of t^k in

$$(1 + t)^d(1 + t)^e = (1 + t)^{d+e}.$$

111. An English lady claims to be able to detect whether tea was made by adding the milk first and the tea second, or vice versa. She says she is correct 80% of the time. As a test, we give her ten cups to classify, and we agree to grant her claim if she is correct at least seven times.

(a)   What is the probability we are convinced of her claim, even though she has no ability (that is, her probability of being correct is $\frac{1}{2}$)?

(b)   What is the probability we deny her claim, even though, in the long run, she *is* correct 80% of the time?

112. Suppose Y has a binomial distribution with

$$P(Y = k) = \binom{n}{k}p^k(1 - p)^{n-k}, \qquad k = 0, 1, 2, \ldots, n.$$

Prove that $P(Y = k)$ first increases monotonically and then decreases monotonically, achieving its maximum when k is the smallest integer less than or equal to $(n + 1)p$. Hint: Consider the ratio,

$$\frac{P(Y = k)}{P(Y = k - 1)}.$$

113. Two baseball teams are negotiating a format for deciding how to determine a division champion. Two possibilities have been mentioned: a "best two of three" playoff series or a "best three of five" series. Suppose the members of one of the teams estimate that they have a 55% chance of defeating their opponent on any given day. Which of the two playoff schemes should they support? Assume each game can be considered an independent Bernoulli trial.

114. Suppose the engines on propeller-driven airplanes fail independently with probability p. Assume such a plane will have a safe flight if at least half its engines remain operable. For what values of p will a two-engine plane be safer than a four-engine plane?

115. Huygens' *De Ratiociniis in Aleae Ludo* (recall Section 1.1) concludes with the following five problems. [See (104) for some historical comments on these exercises].

(a)   A and B play with two dice on the condition that A gains if he throws 6, and B gains if he throws 7. First A has one throw, then B has two throws, then A two throws, and so on until one or the other wins. Show that A's chance is to B's as 10,355 is to 12,276.

(b)   Three gamblers, A, B, and C, take 12 balls, of which four are white and eight black. They play with the rules that the drawer is blindfolded, A is to draw first, then B, and then C; the winner is the one who first draws a white ball. What is the ratio of their chances?

(c) Player A wagers B that, given 40 cards of which ten are of one color, ten of another, ten of another, and ten of yet another, he will draw four so as to have one of each color. Here A's chance is to B's as 1000 is to 8139. (Note: The correct solution is 1000 to 9139.)

(d) Twelve balls are taken, eight of which are black and four white. A plays with B and undertakes in drawing seven balls blindfolded to obtain three white balls. Compare the chances of A and B.

(e) A and B take each 12 counters and play with three dice on the condition that, if 11 is thrown, A gives a counter to B, and if 14 is thrown, B gives a counter to A: and he wins the game who first obtains all the counters. Show that A's chance is to B's as 244,140,625 is to 282,429,536,481.

CHAPTER THREE
Random Variables

JAKOB (JACQUES) BERNOULLI (1654–1705)

One of a Swiss family producing eight distinguished scientists, Jakob was forced by his father to pursue theological studies, but his love of mathematics eventually led him to a university career. He and his brother, Johann, were the most prominent champions of Leibniz' calculus on continental Europe, the two using the new theory to solve numerous problems in physics and mathematics. Bernoulli's main work in probability, *Ars Conjectandi*, was published after his death by his nephew, Nikolaus, in 1713.

3.1 INTRODUCTION

Throughout most of Chapter 2, probability functions were defined directly in terms of the elementary outcomes making up the sample space. Thus, if two fair dice were tossed, a P value was assigned to each of the 36 possible pairs of upturned faces: $P((2, 3)) = \frac{1}{36}$, $P((3, 2)) = \frac{1}{36}$, $P((4, 6)) = \frac{1}{36}$, and so on. We have already seen, though, that in certain situations some attribute of an outcome may hold more interest for the experimenter than the outcome itself. A craps player, for example, may be concerned only that he throws a 7, and not whether the 7 was the result of a 5 and a 2, a 4 and a 3, or a 6 and a 1. Similarly, a virologist conducting a clinical trial in which n subjects are inoculated with a new influenza vaccine would most likely focus on simply the *number* of those subjects developing flulike symptoms: knowing that, say, Mr. R and Ms. W became ill while Mr. T remained healthy would probably be peripheral to the evaluation of the vaccine's effectiveness.

In cases such as these, it could be argued that the original sample space is needlessly complicated and not appropriately attuned to the experiment's objectives. For craps, why not replace the 36-member sample space of (x, y) pairs with the less complicated 11-member set of all possible two-dice *sums*, $\{2, 3, \ldots, 12\}$? Likewise, the analysis of the vaccine trial would be facilitated if the original sample space of all possible 2^n success-failure n-tuples were replaced by the set of integers, $\{0, 1, \ldots, n\}$, giving the $n + 1$ possible numbers of successes.

In more general terminology, what these examples suggest is that certain problems can be clarified by defining a suitable function over an experiment's original sample space, a function whose domain is the members of that sample space and whose range is some relevant subset of the real numbers. Such functions are called *random variables*. [For the dice example, the "sum" random variable, call it X, would map, say, the outcome $(3, 6)$ into the real number 9, and we would write $X((3, 6)) = 9$; likewise, if Y denoted the number of successes among five vaccinated subjects, $Y((s, s, f, f, s))$ would equal 3.]

Comment. It should be acknowledged that there are probably more suitable names for what we are calling random variables. The nomenclature as it stands, though, is sufficiently entrenched to resist any attempts at reform that might be mounted here. Suffice it to say that no heuristic interpretation should be attached to the name "random variable": the words should be treated simply as technical jargon. (As Lewis Carroll's Humpty Dumpty would insist, the words mean exactly what we say they mean.) ∎

Although no formal definition was given earlier, this is not our first encounter with the notion of a random variable. Examples 2.32, 2.39, and 2.40 were all concerned with sequences of Bernoulli trials for which the most relevant information was the total number of successes. Example 3.1 describes a similar sort of situation, and while it, too, could be handled by using the ideas presented in Sections 2.8 and 2.10, we will find it helpful to start thinking of these sorts of problems in terms of random variables.

> EXAMPLE 3.1. Consider again the evidence presented in *People* v. *Collins*, as outlined in Example 2.29. The prosecution's case rested on the unlikelihood of a given couple's matching up with the six characteristics reported by the several eyewitnesses of the crime. It was estimated that the joint occurrence of a white female with blond hair combed in a ponytail riding in a yellow car with a black male having a beard and a mustache was on the order of 1 in 12 million—a number so small, the prosecution contended, that Ms. Collins and her male friend were clearly guilty, a classic open-and-shut case. Not so, argued the counsel for the defense. By approaching the same data from a slightly different perspective, they were able to show that "reasonable doubt" had not really been eliminated, despite the apparent 12-million-to-1 odds.
>
> Suppose N is taken to be the total number of couples who could conceivably have been in the area and perpetrated the crime, and p, the probability that any such couple would share the six characteristics introduced by the prosecution as evidence. Define Y (a random variable) to be the number of couples matching up with the eyewitness accounts. It is not unreasonable, here, to assume that Y is binomial, in which case,

$$P(Y = k) = \binom{N}{k} p^k (1 - p)^{N-k}, \qquad k = 0, 1, \ldots, N.$$

Therefore,

$$P(Y = 1) = Npq^{N-1}$$

and

$$P(Y \geq 1) = 1 - P(Y = 0) = 1 - (1 - p)^N,$$

from which it follows that

$$P(Y > 1) = 1 - (1 - p)^N - Npq^{N-1}.$$

Now, consider the ratio

$$\frac{P(Y > 1)}{P(Y \geq 1)} = \frac{1 - (1 - p)^N - Npq^{N-1}}{1 - (1 - p)^N} \qquad (3.1)$$

$$= P(\text{more than one of the } N \text{ couples}$$
$$\text{fit the description given that at}$$
$$\text{least one does})$$

$$= P(\text{there is at least one other couple}$$
$$\text{who could have committed the crime}).$$

If $P(Y > 1)/P(Y \geq 1)$ is anything other than a very small number, we would have to accept the possibility that Ms. Collins and her friend have a pair of lookalikes and that perhaps *they* were the culprits.

Table 3.1 shows the value of the probability ratio (Equation 3.1) for various values of N befitting a large metropolitan area and for the prosecutor's estimate of p (1/12,000,000). What the last column makes clear is that $P(Y > 1)/P(Y \geq 1)$ is *not* a particularly small number.

TABLE 3.1 Probability ratios

p	N	$P(Y > 1)/P(Y \geq 1)$
1/12,000,000	1,000,000	0.0402
1/12,000,000	2,000,000	0.0786
1/12,000,000	5,000,000	0.1875
1/12,000,000	10,000,000	0.3479

Looked at in this way, the data are certainly not as incriminating as the "1-in-12,000,000" argument would have us believe. At least that was the opinion of the California Supreme Court: based on the probability argument just presented, they overturned the initial verdict of "guilty" that had been handed down by the Superior Court of Los Angeles County.

QUESTION 3.1.1   Show that $P(Y > 1)/P(Y \geq 1)$ approaches a limiting value of approximately 0.42 for the special case where $N \longrightarrow \infty$ and $p = 1/N$. Hint: Recall the definition of e as the limit of a sequence.

As a conceptual framework, random variables are of fundamental importance in the theory of statistics: they provide a single rubric under which *all* probability problems may be brought. Even in cases where the original sample space needs no redefinition—that is, when the measurement recorded is the measurement of interest (see, for instance, Example 2.14)—the concept still applies: we simply take the random variable to be the identity mapping.

On the whole, this is a nuts-and-bolts chapter. Its primary purpose is to introduce many of the bits and pieces of random variable terminology and technique that make up the mathematics of statistics: application of these notions to real-world problems will, for the most part, come later.

3.2 DENSITIES AND DISTRIBUTIONS

If a random variable does, as claimed, measure some important aspect of a sample observation, it seems only reasonable that we would want to know the probability that that variable takes on certain values. What are the chances, for example, that the honor count in a bridge hand (4 · number of aces + 3 · number of kings + 2 · number of queens + 1 · number of jacks) equals, say, 16—or the probability that a craps player rolls a 7? In point of fact, such questions arise sufficiently often that to deal with them conveniently requires some special notation. The effect of the three definitions in this section is to transfer our attention, once and for all, from the experiment's original sample space to the sample space of the associated random variable.

DEFINITION 3.1   Let Y be a random variable defined on a sample space S. For any real number t, the *cumulative distribution function of Y* [hereafter written $F_Y(t)$ and referred to as the *cdf*] is the probability of the set of all the sample points in S whose Y-values are less than or equal to t. That is,

$$F_Y(t) = P(\{s \in S \mid Y(s) \leq t\}),$$

where P is the probability function defined on S.

We will often have occasion to consider expressions such as the following:

$$1 - F_Y(t) = P(\{s \in S \mid Y(s) > t\}) \tag{3.2}$$

and

$$F_Y(b) - F_Y(a) = P(\{s \in S \mid a < Y(s) \leq b\}). \tag{3.3}$$

(That Equations 3.2 and 3.3 are true follows immediately from the properties of P.) Also, to simplify the cdf notation, we will suppress the implicit relationship between Y and S that was set forth in Definition 3.1. Thus, the event $\{s \in S \mid Y(s) \leq t\}$ will be abbreviated to $Y \leq t$; $\{s \in S \mid a < Y(s) \leq b\}$, to $a < Y \leq b$; and so on. It is doubtful that this will lead to any confusion.

The functional properties of a cdf, beyond those self-evident from Definition 3.1, depend on the "nature" of the random variable Y. Before elaborating on that statement mathematically, we consider two examples that will prove useful as stereotypes of cdf behavior.

EXAMPLE 3.2.   Let Y be a binomial random variable defined on $n = 4$ independent Bernoulli trials where $p = P(\text{success})$. Then

$$P(Y = k) = \binom{4}{k} p^k q^{4-k} \qquad \text{for } k = 0, 1, 2, 3, 4.$$

For any t in the semiopen interval $[0, 1)$, $F_Y(t) = P(Y \leq t) = P(Y = 0) = q^4$. Similarly, for any $1 \leq t < 2$, $F_Y(t) = P(Y = 0) + P(Y = 1) = q^4 + 4pq^3$. Continuing this argument establishes $F_Y(t)$ to be a step function with jumps at the points $t = 0, 1, 2, 3,$ and 4.

$$F_Y(t) = \begin{cases} 0, & t < 0 \\ q^4, & 0 \leq t < 1 \\ q^4 + 4pq^3, & 1 \leq t < 2 \\ q^4 + 4pq^3 + 6p^2q^2, & 2 \leq t < 3 \\ q^4 + 4pq^3 + 6p^2q^2 + 4p^3q, & 3 \leq t < 4 \\ 1, & 4 \leq t. \end{cases}$$

Figure 3.1 is a graph of $F_Y(t)$ for the particular Y encountered by the Chevalier de Mere in Example 2.32: Y is the number of 6's in four rolls of a fair die and $p = P(6 \text{ is rolled}) = \frac{1}{6}$.

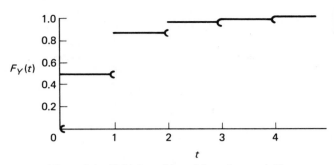

Figure 3.1 $F_Y(t)$ for a binomial random variable.

QUESTION 3.2.1   An urn contains seven chips, four red and three black. Three are drawn at random without replacement. Let Y denote the number of black chips in the sample. Graph $F_Y(t)$.

EXAMPLE 3.3.   In Example 2.15, the lifetime (in hours) of V805 radar tubes was described by the exponential probability function, $\frac{1}{179}e^{-y/179}$, $y > 0$. (What are the range and domain, here, of the random variable Y?) Note that for any $t < 0$, $F_Y(t) = 0$, but for $t \geq 0$,

$$F_Y(t) = P(Y \leq t) = \int_0^t \frac{1}{179}e^{-y/179}\, dy$$

$$= \int_0^{t/179} e^{-u}\, du = -e^{-u}\Big|_0^{t/179}$$

$$= 1 - e^{-t/179}.$$

Having derived $F_Y(t)$, we find it a simple matter to calculate the probabilities of events related to the lifetime of a randomly selected V805. For example, there is a 33% chance that such a tube will last more than 200 hours and an 11% chance that it will last longer than 144 hours but not survive past 194 hours:

$$P(Y > 200) = 1 - P(Y \leq 200) = 1 - F_Y(200)$$

$$= 1 - (1 - e^{-200/179}) = e^{-1.12}$$

$$= 0.33,$$

$$P(144 < Y \leq 194) = F_Y(194) - F_Y(144)$$

$$= (1 - e^{-194/179}) - (1 - e^{-144/179})$$

$$= (1 - 0.34) - (1 - 0.45)$$

$$= 0.11.$$

Figure 3.2 is a graph of $F_Y(t)$. The probabilities of events $Y > 200$ and $144 < Y \leq 194$ appear as lengths of intervals along the vertical axis. Of course, $F_Y(t) \longrightarrow 1$ as $t \longrightarrow \infty$.

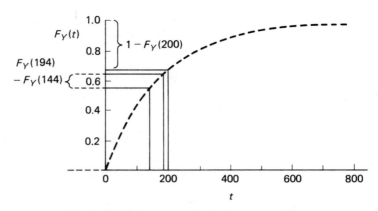

Figure 3.2 $F_Y(t)$ for an exponential random variable.

QUESTION 3.2.2   Find and graph the cdf for the missile-to-pipeline distance described in Example 2.14. Use the cdf to determine the probability that the missile will land within 20 feet of the pipeline.

QUESTION 3.2.3   Find and graph the cdf for a uniform random variable (see Section 2.5) defined over the interval (a, b).

We return at this point to the question of random-variable "types." Recall from Sections 2.4 and 2.5 that, depending on the nature of an experiment's sample space, probability functions fall into one of two categories—they are either discrete or continuous. The very same dichotomy is applied to random variables, with the distinction now resting on the properties of the domain over which Y is defined. (For the sake of comparison, the random variable in Example 3.2 is discrete; the one in Example 3.3 is continuous.)

The remainder of this section examines the relationship between these two kinds of random variables, as well as the relationship between a random variable's cdf and what we will define to be its *pdf*.

> **DEFINITION 3.2**   A *discrete random variable*, Y, is a real-valued function that can assume at most a countably infinite set of values. The *probability density function* for Y, written $f_Y(y)$ and abbreviated *pdf*, is the probability that Y takes on the value y:
> $$f_Y(y) = P(Y = y).$$

As might be expected, there is an equivalence between a discrete random variable's probability density function and its cdf: if we know one, we can determine the other. Specifically,

$$F_Y(t) = \sum_{y \le t} f_Y(y) = P(Y \le t) \tag{3.4}$$

and

$$f_Y(t) = F_Y(t) - \lim_{y \to t^-} F_Y(y). \tag{3.5}$$

We have seen the utility of Equation 3.4 in finding the cdf described in Example 3.2. But suppose what is given *is* the cdf: then the pdf can be reconstructed using Equation 3.5. Refer again to Example 3.2. For any y-value that is not a jump point, say, $y = 1.5$, the probability density function, according to Equation 3.5, will be 0 (as it should be):

$$f_Y(1.5) = F_Y(1.5) - \lim_{y \to 1.5^-} F_Y(y)$$
$$= F_Y(1.5) - F_Y(1.5) = 0.$$

On the other hand, for y equal to, say, 2,

$$f_Y(2) = F_Y(2) - \lim_{y \to 2^-} F_Y(y)$$
$$= F_Y(2) - F_Y(1)$$
$$= q^4 + 4pq^3 + 6p^2q^2 - q^4 - 4pq^3$$
$$= 6p^2q^2 = \binom{4}{2} p^2 q^2,$$

which we recognize to be the correct expression for the probability that a binomial random variable with $n = 4$ takes on the value $Y = 2$.

The next example shows a slightly less trivial application of this pdf-cdf relationship.

EXAMPLE 3.4.   An urn contains n chips, numbered 1 through n. Suppose we draw, without replacement, a sample of k chips. Let Y denote the highest-numbered chip among those drawn. We wish to find $f_Y(t)$. We can do this directly (see Question 3.2.4) or indirectly by applying Equation 3.5. Note that for any integer $1 \leq t \leq n$,

$$F_Y(t) = P(Y \leq t) = \frac{\binom{t}{k}}{\binom{n}{k}}.$$

Therefore,

$$f_Y(t) = \frac{\binom{t}{k} - \lim_{y \to t^-} F_Y(y)}{\binom{n}{k}} = \frac{\binom{t}{k} - \binom{t-1}{k}}{\binom{n}{k}}.$$

Of course, $f_Y(t) = 0$ for any other t.

QUESTION 3.2.4   Find $f_Y(t)$ for the urn problem of Example 3.4 *directly*—that is, without first determining $F_Y(t)$.

DEFINITION 3.3 A random variable Y is said to be *continuous if $F_Y(y)$* is continuous and $F_Y'(y)$ exists at all but a finite number of points. The probability density function for $Y, f_Y(y)$, is defined to be $F_Y'(y)$.

The analog of Equation 3.4 for a continuous random variable is the *fundamental theorem of calculus*:

$$F_Y(t) = \int_{-\infty}^{t} f_Y(y)\, dy. \tag{3.6}$$

As was the situation in the discrete case, $f_Y(y)$ may be zero over certain intervals, so the integral in Equation 3.6 may not need to extend to $-\infty$. Recall the V805 example. There, half the real line could be eliminated from the integration:

$$F_Y(t) = \int_{-\infty}^{t} f_Y(y)\, dy$$

$$= \int_{-\infty}^{0} 0 \cdot dy + \int_{0}^{t} \tfrac{1}{179} e^{-y/179}\, dy$$

$$= \int_{0}^{t} \tfrac{1}{179} e^{-y/179}\, dy,$$

since $f_Y(y)$ was nonzero only for $y > 0$.

Comment. It may be helpful to think of the pdf, $f_Y(y)$, for a continuous random variable as being that function which, when integrated between any two values a and b, gives the probability that Y assumes a value somewhere between those two points. That is, $f_Y(y)$ is the function that describes the behavior of Y in the sense that

$$P(a \leq Y \leq b) = \int_{a}^{b} f_Y(y)\, dy$$

for any a and b. Keep in mind that, unlike what is true for discrete random variables, $f_Y(y)$ is *not* the probability that $Y = y$ if Y is continuous. ▮

3.3 JOINT DENSITIES

Section 3.2 introduced the basic terminology for describing the behavior of a *single* random variable, whether that variable was discrete or continuous. Such information, while adequate for many problems, is insufficient in situations where the number of random variables affecting the outcome of an experiment is two or more. For example, consider an electronic system containing two tubes, one for backup, but both under load. Suppose the only way the system will fail is if both tubes cease to function. Then the probable life, Z, of the system depends *jointly* on the probable lives, X and Y, of the two tubes. Unfortunately, knowing only $F_X(x)$ and $F_Y(y)$ will not necessarily provide us with enough information to determine $F_Z(z)$. What we need is a probability function giving the "simultaneous" behavior of X and Y.

DEFINITION 3.4 Suppose X and Y are two random variables defined on the same sample space S. The *joint cdf of X and Y* is the function $F_{X,Y}(t, u)$, where

$$F_{X,Y}(t, u) = P(X \le t, \, Y \le u).$$

The domain of $F_{X,Y}(t, u)$ is the set of all pairs of real numbers.

The mathematical properties of single-variable cdf's all have their counterparts in the bivariate situation. For example, if X and Y are both discrete, we define the *joint probability density function of X and Y*, at the point (x, y), to be the probability that $X = x$ *and* $Y = y$:

$$f_{X,Y}(x, y) = P(X = x, \, Y = y).$$

From this, it follows immediately that

$$F_{X,Y}(t, u) = \sum_{x \le t} \sum_{y \le u} f_{X,Y}(x, y). \tag{3.7}$$

For X and Y both continuous, it seems reasonable (by analogy with the single-variable case) to try to define $f_{X,Y}(x, y)$ in terms of differentiability conditions on $F_{X,Y}(x, y)$. What proves to be the proper formulation should come as no surprise:

$$f_{X,Y}(x, y) = \frac{\partial^2}{\partial x \, \partial y} F_{X,Y}(x, y).$$

Equivalently, $f_{X,Y}(x, y)$ can be thought of as that function for which

$$F_{X,Y}(t, u) = \int_{-\infty}^{t} \int_{-\infty}^{u} f_{X,Y}(x, y) \, dy \, dx$$

(see Equation 3.7).

Joint pdf's and single-variable pdf's are, themselves, intimately related: given an $f_{X,Y}(x, y)$, we can "recover" the individual pdf's for X and Y by integrating out (or summing over) the unwanted variable. The necessity to do this arises quite often in some of the proofs we will see in later chapters.

THEOREM 3.1 Let X and Y be discrete random variables with joint pdf $f_{X,Y}(x, y)$. Then the probability density functions for X and for Y are given by

$$f_X(x) = \sum_{\text{all } y} f_{X,Y}(x, y), \qquad f_Y(y) = \sum_{\text{all } x} f_{X,Y}(x, y).$$

If X and Y are both continuous [with joint pdf $f_{X,Y}(x, y)$],

$$f_X(x) = \int_{-\infty}^{\infty} f_{X,Y}(x, y) \, dy, \qquad f_Y(y) = \int_{-\infty}^{\infty} f_{X,Y}(x, y) \, dx.$$

PROOF The proof will be given for the discrete case only. The analogous result for continuous X and Y follows similarly.

By definition,

$$f_X(x) = P(X = x) = P(\bigcup_{\text{all } y} (X = x, Y = y)).$$

But the events $(X = x, Y = y_i)$ and $(X = x, Y = y_j)$ are mutually exclusive for $i \neq j$, implying that

$$P(\bigcup_{\text{all } y} (X = x, Y = y)) = \sum_{\text{all } y} P(X = x, Y = y).$$

Therefore,

$$f_X(x) = \sum_{\text{all } y} P(X = x, Y = y) = \sum_{\text{all } y} f_{X,Y}(x, y).$$

The next example offers a simple application of the first part of Theorem 3.1.

EXAMPLE 3.5.   Suppose an experiment consists of three flips of a fair coin, with each outcome being equally likely. Let X denote the number of heads on the last flip; Y, the total number of heads for the three tosses. Table 3.2 shows the (x, y) value associated with each of the eight possible outcomes.

TABLE 3.2   Coin-toss experiment

Outcome	(x, y)	$P(X = x, Y = y)$
(H, H, H)	(1, 3)	$\frac{1}{8}$
(T, H, H)	(1, 2)	$\frac{1}{8}$
(H, T, H)	(1, 2)	$\frac{1}{8}$
(H, H, T)	(0, 2)	$\frac{1}{8}$
(T, T, H)	(1, 1)	$\frac{1}{8}$
(T, H, T)	(0, 1)	$\frac{1}{8}$
(H, T, T)	(0, 1)	$\frac{1}{8}$
(T, T, T)	(0, 0)	$\frac{1}{8}$

By appropriately combining the second and third columns of Table 3.2, we can write the joint pdf of X and Y as a 2×4 matrix (Table 3.3).

TABLE 3.3   Matrix for joint pdf

		Y			
	0	1	2	3	$f_X(x)$
0	$\frac{1}{8}$	$\frac{1}{4}$	$\frac{1}{8}$	0	$\frac{1}{2}$
X 1	0	$\frac{1}{8}$	$\frac{1}{4}$	$\frac{1}{8}$	$\frac{1}{2}$
$f_Y(y)$	$\frac{1}{8}$	$\frac{3}{8}$	$\frac{3}{8}$	$\frac{1}{8}$	

Note that summing across the rows gives $f_X(x)$, while summing down the columns yields $f_Y(y)$.

$$\int 8xy \quad 8xy^2 =$$

Comment. When $f_{X,Y}(x, y)$ is written as a matrix, the individual densities will appear, as they do in Table 3.3, as "margins." For this reason, $f_x(x)$ and $f_Y(y)$, in the context of a joint pdf, are often referred to as *marginal densities.* Keep in mind, though, that the use of the word "marginal" is solely for emphasis and clarity—there is absolutely no difference between a marginal density and a density.

∎

QUESTION 3.3.1   Suppose two fair dice are tossed one time. Let X denote the number of 2's that appear, and Y the number of 3's. Write out the matrix giving the joint probability density function for X and Y. Suppose a third random variable, Z, is defined, where $Z = X + Y$. Use $f_{X,Y}(x, y)$ to find $f_Z(z)$. (We will examine this sort of problem—finding the pdf for a sum—in considerable detail in Section 3.5.)

Let X and Y be discrete random variables and R, some subset of the xy-plane. It is an immediate consequence of the definition of a joint probability density function that

$$P((X,\ Y) \in R) = \sum_{(x,y) \in R} f_{X,Y}(x, y).$$

It would be reassuring if the analogous statement

$$P((X,\ Y) \in R) = \int_R \int f_{X,Y}(x, y)\ dy\ dx$$

were true for continuous random variables—and it is, provided suitable restrictions are placed on R. A careful analysis of what those restrictions are would be technically tedious and, at this stage, not particularly illuminating. We will look instead at a set of examples that illustrate three of the situations encountered when evaluating double integrals.

EXAMPLE 3.6.   Consider the electronic system mentioned in the beginning of this section (for the jargon lover, a system with "hot redundancy"). Suppose the two tubes operate independently and have identical performance characteristics. Let X and Y be random variables denoting their life spans. Experience with tube wearout times suggests that in some cases a good choice for $f_{X,Y}(x, y)$ would be

$$f_{X,Y}(x, y) = \begin{cases} \lambda^2 e^{-\lambda(x+y)}, & x \geq 0, y \geq 0 \\ 0, & \text{otherwise} \end{cases}$$

where λ is some constant greater than 0.

Suppose the manufacturer advertises a money-back guarantee if the system fails to last for more than 1000 hours. What are the chances of a given system's being returned for a refund? Since the system fails only if both tubes fail,

$$P(\text{refund}) = P(X \leq 1000,\ Y \leq 1000)$$

$$= \int_0^{1000} \int_0^{1000} \lambda^2 e^{-\lambda(x+y)}\ dy\ dx$$

$$= \int_0^{1000} \left(\int_0^{1000} \lambda^2 e^{-\lambda x} e^{-\lambda y}\ dy \right) dx.$$

The integration in parentheses is done with respect to y, with x being treated as a constant. Consequently, we can factor out $\lambda e^{-\lambda x}$, in which case

$$\int_0^{1000} \left(\int_0^{1000} \lambda^2 e^{-\lambda x} e^{-\lambda y} \, dy \right) dx = \int_0^{1000} \lambda e^{-\lambda x} \left(\int_0^{1000} \lambda e^{-\lambda y} \, dy \right) dx$$

$$= \int_0^{1000} \lambda e^{-\lambda x} (1 - e^{-\lambda \cdot 1000}) \, dx$$

$$= (1 - e^{-\lambda \cdot 1000}) \int_0^{1000} \lambda e^{-\lambda x} \, dx.$$

Evaluating this final integral gives

$$P(\text{refund}) = (1 - e^{-1000\lambda})^2.$$

QUESTION 3.3.2   Suppose X and Y have a bivariate uniform density over the unit square:

$$f_{X,Y}(x, y) = \begin{cases} c, & 0 < x < 1, 0 < y < 1 \\ 0, & \text{elsewhere.} \end{cases}$$

(a) Find c.
(b) Find $P(0 < X < \frac{1}{2}, 0 < Y < \frac{1}{4})$.

Example 3.6 was particularly pleasant, as double integrals go, because $f_{X,Y}(x, y)$ factored into a product, with one factor involving only x, the other depending solely on y. (Fortunately, this condition—"independence," as we shall see in Section 3.4—is quite common.) Another simplifying feature was the rectangularity of the region of integration. The next example redefines the roles of the two tubes and, in so doing, illustrates the adjustments that must be made when the region of integration is *not* rectangular.

EXAMPLE 3.7.   Let X, Y, and $f_{X,Y}(x, y)$ be defined as they were in Example 3.6. However, suppose the system itself is modified so that one tube is kept unused to serve as a replacement for the other. As before, the system fails only when both tubes burn out, but now the tube lives are cumulative. (This is known as "cold redundancy.") Again we seek the probability that the system fails in 1000 hours or less. Translated into a statement about X and Y, the probability of that event can be written

$$P(X + Y \leq 1000) = \int_R \int \lambda^2 e^{-\lambda(x+y)} \, dy \, dx,$$

where R is the shaded region shown in Figure 3.3.

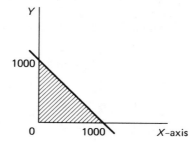

Figure 3.3   Graph showing R.

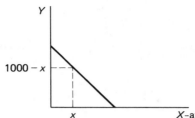

Figure 3.4 Variation of y.

As in the previous example, the double integral is evaluated by first holding x constant. But note that, here, the upper limit of integration for y depends on x: for x fixed, y varies between 0 and $1000 - x$ (Figure 3.4).

Thus,

$$
\int_R \int f_{X,Y}(x, y)\, dy\, dx = \int_0^{1000} \left(\int_0^{1000-x} \lambda^2 e^{-\lambda x} e^{-\lambda y}\, dy \right) dx
$$

$$
= \int_0^{1000} \lambda e^{-\lambda x} \left(\int_0^{1000-x} \lambda e^{-\lambda y}\, dy \right) dx
$$

$$
= \int_0^{1000} \lambda e^{-\lambda x} (1 - e^{-\lambda(1000-x)})\, dx
$$

$$
= \int_0^{1000} (\lambda e^{-\lambda x} - \lambda e^{-\lambda \cdot 1000})\, dx
$$

$$
= 1 - e^{-\lambda \cdot 1000} - 1000\lambda e^{-\lambda \cdot 1000}.
$$

Collecting terms gives

$$
P(\text{system fails in 1000 hours or less}) = 1 - (1 + 1000\lambda)e^{-1000\lambda}.
$$

QUESTION 3.3.3   Let X and Y have the bivariate uniform density as defined in Question 3.3.2. Let $Z = X + Y$. Find $F_Z(z)$. Hint: Consider two cases, $0 < z \leq 1$ and $1 < z < 2$. For each case, draw a diagram showing the region of integration.

The last example in this section deals with a joint pdf that does *not* factor into a function of x and a function of y.

EXAMPLE 3.8.   Suppose X and Y are two random variables jointly distributed over the first quadrant of the xy-plane according to the joint pdf,

$$
f_{X,Y}(x, y) = \begin{cases} y^2 e^{-y(x+1)}, & x \geq 0, y \geq 0 \\ 0, & \text{elsewhere.} \end{cases}
$$

We will show in this case that $f_{X,Y}(x, y) \neq f_X(x) \cdot f_Y(y)$.

First, consider $f_X(x)$:

$$
f_X(x) = \int_{-\infty}^{\infty} f_{X,Y}(x, y)\, dy = \int_0^{\infty} y^2 e^{-y(x+1)}\, dy.
$$

In the integrand, substitute,

$$
u = y(x + 1),
$$

making $du = (x + 1)\, dy$. This gives

$$f_X(x) = \frac{1}{x+1} \int_0^{\infty} \frac{u^2}{(x+1)^2} e^{-u} \, du = \frac{1}{(x+1)^3} \int_0^{\infty} u^2 e^{-u} \, du.$$

After applying integration by parts (twice) to $\int_0^{\infty} u^2 e^{-u} \, du$, we obtain

$$f_X(x) = \frac{1}{(x+1)^3} \left[-u^2 e^{-u} - 2u e^{-u} - 2e^{-u} \Big|_0^{\infty} \right]$$

$$= \frac{1}{(x+1)^3} \left[2 - \lim_{u \to \infty} \left(\frac{u^2}{e^u} + \frac{2u}{e^u} + \frac{2}{e^u} \right) \right]$$

$$= \frac{2}{(x+1)^3}.$$

Finding $f_Y(y)$ is a little easier than finding $f_X(x)$:

$$f_Y(y) = \int_{-\infty}^{\infty} f_{X,Y}(x, y) \, dx = \int_0^{\infty} y^2 e^{-y(x+1)} \, dx$$

$$= y^2 e^{-y} \int_0^{\infty} e^{-yx} \, dx = y^2 e^{-y} \left(\frac{1}{y} \right) \left[-e^{-yx} \Big|_0^{\infty} \right]$$

$$= y e^{-y}.$$

So, by inspection, $f_{X,Y}(x, y) \neq f_X(x) \cdot f_Y(y)$.

QUESTION 3.3.4   Let X and Y be random variables having the joint probability density function

$$f_{X,Y}(x, y) = \begin{cases} 1/x, & 0 < y < x, 0 < x < 1 \\ 0, & \text{elsewhere.} \end{cases}$$

Verify that the double integral over $f_{X,Y}(x, y)$ is 1, and find $f_X(x)$ and $f_Y(y)$.

The definitions and results in this section extend in a very straightforward way to the case of more than two variables. For example, given *three* random variables, we would define

$$F_{X_1, X_2, X_3}(u, v, w) = P(X_1 \leq u, X_2 \leq v, X_3 \leq w).$$

Likewise, if X_1, X_2, and X_3 were continuous, their joint pdf, $f_{X_1, X_2, X_3}(x_1, x_2, x_3)$, would be defined as that function whose triple integral is the joint cdf:

$$F_{X_1, X_2, X_3}(u, v, w) = \int_{-\infty}^{u} \int_{-\infty}^{v} \int_{-\infty}^{w} f_{X_1, X_2, X_3}(x_1, x_2, x_3) \, dx_1 \, dx_2 \, dx_3.$$

The notion of a marginal density also carries over in a natural way: the marginal pdf of X_1, for instance, would be the *double* integral (or double sum) of $f_{X_1, X_2, X_3}(x_1, x_2, x_3)$ over the variables x_2 and x_3.

3.4 INDEPENDENT RANDOM VARIABLES

The concept of independent events that was introduced in Section 2.7 leads quite naturally to a similar definition for independent random variables.

> **DEFINITION 3.5**   Random variables X and Y are said to be *independent* if for any intervals A and B,
>
> $$P(X \in A, Y \in B) = P((X, Y) \in A \otimes B) = P(X \in A) \cdot P(Y \in B).$$

Although it *defines* independence, Definition 3.5 is of little value in *establishing* independence. Given a continuous X and a continuous Y, it would be impossible to check all intervals A and B to verify the relationship between $P(X \in A, Y \in B)$ and $P(X \in A) \cdot P(Y \in B)$. A more workable characterization of this property is provided in Theorem 3.2.

> **THEOREM 3.2** Two random variables X and Y are independent if and only if
>
> $$f_{X,Y}(x, y) = f_X(x) \cdot f_Y(y)$$
>
> for all x and y.

PROOF

We will show that independence implies that the joint pdf factors into a product of the marginals. The converse is left as an exercise.

The proof for discrete X and Y is trivial. Suppose X and Y are discrete and independent; let a and b be any two numbers. Then

$$f_{X,Y}(a, b) = P(X = a, Y = b) = P(X = a) \cdot P(Y = b) = f_X(a) \cdot f_Y(b).$$

Now, let both X and Y be continuous (and independent), and take A and B to be two arbitrary intervals. We can write

$$\int_A \left(\int_B f_{X,Y}(x, y)\, dy \right) dx = \iint_{A \otimes B} f_{X,Y}(x, y)\, dy\, dx = P(X \in A, Y \in B)$$

$$= P(X \in A) \cdot P(Y \in B)$$

$$= \int_A f_X(x)\, dx \cdot \int_B f_Y(y)\, dy$$

$$= \int_A \left(\int_B f_X(x) \cdot f_Y(y)\, dy \right) dx.$$

Note that the functions

$$\int_B f_{X,Y}(x, y)\, dy \quad \text{and} \quad \int_B f_X(x) f_Y(y)\, dy$$

have equal integrals over every interval A; therefore, by a theorem of calculus, the two are equal. But then it follows that $f_{X,Y}(x, y)$ and $f_X(x) \cdot f_Y(y)$ have equal integrals over every interval B, so by the same theorem, *they* are equal.

In Chapter 2, extending the notion of independence from *two* events to n events proved to be something of a problem: the independence of each subset of the n events had to be checked separately (recall Definition 2.6). This is not necessary in the case of random variables: to generalize Theorem 3.2, we need only establish that the joint pdf factors into a product of the n marginals.

DEFINITION 3.6 The n random variables $X_1, X_2, \ldots, X_n$ are said to be *independent* if, for all $x_1, x_2, \ldots, x_n$,

$$f_{X_1, X_2, \ldots, X_n}(x_1, x_2, \ldots, x_n) = f_{X_1}(x_1) \cdot f_{X_2}(x_2) \cdots f_{X_n}(x_n).$$

An important special case of Definition 3.6 arises when all n independent random variables have the same pdf, $f_X(x)$. The set $X_1, X_2, \ldots, X_n$ is then called a *random sample of size n*. If S is the sample space for each of the X_i's, the sample space T for the random sample is the Cartesian product introduced in Section 2.7: $T = S \otimes S \otimes \cdots \otimes S$ (n times). Thus, for each i, X_i is defined on T according to the relationship

$$X_i((s_1, s_2, \ldots, s_i, \ldots, s_n)) = X(s_i). \tag{3.8}$$

Example 3.9 illustrates Equation 3.8 for a Bernoulli random sample of size 2.

EXAMPLE 3.9. Let $S = (s, f)$ be the usual Bernoulli sample space with $P((s)) = p$. Define $X(s) = 1$ and $X(f) = 0$. Suppose $n = 2$ such trials are observed, with the number of successes on the ith trial being denoted by the random variable X_i, $i = 1, 2$. Set $T = S \otimes S$ and let the product probability function be defined as it was in Section 2.7—that is, $P((s, f)) = pq$, $P((s, s)) = p^2$, and so on. Let X_1 be defined on T by $X_1((s, f)) = X_1((s, s)) = X(s) = 1$ and $X_1((f, s)) = X_1((f, f)) = X(f) = 0$. Similarly, define X_2 by $X_2((f, s)) = X_2((s, s)) = X(s) = 1$ and $X_2((s, f)) = X_2((f, f)) = X(f) = 0$. It is a simple matter to verify that

$$f_{X_1}(x) = f_{X_2}(x) = f_X(x)$$

and

$$f_{X_1, X_2}(x_1, x_2) = f_{X_1}(x_1) \cdot f_{X_2}(x_2) = f_X(x_1) \cdot f_X(x_2).$$

QUESTION 3.4.1   Write down the joint probability density function for a random sample of size n drawn from the exponential pdf, $f_Y(y) = (1/\lambda)e^{-y/\lambda}$, $y > 0$.

3.5 COMBINING AND TRANSFORMING RANDOM VARIABLES

Frequently a measurement of interest comes about as a *combination* of random variables. The cold redundancy mentioned in Example 3.7 is not atypical: there the life, Z, of an electronic system depended on the *sum*, $X + Y$, of two tube lives. At issue, of course, was $f_Z(z)$. Other situations may require a change in the *scale* of a random variable—for example, suppose X is originally calibrated in degrees Fahrenheit but the forces of metric conversion demand its reexpression in degrees Celsius. How does the pdf of Y, the Celsius random variable, compare to that of X?

In most cases such as these it is inefficient to compute the pdf's of the "new" random variables by returning to elementary principles. Easier methods are available. If Y is a given function of X—say, $Y = u(X)$—and $f_X(x)$ is known, we can readily find $f_Y(y)$ by first expressing Y's cdf, $F_Y(y)$, in terms of $F_X(u^{-1}(y))$ and then differentiating the latter. Examples 3.10 through 3.14 explore this very useful technique, as well as its extension to bivariate problems where $Z = u(X, Y)$ and we seek $f_Z(z)$.

EXAMPLE 3.10. Suppose a random variable X has pdf

$$f_X(x) = \begin{cases} 6x(1 - x), & 0 < x < 1 \\ 0, & \text{elsewhere.} \end{cases}$$

Let Y be a second random variable functionally related to X according to Equation 3.9:

$$Y = 2X + 1. \tag{3.9}$$

What is the pdf for Y?

Note, first of all, that the random variable Y maps the range of X-values, $(0, 1)$, onto the interval $(1, 3)$. (See Figure 3.5.)

Let $1 < y < 3$. Then

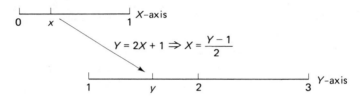

Figure 3.5 Mapping by random variable.

$$F_Y(y) = P(Y \le y) = P(2X + 1 \le y) = P\left(X \le \frac{y-1}{2}\right)$$

$$= F_X\left(\frac{y-1}{2}\right).$$

$6y - 6y^2$

But

$$F_X(t) = \begin{cases} 0, & t \le 0 \\ \int_0^t 6x(1 - x)\, dx = 3t^2 - 2t^3, & 0 < t < 1 \\ 1, & t \ge 1. \end{cases}$$

$6y - 6t^2$

Therefore,

$$F_Y(y) = F_X\left(\frac{y-1}{2}\right) = \begin{cases} 0, & y \le 1 \\ 3\left(\frac{y-1}{2}\right)^2 - 2\left(\frac{y-1}{2}\right)^3 \\ \quad = -\frac{y^3}{4} + \frac{3y^2}{2} - \frac{9y}{4} + 1, & 1 < y < 3 \\ 1, & y \ge 3. \end{cases}$$

Differentiating $F_Y(y)$ gives $f_Y(y)$:

$$f_Y(y) = \begin{cases} 0, & y \le 1 \\ -\frac{3y^2}{4} + 3y - \frac{9}{4}, & 1 < y < 3 \\ 0, & y \ge 3. \end{cases}$$

QUESTION 3.5.1   For X and Y as defined in Example 3.10, find $P(\frac{1}{2} < X < \frac{3}{4})$ *two* different ways.

QUESTION 3.5.2   Let X have the uniform density over $(0, 1)$. Define $Y = 3X - 3$. Find $f_Y(y)$.

   In doing transformation problems, extreme care must be exercised in identifying the proper set of x's that get mapped into the event $Y \leq y$. The best approach is always to sketch a diagram like the one in Figure 3.5; the next two examples show why.

   EXAMPLE 3.11.   Let X have the uniform density over the interval $(-1, 2)$:

$$f_X(x) = \begin{cases} \frac{1}{3}, & -1 < x < 2 \\ 0, & \text{elsewhere.} \end{cases}$$

Find the density function for Y, where $Y = X^2$.
   Here, the set of the x's that get mapped into the interval $Y \leq y$ has a different form depending on whether $0 \leq y < 1$ or $1 \leq y < 4$. To see this, suppose, first of all, that $0 \leq y < 1$. Then the set of x's that get mapped into $0 \leq Y \leq y$ are those from $-\sqrt{y}$ to $+\sqrt{y}$ (see Figure 3.6). Therefore,

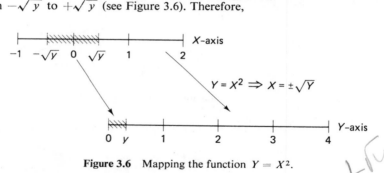

Figure 3.6   Mapping the function $Y = X^2$.

$$F_Y(y) = P(Y \leq y) = P(-\sqrt{y} \leq X \leq +\sqrt{y})$$

$$= \int_{-\sqrt{y}}^{\sqrt{y}} \left(\frac{1}{3}\right) dx = \frac{2\sqrt{y}}{3}. \tag{3.10}$$

On the other hand, suppose $1 \leq y < 4$. Now, from Figure 3.6, the corresponding x's range from -1 to $\sqrt{y}$, and

$$F_Y(y) = P(Y \leq y) = P(-1 < X \leq \sqrt{y})$$

$$= \int_{-1}^{\sqrt{y}} \left(\frac{1}{3}\right) dx = \frac{\sqrt{y} + 1}{3}. \tag{3.11}$$

Equations 3.10 and 3.11, together with the obvious "boundary" cases (when $y < 0$ and $y \geq 4$), define $F_Y(y)$:

$$F_Y(y) = \begin{cases} 0, & y < 0 \\ \dfrac{2\sqrt{y}}{3}, & 0 \leq y < 1 \\ \dfrac{\sqrt{y} + 1}{3}, & 1 \leq y < 4 \\ 1, & y \geq 4. \end{cases}$$

Then, by differentiating $F_Y(y)$, we get

$$f_Y(y) = \begin{cases} \dfrac{1}{3\sqrt{y}}, & 0 \le y < 1 \\ \dfrac{1}{6\sqrt{y}}, & 1 \le y < 4 \\ 0, & \text{elsewhere.} \end{cases}$$

QUESTION 3.5.3   Draw graphs of $f_X(x)$, $F_Y(y)$, and $f_Y(y)$. Why does $f_Y(y)$ have a discontinuity at $y = 1$?

QUESTION 3.5.4   Suppose X has pdf $f_X(x) = e^{-x}$, $x > 0$. Find the pdf for $Y = 1/X$.

QUESTION 3.5.5   A random variable X has density function $f_X(x) = \frac{3}{8}x^2$ over the interval $0 < x < 2$ (and 0, elsewhere). Suppose a circle is generated by a radius whose length is the value of X. Find the density function for Y, the area of the circle.

EXAMPLE 3.12.   Suppose X and Y have a joint uniform density over the unit square:

$$f_{X,Y}(x, y) = \begin{cases} 1, & 0 < x < 1, 0 < y < 1 \\ 0, & \text{elsewhere.} \end{cases}$$

Find the pdf for their product—that is, find $f_Z(z)$, where $Z = XY$.

For $0 < z < 1$, $F_Z(z)$ is the volume above the shaded region in Figure 3.7. Specifically,

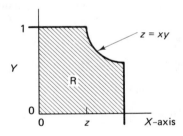

Figure 3.7 Region of integration for $F_Z(z)$.

$$F_Z(z) = P(Z \le z) = P(XY \le z) = \iint_R f_{X,Y}(x, y)\, dy\, dx.$$

By inspection, we see that the double integral over R can be split up into *two* double integrals—one letting x range from 0 to z, the other having x-values extend from z to 1:

$$F_Z(z) = \int_0^z \left(\int_0^1 1\, dy \right) dx + \int_z^1 \left(\int_0^{z/x} 1\, dy \right) dx.$$

But

$$\int_0^z \left(\int_0^1 1\, dy \right) dx = \int_0^z 1\, dx = z$$

and

$$\int_z^1 \left(\int_0^{z/x} 1\, dy \right) dx = \int_z^1 \left(\frac{z}{x} \right) dx = z \ln x \Big|_z^1 = -z \ln z.$$

It follows that

$$F_Z(z) = \begin{cases} 0, & z \le 0 \\ z - z \ln z, & 0 < z < 1 \\ 1, & z \ge 1, \end{cases}$$

in which case

$$f_Z(z) = \begin{cases} -\ln z, & 0 < z < 1 \\ 0 & \text{elsewhere.} \end{cases}$$

QUESTION 3.5.6   Suppose the random variables X and Y have the joint uniform density over the unit square. Find (a) the cdf and (b) the density function for $Z = X/Y$.

The next several examples illustrate a general procedure for finding the density function for the *sum* of two random variables.

EXAMPLE 3.13.   Suppose X and Y are two independent binomial random variables, each with the same success probability but defined on m and n trials, respectively. Specifically,

$$f_X(k) = \binom{m}{k} p^k q^{m-k}, \qquad k = 0, 1, \ldots, m$$

and

$$f_Y(k) = \binom{n}{k} p^k q^{n-k}, \qquad k = 0, 1, \ldots, n.$$

Let $Z = X + Y$; our objective is $f_Z(z)$.

Being a sum of x- and y-values, any given z can come about in a number of different ways, all mutually exclusive: $z = (0) + (z) = (1) + (z - 1)$, and so on. Therefore,

$$P(Z = z) = P\{(X = 0, Y = z) \cup (X = 1, Y = z - 1) \cup \cdots \cup (X = z, Y = 0)\}$$

$$= \sum_{k=0}^{z} P(X = k, Y = z - k),$$

and, since X and Y are independent,

$$P(Z = z) = \sum_{k=0}^{z} P(X = k) \cdot P(Y = z - k)$$

$$= \sum_{k=0}^{z} \binom{m}{k} p^k q^{m-k} \binom{n}{z - k} p^{z-k} q^{n-z+k}$$

$$= \sum_{k=0}^{z} \binom{m}{k} \binom{n}{z - k} p^z q^{m+n-z}.$$

Recall the statement of Review Exercise 110 in Chapter 2,

$$\sum_{k=0}^{z} \binom{m}{k} \binom{n}{z - k} = \binom{m + n}{z},$$

from which it follows that

$$P(Z = z) = f_Z(z) = \binom{m + n}{z} p^z q^{m+n-z}, \qquad z = 0, 1, \ldots, m + n$$

Notice that the binomial distribution "reproduces" itself—that is, X and Y were binomial and their sum, $Z = X + Y$, also proves to be binomial. Not all random variables share this property. The sum of two independent *uniform* random variables, for example, is not itself uniform (recall Question 3.3.3).

QUESTION 3.5.7   Let X and Y be two independent random variables with probability density functions

$$f_X(x) = \frac{e^{-r}r^x}{x!}, \qquad x = 0, 1, 2, \ldots ,$$

and

$$f_Y(y) = \frac{e^{-s}s^y}{y!}, \qquad y = 0, 1, 2, \ldots .$$

(a) Verify that $\sum_{x=0}^{\infty} f_X(x) = 1$.

(b) Let $Z = X + Y$. Find $f_Z(z)$. Do X and Y reproduce themselves?

To find the density function for the sum of two *continuous* random variables, we first derive the sum's cdf and then differentiate it.

EXAMPLE 3.14.   Let X and Y be two independent random variables with pdf's

$$f_X(x) = e^{-x}, \qquad x > 0$$

and

$$f_Y(y) = e^{-y}, \qquad y > 0.$$

Define $Z = X + Y$. Then,

$$F_Z(z) = P(Z \le z) = P(X + Y \le z).$$

Here $F_Z(z)$ is the volume above the shaded triangular region, R, shown in Figure 3.8. For $z > 0$,

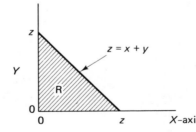

Figure 3.8 Region of integration for $F_Z(z)$.

$$F_Z(z) = \iint_R f_{X,Y}(x, y)\, dy\, dx = \int_0^z e^{-x} \left(\int_0^{z-x} e^{-y}\, dy \right) dx$$

$$= \int_0^z e^{-x}(1 - e^{x-z})\, dx = \int_0^z (e^{-x} - e^{-z})\, dx$$

$$= -e^{-x} \Big|_0^z - xe^{-z} \Big|_0^z$$

$$= 1 - e^{-z} - ze^{-z}$$

and the pdf for the sum readily follows:

$$f_Z(z) = F'_Z(z) = \begin{cases} ze^{-z}, & z > 0 \\ 0, & \text{elsewhere.} \end{cases}$$

QUESTION 3.5.8   Let X_1, X_2, and X_3 each be independent random variables with densities

$$f_{X_i}(x) = e^{-x}, \qquad x > 0, i = 1, 2, 3.$$

Let $Z = X_1 + X_2 + X_3$. Find $f_Z(z)$.

We conclude this section with a set of theorems that give the pdf's for three specific kinds of transformed random variables. Each of these transformations ($Y = aX + b$, $Y = X^2$, and $Z = X + Y$) occurs in later chapters, and it will be convenient not to have to rederive $f_Y(y)$ or $f_Z(z)$ on each occasion. The proofs are straightforward and will be left as exercises. (Theorem 3.3 could have been used for Example 3.10, Theorem 3.4 for Example 3.11, and Theorem 3.5 for Examples 3.13 and 3.14.) For a more thorough treatment of the properties of one-to-one transformations, the reader is referred to (3) or any textbook on advanced calculus.

THEOREM 3.3 Let X be a random variable with probability density function $f_X(x)$. Let $a \neq 0$ and b be constants, and define $Y = aX + b$.

(a) If X is discrete,

$$f_Y(y) = f_X\left(\frac{y - b}{a}\right).$$

(b) If X is continuous,

$$f_Y(y) = \frac{1}{|a|}f_X\left(\frac{y - b}{a}\right).$$

THEOREM 3.4 Let X be a continuous random variable with pdf $f_X(x)$. Let $Y = X^2$. For $y > 0$,

$$f_Y(y) = \frac{1}{2\sqrt{y}}(f_X(\sqrt{y}) + f_X(-\sqrt{y})).$$

THEOREM 3.5 Let X and Y be independent random variables with pdf's $f_X(x)$ and $f_Y(y)$, respectively. Let $Z = X + Y$.

(a) If X and Y are discrete,

$$f_Z(z) = \sum_{\text{all } x} f_X(x)f_Y(z - x).$$

(b) If X and Y are continuous,

$$f_Z(z) = \int_{-\infty}^{\infty} f_X(x)f_Y(z - x)\,dx.$$

Comment. An integral of the form

$$\int_{-\infty}^{\infty} f_X(x)f_Y(z - x)\,dx$$

is referred to as the *convolution* of the functions f_X and f_Y. Besides their frequent appearances in random-variable problems, convolutions turn up in many areas of applied and theoretical mathematics. ∎

3.6 ORDER STATISTICS

The single-variable transformations taken up in Section 3.5 all involved some standard mathematical operation, such as $Y = aX + b$ or $Y = X^2$. The bivariate transformations were similarly arithmetic, typically being concerned with either sums or products. In this section we will consider a different sort of transformation, one involving the *ordering* of an entire *set* of random variables. This particular transformation has wide applicability in many areas of statistics, and we will see some of its consequences in later chapters. Here, though, we will limit our discussion to one basic result: a derivation of the marginal pdf for the ith largest observation, $i = 1, 2, \ldots, n$, in a random sample of size n.

DEFINITION 3.7 Let X be a continuous random variable for which $x_1, x_2, \ldots, x_n$ are the values of a random sample of size n. Reorder the x_i's from smallest to largest:

$$x_1' < x_2' < \cdots < x_n'.$$

(No two of the x_i's are equal, except with probability zero, since X is continuous.) Define the random variable X_i' to have the value x_i', $1 \le i \le n$. Then X_i', is called the *i*th *order statistic*. Sometimes X_n' and X_1' are denoted $X_{\max}$ and $X_{\min}$, respectively.

EXAMPLE 3.15. Suppose four measurements are made on the random variable X: $x_1 = 3.4$, $x_2 = 4.6$, $x_3 = 2.6$, and $x_4 = 3.2$. The corresponding ordered sample would be

$$2.6 < 3.2 < 3.4 < 4.6.$$

The random variable representing the smallest observation would be denoted X_1', with its value for this particular sample being 2.6. Similarly, the value for the second-order statistic, X_2', is 3.2, and so on.

QUESTION 3.6.1   Suppose the random variable referred to in Example 3.15 has pdf $f_X(x) = \frac{1}{3}e^{-x/3}$, $x > 0$. Find $P(X_4 \le 5.1)$ and $P(X_4' \le 5.1)$. Hint: Keep in mind that X_4' will be less than or equal to 5.1 only if *all* the X_i's are less than or equal to 5.1.

THEOREM 3.6 Let X be a continuous random variable with probability density function $f_X(x)$. If a random sample of size n is drawn from $f_X(x)$, the marginal pdf for the ith-order statistic is given by

$$f_{X_i'}(y) = \frac{n!}{(i-1)!(n-i)!}[F_X(y)]^{i-1}[1 - F_X(y)]^{n-i}f_X(y)$$

for $1 \le i \le n$.

PROOF

A standard way to verify a theorem of this sort is to take advantage of the fact that we already know the correct answer and give an induction argument on i; here, it should be noted, the induction is slightly unusual in that it proceeds from n to 1 rather than from 1 to n.

The first step, then, is to find $f_{X_n'}(y)$. This can be done easily by differentiating the corresponding cdf. From Question 3.6.1, recall that

$$F_{X_n'}(y) = P(X_n' \leq y) = P(X_1, X_2, \ldots, X_n \leq y)$$

$$= P(X_1 \leq y) \cdot P(X_2 \leq y) \cdot \cdots \cdot P(X_n \leq y)$$

$$= (F_X(y))^n.$$

Therefore, $f_{X_n'}(y) = F_{X_n'}'(y) = n(F_X(y))^{n-1}f_X(y)$, which, according to Theorem 3.6, is the appropriate form for $i = n$.

Next comes the induction step: we assume the theorem to be correct for $f_{X_{i+1}'}(y)$ and seek an expression for $f_{X_i'}(y)$. By definition,

$$F_{X_i'}(y) = P(X_i' \leq y) = P(X_i' \leq y, X_{i+1}' \leq y) + P(X_i' \leq y, X_{i+1}' > y)$$

$$= P(X_{i+1}' \leq y) + P(X_i' \leq y, X_{i+1}' > y).$$

The first summand on the right is just $F_{X_{i+1}'}(y)$; the second is obtained by noticing that exactly i of the X_i's must be less than or equal to y, and this can occur in $\binom{n}{i}$ ways. Thus, appealing to the binomial distribution,

$$F_{X_i'}(y) = F_{X_{i+1}'}(y) + \binom{n}{i}(F_X(y))^i(1 - F_X(y))^{n-i}. \tag{3.12}$$

We then take the derivative of Equation 3.12,

$$f_{X_i'}(y) = \frac{n!}{i!(n-i-1)!}[F_X(y)]^i[1 - F_X(y)]^{n-i-1}f_X(y) + \frac{n!}{i!(n-i)!}$$

$$\cdot \{i[F_X(y)]^{i-1}[1 - F_X(y)]^{n-i} - (n-i)[F_X(y)]^i[1 - F_X(y)]^{n-i-1}\}f_X(y)$$

$$= \frac{n!}{i!(n-i)!}[F_X(y)]^{i-1}[1 - F_X(y)]^{n-i-1}f_X(y)$$

$$\cdot \{(n-i) F_X(y) + i[1 - F_X(y)] - (n-i)F_X(y)\}$$

$$= \frac{n!}{i!(n-i)!}[F_X(y)]^{i-1}[1 - F_X(y)]^{n-i-1}f_X(y)[i(1 - F_X(y))]$$

$$= \frac{n!}{(i-1)!(n-i)!}[F_X(y)]^{i-1}[1 - F_X(y)]^{n-i}f_X(y).$$

Comparing this latter expression with the statement of Theorem 3.6 completes the induction.

EXAMPLE 3.16. Let X_i' be the ith-order statistic in a random sample of size n drawn from $f_X(x)$. From Theorem 3.6,

$$F_{X_i'}(t) = P(X_i' \leq t)$$

$$= \int_{-\infty}^{t} \frac{n!}{(i-1)!\,(n-i)!} f_X(x)[F_X(x)]^{i-1}[1 - F_X(x)]^{n-i}\, dx.$$

Making a change of variables, let $y = F_X(x)$ so that $dy = f_X(x)\, dx$. Then

$$F_{X_i'}(t) = \int_{0}^{F_Y(t)} \frac{n!}{(i-1)!(n-i)!} y^{i-1}(1-y)^{n-i}\, dy. \qquad (3.13)$$

The right-hand side of Equation 3.13 is called an *incomplete beta integral*.
Note, also, that

$$F_{X_i'}(t) = P(i \text{ or more } X_i\text{'s} \le t)$$

$$= \sum_{j=i}^{n} \binom{n}{j} (F_X(t))^{j}(1 - F_X(t))^{n-j}.$$

Thus, the upper tail of a binomial distribution is equivalent to an incomplete beta integral.

QUESTION 3.6.2   Let $X_1, X_2, \ldots, X_n$ be a random sample from $f_{X_i}(x)$, where

$$f_{X_i}(x) = \begin{cases} e^{-x}, & x > 0 \\ 0, & \text{elsewhere.} \end{cases}$$

Let $Y = X_1'$. Find $f_Y(y)$, the probability density function for the smallest exponential order statistic.

EXAMPLE 3.17.   Suppose that many years of observation have confirmed that the annual maximum flood tide X (in feet) for a certain river has pdf

$$f_X(x) = \tfrac{1}{20}, \qquad 20 < x < 40.$$

(Note: It is unlikely that flood tides would be described by anything as simple as a uniform pdf. We are making that choice here solely to facilitate the mathematics.) The Army Corps of Engineers are planning to build a levee along a certain portion of the river, and they want to build it high enough so that there is only a 30% chance that the second worst flood in the next 33 years will overflow the embankment. How high should the levee be? (We assume that there will be only one potential flood per year.)

Let h be the desired height. If $X_1, X_2, \ldots, X_{33}$ denote the flood tides for the next $n = 33$ years, what we require of h is that

$$P(X_{32}' > h) = 0.30.$$

As a starting point, notice that for $20 < y < 40$,

$$F_X(y) = \int_{20}^{y} \frac{1}{20}\, dx = \frac{y}{20} - 1.$$

Therefore,

$$f_{X_{32}'}(y) = \frac{33!}{31!\,1!}\left(\frac{y}{20} - 1\right)^{31}\left(2 - \frac{y}{20}\right)^{1} \cdot \frac{1}{20},$$

and h is the solution of the integral equation

$$\int_{h}^{40} (33)(32)\left(\frac{y}{20} - 1\right)^{31}\left(2 - \frac{y}{20}\right)^{1} \cdot \frac{dy}{20} = 0.30. \qquad (3.14)$$

If we make the substitution

$$u = \frac{y}{20} - 1,$$

Equation 3.14 simplifies to

$$P(X'_{32} > h) = 33(32) \int_{(h/20)-1}^{1} u^{31}(1 - u)\, du$$

$$= 1 - 33\left(\frac{h}{20} - 1\right)^{32} + 32\left(\frac{h}{20} - 1\right)^{33}. \tag{3.15}$$

Setting the right-hand side of Equation 3.15 equal to 0.30 and solving for h by trial and error gives

$$h = 39.3 \text{ feet.}$$

QUESTION 3.6.3   How high would the levee have to be in order to give a 70% chance of containing the worst flood tide in the next ten years?

3.7 CONDITIONAL DENSITIES

We have already seen that many of the concepts defined in Chapter 2 relating to the probabilities of *events*—for example, independence—have their random-variable counterparts. Another of these carryovers is the notion of a conditional probability, or, in what will be our present terminology, a *conditional probability density function*. Applications of conditional pdf's are not uncommon. The height and girth of a tree, for instance, can be considered a pair of random variables. While it is easy to measure girth, it can be difficult to determine height; thus it might be of interest to, say, a lumberman to know the probabilities of a Ponderosa pine's attaining certain heights given a known value for its girth. Or consider the plight of a school board member agonizing over which way to vote on a proposed budget increase. Her task would be that much easier if she knew the conditional probability that x additional tax dollars would stimulate an average increase of y points among twelfth-graders taking a standardized proficiency exam.

In the case of discrete random variables, a conditional pdf can be treated in the same way as a conditional probability. Note the similarity between Definitions 3.8 and 2.4.

DEFINITION 3.8   Let X and Y be two discrete random variables. The *conditional probability density function of Y given x*—that is, the probability that Y takes on the value y given that X is equal to x—is denoted $f_{Y|x}(y)$ and given by

$$f_{Y|x}(y) = P(Y = y \mid X = x) = \frac{f_{X,Y}(x, y)}{f_X(x)}$$

for $f_X(x) \neq 0$.

EXAMPLE 3.18. Recall the coin-tossing experiment in Example 3.5; Table 3.4 reproduces the joint pdf, $f_{X,Y}(x, y)$.

TABLE 3.4 Joint pdf for coin-tossing experiment

		Y: 0	1	2	3	$f_X(x)$
X	0	$\frac{1}{8}$	$\frac{1}{4}$	$\frac{1}{8}$	0	$\frac{1}{2}$
	1	0	$\frac{1}{8}$	$\frac{1}{4}$	$\frac{1}{8}$	$\frac{1}{2}$
	$f_Y(y)$	$\frac{1}{8}$	$\frac{3}{8}$	$\frac{3}{8}$	$\frac{1}{8}$	

A simple application of Definition 3.8 gives

$$f_{Y\,|\,0}(0) = \frac{\frac{1}{8}}{\frac{1}{2}} = \frac{1}{4}, \qquad f_{Y\,|\,1}(0) = \frac{0}{\frac{1}{2}} = 0,$$

$$f_{Y\,|\,0}(1) = \frac{\frac{1}{4}}{\frac{1}{2}} = \frac{1}{2}, \qquad f_{Y\,|\,1}(1) = \frac{\frac{1}{8}}{\frac{1}{2}} = \frac{1}{4},$$

and so on.

QUESTION 3.7.1 Prove that $f_{Y\,|\,x}(y)$ as defined in Definition 3.8 is a legitimate probability density function.

If the variables X and Y are continuous, we can still appeal to the quotient $f_{X,Y}(x, y)/f_X(x)$ as the definition of $f_{Y\,|\,x}(y)$ and argue its propriety by analogy. A more satisfying approach, though, is to arrive at the same conclusion by taking the limit of Y's "conditional" cdf.

If X is continuous, a direct definition of $F_{Y\,|\,x}(y) = P(Y \leq y\,|\,X = x)$ would lead to division by 0. To avoid that, we think of $P(Y \leq y\,|\,X = x)$ as being a limit:

$$P(Y \leq y\,|\,X = x) = \lim_{h \to 0} P(Y \leq y\,|\,x \leq X \leq x + h)$$

$$= \lim_{h \to 0} \frac{\int_x^{x+h} \int_{-\infty}^{y} f_{X,Y}(t, u)\, du\, dt}{\int_x^{x+h} f_X(t)\, dt}.$$

Evaluating the quotient of the limits gives 0/0, so l'Hospital's rule is indicated:

$$P(Y \leq y\,|\,X = x) = \lim_{h \to 0} \frac{\frac{d}{dh} \int_x^{x+h} \int_{-\infty}^{y} f_{X,Y}(t, u)\, du\, dt}{\frac{d}{dh} \int_x^{x+h} f_X(t)\, dt}. \tag{3.16}$$

By the fundamental theorem of calculus,

$$\frac{d}{dh} \int_x^{x+h} g(t)\, dt = g(x + h),$$

which simplifies Equation 3.16 to

$$P(Y \leq y \mid X = x) = \lim_{h \to 0} \frac{\int_{-\infty}^{y} f_{X,Y}[(x + h), u] \, du}{f_X(x + h)}$$

$$= \frac{\int_{-\infty}^{y} \lim_{h \to 0} f_{X,Y}(x + h, u) \, du}{\lim_{h \to 0} f_X(x + h)} = \int_{-\infty}^{y} \frac{f_{X,Y}(x, u)}{f_X(x)} \, du$$

provided the limit operation and the integration can be interchanged [see (8) for a discussion of when such an interchange is valid]. It follows from this last expression that $f_{X,Y}(x, u)/f_X(x)$ behaves as a conditional probability density function should, and we are justified in extending Definition 3.8 to the continuous case.

EXAMPLE 3.19. A frequent assumption made in both Chapters 2 and 3 has been that the life of an electronic component can be modeled by an exponential pdf, $f_X(x) = \lambda e^{-\lambda x}$, $x > 0$, where λ is a positive constant. In some situations, though, λ is more properly thought of as being a random variable. This would be the case if the component being tested were selected from a large population of components, where the members of that population had a variety of operating characteristics. In any event, if λ *is* a random variable, what we have assumed about X is no longer its pdf but, rather, its *conditional* pdf:

$$f_{X \mid \lambda}(x) = \lambda e^{-\lambda x}, \qquad x > 0.$$

What can we say, then, about the unconditional $f_X(x)$?

To begin, suppose the pdf for Λ is given by

$$f_\Lambda(\lambda) = \lambda e^{-\lambda}.$$

Therefore, appealing to Definition 3.8, the joint pdf for X and Λ can be written

$$f_{X,\Lambda}(x, \lambda) = f_{X \mid \lambda}(x) \cdot f_\Lambda(\lambda) = \lambda e^{-\lambda x} \cdot \lambda e^{-\lambda}$$

$$= \lambda^2 e^{-\lambda(x+1)}. \tag{3.17}$$

From this it follows that the marginal (or unconditional) pdf for X is the integral

$$f_X(x) = \int_0^\infty \lambda^2 e^{-\lambda(x+1)} \, d\lambda. \tag{3.18}$$

But note that Equation 3.17 is the same joint pdf that was given in Example 3.8, and it was shown there that the solution to Equation 3.18 is

$$f_X(x) = \frac{2}{(x + 1)^3}.$$

QUESTION 3.7.2   Let X and Y be continuous random variables with joint pdf

$$f_{X,Y}(x, y) = \begin{cases} \frac{1}{8}(6 - x - y), & 0 < x < 2, 2 < y < 4 \\ 0, & \text{elsewhere.} \end{cases}$$

Find $f_X(x)$ and $f_{Y \mid x}(y)$. Also, draw a diagram showing $f_{Y \mid 1}(y)$.

3.8 EXPECTED VALUES

Probability density functions, as we have seen, provide us with an overview of a random variable's entire behavior pattern. If X is discrete, its pdf gives $P(X = x)$ for any x; if X is continuous, and A is any interval,

$$P(X \in A) = \int_A f_X(x) \, dx.$$

This sort of detailed information is not always what we are looking for, though. Sometimes what is of most interest is not the entire range of X but, rather, some indication of where, *on the average*, the random variable tends to lie. We will call the numerical measure of this "central tendency" of a random variable its *expected value*.

Gambling affords a familiar illustration of the notion of an expected value. Consider the game of roulette. After bets are placed, the croupier spins the wheel, and declares one of 38 numbers, 00, 0, 1, 2, . . . , 36, to be the winner. Disregarding what seems to be a perverse tendency of many roulette wheels to land on numbers for which no money has been wagered, we will assume that each of these 38 numbers is equally likely. Suppose that our particular bet is $1 on "odds." If X denotes our winnings, then X takes on the value 1 if an odd number occurs, and -1 otherwise. By the uniformity assumption,

$$f_X(1) = P(X = 1) = \tfrac{18}{38} = \tfrac{9}{19}$$

and

$$f_X(-1) = P(X = -1) = \tfrac{20}{38} = \tfrac{10}{19}.$$

It follows that we will win $1 $\tfrac{9}{19}$ of the time and lose $1 $\tfrac{10}{19}$ of the time. Intuitively, then, if we persist in this foolishness, we stand to *lose*, on the average, a little more than 5 cents each time we play the game:

$$\text{"expected" winnings} = \$1 \cdot \tfrac{9}{19} + (-\$1) \cdot \tfrac{10}{19}$$
$$= -\$0.053 \doteq -5\cent.$$

The number -0.053 will be called the expected value of X.

If X is a discrete random variable taking on each of its values, $x_1, x_2, \ldots, x_n$, with the same probability, the expected value of X is simply the everyday notion of an arithmetic average or mean:

$$\text{expected value of } X = \sum_{i=1}^{n} x_i \cdot \frac{1}{n} = \frac{1}{n} \sum_{i=1}^{n} x_i.$$

Extending this idea to a discrete X described by an arbitrary pdf, $f_X(x)$, gives

$$\text{expected value of } X = \sum_{i=1}^{\infty} x_i \cdot f_X(x_i). \tag{3.19}$$

Thus, Equation 3.19 can be viewed as a sort of generalized mean, a number marking off, in some sense, the "center" of $f_X(x)$. Although the idea of an average predates Pythagoras (the circle of Pythagoreans knew of more than nine different kinds of means), the probabilistic concept of an expected value made its debut in 1657 in Huygens' *De Ratiociniis in Aleae Ludo.*

> **DEFINITION 3.9**   Let X be a random variable with probability density function $f_X(x)$. The *expected value of X* is denoted $E(X)$, or μ, and is given by
>
> (a) $E(X) = \sum_{\text{all } x} x f_X(x)$ if X is discrete,
>
> (b) $E(X) = \int_{-\infty}^{\infty} x f_X(x)\, dx$ if X is continuous.

Comment. It is assumed that both the sum and the integral in Definition 3.9 converge absolutely:

$$\sum_{\text{all } x} |x| f_X(x) < \infty, \qquad \int_{-\infty}^{\infty} |x| f_X(x)\, dx < \infty.$$

If not, we say that X has no finite expected value. One immediate reason for requiring *absolute* convergence is that a convergent sum that is not absolutely convergent depends on the order in which the terms are added, and, of course, order should not be a consideration when defining a mean.   ∎

EXAMPLE 3.20.   An urn contains nine chips, five red and four white. Three are drawn out at random without replacement. Let X denote the number of red chips in the sample. Find $E(X)$.

From Section 2.10, we recognize X to be a hypergeometric random variable, where

$$P(X = x) = f_X(x) = \frac{\binom{5}{x}\binom{4}{3-x}}{\binom{9}{3}}, \qquad x = 0, 1, 2, 3.$$

Therefore,

$$E(X) = \sum_{x=0}^{3} x \cdot \frac{\binom{5}{x}\binom{4}{3-x}}{\binom{9}{3}}$$

$$= (0)\left(\frac{4}{84}\right) + (1)\left(\frac{30}{84}\right) + (2)\left(\frac{40}{84}\right) + (3)\left(\frac{10}{84}\right)$$

$$= \frac{5}{3}.$$

QUESTION 3.8.1   Find $E(X)$ for the urn problem of Example 3.20 if the three chips are drawn out *with* replacement.

QUESTION 3.8.2   Suppose X is a "constant" random variable—that is, it always takes on the same value c. Find $E(X)$.

EXAMPLE 3.21.   Let X be a binomial random variable defined on n trials, where $p = P(\text{success})$. By definition,

$$E(X) = \sum_{k=0}^{n} k \cdot f_X(k) = \sum_{k=0}^{n} k \binom{n}{k} p^k (1-p)^{n-k}$$

$$= \sum_{k=0}^{n} \frac{k \cdot n!}{k!(n-k)!} p^k (1-p)^{n-k}$$

$$= \sum_{k=1}^{n} \frac{n!}{(k-1)!(n-k)!} p^k (1-p)^{n-k}. \tag{3.20}$$

At this point, a "trick" is called for. If $E(X) = \sum_{\text{all } k} g(k)$ can be factored in such a way that $E(X) = h \sum_{\text{all } k} f_Y(k)$, where $f_Y(k)$ is the pdf for some random variable Y, then $E(X) = h$, since the sum of a pdf over its entire range is 1. Here, suppose np is factored out of Equation 3.20. Then

$$E(X) = np \sum_{k=1}^{n} \frac{(n-1)!}{(k-1)!(n-k)!} p^{k-1}(1-p)^{n-k}$$

$$= np \sum_{k=1}^{n} \binom{n-1}{k-1} p^{k-1}(1-p)^{n-k}.$$

Now, let $j = k - 1$. It follows that

$$E(X) = np \sum_{j=0}^{n-1} \binom{n-1}{j} p^j (1-p)^{n-j-1}.$$

Finally, letting $m = n - 1$ gives

$$E(X) = np \sum_{j=0}^{m} \binom{m}{j} p^j (1-p)^{m-j},$$

and, since the value of the sum is 1 (why?),

$$E(X) = np. \tag{3.21}$$

Equation 3.21 is hardly an unexpected result. If a multiple-choice test, for example, has 100 questions, each with five possible answers, we would "expect" to get 20 correct if we were just guessing. But $20 = E(X) = 100(\frac{1}{5}) = np$. (Compare Equation 3.21 with your answer to Question 3.8.1.)

EXAMPLE 3.22. Consider the following game. A fair coin is flipped until the first tail appears; we win \$2 if it appears on the first toss, \$4 if it first appears on the second toss, and, in general, \$$2^k$ if it first occurs on the kth toss. Let the random variable X denote our winnings. How much should we have to pay in order for this to be a fair game? [Note: A fair game is one where the difference between the ante and $E(X)$ is 0.]

Known as the St. Petersburg paradox, this problem has a rather surprising answer. First of all, note that

$$f_X(2^k) = P(X = 2^k) = \frac{1}{2^k}, \qquad k = 1, 2, \ldots.$$

Therefore,

$$E(X) = \sum_{\text{all } x} x f_X(x) = \sum_{k=1}^{\infty} 2^k \cdot \frac{1}{2^k} = 1 + 1 + 1 + \cdots,$$

which is a divergent sum. That is, X does not have a finite expected value, so in order for this game to be fair, our ante would have to be an infinite amount of money!

Comment. Mathematicians and statisticians have been trying to "explain" the St. Petersburg paradox for almost 200 years. The answer seems clearly absurd—no rational gambler would consider paying even \$25 to play such a game, much less an infinite amount—yet the computations involved in getting $E(X)$ are unassailably correct. Where the difficulty lies, according to one common theory, is with our inability to put in perspective the very small probabilities of winning very large payoffs. Furthermore, the problem assumes that our opponent has infinite capital, which is an impossible state of affairs. We get a much more reasonable answer for $E(X)$ if the stipulation is added that our opponent can pay at most, say, \$1,000,000 (see Review Exercise 77 at the end of this chapter). For a detailed discussion of the St. Petersburg paradox, see (52). ∎

QUESTION 3.8.3   For the St. Petersburg problem, find $E(X)$ if the payoff for a tail appearing for the first time on the kth trial is log 2^k. This was a modification suggested by D. Bernoulli (a nephew of James Bernoulli) to take into account the decreasing marginal utility of money—that is, the more one has, the less useful a bit more is.

EXAMPLE 3.23.   In Example 3.3, the lifetimes of radar tubes were said to follow an exponential pdf,

$$f_X(x) = \tfrac{1}{179}e^{-x/179}, \qquad x > 0.$$

Finding $E(X)$ in this case requires that $x \cdot f_X(x)$ be integrated:

$$E(X) = \int_{-\infty}^{\infty} x f_X(x)\, dx = \int_{0}^{\infty} x(\tfrac{1}{179})e^{-x/179}\, dx$$

$$= \int_{0}^{\infty} 179 u e^{-u}\, du$$

$$= 179\left[-u e^{-u} - e^{-u}\right]\Big|_{0}^{\infty}$$

$$= 179.$$

Notice that the probability of a tube's wearing out before its average life is *not* 0.5:

$$\int_{0}^{179} \tfrac{1}{179}e^{-x/179}\, dx = 0.632.$$

Thus, in general, the expected value differs from the *median*, that number m for which $P(X < m) = P(X > m)$.

QUESTION 3.8.4   What must be true of $f_X(x)$ in order for $E(X)$ to be equal to the median for X?

QUESTION 3.8.5   Find $E(Y)$ for the transformed random variable derived in Example 3.11.

Next we consider several properties of expected values that will prove useful in theoretical contexts as well as for computational purposes.

THEOREM 3.7   For any random variables X and Y and any numbers a and b,
$$E(aX + bY) = aE(X) + bE(Y).$$

PROOF   We will prove the theorem by establishing that $E(aX) = aE(X)$ and $E(X + Y) = E(X) + E(Y)$. Only the continuous case will be considered.

If $f_{aX}(t)$ is the probability density function for aX, then

$$E(aX) = \int_{-\infty}^{\infty} t f_{aX}(t)\, dt = \int_{-\infty}^{\infty} t\, \frac{1}{|a|} f_X\!\left(\frac{t}{a}\right) dt,$$

the last equality being a consequence of Theorem 3.3. Assume $a > 0$ so that $a = |a|$; the other case is similar. Make the substitution $x = t/a$, and $dx = (1/a)\, dt$. Then

$$E(aX) = \int_{-\infty}^{\infty} ax \left(\frac{1}{a}\right) f_X(x)\, a\, dx = a \int_{-\infty}^{\infty} x f_X(x)\, dx = aE(X).$$

The crux of the second part of the proof is to find an expression for $E(X + Y)$. One approach would be to derive a general formula for $f_{X+Y}(t)$ and then apply Definition 3.9. It is simpler and more illuminating, though, to return to the basic concept of an expected value. The random variable $X + Y$ takes on values of the form $x + y$, with the "likelihood" of such a value being $f_{X,Y}(x, y)$. Thus, the expected value should be

$$\sum_{\text{all } x} \sum_{\text{all } y} (x + y) f_{X,Y}(x, y)$$

or

$$\int_{-\infty}^{\infty} \int_{-\infty}^{\infty} (x + y) f_{X,Y}(x, y)\, dy\, dx,$$

depending on the nature of X and Y.

Therefore, to verify that $E(X + Y) = E(X) + E(Y)$ in the continuous case, observe that

$$\begin{aligned}
E(X + Y) &= \int_{-\infty}^{\infty} \int_{-\infty}^{\infty} (x + y) f_{X,Y}(x, y)\, dy\, dx \\
&= \int_{-\infty}^{\infty} \int_{-\infty}^{\infty} x f_{X,Y}(x, y)\, dy\, dx + \int_{-\infty}^{\infty} \int_{-\infty}^{\infty} y f_{X,Y}(x, y)\, dy\, dx \\
&= \int_{-\infty}^{\infty} x \left(\int_{-\infty}^{\infty} f_{X,Y}(x, y)\, dy \right) dx + \int_{-\infty}^{\infty} y \left(\int_{-\infty}^{\infty} f_{X,Y}(x, y)\, dx \right) dy \\
&= \int_{-\infty}^{\infty} x f_X(x)\, dx + \int_{-\infty}^{\infty} y f_Y(y)\, dy = E(X) + E(Y).
\end{aligned}$$

An easy induction argument extends Theorem 3.7 to the case of n variables.

COROLLARY   Given random variables $X_1, X_2, \ldots, X_n$ and constants $a_1, a_2, \ldots, a_n$,

$$E(a_1 X_1 + a_2 X_2 + \cdots + a_n X_n) = a_1 E(X_1) + a_2 E(X_2) + \cdots + a_n E(X_n).$$

Comment. The reader who speaks linear algebra will immediately recognize the statement of the corollary as English for "E is a linear transformation." ∎

An excellent application of the corollary is finding the expected value of a binomial random variable. Contrast the simplicity of this approach with the intricacies of the direct summation done in Example 3.21.

EXAMPLE 3.24. Consider a series of n independent Bernoulli trials. Define the random variable X_i, $i = 1, 2, \ldots, n$, to be 1 if the ith trial is a success, and 0 if it is a failure. Then $f_{X_i}(1) = p$ and $f_{X_i}(0) = 1 - p = q$, giving $E(X_i) = 1 \cdot p + 0 \cdot q = p$. Now, let X be the usual binomial random variable. Since $X = X_1 + \cdots + X_n$, $E(X)$, by the corollary, is simply $E(X) = E(X_1) + \cdots + E(X_n) = np$.

QUESTION 3.8.6   Suppose X_i is a random variable for which $E(X_i) = \mu$, $i = 1, 2, \ldots, n$. Under what conditions will

$$E\left(\sum_{i=1}^{n} a_i X_i\right) = \mu\,?$$

The technique introduced in the proof of Theorem 3.7 for finding the expected value of a function works in a very general setting. The next two theorems formalize the idea. The proofs will not be given: while technically demanding, they are not particularly enlightening.

THEOREM 3.8

(a) If X is a discrete random variable with pdf $f_X(x)$ and if $g(x)$ is any real-valued function of a real variable,

$$E(g(X)) = \sum_{\text{all } x} g(x) \cdot f_X(x).$$

(b) If X is a continuous random variable and if $g(x)$ is a continuous real-valued function,

$$E(g(X)) = \int_{-\infty}^{\infty} g(x) \cdot f_X(x)\, dx.$$

EXAMPLE 3.25. Suppose X is a random variable with $f_X(-2) = \frac{5}{8}$, $f_X(1) = \frac{1}{8}$, and $f_X(2) = \frac{2}{8}$. Let $g(X) = X^2$. To compute $E(g(X)) = E(X^2)$ from first principles—that is, using Definition 3.9—we need to find f_{X^2}, easily seen in this case to have the values $f_{X^2}(1) = \frac{1}{8}$ and $f_{X^2}(4) = \frac{7}{8}$. Then

$$E(g(X)) = 1 \cdot \tfrac{1}{8} + 4 \cdot \tfrac{7}{8} = \tfrac{29}{8}.$$

Theorem 3.8, though, offers a simpler solution:

$$E(g(X)) = (-2)^2 \cdot \tfrac{5}{8} + 1^2 \cdot \tfrac{1}{8} + 2^2 \cdot \tfrac{2}{8} = \tfrac{29}{8}.$$

QUESTION 3.8.7   Let X have the probability density function

$$f_X(x) = \begin{cases} 2(1 - x), & 0 < x < 1 \\ 0, & \text{elsewhere.} \end{cases}$$

Let $Y = g(X) = X^2$. Find $E(g(X))$ two ways: first, by solving for $f_Y(y)$ and integrating $y \cdot f_Y(y)$, and, second, by using Theorem 3.8.

THEOREM 3.9

(a) Suppose X and Y are discrete random variables with joint density $f_{X,Y}(x, y)$. Let $g(x, y)$ be any real-valued function of two real variables. Then

$$E(g(X, Y)) = \sum_{\text{all } x} \sum_{\text{all } y} g(x, y) \cdot f_{X,Y}(x, y).$$

(b) Suppose X and Y are continuous random variables and $g(x, y)$ is a continuous function of two real variables. Then

$$E(g(X, Y)) = \int_{-\infty}^{\infty} \int_{-\infty}^{\infty} g(x, y) \cdot f_{X,Y}(x, y) \, dy \, dx.$$

EXAMPLE 3.26.   Let X and Y be defined as in Example 3.5 and let $g(x, y) = 3x - 2xy + y$. One way to find $E(g(X, Y))$ is first to compute the pdf for $Z = 3X - 2XY + Y$ from Table 3.3, as shown in Table 3.5.

TABLE 3.5   The pdf for $Z = 3X - 2XY + Y$

z	0	1	2	3
$f_Z(z)$	$\frac{1}{4}$	$\frac{1}{2}$	$\frac{1}{4}$	0

From Table 3.5, then, we get that $E(Z)$ is equal to 1:

$$E(Z) = 0 \cdot \tfrac{1}{4} + 1 \cdot \tfrac{1}{2} + 2 \cdot \tfrac{1}{4} + 3 \cdot 0 = 1.$$

An alternate solution using Theorem 3.9 avoids the intermediate step of determining $f_Z(z)$. Directly from Table 3.3 we get

$$E(g(X, Y)) = E(Z) = 0 \cdot \tfrac{1}{8} + 1 \cdot \tfrac{1}{4} + 2 \cdot \tfrac{1}{8} + 3 \cdot 0$$
$$+ 3 \cdot 0 + 2 \cdot \tfrac{1}{8} + 1 \cdot \tfrac{1}{4} + 0 \cdot \tfrac{1}{8}$$
$$= 1.$$

QUESTION 3.8.8   Consider the joint density described in Question 3.3.1. Let $Z = g(X, Y) = XY^2$. Find $E(Z)$ by using Definition 3.9 and also by using Theorem 3.9.

A simple yet very important application of Theorem 3.9 is the following statement about the expected value of a product of independent random variables.

THEOREM 3.10   If X and Y are independent,

$$E(XY) = E(X) \cdot E(Y).$$

PROOF   For variety, we will prove the discrete case and leave the continuous case as an exercise. Using Theorems 3.2 and 3.9,

$$E(XY) = \sum_{\text{all } x} \sum_{\text{all } y} xy f_{X,Y}(x, y)$$
$$= \sum_{\text{all } x} \sum_{\text{all } y} xy f_X(x) \cdot f_Y(y)$$
$$= \sum_{\text{all } x} x f_X(x) \cdot [\sum_{\text{all } y} y f_Y(y)]$$
$$= E(X) \cdot E(Y).$$

We conclude this section with a nontrivial, nongambling example showing the utility of the concept of expected value.

EXAMPLE 3.27. Suppose N people are to be tested for some rare condition—for example, whether or not they have a certain disease or, in the case of hospital workers, whether or not they have an overexposed radiation badge. A familiar example of the former is a blood test. If every person is tested, obviously N tests will be done. An alternate strategy is to pool the blood samples into groups of k. Each pooled sample is then tested. If it is negative, all k in the group are free from the disease. If it is positive, each person must be retested, ultimately resulting in $k + 1$ tests being done on that particular group. Suppose the probability of any one person's having a positive test is p. What will be the expected number of tests under the "pooling" plan—and how does it compare to N?

Let X be the random variable counting the number of tests done on a pooled sample. Then X takes on only two values, either 1 or $k + 1$, where

$$f_X(1) = P(\text{none of the } k \text{ gives a positive test})$$
$$= (1 - p)^k$$

and

$$f_X(k + 1) = 1 - (1 - p)^k.$$

Thus,

$$E(X) = 1 \cdot (1 - p)^k + (k + 1)[1 - (1 - p)^k]$$
$$= (k + 1) - k(1 - p)^k.$$

Since there are approximately N/k groups of size k, the total expected number of tests is roughly

$$\frac{N}{k} \cdot E(X) = \frac{N}{k}[(k + 1) - k(1 - p)^k] = N[1 - (1 - p)^k + (1/k)].$$

Table 3.6 gives $(N/k) \cdot E(X)$ for p values of $\frac{1}{100}$ and $\frac{1}{1000}$ and for $k = 2, 5, 10, 20, 40,$ and 100. Notice that for small p, especially, pooling can reduce considerably the number of tests that need to be performed.

TABLE 3.6   Reduction of tests by pooling

p	k	Expected Number of Tests
$\frac{1}{100}$	2	$0.52N$
	5	$0.25N$
	10	$0.20N$
	20	$0.23N$
	40	$0.36N$
	100	$0.64N$
$\frac{1}{1000}$	2	$0.50N$
	5	$0.20N$
	10	$0.11N$
	20	$0.07N$
	40	$0.06N$
	100	$0.10N$

QUESTION 3.8.9   Find the optimal pooling size, k, if $p = \frac{1}{50}$.

3.9 THE VARIANCE AND HIGHER MOMENTS

The expected value is a good enough measure of central tendency, but it still leaves out some critical information about a random variable's behavior. Unless we are also provided with some indication of how *spread out* a random variable's probability density function is, the expected value by itself can be misleading, and we are prey to absurdities such as, "A person with his head in the freezer and feet in the oven is *on the average* quite comfortable." More quantitatively, a football team may have an interior line averaging 200 pounds if it has five players all weighing close to that figure, or if it has four 150-pounders and a 400-pound behemoth at left tackle. One would expect the opposition to recognize the tactical significance of such a difference in weight dispersion rather quickly. Finally, a random variable representing a game where $1 is won or lost with equal probability has expected value zero, as does one where $100,000 is the stake; yet, surely, these are different games (or, at the very least, played by different people).

Given, then, that the assessment of a random variable's dispersion, or variability, is a worthwhile pursuit, how should we proceed? One seemingly reasonable approach would be to average, in the generalized sense, the deviations of the values of X from their expected value. Unfortunately, this proves to be futile, since the numerical value of such an average will always be 0:

$$E(X - \mu) = E(X) - \mu = \mu - \mu = 0. \tag{3.22}$$

Another possibility would be to modify Equation 3.22 by making all the deviations positive—that is, replace $E(X - \mu)$ with $E(|X - \mu|)$. This does work, and it *is* sometimes used to measure dispersion, but the absolute value is somewhat troublesome mathematically: it does not have a simple arithmetic formula, nor is it a differentiable function. *Squaring* the deviations proves to be a much more attractive solution.

> **DEFINITION 3.10** The *variance* of a random variable X, denoted Var(X), is the expected value of its squared deviations from μ:
>
> $$\text{Var}(X) = E((X - \mu)^2)$$
>
> where $\mu = E(X)$.

Comment. One unfortunate consequence of Definition 3.10 is that the units for the variance are the square of the units for X: if X is measured in inches, the units for Var(X) are inches². This causes obvious problems in relating the variance back to the sample values. In applied statistics, the variance is often replaced as a measure of dispersion by the *standard deviation*, where

$$\text{standard deviation of } X = \sqrt{\text{Var}(X)}. \quad \blacksquare$$

EXAMPLE 3.28.   Consider an urn containing two red chips and three black chips. We draw two chips at random, *without replacement*. Let X denote the number of red chips in the sample. Find Var(X).

Since the sampling is done without replacement, the appropriate pdf for X is the hypergeometric:

$$f_X(x) = \frac{\binom{2}{x}\binom{3}{2-x}}{\binom{5}{2}}, \qquad x = 0, 1, 2.$$

Therefore,

$$E(X) = 0 \cdot \frac{\binom{2}{0}\binom{3}{2}}{\binom{5}{2}} + 1 \cdot \frac{\binom{2}{1}\binom{3}{1}}{\binom{5}{2}} + 2 \cdot \frac{\binom{2}{2}\binom{3}{0}}{\binom{5}{2}} = 0.8.$$

By Definition 3.10,

$$\text{Var}(X) = \sum_{\text{all } x} (x - \mu)^2 \cdot f_X(x)$$

$$= (0 - 0.8)^2 \cdot \frac{\binom{2}{0}\binom{3}{2}}{\binom{5}{2}} + (1 - 0.8)^2 \cdot \frac{\binom{2}{1}\binom{3}{1}}{\binom{5}{2}} + (2 - 0.8)^2 \cdot \frac{\binom{2}{2}\binom{3}{0}}{\binom{5}{2}}$$

$$= 0.36.$$

QUESTION 3.9.1   Find Var (X) for the urn problem of Example 3.28 if the sampling was done *with* replacement.

Definition 3.10 does not provide a particularly convenient *computing* formula for Var (X). Theorem 3.11 offers an improvement.

THEOREM 3.11   Var $(X) = E(X^2) - \mu^2$.

PROOF   The proof is a simple application of the distributive property of expected values:

$$\begin{aligned}\text{Var}(X) &= E((X - \mu)^2) \\ &= E(X^2 - 2\mu X + \mu^2) \\ &= E(X^2) - 2\mu E(X) + \mu^2 \\ &= E(X^2) - 2\mu^2 + \mu^2 \\ &= E(X^2) - \mu^2.\end{aligned}$$

EXAMPLE 3.29.   For the urn problem of Example 3.28, $E(X) = 0.8$. Also,

$$E(X^2) = \sum_{\text{all } x} x^2 \cdot f_X(x)$$

$$= 0^2 \cdot \frac{\binom{2}{0}\binom{3}{2}}{\binom{5}{2}} + 1^2 \cdot \frac{\binom{2}{1}\binom{3}{1}}{\binom{5}{2}} + 2^2 \cdot \frac{\binom{2}{2}\binom{3}{0}}{\binom{5}{2}}$$

$$= 1.00.$$

Therefore, by Theorem 3.11,

$$\text{Var}(X) = 1.00 - (0.8)^2 = 0.36$$

the same answer obtained in Example 3.28 by using Definition 3.10.

QUESTION 3.9.2   Use Theorem 3.11 to find the variance of a uniform random variable defined over the unit interval.

If a random variable is subjected to a change of scale, its variance is also changed, as our intuition should lead us to suspect. Theorem 3.12 relates the notion of variance back to the linear transformations discussed in Section 3.5.

THEOREM 3.12 Let X and Y be random variables and a and b be constants such that $Y = aX + b$. Then

$$\text{Var}(Y) = a^2 \text{Var}(X).$$

PROOF   Since $E(aX + b) = a\mu + b$,

$$\text{Var}(Y) = E([(aX + b) - (a\mu + b)]^2)$$

$$= E(a^2(X - \mu)^2)$$

$$= a^2 E((X - \mu)^2)$$

$$= a^2 \text{Var}(X).$$

In general, the variance of a sum (of variables) is not equal to the sum of the variances. Equality does hold, though, in the important special case of independent random variables.

THEOREM 3.13 Let $X_1, X_2, \ldots, X_n$ be independent random variables and let $Y = X_1 + X_2 + \cdots + X_n$. Then

$$\text{Var}(Y) = \text{Var}(X_1) + \text{Var}(X_2) + \cdots + \text{Var}(X_n).$$

PROOF   We give a proof for $Y = X_1 + X_2$; an easy induction completes the argument for general n. From Theorems 3.7 and 3.11,

$$\text{Var}(Y) = E((X_1 + X_2)^2) - (E(X_1) + E(X_2))^2.$$

Writing out the squares gives

$$\text{Var}(Y) = E(X_1^2 + 2X_1 X_2 + X_2^2) - (E(X_1))^2 - 2E(X_1)E(X_2) - (E(X_2))^2$$

$$= E(X_1^2) - (E(X_1))^2 + E(X_2^2) - (E(X_2))^2$$

$$+ 2(E(X_1 X_2) - E(X_1)E(X_2)). \quad (3.23)$$

By the independence of X_1 and X_2, $E(X_1 X_2) = E(X_1)E(X_2)$, making the last term in Equation 3.23 vanish. The remaining terms combine to give the desired result: $\text{Var}(Y) = \text{Var}(X_1) + \text{Var}(X_2)$.

The binomial random variable, X, is an obvious candidate for Theorem 3.13. Since X is the sum of n independent Bernoulli random variables, the variance of X should be the sum of the variances of the n Bernoullis. Theorem 3.14 gives the formal statement.

THEOREM 3.14   Let X denote the number of successes in n independent trials, where p is the probability of success at any given trial. Then

$$\text{Var}(X) = npq.$$

PROOF   As we did in Example 3.24, write $X = X_1 + X_2 + \cdots + X_n$. Recall that $E(X_i) = p$. Since X_i takes on only the values 0 and 1, X_i^2 and X_i are same. Therefore,

$$\text{Var}(X_i) = E(X_i^2) - (E(X_i))^2 = E(X_i) - (E(X_i))^2$$
$$= p - p^2 = p(1 - p) = pq$$

and

$$\text{Var}(X) = \text{Var}(X_1) + \text{Var}(X_2) + \cdots + \text{Var}(X_n) = npq.$$

In more general terminology, what we have called the expected value of X is often referred to as the *first moment of X about the origin*, and the variance of X as the *second moment of X about the mean*. While these are certainly the two most important moments in statistics, we will encounter others as we go along. These "higher moments" are defined here for later reference.

DEFINITION 3.11   For any positive integer r, the *rth moment of X about the origin* is given by $E(X^r)$; the *rth moment of X about the mean*, by $E((X - \mu)^r)$.

We conclude this section with two well-known theorems. The first has theoretical applications in later chapters but is worth introducing here because it makes more precise the meaning of $\text{Var}(X)$ as a measure of dispersion. It shows that the probabilities of deviations of a random variable from its mean are bounded by a function of $\text{Var}(X)$. The second theorem is an application of the first and considers the behavior of the *sample mean*, $(1/n) \sum_{i=1}^{i} X_i$, a quantity that will prove to be of paramount importance when we begin our discussion of statistics.

THEOREM 3.15 (Chebyshev's inequality)   Suppose X is a random variable with mean μ and variance $\text{Var}(X)$. For any number $\epsilon > 0$,

$$P(|X - \mu| < \epsilon) > 1 - \frac{\text{Var}(X)}{\epsilon^2}$$

or, equivalently,

$$P(|X - \mu| \geq \epsilon) \leq \frac{\text{Var}(X)}{\epsilon^2}.$$

PROOF   In the continuous case,

$$\text{Var}(X) = \int_{-\infty}^{\infty} (x - \mu)^2 f_X(x) \, dx$$

$$= \int_{-\infty}^{\mu-\epsilon} (x - \mu)^2 f_X(x) \, dx + \int_{\mu-\epsilon}^{\mu+\epsilon} (x - \mu)^2 f_X(x) \, dx + \int_{\mu+\epsilon}^{\infty} (x - \mu)^2 f_X(x) \, dx.$$

Omitting the nonnegative middle integral gives an inequality:

$$\text{Var}(X) \geq \int_{-\infty}^{\mu-\epsilon} (x - \mu)^2 f_X(x) \, dx + \int_{\mu+\epsilon}^{\infty} (x - \mu)^2 f_X(x) \, dx$$

$$\geq \int_{|x-\mu|\geq\epsilon} (x - \mu)^2 f_X(x) \, dx$$

$$\geq \int_{|x-\mu|\geq\epsilon} \epsilon^2 f_X(x) \, dx$$

$$= \epsilon^2 P(|X - \mu| \geq \epsilon).$$

Division by ϵ^2 completes the proof. (If X is discrete, replace the integrals with summations.)

> EXAMPLE 3.30. A student makes 100 check transactions between receiving two consecutive bank statements. Rather than subtract the amounts exactly, he rounds each entry off to the nearest dollar. Use Chebyshev's inequality to get an upper bound for the probability that the student's accumulated error (either positive or negative) is \$5 or more.
>
> Let Y_i denote the roundoff error associated with the ith transaction, $i = 1, 2, \ldots,$ 100. It can be assumed that the Y_i's are independent random variables and follow a uniform pdf over the interval $(-\frac{1}{2}\$, +\frac{1}{2}\$)$. Clearly,
>
> $$E(Y_i) = 0$$
>
> and
>
> $$\text{Var}(Y_i) = \tfrac{1}{12} \qquad \text{(see Question 3.9.2).}$$
>
> Let $Y = Y_1 + Y_2 + \cdots + Y_{100}$ denote the total accumulated error. By the corollary to Theorem 3.7,
>
> $$E(Y) = \mu = 0$$
>
> and by Theorem 3.13,
>
> $$\text{Var}(Y) = \tfrac{100}{12} = 8.3.$$
>
> It follows, then, from Chebyshev's inequality that
>
> $$P(|Y - \mu| \geq \epsilon) = P(|Y| \geq \$5) \leq \frac{8.3}{25} = 0.33.$$
>
> That is, there is less than a 33% chance that the student's total will be off by as much as \$5.

QUESTION 3.9.3   Suppose the exact pdf, $f_Y(y)$, for the random variable described in Example 3.30 was known. Would we expect

$$\int_{-5}^{5} f_Y(y) \, dy$$

to be greater than, less than, or equal to 0.33? Explain.

We conclude this section with a statement of the so-called *weak law of large numbers*. In words, it says that the average of *n* identically distributed random variables, each having mean μ, converges to the common mean in the sense that the probability that the average falls outside an ϵ-neighborhood of μ decreases to 0 as *n* goes to infinity. The proof follows immediately from Chebyshev's inequality.

THEOREM 3.16   If $X_1, X_2, \ldots, X_n$ are continuous random variables with the same mean μ and variance Var (X),

$$\lim_{n \to \infty} P\left(\left|\frac{X_1 + \cdots + X_n}{n} - \mu\right| > \epsilon\right) = 0.$$

PROOF   Let $Y = (X_1 + X_2 + \cdots + X_n)/n$. From Theorems 3.12 and 3.13,

$$\text{Var}(Y) = \frac{1}{n^2} \sum_{i=1}^{n} \text{Var}(X_i) = \frac{\text{Var}(X)}{n}.$$

Applying Theorem 3.15, then, gives the result:

$$P\left(\left|\frac{X_1 + \cdots + X_n}{n} - \mu\right| > \epsilon\right) \leq \frac{\text{Var}(X)}{n\epsilon^2} \longrightarrow 0 \qquad \text{as } n \longrightarrow \infty.$$

Comment. There are much stronger results similar to Theorem 3.16; the weak law is mainly a historical curiosity. The basic idea was developed by Bernoulli for the case of 0, 1 variables and was included in his *Ars Conjectandi*. ∎

3.10 MOMENT-GENERATING FUNCTIONS

Finding moments of random variables, particularly the higher moments defined in Section 3.9, is conceptually straightforward but can be difficult to accomplish in practice: depending on the nature of $f_X(x)$, integrals of the form $\int_{-\infty}^{\infty} x^r f_X(x)\, dx$ are not always easy to evaluate. Fortunately, an alternate method is available. For some densities, we can find a *moment-generating function* (or *mgf*), $M_X(t)$, one of whose properties is that its *r*th derivative evaluated at 0 is equal to $E(X^r)$. If $M_X(t)$ can be found, $M_X^{(r)}(t)$ will often be easier to evaluate than $\int_{-\infty}^{\infty} x^r f_X(x)\, dx$.

Moment-generating functions can also be extremely useful in deriving the distribution of a *sum* of independent random variables. Such problems are important in statistics and typically difficult: recall, for instance, Example 3.13, where it took considerable effort to prove that the sum of two binomials is also binomial. Using moment-generating functions, that particular problem is trivial.

> **DEFINITION 3.12** Let X be a random variable. The *moment-generating function for X*, denoted $M_X(t)$, is given by
> $$M_X(t) = E(e^{tX})$$
> at all values of t for which the expected value exists.

Comment. It is not necessarily true that the moment-generating function exists for any t other than 0. Fortunately, in many cases of interest there *is* some interval about 0 on which the function is defined. ▮

Before investigating the properties of moment-generating functions, we will find $M_X(t)$ for two of the densities already encountered, the binomial and the exponential. The computations here are simply an application of Theorem 3.8.

EXAMPLE 3.31.   Let X be a binomial random variable with pdf

$$f_X(x) = \binom{n}{x} p^x q^{n-x}, \quad x = 0, 1, \ldots, n.$$

Then

$$M_X(t) = E(e^{tX}) = \sum_{x=0}^{n} e^{tx} \binom{n}{x} p^x q^{n-x} = \sum_{x=0}^{n} \binom{n}{x} (pe^t)^x q^{n-x}$$

$$= (pe^t + q)^n.$$

In this case, $M_X(t)$ is defined for all values of t.

EXAMPLE 3.32.   Let X have the exponential pdf, $f_X(x) = \lambda e^{-\lambda x}$, $x > 0$. Then

$$M_X(t) = E(e^{tX}) = \int_0^\infty e^{tx} \lambda e^{-\lambda x} \, dx$$

$$= \int_0^\infty \lambda e^{(t-\lambda)x} \, dx$$

$$= \frac{\lambda}{t - \lambda} e^{(t-\lambda)x} \Big|_0^\infty$$

$$= \frac{\lambda}{t - \lambda} [\lim_{x \to \infty} e^{(t-\lambda)x} - 1].$$

Note that this limit exists (and equals 0) only if $t - \lambda < 0$. Thus, for $t < \lambda$,

$$M_X(t) = \frac{\lambda}{\lambda - t}.$$

QUESTION 3.10.1   Let X be a discrete random variable with

$$f_X(x) = \begin{cases} pq^{x-1}, & x = 1, 2, \ldots \\ 0, & \text{elsewhere.} \end{cases}$$

Show that $M_X(t) = (pe^t)/(1 - qe^t)$. Hint: Recall the formula for the sum of a geometric series,

$$\sum_{t=0}^{\infty} r^t = \frac{1}{1 - r}, \quad \text{for } |r| < 1.$$

The next theorem shows that $M_X(t)$ does, indeed, generate moments.

THEOREM 3.17 Let X be a random variable with probability density function $f_X(x)$. (If X is continuous, f_X must be sufficiently smooth to allow the order of differentiation and integration to be interchanged.) If $M_X(t)$ is the moment-generating function for X,

$$M_X^{(r)}(0) = E(X^r).$$

PROOF We will verify the theorem for the continuous case where r is either 1 or 2. The extensions to discrete random variables and to an arbitrary positive integer r are straightforward.

For $r = 1$,

$$M_X^{(1)}(0) = \frac{d}{dt} \int_{-\infty}^{\infty} e^{tx} f_X(x)\, dx \bigg|_{t=0} = \int_{-\infty}^{\infty} \frac{d}{dt} e^{tx} f_X(x)\, dx \bigg|_{t=0}$$

$$= \int_{-\infty}^{\infty} x e^{tx} f_X(x)\, dx \bigg|_{t=0} = \int_{-\infty}^{\infty} x e^{0 \cdot x} f_X(x)\, dx$$

$$= \int_{-\infty}^{\infty} x f_X(x)\, dx = E(X).$$

For $r = 2$,

$$M_X^{(2)}(0) = \frac{d^2}{dt^2} \int_{-\infty}^{\infty} e^{tx} f_X(x)\, dx \bigg|_{t=0} = \int_{-\infty}^{\infty} \frac{d^2}{dt^2} e^{tx} f_X(x)\, dx \bigg|_{t=0}$$

$$= \int_{-\infty}^{\infty} x^2 e^{tx} f_X(x)\, dx \bigg|_{t=0} = \int_{-\infty}^{\infty} x^2 e^{0 \cdot x} f_X(x)\, dx \bigg|_{t=0}$$

$$= \int_{-\infty}^{\infty} x^2 f_X(x)\, dx = E(X^2).$$

EXAMPLE 3.33. Let X be a binomial random variable defined over n trials. We wish to find $E(X)$ using $M_X(t)$. From Example 3.31,

$$M_X(t) = (pe^t + q)^n.$$

Therefore,

$$M_X^{(1)}(t) = n(pe^t + q)^{n-1} pe^t,$$

so that

$$E(X) = M_X^{(1)}(0) = n(p + q)^{n-1} p = np,$$

the same result obtained in Examples 3.21 and 3.24.

QUESTION 3.10.2   Use Theorem 3.17 to find $E(X^2)$ for a binomial random variable.

If the Taylor series expansion for $M_X(t)$ is known, then $E(X^r)$ is simply the coefficient of $t^r/r!$. This follows immediately if we expand $M_X(t)$ about the point 0:

$$M_X(t) = \sum_{r=0}^{\infty} \frac{M_X^{(r)}(0)}{r!} t^r. \tag{3.24}$$

The next example applies Equation 3.24 to the problem of finding moments for an exponential random variable.

EXAMPLE 3.34.   Let X have the exponential pdf $f_X(x) = \lambda e^{-\lambda x}$, $x > 0$. From Example 3.32,

$$M_X(t) = \frac{\lambda}{\lambda - t} = \frac{1}{1 - (t/\lambda)}$$

$$= \sum_{r=0}^{\infty} \left(\frac{t}{\lambda}\right)^r = \sum_{r=0}^{\infty} \frac{1}{\lambda^r} \cdot t^r.$$

Therefore,

$$\frac{M_X^{(r)}(0)}{r!} = \frac{1}{\lambda^r},$$

implying that

$$E(X^r) = \frac{r!}{\lambda^r}.$$

QUESTION 3.10.3   Suppose X is a random variable for which

$$M_X(t) = e^{t^2/2}.$$

Write out the series expansion for $e^{t^2/2}$, compare it to the Taylor series for $M_X(0)$, and deduce a general formula for the rth moment of X.

As mentioned earlier, moment-generating functions do more than generate moments: they also characterize probability density functions and facilitate the derivation of $f_Z(z)$, where $Z = X_1 + X_2 + \cdots + X_n$. The next two theorems state the basic results we need. The first gives the uniqueness property of moment-generating functions: if X and Y have the same mgf's, then they must necessarily have the same pdf's. The second gives two results that are of particular help in manipulating moment-generating functions. The proof of Theorem 3.18 is beyond the scope of this text. The proof of Theorem 3.19 will be left as an exercise.

THEOREM 3.18 Suppose X and Y are random variables for which $M_X(t) = M_Y(t)$ for some interval of t's containing 0. Then $f_X = f_Y$.

THEOREM 3.19
(a) Let X be a random variable with moment-generating function $M_X(t)$. Let $Y = aX + b$. Then

$$M_Y(t) = e^{bt} M_X(at).$$

(b) Let $X_1, X_2, \ldots, X_n$ be independent random variables with moment-generating functions $M_{X_1}(t), M_{X_2}(t), \ldots,$ and $M_{X_n}(t)$, respectively. Let $Y = X_1 + X_2 + \cdots + X_n$. Then

$$M_Y(t) = M_{X_1}(t) \cdot M_{X_2}(t) \cdots M_{X_n}(t).$$

For the final example in this chapter we turn Theorems 3.18 and 3.19 loose on the transformation problem described in Example 3.13.

EXAMPLE 3.35.   Let X and Y be independent binomial random variables defined on m and n trials, respectively. Let p be the same for both X and Y. We wish to find the probability density function for $Z = X + Y$.

From the second part of Theorem 3.19,

$$M_Z(t) = M_X(t) \cdot M_Y(t)$$
$$= (pe^t + q)^m(pe^t + q)^n$$
$$= (pe^t + q)^{m+n}. \tag{3.25}$$

But we recognize Equation 3.25 to be the moment-generating function for a binomial random variable defined on $m + n$ trials (and with success probability p). Thus, from Theorem 3.18, we conclude that Z is binomial—and, in particular,

$$f_Z(z) = \binom{m + n}{z} p^z q^{m+n-z}, \qquad z = 0, 1, \ldots, m + n.$$

QUESTION 3.10.4   Let X and Y be two independent exponential random variables:

$$f_X(x) = \lambda e^{-\lambda x}, \quad x > 0$$
$$f_Y(y) = \lambda e^{-\lambda y}, \quad y > 0.$$

Let $Z = X + Y$. Does Z have an exponential distribution?

Comment. The reader may find the moment-generating-function approach for finding pdf's of sums preferable to the convolution technique introduced in Example 3.13. Indeed, in many situations in later chapters a very simple argument could be given using moment-generating functions, but we choose to follow the more tedious convolution approach and relegate $M_X(t)$ to the exercises. There is a reason for this: it is the authors' firm conviction that there is considerable peda-gogical merit in using basic statistical principles and mathematical tools rather than moment-generating functions. The essence of a subject can go unnoticed if proofs are too "slick." One is reminded of the famous sword duel where one duelist took a mighty swipe at the other. "Aha! You missed," cried the second swordsman. "Wait till you try to turn your head," was the rejoinder.   ∎

APPENDIX 3.1 ERROR PROPAGATION

It is not an uncommon problem for a laboratory scientist to have to measure several quantities, each subject to a certain amount of "error," in order to calculate a final desired result. For example, a physics student trying to determine the accel-eration due to gravity, G, knows that the distance, D, traveled by a freely falling body in time, T, is related to G by the equation,

$$D = \tfrac{1}{2}GT^2$$

(assuming the body is initially at rest) or, equivalently,

$$G = \frac{2D}{T^2}.$$

Suppose distance and time are to be measured directly with a yardstick and a stopwatch. The values obtained, D and T, will not be exactly correct; rather, we can think of them as being realizations of random variables, with those variables having "true" values μ_D and μ_T and variances Var (D) and Var (T), the latter two numbers reflecting the lack of precision in the measuring process. Suppose we know from past experience the precisions characteristic of the distance and time measurements—what can we then conclude about the precision in the calculated value for G? That is, knowing Var (D) and Var (T), can we find Var (G)?

By way of background, we have already seen one result that bears directly on this sort of "error-propagation" problem. If the quantity to be calculated, Y, is the *sum* of n independently measured quantities, $X_1, X_2, \ldots, X_n$, and if the variance associated with each of the X_i's is known, we can appeal to Theorem 3.13 and say that

$$\text{Var}(Y) = \text{Var}(X_1) + \text{Var}(X_2) + \cdots + \text{Var}(X_n). \tag{3.26}$$

In general, extending Equation 3.26 in any *exact* way to situations where Y is some arbitrary function of a set of X_i's—say, $Y = g(X_1, X_2, \ldots, X_n)$—is extremely difficult. It is a relatively simple matter, though, to get an approximation for the variance of Y by considering a Taylor expansion of the function $g(X_1, X_2, \ldots, X_n)$. We will look at a specific example of this technique before taking up the general formulation.

EXAMPLE 3.36. Suppose X and Y are independent random variables with means μ_X and μ_Y and variances Var (X) and Var (Y). Let $Z = XY$. We wish to find an approximation for Var (Z).

Using the first-order terms in a Taylor expansion of the function $g(X, Y) = XY$ around the point (μ_X, μ_Y) gives

$$Z = g(X, Y) \doteq \mu_X \mu_Y + (X - \mu_X) \left[\frac{\partial g}{\partial X} \bigg|_{(\mu_X, \mu_Y)} \right] + (Y - \mu_Y) \left[\frac{\partial g}{\partial Y} \bigg|_{(\mu_X, \mu_Y)} \right]$$

$$= \mu_X \mu_Y + (X - \mu_X) \left[Y \big|_{(\mu_X, \mu_Y)} \right] + (Y - \mu_Y) \left[X \big|_{(\mu_X, \mu_Y)} \right]$$

$$= \mu_X \mu_Y + (X - \mu_X) \mu_Y + (Y - \mu_Y) \mu_X.$$

It follows, then, by an application of Theorems 3.12 and 3.13 that

$$\text{Var}(Z) \doteq \mu_Y^2 \,\text{Var}(X) + \mu_X^2 \,\text{Var}(Y). \tag{3.27}$$

Comment. In situations where μ_X and μ_Y are unknown, they would be replaced by the actual x- and y-values recorded. ■

A physics application of this particular functional relationship would be the familiar equation relating distance, D, rate, R, and time, T:

$$D = RT.$$

Suppose an experiment were performed where both R and T were measured, with the values obtained being $R = 2.1$ feet/second and $T = 6.0$ seconds. Furthermore, suppose it were known that the standard deviations associated with R and T are 0.1 feet/second and 0.2 seconds, respectively. Then the calculated value for the distance would be

$$D = (2.1 \text{ ft/sec})(6.0 \text{ sec}) = 12.6 \text{ feet},$$

and it would have an approximate standard deviation, from Equation 3.27, of 0.7 feet:

$$\text{standard deviation of } D = \sqrt{\text{Var}(D)}$$
$$\doteq \sqrt{(6.0)^2(0.1)^2 + (2.1)^2(0.2)^2}$$
$$= 0.7 \text{ feet}.$$

Consider, now, the general case where Y is a function of n independent random variables, $Y = g(X_1, X_2, \ldots, X_n)$. Assume that each X_i has mean μ_i and variance $\text{Var}(X_i)$. By expanding $g(X_1, X_2, \ldots, X_n)$ around the point $(\mu_1, \mu_2, \ldots, \mu_n)$—and ignoring higher-order terms—we have that

$$Y \doteq g(\mu_1, \mu_2, \ldots, \mu_n) + (X_1 - \mu_1)\left[\frac{\partial g}{\partial X_1}\bigg|_{(\mu_1, \ldots, \mu_n)}\right]$$

$$+ (X_2 - \mu_2)\left[\frac{\partial g}{\partial X_2}\bigg|_{(\mu_1, \ldots, \mu_n)}\right] + \cdots + (X_n - \mu_n)\left[\frac{\partial g}{\partial X_n}\bigg|_{(\mu_1, \ldots, \mu_n)}\right].$$

Therefore,

$$\text{Var}(Y) \doteq \left[\frac{\partial g}{\partial X_1}\bigg|_{(\mu_1, \mu_2, \ldots, \mu_n)}\right]^2 \text{Var}(X_1) + \left[\frac{\partial g}{\partial X_2}\bigg|_{(\mu_1, \mu_2, \ldots, \mu_n)}\right]^2 \text{Var}(X_2)$$

$$+ \cdots + \left[\frac{\partial g}{\partial X_n}\bigg|_{(\mu_1, \mu_2, \ldots, \mu_n)}\right]^2 \text{Var}(X_n). \tag{3.28}$$

We will apply Equation 3.28 in Case Study 3.1 to a problem encountered by X-ray equipment inspectors.

CASE STUDY

3.1

In a typical dental X-ray unit, electrons from the cathode of the X-ray tube are decelerated by nuclei in the anode, thereby producing Bremsstrahlung radiation (X rays). These emissions, when collimated by a lead-lined tube, effect the desired image on a sheet of film.

Tennessee state regulations (172) require that the distance, Q, from the focal spot on the anode of an X-ray tube to the patient's skin be at least 18 cm. On some equipment, particularly older units, that distance cannot be measured directly because the exact location of the focal spot cannot be determined just by looking at the tube's outer housing. When this is the case, state inspectors resort to an indirect measuring procedure. Two films are exposed, one at the unknown distance Q and a second at a distance $Q + Z$. The two diameters, X and Y, of the resulting circular images are then measured (see Figure 3.9).

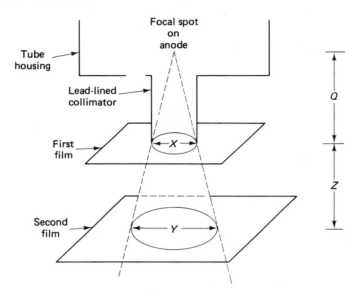

Figure 3.9   Indirect measuring procedure.

By similar triangles,

$$\frac{X}{Q} = \frac{Y}{Q + Z} \qquad \text{or} \qquad Q = \frac{XZ}{Y - X}. \qquad (3.29)$$

Phrased in the context of our previous notation,

$$Q = g(X, Y, Z) = XZ(Y - X)^{-1}.$$

During the course of one such inspection (96), values measured for the two diameters X and Y and the backoff distance Z were 6.4 cm, 9.7 cm, and 10.2 cm, respectively. From Equation 3.29, then, the anode-to-patient distance is estimated to be

$$Q = \frac{(6.4)(10.2)}{9.7 - 6.4} = 19.8 \text{ cm},$$

indicating the unit is in compliance. If the error in Q, though, were sufficiently large, there might still be a sizable probability that the *true* Q was less than 18 cm, meaning the unit was, in fact, out of compliance. It is not unreasonable, therefore, to inquire about the magnitude of Var (Q).

To apply Equation 3.28, we first need to compute the partial derivatives of $g(X, Y, Z)$. In this case,

$$\frac{\partial g}{\partial X} = \frac{XZ}{(Y - X)^2} + \frac{Z}{(Y - X)}$$

$$\frac{\partial g}{\partial Y} = \frac{-XZ}{(Y - X)^2}$$

and

$$\frac{\partial g}{\partial Z} = \frac{X}{(Y - X)}.$$

Furthermore, inspectors feel that the standard deviation in any of their measurements is on the order of 0.08 cm, so Var (X) = Var (Y) = Var (Z) = $(0.08)^2$. Substituting the variance estimates and the partial derivatives, evaluated at the point (μ_X, μ_Y, μ_Z) = (6.4, 9.7, 10.2) into Equation 3.28 gives

$$\text{Var}(Q) \doteq \left[\frac{(6.4)(10.2)}{(9.7 - 6.4)^2} + \frac{10.2}{(9.7 - 6.4)} \right]^2 (0.08)^2$$

$$+ \left[\frac{-(6.4)(10.2)}{(9.7 - 6.4)^2} \right]^2 (0.08)^2 + \left[\frac{6.4}{(9.7 - 6.4)} \right]^2 (0.08)^2$$

$$= 0.782.$$

Therefore, the estimated standard deviation associated with the calculated value of Q is $\sqrt{0.782}$, or 0.88 cm. (See Review Exercise 101 for the "conclusion" of this problem.)

REVIEW EXERCISES FOR CHAPTER 3

1. Hemophilia is a recessive sex-linked disease. If a woman who is a *carrier*—that is, who has the recessive hemophilia gene but not the symptoms of the disease—has a child with a man who is normal, the probability is $\frac{1}{4}$ that their next child will be hemophilic. Suppose the couple intend to have three children. Let the random variable X denote the number who will be hemophilic. Find $f_X(x)$.

2. Listed below is the length distribution of World Series competition for the 50 years from 1926 to 1975.

World Series lengths

Number of Games, X	Number of Years
4	9
5	11
6	8
7	22
	50

If it is assumed that each World Series game is an independent event and that the probability of either team's winning any particular contest is 0.5, find the probability model describing the series length, X—that is, find $f_X(x)$, $x = 4, 5, 6, 7$. How well does the model fit the data? (Compute the "expected" frequencies.)

3. Suppose the length of time, X, that a customer has to stand in line in front of a bank teller's window before being served can be described quite well by the pdf

$$f_X(x) = \tfrac{1}{5}e^{-x/5}, \qquad x > 0,$$

where X is in minutes. However, if he is not waited on within 10 minutes of his arrival in line, the customer will leave. What is the probability of that event? Suppose the customer intends to make five trips to the bank in the next month. Let Y denote the number of times he will leave before getting waited on. Find $f_Y(2)$. What is the probability he will leave the line at least once?

4. Among the most important of all probability models is the *Poisson*, which will be discussed in detail in Chapter 4. Example 2.13 offered a preview of this particular model: for the 12 years from 1948 to 1959 the number of major labor strikes, X, per week in the United Kingdom can be accurately described by the particular Poisson pdf

$$f_X(x) = \frac{e^{-0.90}(0.90)^x}{x!}, \qquad x = 0, 1, 2, \ldots.$$

In how many weeks during a year would we expect to have exactly three major labor strikes? In how many weeks during a year would we expect to have fewer than three major labor strikes?

5. A basketball player has a 45% shooting percentage. Let X denote the field-goal attempt at which he makes his third basket of the game. Find $f_X(5)$. Write down a general expression for $f_X(x)$.

6. If the random variable X denotes the speed of a molecule in a uniform gas at equilibrium, then

$$f_X(x) = ax^2 e^{-bx^2}, \qquad x > 0.$$

Here, $b = m/(2kT)$, where k is Boltzmann's constant, T is absolute temperature, and m denotes the molecule's mass. Find a.

7. In the game of odd man out, three players each toss a fair coin. The first to get an outcome different from that of the other two, wins. Let X denote the number of tosses required to determine a winner, $x = 1, 2, \ldots$. Find $f_X(x)$. What is the probability that more than four tosses will be required?

8. In the past, persons afflicted with a certain neurological disease have had a 30% chance of complete recovery. A radically different therapy has been tested recently on ten patients, and eight have recovered. Let the random variable X denote the number who recover when given the new therapy. Find $1 - F_X(7)$ under the assumption that the new therapy is no better than the old. In very general terms, how would the magnitude of $1 - F_X(7)$ be used to draw an inference about the efficacy of the new treatment?

9. Six fair dice are tossed. Give an expression for the pdf of X, the number of 1's and 2's that appear.

10. In the first game of the season, Vanderbilt's quarterback throws 11 passes, five in the direction of his halfback, four toward his tight end, and the remaining two in the general vicinity of the other team's middle linebacker. Suppose four are caught and each "receiver" has the same chance as the others of catching any of the passes thrown in his direction. Let the random variable X denote the number of passes caught by the tight end. Find $f_X(x)$. What is the probability that each team catches two passes?

11. Suppose that in a certain third-world nation the distribution of per family disposable income, X, is described by the pdf

$$f_X(x) = xe^{-x}, \qquad x > 0,$$

where x is measured in thousands of dollars. Find the first, second, and third quartiles of the income distribution. That is, find the values of x for which $F_X(x) = 0.25$, $F_X(x) = 0.50$, and $F_X(x) = 0.75$.

12. The electronic timing device used in a bobsled race is accurate to the nearest 0.1 second: if the recorded time for a team to complete the course is 17.2 seconds, their "true" time might be anything from 17.15 to 17.25 seconds. Let X represent the timing *error*—that is, X is the difference between the recorded time and the actual time. Assuming that X is a uniform random variable, find the probability that the recorded time is in error by more than 0.03 seconds.

13. Urns I and II each contain four chips. An urn is chosen at random and one of its chips removed. The same procedure, choosing an urn and drawing out a chip, is repeated until one of the urns is empty. Let X denote the number of chips remaining in the *other* urn. Find $f_X(x)$.

14. Let X be a random variable with pdf

$$f_X(x) = \begin{cases} \frac{3}{4}, & 0 \leq x \leq 1 \\ \frac{1}{4}, & 2 \leq x \leq 3 \\ 0 & \text{elsewhere.} \end{cases}$$

(a) Graph $f_X(x)$.   (b) Find and graph $F_X(x)$.

15. The Admissions Committee of a medical school has ten applicants, eight white and two who are members of ethnic minorities, from which the remaining n positions in the first-year class are to be filled. Assume all candidates are equally qualified and the committee decides to choose the n at random. What is the smallest value for n that will insure a probability greater than $\frac{1}{2}$ of admitting at least one minority applicant?

16. If $f_X(x)$ is the pdf of the random variable X, find and graph the cdf in the following three cases:
(a) $f_X(x) = \frac{1}{3}, x = 0, 1, 2$.
(b) $f_X(x) = 3(1 - x)^2, 0 < x < 1$.
(c) $f_X(x) = 1/x^2, x > 1$.

17. Suppose the pdf describing X, the proportion of right answers a student gets on a certain test, is given by

$$f_X(x) = 6x(1 - x), \qquad 0 < x < 1.$$

Anyone getting less than 60% fails.
(a) Find $F_X(0.60)$.
(b) If ten students take the test, what is the probability that at least eight pass?

18. In England from 1875 to 1951 the interval T (in days) between consecutive mining accidents resulting in at least ten fatalities is described by the exponential pdf

$$f_T(t) = \tfrac{1}{241}e^{-t/241}, \qquad t > 0$$

(103). Find $F_T(t)$. Also, use $F_T(t)$ to determine the probability that the gap between consecutive accidents is between 50 days and 100 days, inclusive.

19. Let X have pdf

$$f_X(x) = \begin{cases} 0, & |x| > 1 \\ 1 - |x|, & |x| \le 1. \end{cases}$$

Find $F_X(x)$.

20. Suppose $f_X(x)$ is a symmetric continuous function, $f_X(x) = f_X(-x)$, for $x > 0$. Show that $P(-a < X < a) = 2 \cdot F_X(a) - 1$, $a > 0$.

21. Let X be a random variable with pdf

$$f_X(x) = \begin{cases} \tfrac{1}{2}, & \tfrac{1}{2} \le x < 1 \\ x/2, & 1 \le x < 2 \\ 0 & \text{elsewhere.} \end{cases}$$

(a) Graph $f_X(x)$.
(b) Find and graph $F_X(x)$.
(c) What is the probability that X takes on a value between $\tfrac{3}{4}$ and $\tfrac{5}{4}$?

22. For persons in a certain state convicted of grand theft auto and given three-year sentences, the length of time, X, they actually serve is described by the pdf $f_X(x) = \tfrac{1}{9}x^2$, $0 \le x \le 3$. Find $F_X(x)$ and use it to find the probability that a car thief will actually spend less than one year in jail. Suppose the records of three such prisoners are reviewed independently by the parole board. What is the probability that exactly two will be released in less than a year?

23. If the random variable X has cdf,

$$F_X(x) = \begin{cases} 0, & x \le 1 \\ \ln x, & 1 < x \le e \\ 1, & e < x, \end{cases}$$

find
(a) $P(X < 2)$.
(b) $P(2 < X < 2\tfrac{1}{2})$.
(c) $P(2 < X \le 2\tfrac{1}{2})$.
(d) $f_X(x)$.

24. Suppose n batteries are put into n toy robots, and the robots are left on for five hours. If the battery life, X, is described by the pdf $f_X(x) = (1/\theta)e^{-x/\theta}$, $\theta > 0$, find an expression for the probability that exactly r robots are still working after the five hours have elapsed.

25. An insurance company has records showing that the length of time, X, that relatives take to file for death benefits is described by the pdf $f_X(x) = 4x^2e^{-2x}$, $x > 0$, where x is measured in days.
(a) Find $F_X(x)$.
(b) Verify that $\lim_{t \to \infty} F_X(t) = 1$ and $\lim_{t \to 0} F_X(t) = 0$.
(c) What proportion of clients file within three days of their relative's death?

26. A point is chosen at random from the interior of a circle whose equation is $x^2 + y^2 \leq 4$. Let the random variables X and Y denote the x- and y-coordinates of the sampled point. Find $f_{X,Y}(x, y)$.

27. A senior math major needs to take three more math courses in his final semester to complete his major. The math department at his school has a faculty of 15, of whom four are incompetent, six are hacks, and the remaining five are completely worthless. Suppose he chooses his final three courses at random. Let X denote the number of incompetents he has for teachers and Y the number of hacks. Find $f_{X,Y}(x, y)$.

28. For each of the following joint pdf's, $f_{X,Y}(x, y)$, find $f_X(x)$ and $f_Y(y)$.
    (a) $f_{X,Y}(x, y) = \dfrac{1}{36x!y!(3 - x - y)!} 4^{3-x-y}$; $x = 0, 1, 2, 3$; $y = 0, 1, 2, 3$.
    (b) $f_{X,Y}(x, y) = \frac{1}{2}$; $0 < x < 2$; $0 < y < 1$.
    (c) $f_{X,Y}(x, y) = xye^{-(x+y)}$; $x > 0$; $y > 0$.
    (d) $f_{X,Y}(x, y) = 2e^{-(x+y)}$; $0 < x < y$; $0 < y$.

29. Four cards are drawn from a standard poker deck without replacement. Let X be the number of kings drawn and Y the number of queens. Find $f_{X,Y}(x, y)$.

30. If $f_{X,Y}(x, y) = cxy$ at the points $(1, 1)$, $(2, 1)$, $(2, 2)$, and $(3, 1)$, and equals 0 elsewhere, find c. Also, find $f_X(x)$.

31. Suppose $f_{X,Y}(x, y) = xye^{-(x+y)}$, $x > 0$, $y > 0$. Prove for any real numbers a, b, c, and d that

$$P(a < X < b, c < Y < d) = P(a < X < b) \cdot P(c < Y < d),$$

thereby establishing the independence of X and Y.

32. Given the joint pdf $f_{X,Y}(x, y) = 2x + y - 2xy, 0 < x < 1, 0 < y < 1$, find numbers $a, b, c,$ and d such that

$$P(a < X < b, c < Y < d) \neq P(a < X < b) \cdot P(c < Y < d),$$

thus demonstrating that X and Y are not independent.

33. Prove that if X and Y are two independent random variables, then $U = g(X)$ and $V = h(Y)$ are also independent.

34. If two random variables X and Y are defined over a region in the XY-plane that is *not* a rectangle (possibly infinite) with sides parallel to the coordinate axes, can X and Y be independent?

35. Find the pdf of $R = A \sin \theta$, where A is a constant and θ is uniformly distributed on $(-\pi/2, \pi/2)$. Variables like R arise often in the theory of ballistics. If a projectile is fired at an angle α from the earth with a velocity v, then the distance R at which it returns to the earth can be expressed as $R = (v^2/g)(\sin 2\alpha)$, where g is the gravitational constant.

36. The temperature, X, achieved in a certain chemical reaction varies considerably from experiment to experiment but appears to be described quite well by a pdf of the form

$$f_X(x) = xe^{-x^2/2}, \qquad x \geq 0,$$

where x is measured in degrees Fahrenheit. The conversion formula for going from degrees Fahrenheit (X) to degrees Centigrade (Y) is

$$Y = \tfrac{5}{9}(X - 32).$$

Describe the distribution of temperatures in terms of degrees Centigrade. That is, find $f_Y(y)$.

37. Let X have pdf $f_X(x) = x^2/9$, $0 < x < 3$. Let $Y = X^2$. Find $f_Y(y)$.

38. Let $f_X(x) = e^{-x}$, $x \geq 0$. Find $f_Y(y)$ for $Y = \ln X$.

39. Let X be any continuous random variable with cdf $F_X(x)$. Define $Y = F_X(x)$. Show that Y has a uniform distribution over the unit interval.

40. A random observation from an exponential pdf, $f_X(x) = (1/\lambda)e^{-x/\lambda}$, $x > 0$, can be generated by multiplying minus λ times the ln of a number chosen at random from a uniform pdf over the unit interval. Prove why that should be true.

41. The velocity of a gas molecule of mass m is a random variable X with pdf $f_X(x) = ax^2e^{-bx^2}$, $x > 0$, where a and b are constants depending on the gas. The kinetic energy of the molecule is a random variable Y, where $Y = (m/2)X^2$. Find $f_Y(y)$.

42. Suppose you were given a random sample $Y_1, Y_2, \ldots, Y_n$ drawn from the uniform pdf over $(0, 1)$. How could those numbers be used to generate a random observation from a binomial pdf with parameters n and p?

43. To become an agent for a certain government intelligence agency, a candidate must take two personality-profile tests. On the basis of the first, he is assigned a Furtiveness Quotient, X; on the basis of the second, an Appearance Index, Y (both X and Y are measured in Bogart units). Among the general public, Furtiveness Quotients vary according to the pdf $f_X(x) = e^{-x}$, $x > 0$. Appearance Indices are described by $f_Y(y) = ye^{-y}$, $y > 0$. Past experience has shown that a candidate will be a successful agent only if he achieves a total of at least 4.5 Bogart units on the two tests. Find the proportion of potentially successful agents in the general population.

44. Prove the following intuitively obvious proposition: If a random variable X is independent of random variables Y and Z, then it is independent of $Y + Z$.

45. Let X and Y have the joint pdf $f_{X,Y}(x, y) = e^{-x-y}$, $x > 0$, $y > 0$. Find the pdf for Z, where $Z = X + Y$.

46. A number is chosen at random from the interval $(0, 3)$. A second number is chosen independently and at random from the interval $(0, 4)$. What is the eightieth percentile of the sum of the two numbers?

47. Suppose X is a continuous random variable with only positive values. Let $Y = \sqrt{X}$. Find an expression for f_Y in terms of f_X.

48. Let X and Y be continuous random variables for which $f_{X,Y}(x, y) = x + y$, $0 < x < 1$, $0 < y < 1$. Find $f_Z(z)$, where $Z = XY$.

49. Suppose the length of time, in minutes, that you have to wait at a bank teller's window is uniformly distributed over the interval $(0, 10)$. Suppose you go to the bank four times during the next month. What is the probability your second longest wait will be less than 5 minutes?

50. What is the probability that the larger of two random variables from any continuous pdf will exceed the $(100p)$th percentile?

51. In a certain large metropolitan area the proportion, X, of students bused varies widely from school to school. The distribution of bussing proportions is roughly described by the following pdf:

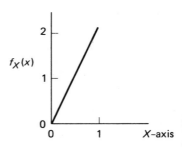

Suppose an HEW official examines the enrollment figures for five schools selected at random. What is the probability that the school with the fourth highest proportion of bused children will actually have an X-value in excess of 0.75? What is the probability that none of the schools will have fewer than 10% of their students bussed?

52. Let $X'_1 < X'_2 < \cdots < X'_n$ be a set of order statistics based on a random sample of size n taken from a uniform pdf over the unit interval. Let R denote the sample range, $R = X'_n - X'_1$. Find the pdf for R. If n equals 4, find the median for R. Find the probability that R is less than $\frac{1}{2}$.

53. An automobile manufacturer road tests a large sample of its new subcompacts to determine their gas consumption, both on the highway (X) and in the city (Y). They have found that if X and Y are given in fractions of the best possible mileages predicted by the car's designers, then

$$f_{X,Y}(x, y) = 8xy, \qquad 0 \le x \le 1, 0 \le y \le x.$$

(a) Verify that $f_{X,Y}(x, y)$ is a pdf.
(b) Find the pdf's describing highway mileage and city mileage, respectively.
(c) Find the conditional pdf describing the city mileage gotten by cars whose highway mileage is 80% of what the designers predicted.
(d) Find $P(Y > \frac{1}{4} | \frac{1}{4} < X < \frac{1}{2})$.

54. Let X have pdf

$$f_X(x) = \binom{n}{x} p^x(1 - p)^{n-x}, \qquad x = 0, 1, \ldots, n,$$

and let Y have pdf

$$f_Y(y) = \binom{n}{y} p^y(1 - p)^{n-y}, \qquad y = 0, 1, \ldots, n.$$

Let $Z = X + Y$. Show that the conditional pdf, $f_{X|z}(x)$, is hypergeometric.

55. Suppose the random variables X and Y are jointly distributed according to the pdf

$$f_{X,Y}(x, y) = \frac{6}{7}\left(x^2 + \frac{xy}{2}\right), \qquad 0 < x < 1, 0 < y < 2.$$

Find
(a) $f_X(x)$.
(b) $P(X > 2Y)$.
(c) $P(Y > 1 | X > \frac{1}{2})$.

56. Suppose the joint demands, X and Y, for hula hoops and yo-yo's, respectively, are described by the pdf

$$f_{X,Y}(x, y) = \frac{e^{-x/y}e^{-y}}{y}, \qquad 0 < x < \infty, 0 < y < \infty,$$

where x and y are given in terms of units sold per year. Find the probability that at least 2000 hula hoops will be sold given that the sales for yo-yo's fall somewhere between 5000 and 10,000.

57. Let X be a nonnegative random variable. We say that X is *memoryless* if

$$P(X > s + t \,|\, X > t) = P(X > s) \qquad \text{for all } s, t \geq 0.$$

Show that a random variable with pdf $f_X(x) = (1/\lambda)e^{-x/\lambda}$, $x > 0$, is memoryless.

58. Unfortunately it is not true, but suppose the life span, X, of a new car were described by an exponential pdf with λ equal to 10,

$$f_X(x) = \tfrac{1}{10}e^{-x/10}, \qquad x > 0,$$

where x is given in years. What would be the probability of a two-year-old car's lasting at least ten more years? of a ten-year-old car's lasting at least ten more years?

59. Given the joint pdf

$$f_{X,Y}(x, y) = 2e^{-(x+y)}, \qquad 0 < x < y, y > 0,$$

find
(a) $P(Y < 1 \,|\, X < 1)$.   (b) $P(Y < 1 \,|\, X = 1)$.
(c) $f_{Y|x}(y)$.

60. Suppose that among a certain socioeconomic class the number of hours (X) per day that a teenage boy watches television and the number of hours (Y) per day he spends studying are described by the joint pdf

$$f_{X,Y}(x, y) = xye^{-(x+y)}, \qquad x > 0, y > 0.$$

(a) Find the probability that he spends less than two hours studying given that he spends less than one hour watching television.
(b) Find $P(Y < 1 \,|\, X = 1)$.   (c) Find $f_{Y|x}(y)$.
(d) Sketch the surface described by $f_{X,Y}(x, y)$. Is this a reasonable model for approximating the joint behavior of TV watching and studying? Explain.

61. Suppose

$$f_{Y|x}(y) = \frac{2y + 4x}{1 + 4x} \quad \text{and} \quad f_X(x) = \tfrac{1}{3}(1 + 4x)$$

for $0 < x < 1$, $0 < y < 1$. Find the marginal pdf for Y.

62. A single elimination tennis tournament has eight players. Assume the players can be ranked from 1 to 8, with player 1 always being able to defeat player 2, player 2 always superior to player 3, and so on. If the initial "draw" for the tournament is done at random, what is the expected rank of the runner-up? [See (198) for a proof showing that as the number of players increases, the expected rank of the runner-up converges to 3.]

63. A disgruntled secretary is upset about having to stuff envelopes. Handed a box of n letters and n envelopes, she retaliates by putting the former into the latter *at random*. What is the expected number of envelopes receiving their proper letter?

64. A fair die is rolled three times. Let X denote the number of different faces showing, $X = 1, 2, 3$. Find $E(X)$.

65. For two continuous random variables, X and Y, the conditional expectation of Y given $X = x$ is written $E(Y \mid X = x)$ and given by the integral over y of the conditional pdf:

$$E(Y \mid X = x) = \int_{-\infty}^{\infty} y \cdot f_{Y \mid x}(y) \, dy.$$

The definition for discrete X and Y is analogous. Let

$$f_{X,Y}(x, y) = \frac{6}{7}\left(x^2 + \frac{xy}{2}\right), \qquad 0 < x < 1, 0 < y < 2.$$

Find $E(Y \mid X = 1)$.

66. A local tavern has six bar stools. The bartender predicts that if two strangers come into the bar they will sit in such a way as to leave at least two stools between them.
   (a) If two strangers do come in, but choose their seats at random, what is the probability of the bartender's prediction coming true?
   (b) Compute the expected value of the number of stools between the two customers, under the assumption that they seat themselves randomly.

67. Find the expected length of time a person convicted of grand theft auto actually spends in jail (see Review Exercise 22).

68. Ten equally qualified applicants, six men and four women, apply for three lab technician positions. Unable to justify choosing any of the applicants over all the others, the personnel director decides to select the three at random. Let Y denote the number of men hired. Find $E(Y)$. Also, if the positions are filled by either all men or all women, a suit charging unfair labor practices will be filed. What is the probability of that event?

69. Two distinct integers are chosen at random from the first five positive integers. Compute the expected value of the absolute value of the difference of the two numbers.

70. A baseball player with a .320 average has five official at-bats during a game. Let X denote the number of hits he makes. Find $E(X)$.

71. The *skewness*, B, of a random variable is defined to be the third moment about its mean: $B = E[(X - \mu)^3]$. Find the skewness of the exponential pdf, $f_X(x) = e^{-x}$, $x > 0$.

72. The president of a local civic club is trying to set up a "Bingo Night" that will yield a profit of $500, the proceeds going to a scholarship fund. Each of the 25 games is to be played with 75 numbers, ten of which will be drawn from a bowl by a representative of the club. Before the drawing, players will have written down four numbers. If all a player's numbers match up with those drawn from the bowl, he wins a $300 TV set. Find an expression for the entry fee that should be paid by each contestant for each game, as a function of n, the *number* of contestants. (Assume that n is constant for all 25 games.)

73. Find $E(X)$ for the following pdf's:
   (a) $f_X(x) = \lambda e^{-\lambda x}$, $x > 0$.   (b) $f_X(x) = \frac{1}{3}$, $0 \le x \le 3$.

(c) $f_X(x) = 1/[\pi(1 + x^2)]$, $-\infty < x < \infty$.

(d) $f_X(x) = 4xe^{-2x}$, $x > 0$.   (e) $f_X(x) = \sin x$, $0 \leq x \leq \pi/2$.

74. Let M denote the median of a random sample of size three drawn from $f_X(x) = 1/\theta$, $0 < x < \theta$. Find $E(M)$ and $E(\bar{X})$, where $\bar{X} = (\frac{1}{3})(X_1 + X_2 + X_3)$.

75. An example of a Poisson random variable was described in Review Exercise 4. If X denotes the number of major labor strikes (per week) in the United Kingdom, then

$$f_X(x) = \frac{e^{-0.90}(0.90)^x}{x!}, \qquad x = 0, 1, 2, \ldots .$$

Find $E(X)$. Hint: Factor 0.90 out of the expression $\sum_{x=0}^{\infty} xf_X(x)$ and show that the remaining summation is 1.

76. Find the expected payoff for the St. Petersburg problem (see Example 3.22) if the amounts won are c^k instead of 2^k, where $0 < c < 2$.

77. How much would you have to ante to make the St. Petersburg problem "fair" (see Example 3.22) if the most you could win was $1000? That is, you win 2^k for $1 \leq k \leq 9$, and $1000 for $k \geq 10$.

78. Let X be a random variable with finite mean μ. Define for every real number a, $g(a) = E[(X - a)^2]$. Show that

$$g(a) = E[(X - \mu)^2] + (\mu - a)^2.$$

What is another name for $\min_a g(a)$?

79. Find the expected values of the following random variables defined on the sample space of throws of a pair of dice:
   (a) The sum of the two numbers showing.
   (b) The larger of the two numbers showing.
   (c) The number of sixes showing.

80. Suppose $f_{X,Y}(x, y) = e^{-x-y}$, $x > 0$, $y > 0$. Find $E(X + Y)$.

81. Let X_1, X_2, $X_{,3}$ X_4 be a random sample from $f_X(x) = 2x$, $0 < x < 1$. Find the mean and variance of the second-order statistic.

82. An urn has two red chips and three black chips. Two chips are drawn at random and without replacement. Let X denote the number of red chips in the sample. Find Var (X).

83. If X_1 is a binomial random variable based on n trials and a success probability of p_1, and if X_2 is an independent binomial random variable based on m trials and a success probability of p_2, find $E(Y)$ and Var (Y), where $Y = 4X_1 + 6X_2$.

84. Find the variance of the time spent in jail by a person convicted of grand theft auto (see Review Exercise 22).

85. Find the variance of X if

   (a) $f_X(x) = 6x(1 - x)$, $0 \leq x \leq 1$.   (b) $f_X(x) = \begin{cases} \frac{3}{4}, & 0 \leq x \leq 1. \\ \frac{1}{4}, & 2 \leq x \leq 3 \\ 0 & \text{elsewhere.} \end{cases}$

(c) $f_X(x) = 3(1 - x)^2, 0 \le x \le 1.$ (d) $f_X(x) = \begin{cases} x, & 0 \le x \le 1 \\ 2 - x, & 1 < x \le 2. \end{cases}$

86. A gambler plays n hands of poker. If he wins the kth hand, he collects k dollars; if he loses the kth hand, he collects nothing. Let T denote his total winnings in n hands. Assuming his chances of winning each hand are constant (say, equal to p) and are independent of his success or failure at any other hand, find $E(T)$ and Var(T).

87. Use Chebyshev's inequality to determine how many times a fair coin must be tossed in order for the probability to be at least 0.90 that the ratio of the observed number of heads to the number of tosses will be between 0.4 and 0.6.

88. If X_1 and X_2 are independent with means μ_1 and μ_2 and variances σ_1^2 and σ_2^2, respectively, show that the mean and variance of $Y = X_1 X_2$ are $\mu_1 \mu_2$ and $\sigma_1^2 \sigma_2^2 + \mu_1^2 \sigma_2^2 + \mu_2^2 \sigma_1^2$, respectively.

89. A fair die is tossed 100 times. Let X_k denote the outcome of the kth roll. Use Chebyshev's inequality to get a lower bound for the probability that $Y = X_1 + \cdots + X_{100}$ is between 300 and 400.

90. Prove Schwarz's inequality for random variables: if X and Y are random variables with finite variances, then

$$[E(XY)]^2 \le E(X^2)E(Y^2).$$

91. Find the variances for the random variables listed in parts (a), (b), and (d) of Review Exercise 73.

92. Suppose a certain IQ test has mean score 100 and standard deviation 16. Show that the probability of a score more than 148 or less than 52 is at most $\frac{1}{9}$.

93. Show that the moment-generating function of the random variable X having pdf $f_X(x) = \frac{1}{3}, -1 < x < 2$, is

$$M_X(t) = \begin{cases} \dfrac{e^{2t} - e^{-t}}{3t}, & t \ne 0 \\ 1, & t = 0. \end{cases}$$

94. Let $C_X(t) = \ln M_X(t)$, where $M_X(t) = E(e^{tX})$. Prove that $C'_X(0) = \mu$ and $C''_X(0) = \sigma^2$.

95. Let X have pdf

$$f_X(x) = \begin{cases} x, & 0 \le x \le 1 \\ 2 - x, & 1 \le x \le 2 \\ 0 & \text{elsewhere.} \end{cases}$$

Find $M_X(t)$.

96. Find the moment-generating function for $Y = (X/2) + 3$, where X has the pdf $f_X(x) = pq^{x-1}, x = 1, 2, \ldots; p + q = 1$. Also, use $M_Y(t)$ to find the mean and variance of Y.

97. A random variable, X, is said to have a *normal* pdf with mean μ and variance σ^2 if

$$f_X(x) = \frac{1}{\sqrt{2\pi}\sigma} e^{-(1/2)[(x-\mu)/\sigma]^2}, \qquad -\infty < x < \infty$$

(see Chapter 4). Also, the moment-generating function for X can be shown to equal

$$M_X(t) = e^{\mu t + (\sigma^2 t^2 / 2)}.$$

Suppose X and Y are two independent normal random variables with means μ_X and μ_Y and variances σ_X^2 and σ_Y^2, respectively. Find the pdf of Z, where $Z = X + Y$.

98. Let X be a binomial random variable with pdf

$$f_X(x) = \binom{n}{x} p^x (1 - p)^{n-x}, \qquad x = 0, 1, 2, \ldots, n.$$

Find the skewness of the binomial distribution (see Review Exercise 71) by differentiating $M_X(t)$.

99. A physics student is trying to determine the gravitational constant, G, using the expression

$$G = \frac{2D}{T^2},$$

where both distance (D) and time (T) are to be measured (see Appendix 3.1). Suppose that the standard deviation of the measurement errors in D is 0.0025 feet and in T, 0.045 seconds. If the experimental apparatus is set up so that D will be 4 feet, then T will be approximately $\frac{1}{2}$ second. If D is set at 16 feet, T will be close to 1 second. Which of these two sets of values for D and T will give a smaller variance for the calculated G?

100. Find a second-order approximation for the variance of D in the relationship $D = RT$ (see Appendix 3.1 on page 122).

101. Refer to the dental X-ray inspection data described in Case Study 3.1. Use Chebyshev's inequality to find an upper bound for the probability that $|Q - 18 \text{ cm}| > 1.8$ cm, where Q is the distance (in cm) from the focal spot on the anode of the X-ray tube to the patient. What does this probability represent in terms of the unit's being in compliance with state regulations?

102. Compute the variance of Z, where $Z = XY$ and X and Y have a bivariate uniform pdf (see Example 3.12). Also, approximate Var (Z) using (a) a first-order Taylor expansion and (b) a second-order Taylor expansion. Compare the approximations to the exact value. Note:

$$\int_0^1 z^p \ln z \, dz = -\frac{1}{(p + 1)^2}.$$

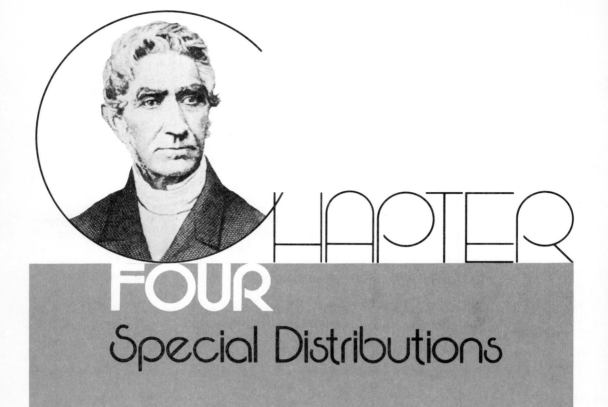

CHAPTER FOUR

Special Distributions

LAMBERT ADOLPHE JULES QUETELET (1796–1874)

Although he maintained lifelong literary and artistic interests, Quetelet's mathematical talents led him to a doctorate from the University of Ghent and from there to a college teaching position in Brussels. In 1833 he was appointed astronomer at the Brussels Royal Observatory, after having been largely responsible for its founding. His work with the Belgian census marked the beginning of his pioneering efforts in what today would be called mathematical sociology. Quetelet was well known throughout Europe in scientific and literary circles: at the time of his death he was a member of more than 100 learned societies.

4.1 INTRODUCTION

Certain probability functions occur sufficiently often, in theoretical as well as applied contexts, to make their study both useful and rewarding. Almost invariably these functions occur in *families*, where they are indexed by one or more unknown parameters. To single out a particular member of such a family, numerical values must be assigned to each of the function's parameters. We have already encountered the one-parameter exponential family, $f_X(x) = (1/\lambda)e^{-x/\lambda}$, in this regard: recall that $f_X(x) = \frac{1}{179}e^{-x/179}$ was used as a model for the life distribution of aircraft radar tubes (see Example 2.15). Another example is the binomial distribution,

$$P(X = k) = \binom{n}{k} p^k(1-p)^{n-k}, \ k = 0, 1, \ldots, n: \text{this is a family indexed by } two$$

parameters, n and p.

For the statistician, many families of probability functions merit special attention—far more than can be discussed in a text of this length. Hence, only those finding particularly extensive application will be dealt with. The reader is referred to (71) or the Johnson and Kotz series (88, 89, 90) for a more extensive survey.

Altogether, a total of five families of probability functions—the Poisson, normal, geometric, negative binomial, and gamma—will be introduced in the next several sections. In certain cases it will prove enlightening to recount briefly the history of a distribution—how it came to be discovered and what its first applications were. Some of the results presented here date back more than 150 years and are now part of the bedrock of modern statistical theory.

4.2 THE POISSON DISTRIBUTION

Simeon Denis Poisson (1781–1840) was an eminent French mathematician and physicist, an academic administrator of some note, and, according to an 1826 letter from the mathematician Abel to a friend, a man who knew "how to behave with a great deal of dignity." One of Poisson's many interests was the application of probability to the law, and in 1837 he wrote *Recherches sur la Probabilite de Judgements*. This text contained a good deal of mathematics, including a limit theorem for the binomial distribution.[1] Although initially viewed as little more than a welcome approximation for hard-to-compute binomial probabilities, this particular result was destined for bigger things: it was the analytical seed out of which grew what is now one of the most important of all probability models, the Poisson distribution.

---

[1] Although credit for this theorem is given to Poisson, there is some evidence that DeMoivre may have discovered it almost a century earlier (35).

THEOREM 4.1   Let λ be a fixed number and n an arbitrary positive integer. For each nonnegative integer x,

$$\lim_{n \to \infty} \binom{n}{x} p^x (1 - p)^{n-x} = \frac{e^{-\lambda} \lambda^x}{x!},$$

where $p = \lambda/n$.

PROOF

We begin by rewriting the binomial probability in terms of λ:

$$\lim_{n \to \infty} \binom{n}{x} p^x (1 - p)^{n-x} = \lim_{n \to \infty} \binom{n}{x} \left(\frac{\lambda}{n}\right)^x \left(1 - \frac{\lambda}{n}\right)^{n-x}$$

$$= \lim_{n \to \infty} \frac{n!}{x!(n-x)!} \lambda^x \left(\frac{1}{n^x}\right) \left(1 - \frac{\lambda}{n}\right)^{-x} \left(1 - \frac{\lambda}{n}\right)^n$$

$$= \frac{\lambda^x}{x!} \lim_{n \to \infty} \frac{n!}{(n-x)!} \frac{1}{(n-\lambda)^x} \left(1 - \frac{\lambda}{n}\right)^n.$$

But since $[1 - (\lambda/n)]^n \longrightarrow e^{-\lambda}$ as $n \longrightarrow \infty$, we need only show that

$$\frac{n!}{(n-x)! \, (n-\lambda)^x} \longrightarrow 1$$

to prove the theorem. However, note that

$$\frac{n!}{(n-x)! \, (n-\lambda)^x} = \frac{n(n-1) \cdots (n-x+1)}{(n-\lambda)(n-\lambda) \cdots (n-\lambda)},$$

a quantity which, indeed, tends to 1 as $n \longrightarrow \infty$.

EXAMPLE 4.1.   Tables 4.1 and 4.2 give an indication of the *rate* at which

$$\binom{n}{x} p^x (1 - p)^{n-x}$$

converges to

$$\frac{e^{-np} (np)^x}{x!}.$$

TABLE 4.1   Binomial probabilities and Poisson limits; $n = 5$ and $p = \frac{1}{5}$ ($\lambda = 1$)

x	$\binom{5}{x}(0.2)^x(0.8)^{5-x}$	$\dfrac{e^{-1}(1)^x}{x!}$
0	0.328	0.368
1	0.410	0.368
2	0.205	0.184
3	0.051	0.061
4	0.006	0.015
5	0.000	0.003
6+	0	0.001
	1.000	1.000

TABLE 4.2   Binomial probabilities and Poisson limits;
$n = 100$ and $p = \frac{1}{100}$ ($\lambda = 1$)

x	$\binom{100}{x}(0.01)^x(0.99)^{100-x}$	$\dfrac{e^{-1}(1)^x}{x!}$
0	0.366032	0.367879
1	0.369730	0.367879
2	0.184865	0.183940
3	0.060999	0.061313
4	0.014942	0.015328
5	0.002898	0.003066
6	0.000463	0.000511
7	0.000063	0.000073
8	0.000007	0.000009
9	0.000001	0.000001
10	0.000000	0.000000
	1.000000	0.999999

In both cases $\lambda = np$ is equal to 1, but in the former, n is set equal to 5—in the latter, to 100. We see in Table 4.1 ($n = 5$) that for some x the agreement between the binomial probability and Poisson's limit is not very good. By the time n has reached 100, though (Table 4.2), the agreement is remarkably good for all x.

The next example is typical of how convenient the Poisson limit can be in approximating binomial probabilities. The data concern the incidence of childhood leukemia and whether or not the disease is contagious, a question that has been around for quite a few years but still has medical researchers baffled.

CASE STUDY

4.1   Leukemia is a rare, but highly virulent, form of cancer whose cause and mode of transmission remain largely unknown. While evidence abounds that excessive exposure to radiation can increase a person's risk of contracting the disease, it is at the same time true that most cases occur among persons whose history contains no such over-exposure. A related issue, one maybe even more basic than the cau-sality question, concerns the *spread* of the disease. It is safe to say that the prevailing medical opinion is that most forms of leukemia are not contagious—still, the hypothesis persists that *some* forms of the disease, particularly the childhood variety, are. What continues to fuel this speculation are the discoveries of so-called "leukemia clusters," aggregations in time and space of unusually large numbers of cases.

To date, one of the most frequently cited leukemia clusters in the medical literature occurred during the late 1950s and early 1960s in Niles, Illinois, a suburb of Chicago (73). In the $5\frac{1}{3}$-year period from 1956 to the first four months of 1961, physicians in Niles reported a total of eight cases of leukemia among children less than 15 years of age. The number at risk (that is, the number of residents in that age range) was 7076. To assess the likelihood of that many cases occurring in such a small population, it is necessary to look first at the leukemia incidence in neighboring towns. For all of Cook county, excluding Niles, there were 1,152,695 children less than 15—and among those, 286 diagnosed cases of leukemia. That gives an average $5\frac{1}{3}$-year leukemia rate of 24.8 cases per 100,000:

$$\frac{286 \text{ cases for } 5\frac{1}{3} \text{ years}}{1,152,695 \text{ children}} \times \frac{100,000}{100,000}$$

$$= 24.8 \text{ cases}/100,000 \text{ children}/5\frac{1}{3} \text{ years.}$$

Now, imagine the 7076 children in Niles to be a series of $n = 7076$ (independent) Bernoulli trials, each having a probability of $p = 24.8/100,000 = 0.000248$ of contracting leukemia. The question then becomes, given an n of 7076 and a p of 0.000248, how likely is it that eight "successes" would occur? (The expected number, of course, would be $7076 \times 0.000248 = 1.75$.) Actually, for reasons that will be elaborated on in Chapter 6, it will prove more meaningful to consider the related event, eight *or more* cases occurring in a $5\frac{1}{3}$-year span. If the probability associated with the latter is very small, it could be argued that leukemia did not occur randomly in Niles and that, perhaps, contagion was a factor.

Using the binomial distribution, we can express the probability of eight or more cases as

$$P(8 \text{ or more cases}) = \sum_{k=8}^{7076} \binom{7076}{k} (0.000248)^k (0.999752)^{7076-k}$$

a computation not entirely pleasant to contemplate. With the aid of Poisson's theorem, though, the problem is trivial. By letting $\lambda = np = 7076 \times 0.000248 = 1.75$, we easily obtain 0.00049 as the Poisson limit:

$$P(8 \text{ or more cases}) \doteq \sum_{k=8}^{\infty} \frac{e^{-1.75}(1.75)^k}{k!}$$

$$= 1 - \sum_{k=0}^{7} \frac{e^{-1.75}(1.75)^k}{k!}$$
$$= 1 - 0.99951$$
$$= 0.00049.$$

Comment. Considering the accuracy of the Poisson limit for n as small as 100 (Table 4.2), we should feel very confident here, with an n of 7076, that the 0.00049 is extremely close to the true binomial sum. ∎

Comment. Interpreting this result physically is not as easy as assessing its accuracy mathematically. The fact that the probability is so very small tends to denigrate the hypothesis that leukemia in Niles occurred at random. On the other hand, rare events, such as clusters, *do* happen by chance. The basic difficulty is putting the probability associated with a given cluster in any meaningful perspective, not knowing in how many similar communities leukemia did *not* exhibit a tendency to cluster. That there is no obvious way to do this is one reason the leukemia controversy is still with us. ∎

QUESTION 4.2.1   Given 400 people, estimate the probability that three or more will have a birthday on July 4. Assume there are 365 days in a year, each equally likely to be a randomly chosen person's birthday.

Poisson's theorem certainly sent no shock waves through the scientific community: on the contrary, it languished in mathematical limbo for over 50 years, attracting little interest and finding no real application. Its resurrection finally came in 1898 when Ladislaus von Bortkiewicz, a German professor born in Russia of Polish ancestry, wrote a monograph entitled *Das Gesetz der Kleinen Zahlen* (*The Law of Small Numbers*). Included in this work was Poisson's theorem (with due credit), possibly the first consideration of the Poisson limit as a probability distribution in its own right, and, most importantly, the use of the "Poisson" to model real-world phenomena.

Comment. Bortkiewicz was one of several nineteenth-century scholars (we shall read of others in Section 4.3) who pioneered the use of probability distributions as data models. Before the early 1800s, probability and statistics were considered separate subjects—probability being a branch of pure mathematics and statistics the collection of data. ∎

THEOREM 4.2   Let $p(x, \lambda)$ denote Poisson's limit,

$$p(x, \lambda) = \frac{e^{-\lambda}\lambda^x}{x!}, \qquad \text{for } x = 0, 1, \ldots.$$

Then $p(x, \lambda)$ is a probability function. Also, if X is a random variable with pdf $p(x, \lambda)$, then $E(X) = \lambda$ and $\text{Var}(X) = \lambda$. We will call X a *Poisson random variable*.

PROOF   To show that $p(x, \lambda)$ qualifies as a probability function, note, first of all, that $p(x, \lambda) \geq 0$ for all nonnegative integers x. Also, it sums to 1:

$$\sum_{x=0}^{\infty} p(x, \lambda) = \sum_{x=0}^{\infty} \frac{e^{-\lambda}\lambda^x}{x!} = e^{-\lambda}\sum_{x=0}^{\infty} \frac{\lambda^x}{x!} = e^{-\lambda} \cdot e^{\lambda} = 1,$$

since $\sum_{x=0}^{\infty} \frac{\lambda^x}{x!}$ is the Taylor series expansion of e^{λ}.

The expected value for X can be gotten by direct summation:

$$E(X) = \sum_{x=0}^{\infty} xp(x, \lambda)$$

$$= \sum_{x=0}^{\infty} x\frac{e^{-\lambda}\lambda^x}{x!}$$

$$= \lambda\sum_{x=1}^{\infty} \frac{e^{-\lambda}\lambda^{x-1}}{(x - 1)!}$$

$$= \lambda\sum_{y=0}^{\infty} \frac{e^{-\lambda}\lambda^y}{y!} = \lambda \cdot 1 = \lambda,$$

where $y = x - 1$. The proof that Var$(X) = \lambda$ is similar and will be left as an exercise.

QUESTION 4.2.2   Derive the variance for a Poisson random variable. Hint: Consider the factorial moment, $E(X(X - 1))$.

QUESTION 4.2.3   Show that the moment-generating function for a Poisson random variable is

$$M_X(t) = e^{-\lambda + \lambda e^t},$$

and use $M_X(t)$ to verify the formulas for $E(X)$ and Var(X).

Among the several phenomena that Bortkiewicz successfully "fit" with the Poisson model, the one best remembered comprises the Prussian cavalry data described in the next example.

CASE STUDY

4.2   During the latter part of the nineteenth century, Prussian officials gathered information on the hazards that horses posed to cavalry soldiers. A total of ten cavalry corps were monitored over a period of 20 years (15). Recorded for each year and each corps was X, the number of fatalities due to kicks. Table 4.3 shows the empirical distribution of X for these 200 "corps-years."

TABLE 4.3   Observed horse-kick fatalities

x = Number of Deaths	Observed Number of Corps-Years in Which x Fatalities Occurred
0	109
1	65
2	22
3	3
4	1
	200

Altogether there were 122 fatalities [109(0) + 65(1) + 22(2) + 3(3) + 1(4)], meaning that the observed fatality *rate* was 122/200, or 0.61 fatalities per corps-year. For reasons that will be discussed in Chapter 5, Bortkiewicz set λ equal to 0.61 and proposed as a model for X,

$$P(X = x) = p(x, 0.61) = \frac{e^{-0.61}(0.61)^x}{x!}.$$

Multiplying $p(x, 0.61)$ by 200 would then yield the *expected* number of years in which x fatalities occurred. Clearly, the agreement between the second and third columns of Table 4.4 is excellent. We would infer

TABLE 4.4   Poisson-distribution prediction of horse-kick fatalities

x	Observed Number of Corps-Years	Expected Number of Corps-Years
0	109	108.7
1	65	66.3
2	22	20.2
3	3	4.1
4	1	.6
	200	199.9

from this that the phenomenon of Prussian soldiers' being kicked to death by their horses was modeled very well by the Poisson distribution.

Comment. The first step in fitting any probability function to a set of data is "estimating" the model's unknown parameters (just as, here, the parameter λ was assigned the value 0.61). Methods for doing this will be discussed in considerable detail in Chapter 5. For the present, our emphasis lies with the model's family rather than its parameters. Accordingly, for the remaining examples in this chapter we will simply state what the estimated parameters are, making no attempt to derive them.   ∎

Comment. In Chapter 9 a formal procedure will be introduced for measuring the "goodness-of-fit" between a probability model and an empirical distribution. Until then, any assessment of the appropriateness of a model will have to be a purely subjective one, based on our perception of the extent of agreement, or lack of agreement, between the observed and expected frequencies.   ∎

With Bortkiewicz' work paving the way, not many years passed before the Poisson distribution was firmly entrenched in the repertoire of mathematicians, statisticians, and scientists of all types. What began as a simple limiting form of the binomial has since been shown to be an excellent model for the enumeration of such diverse phenomena as radioactive emission, outbreaks of war, the positioning of stars in space, telephone calls arriving at a central switchboard, traffic accidents along a given stretch of road, mine disasters, and even misprints in books.

We might well ask what causes this distribution to appear so often. Or, turning the question around, what can all these various phenomena, so apparently different, possibly have in common? To answer that, we need to reexamine the implications of Poisson's theorem.

Consider one of the applications just mentioned: calls coming into a switchboard. Suppose the calls are arriving at an average rate of λ per unit time. We will imagine a time interval of length T partitioned into n subintervals, each having length T/n. It will also be supposed that these subintervals are so small that the probability of more than one call's arriving during any particular one is negligible. Finally, we will assume these intervals are independent—that is, a call's arriving during the ith subinterval has no effect on the probability of a call's arriving during the jth subinterval, $i \neq j$.

Now, if p_n denotes the probability of a call's being received during any subinterval, the probability of exactly k calls arriving during the interval of length T is simply the binomial, $b(k, n, p_n)$. It follows that during this time approximately np_n calls will be received (the mean for a binomial random variable); on the other hand, since λ is the call rate per unit time, the number arriving should be approximately λT. Thus, we may assume that $np_n \longrightarrow \lambda T$ as $n \longrightarrow \infty$. Equivalently,

$$b(k, n, p_n) \doteq b\left(k, n, \frac{\lambda T}{n}\right)$$

but, by Poisson's theorem,

$$b\left(k, n, \frac{\lambda T}{n}\right) \longrightarrow p(k, \lambda T). \tag{4.1}$$

Equation 4.1, then, is what all these phenomena have in common. They are all basically a series of independent, Bernoulli trials, where n is large, p is small, and the Poisson limit to the binomial holds.

Comment. In light of the above discussion, it may be helpful to think of Theorem 4.2 in more general terms. If an event is occurring at a constant rate of λ per unit time, and if X denotes the number occurring in a period of length T, then

$$P(X = x) = \frac{e^{-\lambda T}(\lambda T)^x}{x!}, \qquad x = 0, 1, 2, \ldots .$$

By replacing λ with λT, the periods spanned by the random variable and by the assigned parameter are made to agree, as they must. Of course, if X refers to the number of occurrences in "unit" time, as was the case with the cavalry data, T is set equal to 1. ▮

QUESTION 4.2.4   The Brown's Ferry incident of 1975 has focused national attention on the ever-present danger of fires breaking out in nuclear power plants. The Nuclear Regulatory Commission has estimated that with present technology there will be, on the average, one fire for every ten nuclear reactor-years. Suppose a given state intends to put two reactors "on line" by 1983. Assuming the incidence of fires can be described by a Poisson distribution, what is the probability that by 1986 at least two fires will have occurred? Does the Poisson assumption seem reasonable here?

We conclude this section with two other examples illustrating the diversity of Poisson applications—the first from particle physics, the second from sociology.

CASE STUDY

4.3

Among the early research efforts in radioactivity was a 1910 study of alpha emission done by the now-famous Rutherford and Geiger (149). Their experimental setup consisted of a polonium source placed a short distance from a small screen. For each of 2608 eighth-minute intervals, the two physicists recorded the number of α particles impinging on the screen (Columns 1 and 2 of Table 4.5).

The particular Poisson being fitted here (column 3) is

$$P(X = x) = \frac{e^{-3.87}(3.87)^x}{x!}.$$

Note that the large number of observations tends to magnify the discrepancies between the observed and expected columns. The *relative* errors, though, as measured by

$$\frac{\text{observed frequency} - \text{expected frequency}}{\text{expected frequency}}$$

TABLE 4.5   Rutherford-Geiger study

Number of α Particles Recorded, x	Observed Frequency	Expected Frequency
0	57	54
1	203	211
2	383	407
3	525	526
4	532	508
5	408	394
6	273	254
7	139	140
8	45	68
9	27	29
10	10	11
11+	6	6
	2608	2608

are really quite small—this would be considered a very good fit.

CASE STUDY

4.4

In the 432 years from 1500 to 1931, war broke out somewhere in the world a total of 299 times. (By definition, a military action was a war if it either was legally declared, involved over 50,000 troops, or resulted in significant boundary realignments. To achieve greater uniformity from war to war, major confrontations were split into

TABLE 4.6   Outbreaks of war

Number of Wars, x, Beginning in a Given Year	Observed Frequency	Expected Frequency
0	223	217
1	142	149
2	48	52
3	15	12
4+	4	2
	432	432

smaller "subwars": World War I, for example, was treated as five separate wars.) Table 4.6 gives the distribution of the number of years in which x wars broke out (140). The last column gives the expected frequencies based on the Poisson model, $P(X = x) = e^{-0.69}(0.69)^x/x!$.

Comment. One explanation for the excellent fit evident in Table 4.6 is that human society has a "hostility level"—as measured by the number of new wars initiated per year—that is constant. However, a similar breakdown showing the distribution of the number of wars *ending* in a given year also proved to be well described by a Poisson [see (140)]. Putting these two facts together, it could be speculated that what is constant in society is not a desire for war or a desire for peace, but a desire for change. ∎

4.3 THE NORMAL DISTRIBUTION

The limit proposed by Poisson was not the only, or even the first, approximation to the binomial: DeMoivre had already derived a quite different one in his 1718 tract, *Doctrine of Chances*. Like Poisson's work, DeMoivre's theorem did not initially attract the attention it deserved; it did catch the eye of Laplace, though, who generalized it and included it in his influential *Théorie Analytique des Probabilitiés*, published in 1812.

THEOREM 4.3   (DeMoivre-Laplace) Let $X \sim b(n, p)$. For any numbers c and d,

$$\lim_{n \to \infty} P\left(c < \frac{X - np}{\sqrt{npq}} < d\right) = \frac{1}{\sqrt{2\pi}} \int_c^d e^{-x^2/2}\, dx.$$

PROOF   We will prove a generalization of this result in Chapter 7. For a proof of Theorem 4.3 itself, see (45).

Comment. The integral in Theorem 4.3 cannot be expressed in closed form and must be approximated by numerical techniques. As early as 1783, Laplace recognized the pressing need for its tabulation and suggested an appropriate computational formula. The first substantial tables were due to the French physicist, C. Kramp. They were published in 1799. A table of

$$\frac{1}{\sqrt{2\pi}} \int_{-\infty}^d e^{-x^2/2}\, dx$$

for values of d ranging from -3.90 to $+3.90$ is given in the Appendix. ∎

QUESTION 4.3.1   Use Table A.1 in the Appendix to evaluate the integrals

(a) $\dfrac{1}{\sqrt{2\pi}} \displaystyle\int_{1.25}^{2.56} e^{-x^2/2}\, dx.$

(b) $\dfrac{1}{\sqrt{2\pi}} \displaystyle\int_{1.17}^{\infty} e^{-x^2/2}\, dx.$

QUESTION 4.3.2   Using the DeMoivre-Laplace theorem, prove the weak law of large numbers for Bernoulli trials. That is, if $X \sim b(n, p)$ and $d > 0$, show that

$$\lim_{n \to \infty} P\left(\left|\frac{X}{n} - p\right| > d\right) = 0.$$

What does this law mean intuitively?

     Case Study 4.5 shows how the DeMoivre-Laplace theorem is typically put into practice. As with the leukemia clustering problem, we will assess the statistical significance of an observed event—this time the number of correct guesses in an ESP experiment—by computing the probability of getting, by chance, the number actually observed, *or more.*

CASE STUDY
4.5

Research in extrasensory perception has ranged from the slightly unconventional to the downright bizarre. Toward the latter part of the nineteenth century and even well into the twentieth century, much of what was done involved spiritualists and mediums. But beginning around 1910, experimenters moved out of the seance parlors and into the laboratory, where they began setting up controlled studies that could be analyzed statistically. In 1938, Pratt and Woodruff, working out of Duke University, did an experiment that remains to this day a classic in the field (69).

     The investigator and a subject sat at opposite ends of a table. Between them was a screen with a large gap at the bottom. Five blank cards, visible to both participants, were placed side-by-side on the table beneath the screen. On the subject's side of the screen one of the standard ESP symbols (at the top of the next page) was

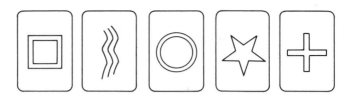

hung over each of the blank cards.

The experimenter shuffled a deck of ESP cards, picked up the top one, and concentrated on it. The subject tried to guess its identity: if he thought it was a *circle*, he would point to the blank card on the table that was beneath the circle card hanging on his side of the screen. The procedure was then repeated. Altogether, a total of 32 subjects, all students, took part in the experiment. They made a total of 60,000 guesses—and were correct 12,489 times.

With five denominations involved, the probability of a subject's making a correct identification just by chance was $\frac{1}{5}$. Assuming a binomial model, the expected number of correct guesses would be $60,000 \times \frac{1}{5}$, or 12,000. The question is, how "near" to 12,000 is 12,489? Should we write off the observed excess of 489 as nothing more than luck, or can we conclude that ESP has been demonstrated?

To effect a resolution, here, between the conflicting "luck" and "ESP" hypotheses, we need to compute the probability of the students' getting 12,489 or more correct answers *under the presumption that $p = \frac{1}{5}$*. Only if that probability is very small, can 12,489 be construed as evidence in support of ESP.

Let the random variable X denote the number of correct responses in 60,000 tries. Then

$$P(X \geq 12{,}489) = \sum_{x=12{,}489}^{60{,}000} \binom{60{,}000}{x} \left(\frac{1}{5}\right)^x \left(\frac{4}{5}\right)^{60{,}000-x}. \quad (4.2)$$

Now, rather than compute the 47,512 binomial probabilities indicated in Equation 4.2 (or the 12,489 making up its complement), we will appeal to the DeMoivre-Laplace theorem:

$$P(X \geq 12{,}489) = P\left(\frac{X - np}{\sqrt{npq}} \geq \frac{12{,}489 - 60{,}000(\frac{1}{5})}{\sqrt{60{,}000(\frac{1}{5})(\frac{4}{5})}}\right)$$

$$= P\left(\frac{X - np}{\sqrt{npq}} \geq 4.99\right)$$

$$\doteq \frac{1}{\sqrt{2\pi}} \int_{4.99}^{\infty} e^{-x^2/2}\, dx$$

$$= 0.0000003,$$

this last value being obtained from a more extensive version of Table A.1 in the Appendix.

Here, the fact that $P(X \geq 12{,}489)$ is so extremely small makes the "luck" hypothesis ($p = \frac{1}{5}$) untenable. It would appear that something other than chance had to be responsible for the occurrence of so many correct guesses.

QUESTION 4.3.3   A theory embraced by certain parapsychologists is that hypnosis can "bring out" a person's ESP ability. In a recent experiment set up to test that hypothesis (22), 15 students were hypnotized and then asked to guess the identity of an ESP card that another person (also hypnotized) was concentrating on. Each of the students made 100 guesses during the course of the experiment. The total number correct (out of 1500) was 326. Write down the binomial expression giving the probability of the event $X \geq 326$ and then use the DeMoivre-Laplace theorem to approximate it numerically. What would your conclusion be regarding the efficacy of hypnosis as a technique for improving a person's ESP? (We will reexamine this problem in Section 12.2.)

QUESTION 4.3.4   Airlines A and B offer identical service on two flights leaving at essentially the same time (meaning that the probability of a passenger's choosing either is $\frac{1}{2}$). Suppose both airlines are competing for the same pool of 400 potential passengers. If Airline A sells tickets to everyone who requests one, and the capacity of its plane is 230, approximate the probability of the plane's being overbooked.

We saw in Theorem 4.2 that Poisson's limit to the binomial was, itself, a probability distribution. The same is true of the DeMoivre-Laplace limit. To verify this, it is necessary to show that

$$\frac{1}{\sqrt{2\pi}} \int_{-\infty}^{\infty} e^{-x^2/2} \, dx = 1.$$

(Clearly, $(1/\sqrt{2\pi})e^{-x^2/2}$ is greater than or equal to 0 for all x.) The proof is not obvious and relies on the trick of changing to polar coordinates to show that the *square* of the integral is 1.

THEOREM 4.4

$$\frac{1}{\sqrt{2\pi}} \int_{-\infty}^{\infty} e^{-x^2/2} \, dx = 1.$$

PROOF   The theorem will be proved if we can show that

$$\int_{-\infty}^{\infty} e^{-x^2/2} \, dx \cdot \int_{-\infty}^{\infty} e^{-y^2/2} \, dy = 2\pi. \tag{4.3}$$

To begin, note that the product of the integrals in Equation 4.3 is equal to a double integral:

$$\int_{-\infty}^{\infty} e^{-x^2/2}\, dx \cdot \int_{-\infty}^{\infty} e^{-y^2/2}\, dy = \int_{-\infty}^{\infty} \int_{-\infty}^{\infty} e^{-(1/2)(x^2+y^2)}\, dx\, dy.$$

Let $x = r\cos\theta$ and $y = r\sin\theta$, making $dx\, dy = r\, dr\, d\theta$. Then

$$\int_{-\infty}^{\infty} \int_{-\infty}^{\infty} e^{-(1/2)(x^2+y^2)}\, dx\, dy = \int_{0}^{2\pi} \int_{0}^{\infty} e^{-r^2/2}\, r\, dr\, d\theta$$

$$= \int_{0}^{\infty} re^{-r^2/2}\, dr \cdot \int_{0}^{2\pi} d\theta$$

$$= -e^{-r^2/2}\big|_{0}^{\infty} \cdot \theta\big|_{0}^{2\pi}$$

$$= 1(2\pi) = 2\pi.$$

The next definition introduces a critical generalization of the density we have just looked at. Not having any parameters to lend it flexibility, $f_X(x) = (1/\sqrt{2\pi})e^{-x^2/2}$ would be of little practical value as a model for data. Transformed in the manner stated in Definition 4.1, though, it suddenly becomes the most useful of *all* probability models.

DEFINITION 4.1 A random variable Z is said to have a *standard normal* distribution if its pdf is given by

$$f_Z(z) = \frac{1}{\sqrt{2\pi}} e^{-z^2/2}, \qquad -\infty < z < \infty.$$

A random variable X is said to have a *normal distribution with parameters μ and σ* if

$$f_X(x) = \frac{1}{\sqrt{2\pi}\,\sigma} e^{-(1/2)[(x-\mu)/\sigma]^2}, \qquad -\infty < x < \infty.$$

We will denote such a random variable by the symbol $N(\mu, \sigma^2)$. [It follows that a standard normal random variable would be denoted $N(0, 1)$.]

Comment. The two pdf's of Definition 4.1 are related through Theorem 3.3. If X is a $N(\mu, \sigma^2)$ random variable, and we define the transformation

$$Z = \frac{X - \mu}{\sigma},$$

then Z's pdf will be the $f_Z(z)$ for the standard normal. Put another way, if $X \sim N(\mu, \sigma^2)$, then $(X - \mu)/\sigma \sim N(0, 1)$. The significance of the relationship between X, $(X - \mu)/\sigma$, and Z will be demonstrated in the examples presented later in this section. ∎

Mathematicians and scientists were quick to recognize the usefulness of the two-parameter normal. One of its first applications was due to Gauss, who used it to model observational errors in astronomy. [It appeared in this context in Gauss' celebrated work of 1809, *Theoria Motus Corporum Coelestium in Sectionibus Conicis Solem Ambientium* (*Theory of the Motion of Heavenly Bodies Moving about the Sun in Conic Sections*).] By the 1830s, the normal distribution was in quite general use by physicists as well. It was Quetelet, though, who established the normal as one of the cornerstones of statistical practice.

Lambert Adolphe Jacques Quetelet (1796–1874) was a Belgian scholar of broad training and interests—mathematics, statistics, astronomy, and writing poetry, just to name a few. In preparation for establishing the royal observatory at Brussels, he studied probability under Laplace in Paris during the winter of 1823–1824. Quetelet had a passion for collecting data, and he was able to demonstrate that, as a model,

$$f_X(x) = \frac{1}{\sqrt{2\pi}\,\sigma} e^{-(1/2)[(x-\mu)/\sigma]^2}$$

described a wide range of sociological and anthropological phenomena. Fortunately, he was sufficiently peripatetic (he belonged to over one hundred scholarly societies) to spread his ideas throughout Europe.

Another scholar of the nineteenth century often associated with the normal distribution is the noted English biologist Sir Francis Galton (although to label him as a biologist does a disservice to the wide scope of his intellectual activities). That Galton was influenced by the work of Quetelet is readily apparent from a statement he made in 1889 (56):

> I need hardly remind the reader that the Law of Error upon which these Normal Values are based, was excogitated for the use of astronomers and others who are concerned with extreme accuracy of measurement, and without the slightest idea until the time of Quetelet that they might be applicable to human measures. But Errors, Differences, Deviations, Divergencies, Dispersions, and individual Variations, all spring from the same kind of causes. Objects that bear the same name, or can be described by the same phrase, are thereby acknowledged to have common points of resemblance, and to rank as members of the same species . . . .

Before discussing one of Quetelet's applications of the normal distribution, it would be well to clarify the roles played by μ and σ. As the notation would suggest, one is related to the distribution's location; the other, to its dispersion.

THEOREM 4.5 Let the random variable X have a normal distribution with parameters μ and σ. Then

$$E(X) = \mu \quad \text{and} \quad \text{Var}(X) = \sigma^2.$$

PROOF   By virtue of Theorems 3.7 and 3.12, it suffices to show that

$$E\left(\frac{X-\mu}{\sigma}\right) = 0 \quad \text{and} \quad \text{Var}\left(\frac{X-\mu}{\sigma}\right) = 1.$$

The statement about μ follows immediately, since the integrand for the expected value is an odd function:

$$E\left(\frac{X-\mu}{\sigma}\right) = \int_{-\infty}^{\infty} y \cdot \frac{1}{\sqrt{2\pi}} e^{-y^2/2}\, dy = 0.$$

To prove the second part of the theorem, we note that

$$\mathrm{Var}\left(\frac{X-\mu}{\sigma}\right) = E\left(\left(\frac{X-\mu}{\sigma}\right)^2\right) - 0^2$$

$$= \int_{-\infty}^{\infty} y^2 \cdot \frac{1}{\sqrt{2\pi}} e^{-y^2/2}\, dy. \qquad (4.4)$$

After an integration by parts, the right-hand side of Equation 4.4 reduces to

$$\int_{-\infty}^{\infty} y^2 \cdot \frac{1}{\sqrt{2\pi}} e^{-y^2/2}\, dy = \frac{1}{\sqrt{2\pi}} y e^{-y^2/2}\Big|_{-\infty}^{\infty} + \frac{1}{\sqrt{2\pi}} \int_{-\infty}^{\infty} e^{-y^2/2}\, dy$$

$$= 0 + 1 = 1,$$

and the result follows.

QUESTION 4.3.5   Show that the moment-generating function for the normal distribution with parameters μ and σ is

$$M_X(t) = e^{\mu t + (1/2)\sigma^2 t^2}.$$

Compute $M'_X(0)$ and $M''_X(0)$ and verify Theorem 4.5.

CASE STUDY

4.6

Among Quetelet's many anthropometric applications of the normal was one where he fitted the distribution to the chest measurements of 5738 Scottish soldiers. Table 4.7 is a facsimile of Quetelet's data as they appeared in an 1846 book of letters to the Duke of Saxe-Cobourg and Gotha (134). Column 1 lists, in inches, the range of possible chest measurements. Column 2 gives the corresponding observed frequencies, which are then divided by 5738 and reappear in Column 3 as *relative* frequencies (the decimal points are omitted).

In the last column are the expected probabilities, assuming a normal distribution. Specifically, the parameter μ was estimated to be 39.8 inches and σ, 2.05 inches, so these final entries were based on the model,

$$f_X(x) = \frac{1}{\sqrt{2\pi}\,(2.05)} e^{-(1/2)[(x-39.8)/2.05]^2}.$$

The expected probability, then, of a Scottish soldier's having a chest

TABLE 4.7   Quetelet's chest measurements of Scottish soldiers

Mea- sures de la poitrine	Nom- bre d'hom- mes	Nom- bre Propor- tionnel	Prob- abilité d'après L'obser- vation	Rang dans La table	Rang d'après le calcul.	Prob- abilité d'après La table	Nombre d'obser- vations calculé
Pouces.							
33	3	5	0,5000			0,5000	7
34	18	31	0,4995	52	50	0,4993	29
35	81	141	0,4964	42,5	42,5	0,4964	110
36	185	322	0,4823	33,5	34,5	0,4854	323
37	420	732	0,4501	26,0	26,5	0,4531	732
38	749	1305	0,3769	18,0	18,5	0,3799	1333
39	1073	1867	0,2464	10,5	10,5	0,2466	1838
			0,0597	2,5	2,5	0,0628	
40	1079	1882	0,1285	5,5	5,5	0,1359	1987
41	934	1628	0,2913	13	13,5	0,3034	1675
42	658	1148	0,4061	21	21,5	0,4130	1096
43	370	645	0,4706	30	29,5	0,4690	560
44	92	160	0,4866	35	37,5	0,4911	221
45	50	87	0,4953	41	45,5	0,4980	69
46	21	38	0,4991	49,5	53,5	0,4996	16
47	4	7	0,4998	56	61,8	0,4999	3
48	1	2	0,5000			0,5000	1
	5738	1,0000					1,0000

measurement, x, of, say, 42 inches would be gotten by integration:

$$P(X \text{ "equals" } 42) = P(41.5 < X < 42.5)$$

$$= P\left(\frac{41.5 - 39.8}{2.05} < \frac{X - \mu}{\sigma} < \frac{42.5 - 39.8}{2.05}\right)$$

$$= P\left(0.829 < \frac{X - \mu}{\sigma} < 1.317\right)$$

$$= \frac{1}{\sqrt{2\pi}} \int_{0.829}^{1.317} e^{-y^2/2} \, dy$$

$$= 0.1096,$$

this latter figure being the tenth entry appearing in the last column.

QUESTION 4.3.6   The Stanford-Binet IQ test is scaled to have a mean score of 100 and a standard deviation of 16. If children with IQ's of less than 80 or greater than 145 are deemed to need special attention, how many out of 2000 children should be expected to fall into this category?

To put the normal distribution in its proper perspective, we should leave the nineteenth century—and Quetelet—and look at an example of more recent vintage. In fields such as education and psychology, a frequent application of the normal— one certainly familiar to students—is the description of test scores: IQ's, ACT's, and CEEB's, just to name a few, are all standardized exams whose distributions are known to strongly resemble the bell-shaped curve of Definition 4.1. The next example is a variation on that same theme, the data coming this time from an experiment in role analysis.

CASE STUDY

4.7

On the basis of which factors influence them, decision makers can be assigned positions along a so-called *motivation scale*. At one end of the scale are the *moralists*, individuals who always do what they believe to be "right," regardless of the consequences. At the other end sit the *expedients*, persons whose decisions reflect more pragmatic concerns—peer-group pressure, short-range gain, the desire to be liked, and so on. Most decision makers, of course, would fall somewhere between these extremes, sometimes acting out of righteousness but other times out of expediency. In the experiment to be described (64), a questionnaire was used to quantify the nature of a person's motivation. The purpose was to investigate—and, hopefully, model—the resulting distribution.

A group of 106 subjects were given a test that consisted of 37 different conflict situations. A possible solution was proposed for each one. The subjects had to decide—if it was up to them—whether *they* would resolve the conflict in the manner that was suggested. But instead of answering just "Yes" or "No," they were required to indicate the strength of their conviction in taking (or not taking) the recommended course of action. Their possible responses were:

1. Absolutely must (take the action).
2. Preferably should (take the action).
3. May or may not (take the action).
4. Preferably should not (take the action).
5. Absolutely must not (take the action).

It was felt that the "Absolutely must" and "Absolutely must not" responses were probably moralistic in origin, and a subject was awarded a "point" each time he gave either one. The total number of points accumulated over the 37 questions was then taken as an index of the extent to which a subject sought moral solutions. The scores are listed in Table 4.8.

TABLE 4.8   Motivation scores

13	12	11	19	24	2	13
17	15	2	17	15	7	15
13	27	4	16	13	9	5
8	19	4	17	12	5	28
7	23	13	13	6	21	20
10	6	10	7	17	18	19
10	2	13	9	27	17	14
21	9	19	12	3	18	11
18	11	25	11	10	12	14
17	5	14	30	7	15	4
19	18	11	19	1	13	8
15	20	4	4	14	13	10
15	24	14	11	22	15	7
23	15	12	18	16	6	23
12	14	23	18	10	25	18
						24

The estimates for μ and σ in this case are 13.8 and 6.5, respectively. Figure 4.1 shows the normal curve,

$$f_X(x) = \frac{1}{\sqrt{2\pi}\,(6.5)}\,e^{-(1/2)[(x-13.8)/6.5]^2}$$

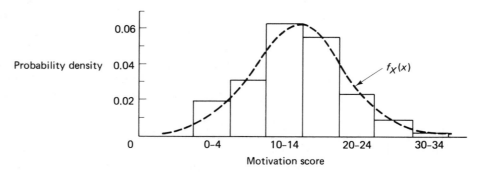

Figure 4.1   Normal curve over histogram.

superimposed over a histogram of the data from Table 4.8. The fit appears to be quite good.

It should be noted that for the normal curve the ordinate is simply $f_x(x)$. For the histogram, though, it is

$$\frac{\text{class frequency}}{\text{total number of observations times class width}}$$

Dividing the relative frequency in this way by the class width scales the area of the histogram to 1, a necessary step if the shape of the histogram is to be compared with a probability model. For example, the observed frequency for the "20–24" class is 12—therefore, the height of the "20–24" bar is 12/(106 × 5), or 0.0226.

Applications of statistical reasoning are not always confined to laboratory research or field surveys, nor do they always appear in technical journals. The next example is a case in point.

CASE STUDY

4.8

The following letter appeared in a well-known, advice-to-the-love-lorn column (178):

> Dear Abby: You wrote in your column that a woman is pregnant for 266 days. Who said so? I carried my baby for ten months and five days, and there is no doubt about it because I know the exact date my baby was conceived. My husband is in the Navy and it couldn't have possibly been conceived any other time because I saw him only once for an hour, and I didn't see him again until the day before the baby was born.
>
> I don't drink or run around, and there is no way this baby isn't his, so please print a retraction about the 266-day carrying time because otherwise I am in a lot of trouble.
>
> San Diego Reader

While the full implications of San Diego Reader's plight are beyond the scope of this text, it *is* possible to assess quantitatively the statistical likelihood of a pregnancy being 310 days long (= 10 months and 5 days). By the same reasoning used in Case Study 4.1, this would be done by computing the probability that, by chance alone, a pregnancy will be 310 days long *or longer*. That is, if Y denotes an arbitrary pregnancy duration, we want to compute $P(Y \geq 310)$—the smaller this probability is, the less credibility San Diego Reader has. Of course, despite our best-intentioned efforts to come up with a relevant and objective probability, the interpretation of that probability will remain somewhat subjective, since the particular value of $P(Y \geq 310)$ that delineates what we choose to accept as the truth from what we reject as nontruth is necessarily arbitrary.

According to well-documented norms, the mean and standard deviation for Y are 266 days and 16 days, respectively. If it can be assumed that the distribution of Y is normal,

$$P(Y \geq 310) = P\left(\frac{Y - 266}{16} \geq \frac{310 - 266}{16}\right) = P(Z \geq 2.75).$$

But, from Table A.1 in the Appendix,

$$P(Z \geq 2.75) = 1 - 0.9970$$
$$= 0.003.$$

Figure 4.2 shows the two equivalent areas involved, the original one under the $N(266, (16)^2)$ distribution and the transformed one under the $N(0, 1)$. It is left to the reader to decide how the 0.003 should be interpreted in this particular setting.

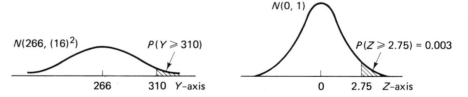

$N(0, 1)$

$N(266, (16)^2)$ $P(Y \geq 310)$ $P(Z \geq 2.75) = 0.003$

266   310 Y-axis   0   2.75 Z-axis

Figure 4.2   Equivalent areas under two normal distributions.

Comment. Insurance companies sometimes use probability computations of this sort in deciding whether to reimburse married couples for maternity costs. Certain policies contain a clause to the effect that if Y is too much *less* than 266 (as measured from the couple's date of marriage) the company can refuse to pay. ∎

The final example in this section is typical of an entire class of problems where tables of the standard normal distribution have to be used *backward*.

EXAMPLE 4.2.   Mensa is an international society whose membership is limited to persons having IQ's above the general population's 98th percentile. It is well known that the average IQ for the general population is 100, the standard deviation is 16, and the distribution, itself, is normal (recall Question 4.3.6). What, then, is the *lowest* IQ that will qualify a person to belong to Mensa?

Let Y denote the IQ of a random person. By definition, the lowest qualifying value for Y, call it Y^*, must satisfy the equation

$$P(Y \geq Y^*) = 0.02.$$

Standardizing Y as prescribed by the corollary gives

$$P(Y \geq Y^*) = P\left(\frac{Y - 100}{16} \geq \frac{Y^* - 100}{16}\right) = P\left(Z \geq \frac{Y^* - 100}{16}\right) = 0.02.$$

From Table A.1 in the Appendix, though, the 98th percentile of the $N(0, 1)$ is 2.05:

$$P(Z \geq 2.05) = 0.02.$$

Therefore, setting the "two" 98th percentiles equal gives

$$\frac{Y^* - 100}{16} = 2.05,$$

from which it follows that a person must have an IQ of at least 133 to be eligible for membership:

$$Y^* = 100 + 2.05(16) = 133.$$

Comment. The material presented in this section is intended to serve as only the briefest of introductions to the normal distribution. In Chapters 7 and 8 we will see much more of this singularly important model, its mathematical properties, and how it is used. ∎

QUESTION 4.3.7   Dental structure provides an effective criterion for classifying certain fossils. Not long ago a baboon skull of unknown origin was discovered in a cave in Angola (112); its third-molar length was 9.0 mm. Speculation arose that the baboon in question might be a "missing link" and belong to the genus *Papio*. Members of this genus have third molars that measure, on the average, 8.18 mm long with a standard deviation of 0.47 mm. Quantify the significance of the 9.0-mm molar. What would your inference be?

QUESTION 4.3.8   A college professor teaches Chemistry 101 each fall to a large class of freshmen. For tests, she uses standardized exams that she knows from past experience produce bell-shaped grade distributions with a mean (μ) of 70 and a standard deviation (σ) of 12. Her philosophy of grading is to impose standards that will yield, in the long run, 14% A's, 20% B's, 32% C's, 20% D's, and 14% F's. Where should the cutoff be between the A's and the B's? between the B's and the C's?

4.4 THE GEOMETRIC DISTRIBUTION

Given a series of independent Bernoulli trials, we are accustomed to thinking of n and p as fixed, and x, the number of successes, as the (binomial) random variable. Suppose the problem is turned around, though, and the question is asked, how many trials will be required in order to achieve the first success? Put this way, n is the random variable and x is what is fixed. Clearly, the first success will occur on the very first trial with probability p, on the second trial with probability $(1 - p)p$, and so on. In general, the probability function for N, the trial number of the first success, will be

$$f_N(n) = p(1 - p)^{n-1} = pq^{n-1}, \qquad n = 1, 2, \ldots.$$

This is called the *geometric distribution* (with parameter p).

Intuitively, the average number of trials required for the first success to appear—that is, $E(N)$—should be inversely proportional to p. The next theorem shows that, in fact, $E(N)$ is *exactly* $1/p$.

THEOREM 4.6   Let N be a geometric random variable with parameter p [$= P(\text{success at any trial})$]. Then

$$E(N) = \frac{1}{p} \quad \text{and} \quad \text{Var}(N) = \frac{q}{p^2}.$$

PROOF   By definition,

$$E(N) = \sum_{n=1}^{\infty} npq^{n-1} = p \sum_{n=1}^{\infty} nq^{n-1}. \tag{4.5}$$

This last sum is not immediately recognizable but it can be finessed by considering it as a function of q and taking the derivative of its integral. Set

$$h(q) = \sum_{n=1}^{\infty} nq^{n-1}.$$

Then

$$\int h(q)\, dq = \sum_{n=1}^{\infty} q^n = \sum_{n=0}^{\infty} q^n - q^0$$

$$= \frac{1}{1-q} - 1$$

$$= \frac{q}{1-q}.$$

By using the rule for the derivative of a quotient, we get that $h(q) = 1/p^2$:

$$h(q) = \frac{d}{dq} \int h(q)\, dq = \frac{(1-q) - (-q)}{(1-q)^2} = \frac{1}{(1-q)^2} = \frac{1}{p^2}.$$

Substituting this value into Equation 4.5 gives the first part of the theorem,

$$E(N) = p\left(\frac{1}{p^2}\right) = \frac{1}{p}.$$

The proof that $\text{Var}(N) = q/p^2$ is left as an exercise.

QUESTION 4.4.1   Show that the moment generating function for the geometric distribution is

$$M_N(t) = \frac{pe^t}{1 - qe^t}$$

and use it to find $E(N)$ and Var (N).

Although not nearly so ecumenical as the Poisson or the normal, the geometric distribution does find a variety of applications: Case Study 4.9 describes one in an ornithological setting.

CASE STUDY

4.9

The basic building block of all bird songs is the *syllable*. A bird such as the cardinal can make at least ten distinct sounds, or syllables. Individual syllables sung in rapid succession are said to belong to the same *utterance*, and a series of consecutive utterances is called a *bout*.

A recent study (100) set out to characterize the variability in the number of utterances per bout for the North American cardinal. The subjects were male cardinals (*Richmondena cardinalis*) nesting near the campus of the University of Western Ontario. Over a period of several months, the songs of these birds were taped and analyzed with a sound spectrograph, a device that graphs the frequency of a bird's call as a function of time. The spectrograph for one of the syllables in a cardinal's repertoire, Song Type D, is shown in Figure 4.3.

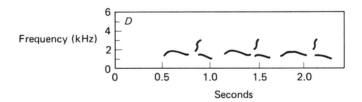

Figure 4.3   Spectrograph for cardinal syllable.

During the course of the study, a total of 250 bouts cf Song Type D were recorded. Table 4.9 shows the distribution of the number of utterances per bout.

If the last utterance in a given bout is considered a success—indeed, the *first* success—then a potential model for the data of Table 4.9 would be the geometric,

$$f_N(n) = pq^{n-1}, \qquad n = 1, 2, \ldots,$$

TABLE 4.9   Distribution of number of utterances per bout

Number of Utterances/Bout, n_i	Frequency, f_i
1	132
2	52
3	34
4	9
5	7
6	5
7	5
8+	6
	250

where n is the number of utterances in a bout. Here, the parameter p would be estimated (using the methods of Chapter 5) as 0.48, so the particular model being proposed is

$$f_N(n) = (0.48)(0.52)^{n-1}, \qquad n = 1, 2, \ldots .$$

The last column of Table 4.10 shows the results. The agreement between the observed and expected frequencies is not as good as what

TABLE 4.10   Fitting a geometric pdf

Number of Utterances/Bout, n_i	Observed Frequency, f_i	Expected Frequency $= 250 f_N(n_i)$
1	132	120.0
2	52	62.4
3	34	32.4
4	9	16.9
5	7	8.8
6	5	4.6
7	5	2.4
8+	6	2.5
	250	250.0

we have seen in previous examples, but still fairly adequate, especially considering the simplicity of the model.

The next example describes the application of an unusual "hybrid" geometric model. The data are meteorological: the lengths (in days) of weather cycles in Tel Aviv.

CASE STUDY

4.10

A *wet spell of x days* is defined to be a "run" of x days on each of which measurable precipitation occurs. Under the assumption that the weather tomorrow depends only on the weather today, there is a probability p_0 that a wet day will be followed by a dry day. From this, the probability of an x-day-long wet spell is $p_0(1 - p_0)^{x-1} = p_0 q_0^{x-1}$—that is, a geometric distribution. We will let X denote the random variable representing the lengths of wet spells.

In a similar way, there is a random variable Y associated with the lengths of *dry* spells; furthermore, $f_Y(y) = p_1 q_1^{y-1}$, where p_1 is the probability of a dry day followed by a wet one. Since a given day's weather is affected only by the previous day's, it follows that X and Y are independent.

Now, a *weather cycle* will be defined as, say, a wet spell followed by a dry spell. If Z denotes the length of such a cycle, then $Z = X + Y$. We leave it to the reader to verify that

$$f_{X+Y}(z) = p_0 p_1 \frac{q_0^{z-1} - q_1^{z-1}}{q_0 - q_1}, \quad z = 2, 3, \ldots.$$

[Hint: Write $P(X + Y = z) = P(X = 1)P(Y = z - 1) + \cdots + P(X = z - 1)P(Y = 1)$ and simplify.]

The lengths of 351 weather cycles in Tel Aviv were recorded during the months of December, January, and February from the winter of 1923–1924 through the winter of 1949–1950 (54). The proposed model's two parameters, p_0 and p_1, were estimated to be 0.338 and 0.250, respectively.

Table 4.11 lists both the observed and expected distributions for Z, the latter being computed from the model

$$f_{X+Y}(z) = (0.338)(0.250) \frac{(0.662)^{z-1} - (0.750)^{z-1}}{0.662 - 0.750}.$$

TABLE 4.11   Weather-cycle distributions

Cycle Length	Observed Frequency	Expected Frequency
2	33	30
3	38	42
4	33	44
5	50	42
6	39	37
7	26	32
8	30	26
9	24	21
10	16	17
11	14	14
12	15	11
13	9	8
14	3	6
15	3	5
16	8	4
17	1	3
18	4	2
19	2	2
20+	3	5

Unlike that of the bird-call data of the previous example, the agreement between the observed and expected frequencies here is very good.

4.5 THE NEGATIVE BINOMIAL DISTRIBUTION

In this section we consider a natural generalization of the geometric distribution. Rather than focus on the number of Bernoulli trials required for the *first* success to occur, we now look at the number required for the occurrence of the *rth* success. Let X denote the number of trials necessary for this occurrence. It should be clear that the only possible scenario for X's being equal to, say, x is for a total of $r - 1$ successes [and $x - 1 - (r - 1)$ failures] to occur somewhere during the first $x - 1$ trials and for one additional success (the *rth*) to occur on the very next (i.e., the *xth*) trial. Assuming all trials to be independent, it follows that the distribution of X should be

$$f_X(x) = \binom{x - 1}{r - 1} p^{r-1} q^{x-1-(r-1)} p$$

$$= \binom{x - 1}{r - 1} p^r q^{x-r}, \qquad x = r, r + 1, \ldots .$$

In practice, the random variable X is often replaced by Y, where Y is the number of trials *in excess of r* required to produce the rth success. That is, $Y = X - r$. Letting n be the argument for Y, we can write

$$f_Y(n) = \binom{n+r-1}{r-1}p^r q^n = \binom{n+r-1}{n}p^r q^n, \qquad n = 0, 1, \ldots.$$

Theorem 4.7 verifies that $f_Y(y)$ is a legitimate probability function.

THEOREM 4.7

$$\sum_{n=0}^{\infty} f_Y(n) = \sum_{n=0}^{\infty}\binom{n+r-1}{n}p^r q^n = 1.$$

PROOF   We begin by establishing a combinatorial identity. Recall that the symbol $\binom{k}{n}$, as a shorthand for $\dfrac{k(k-1)\cdots(k-n+1)}{n!}$, has been used up to this point only in cases where k was a nonnegative integer. In Chapter 2, though, it was defined for *any* real k. Thus, for $k = -r$,

$$\binom{-r}{n} = \frac{-r(-r-1)\cdots(-r-n+1)}{n!}$$

$$= (-1)^n \frac{r(r+1)\cdots(n+r-1)}{n!}$$

$$= (-1)^n \binom{n+r-1}{n}$$

or

$$\binom{n+r-1}{n} = (-1)^n \binom{-r}{n}. \tag{4.6}$$

With Equation 4.6, we can evaluate $\sum_{n=0}^{\infty} f_Y(n)$. Note, first of all, that

$$\sum_{n=0}^{\infty}\binom{n+r-1}{n}p^r q^n = p^r \sum_{n=0}^{\infty}\binom{n+r-1}{n}q^n$$

$$= p^r \sum_{n=0}^{\infty}\binom{-r}{n}(-1)^n q^n$$

$$= p^r \sum_{n=0}^{\infty}\binom{-r}{n}(-q)^n. \tag{4.7}$$

Now, recall Newton's binomial formula: for any real k,

$$(1+t)^k = \sum_{n=0}^{\infty}\binom{k}{n}t^n. \tag{4.8}$$

Applying Equation 4.8 to Equation 4.7, then, gives the result:

$$\sum_{n=0}^{\infty}\binom{n+r-1}{n}p^r q^n = p^r(1-q)^{-r}$$

$$= p^r p^{-r} = 1.$$

DEFINITION 4.2 The random variable Y is said to have the negative binomial distribution with parameters r and p if

$$f_Y(n) = \binom{n + r - 1}{n} p^r q^n, \qquad n = 0, 1, \ldots.$$

THEOREM 4.8 If the random variable Y has the negative binomial distribution with parameters r and p, then

$$E(Y) = \frac{rq}{p} \quad \text{and} \quad \text{Var}(Y) = \frac{rq}{p^2}.$$

PROOF   The theorem is true for arbitrary $r > 0$, but we will consider only the case where r is a positive integer. Let $X_1, X_2, \ldots, X_r$ be independent geometric random variables, each with the same parameter p. Since each X_i can be thought of as the number of trials required for the first success, it seems intuitively reasonable that their sum, $X = X_1 + X_2 + \cdots + X_r$, should have the same distribution as the number of trials required for the rth success—that is, $Y = X - r$ should be negative binomial with parameters r and p (see Review Exercise 55 at the end of this chapter). It follows, then, from Theorem 4.6 and the properties of sums of independent random variables, that

$$E(X) = rE(X_1) = r\left(\frac{1}{p}\right) = \frac{r}{p}$$

and

$$\text{Var}(X) = r \, \text{Var}(X_1) = r\left(\frac{q}{p^2}\right) = \frac{rq}{p^2}.$$

Then the expected value of Y is $E(X - r) = E(X) - r = (r/p) - r = rq/p$, and the variance of Y is $\text{Var}(X - r) = \text{Var}(X)$.

QUESTION 4.5.1   Suppose an underground military installation is fortified to the extent that it can withstand up to four direct air-to-surface missile hits and still function. Enemy aircraft can score a direct hit with this particular missile seven times out of ten. What is the probability that a plane will require fewer than eight shots to destroy the installation? Assume all firings are independent.

QUESTION 4.5.2   Let Y be a negative binomial random variable. Prove directly that the expected value of $X = Y + r$ is r/p by evaluating the sum

$$\sum_{x=r}^{\infty} x \binom{x - 1}{r - 1} p^r q^{x-r}.$$

Hint: Reduce the sum to one involving negative binomial probabilities with parameters $r + 1$ and p.

In fitting a multiparameter probability model to a set of data it is not always obvious what the parameters represent in a physical sense. The next example is a case in point.

CASE STUDY

4.11

An early application of the negative binomial was due to the famous British statistician, Sir Ronald Alymer Fisher. Pursuing a question of somewhat less gravity than was his custom, Fisher was able to show that Y, the number of ticks found on a sheep, could be described remarkably well by a function of the form,

$$f_Y(n) = \binom{n + r - 1}{n} p^r q^n.$$

His data consisted of 60 sheep and their uninvited entourage of some 200 ticks. Table 4.12 shows the observed and expected tick

TABLE 4.12 Distribution of ticks on sheep

No. of Ticks	Observed Frequency	Expected Frequency
0	7	6
1	9	10
2	8	11
3	13	10
4	8	8
5	5	5
6	4	4
7	3	2
8	0	1
9	1	1
10+	2	2
	60	60

distributions (49). The parameters r and p were estimated to be 3.75 and 0.536, respectively, so the entries in column 3 were gotten by multiplying 60 times

$$\binom{n + 3.75 - 1}{n}(0.536)^{3.75}(0.464)^n.$$

Comment. Computing expected frequencies for a negative binomial model can be greatly facilitated by using a recurrence formula. Consider the ratio of two successive negative binomial probabilities:

$$\frac{f_Y(n)}{f_Y(n-1)} = \frac{\binom{n+r-1}{n}p^r q^n}{\binom{n-1+r-1}{n-1}p^r q^{n-1}}$$

$$= q \cdot \frac{(n+r-1) \cdots (r+1)r}{n!} \cdot \frac{(n-1)!}{(n+r-2) \cdots (r+1)r}$$

$$= q \cdot \frac{n+r-1}{n}.$$

For the tick example,

$$\frac{f_Y(n)}{f_Y(n-1)} = (1 - 0.536)\left(\frac{n+3.75-1}{n}\right)$$

$$= 0.464 + \frac{1.276}{n}.$$

Therefore, if the $(n-1)$st expected frequency has been calculated—here, $60 \cdot f_y(n-1)$—the nth is equal to

$$60 \cdot f_Y(n) = \left(0.464 + \frac{1.276}{n}\right)60 \cdot f_Y(n-1).$$

For example, in Table 4.12 the expected value for $Y = 0$ is

$$60\binom{0+3.75-1}{0}(0.536)^{3.75}(0.464)^0 = 60(0.536)^{3.75} \doteq 6.$$

Then the entry for $Y = 1$ becomes

$$6\left(0.464 + \frac{1.276}{1}\right) \doteq 10. \quad \blacksquare$$

QUESTION 4.5.3   Suppose Y_1 and Y_2 are two independent negative binomial random variables, each with parameters r and p. Show that $Y_1 + Y_2$ is negative binomial with parameters $2r$ and p.

                  Hint: Use the identity in Review Exercise 110 at the end of Chapter 2 and recall that

$$\binom{-r}{n} = (-1)^n \binom{n+r-1}{n}.$$

4.6 THE GAMMA DISTRIBUTION

In Section 4.2 the important Poisson distribution was introduced as a limiting form of the binomial. We will see in this section that the limit of a *negative* binomial suggests a useful continuous probability model.

Suppose some event occurs over a time interval of length T at an average rate λ (per unit time). We wish to examine the continuous variable X measuring the length of time required for r events to occur. First, we set up a discrete approximation: divide the time interval of length T into n equal subintervals of length T/n. Choose n so large that the probability of more than one event in a given subinterval is negligible. Assume the subintervals are independent. If p_n denotes the probability of an event's occurring during any particular subinterval, it follows that $p_n = (\lambda T)/n$. Also, from the preceding section, the probability it will take $k(T/n)$ time units for r events to occur is

$$\binom{k-1}{r-1} p_n^r (1 - p_n)^{k-r}.$$

With the above assumptions,

$$f_X(T) = F'_X(T) = \lim_{n \to \infty} \frac{F_X(T) - F_X\left(\dfrac{(n-1)T}{n}\right)}{\dfrac{T}{n}}$$

is approximated by

$$\lim_{n \to \infty} \frac{\binom{n-1}{r-1} p_n^r (1 - p_n)^{n-r}}{\dfrac{T}{n}}$$

$$= \lim_{n \to \infty} \frac{\dfrac{(n-1)!}{(n-r)!(r-1)!} p_n^r (1 - p_n)^{n-r}}{\dfrac{T}{n}}$$

With the substitution $p_n = \lambda T/n$,

$$f_X(T) \doteq \lim_{n \to \infty} \frac{(n-1)!}{(n-r)!(r-1)! \dfrac{T}{n}} \left(\frac{\lambda T}{n}\right)^r \left(1 - \frac{\lambda T}{n}\right)^n \left(1 - \frac{\lambda T}{n}\right)^{-r}$$

$$= \lim_{n \to \infty} \frac{(n-1)(n-2)\cdots(n-(r-1))}{(r-1)!} \lambda^r \left(\frac{T}{n}\right)^{r-1} \left(1 - \frac{\lambda T}{n}\right)^n \cdot 1$$

$$= \lim_{n \to \infty} \frac{\left(1 - \dfrac{1}{n}\right)\left[1 - \dfrac{2}{n}\right]\cdots\left[1 - \dfrac{r-1}{n}\right]}{(r-1)!} \lambda^r T^{r-1}\left(1 - \frac{\lambda T}{n}\right)^n$$

$$= \frac{\lambda^r}{(r-1)!} T^{r-1} e^{-\lambda T}. \tag{4.9}$$

What we have just derived for $f_X(T)$ is a special case of a *gamma* density, a family of probability distributions that will be formally introduced in Definition 4.4. Before examining the gamma in any detail, though, we will generalize Equation 4.9 to include those cases where r is not an integer. To do this, it is necessary to replace $(r-1)!$ with a continuous function of (nonnegative) r, say $\Gamma(r)$, where $\Gamma(r)$ reduces to $(r-1)!$ when r is a positive integer. Such a function was first defined in 1731 by Euler, who reexpressed it fifty years later in the form it takes in Definition 4.3. Legendre is credited with giving it the name "gamma function."

DEFINITION 4.3 For any real number $r > 0$, the gamma function (of r) is given by

$$\Gamma(r) = \int_0^\infty x^{r-1}e^{-x}\,dx.$$

Six properties of the gamma function are stated in Theorem 4.9. The first five will be left as homework exercises; the sixth is proved in Appendix 4.1 on page 177.

THEOREM 4.9 Let $\Gamma(r) = \int_0^\infty x^{r-1}e^{-x}\,dx.$ Then

(a) $\Gamma(1) = 1.$
(b) $\Gamma(\frac{1}{2}) = \sqrt{\pi}.$
(c) $\Gamma(r+1) = r\Gamma(r),$ for any positive real r.
(d) $\Gamma(r+1) = r!,$ if r is a nonnegative integer.
(e) $\dbinom{n+r-1}{n} = \dfrac{\Gamma(n+r)}{\Gamma(n+1)\Gamma(r)}$
(f) $\dfrac{\Gamma(r)\Gamma(s)}{\Gamma(r+s)} = \displaystyle\int_0^1 u^{r-1}(1-u)^{s-1}\,du.$

QUESTION 4.6.1   Show that $\Gamma(\frac{7}{2}) = (15\sqrt{\pi})/8.$

Having found the appropriate generalization of $(r-1)!$ for arbitrary (non-negative) r, we can now define the gamma family of probability functions.

DEFINITION 4.4 Let X be a random variable such that

$$f_X(x) = \frac{\lambda^r}{\Gamma(r)}x^{r-1}e^{-\lambda x}, \qquad x > 0.$$

Then X is said to have a gamma distribution with parameters r and λ. Both r and λ must be greater than 0.

Comment. In the gamma distribution, λ plays the role of a scale parameter. This is apparent from the fact that if X is gamma with parameters r and λ, then λX is gamma with parameters r and 1:

$$f_{\lambda X}(x) = \frac{1}{\lambda}f_X\left(\frac{x}{\lambda}\right)$$

$$= \frac{1}{\lambda}\frac{\lambda^r}{\Gamma(r)}\left(\frac{x}{\lambda}\right)^{r-1}e^{-x}$$

$$= \frac{1}{\Gamma(r)}x^{r-1}e^{-x}. \quad \blacksquare$$

Comment. When $r = 1$, the gamma distribution reduces to the exponential distribution. ∎

QUESTION 4.6.2 Show that $f_X(x)$ as stated in Definition 4.4 is a true probability density: verify that

$$\int_0^\infty \frac{\lambda^r}{\Gamma(r)} x^{r-1} e^{-\lambda x}\, dx = 1.$$

CASE STUDY

4.12

Records of daily rainfall in Sydney, Australia, for the period from October 17 to November 7 were assembled for the years 1859 through 1952 (33). It was postulated that these data might have a gamma distribution, the reason being that precipitation occurs only if water particles can coalesce around dust of sufficient mass, but the accumulation of such dust is similar to the "waiting-time" aspect implicit in the gamma model.

Table 4.13 lists the results. Estimates for r and λ were found to be 0.105 and 0.013, respectively.

TABLE 4.13 Distribution of rainfall

Rainfall (mm)	Observed Frequency	Expected Frequency
0–5	1631	1639
6–10	115	106
11–15	67	62
16–20	42	44
21–25	27	32
26–30	26	26
31–35	19	21
36–40	14	17
41–45	12	14
46–50	18	12
51–60	18	20
61–70	13	15
71–80	13	12
81–90	8	9
91–100	8	7
101–125	16	12
126–150	7	7
151–425	14	13

To compute the expected probability of rainfall measuring between, say, 0 and 5 mm, we compute

$$p = \frac{(0.013)^{0.105}}{\Gamma(0.105)} \int_0^{5.5} x^{0.105-1} e^{-0.013x} \, dx,$$

or, making the substitution $u = 0.013x$,

$$p = \frac{1}{\Gamma(0.105)} \int_0^{0.0715} u^{0.105-1} e^{-u} \, du. \tag{4.10}$$

The integral in Equation 4.10 can be evaluated using Tables of the Incomplete Gamma Function [see (177)]. In this case, $p = 0.793$, so the expected number of days with rainfall between 0 and 5 mm is equal to 2068 × 0.793, or 1639.

Comment. The gamma distribution is a model frequently used in meteorological settings. In Chapter 5 we will see it applied to the amount of rainfall deposited by inland hurricanes. ∎

THEOREM 4.10 Let X be a random variable having a gamma distribution with parameters r and λ. Then

$$E(X) = \frac{r}{\lambda} \quad \text{and} \quad \text{Var}(X) = \frac{r}{\lambda^2}.$$

PROOF   The proof is an easy exercise in substitution. By definition,

$$E(X) = \int_0^\infty x \frac{\lambda^r}{\Gamma(r)} x^{r-1} e^{-\lambda x} \, dx$$

$$= \frac{\lambda^r}{\Gamma(r)} \int_0^\infty x^r e^{-\lambda x} \, dx.$$

Set $u = \lambda x$. Then

$$E(X) = \frac{\lambda^{r-1}}{\Gamma(r)} \int_0^\infty \left(\frac{u}{\lambda}\right)^r e^{-u} \, du$$

$$= \frac{1}{\lambda \Gamma(r)} \int_0^\infty u^r e^{-u} \, du$$

$$= \frac{\Gamma(r+1)}{\lambda \Gamma(r)} = \frac{r}{\lambda}.$$

The proof for Var (X) is similar.

Comment. Notice the similarity between Theorems 4.8 and 4.10. ∎

QUESTION 4.6.3   Find the moment-generating function for a gamma random variable.

   We conclude Chapter 4 with an application taken from psychology that makes use of conditional probabilities. This example brings out an interesting interrelationship among the Poisson, gamma, and negative binomial distributions.

CASE STUDY

4.13

Psychologists use a two-handed coordination test to study performance speed and error liability. Individuals have been found to be quite consistent with respect to the latter, and for any given person the number of errors on this test is a Poisson variable. Thus, for a subject whose error rate is λ,

$$P(x \text{ errors}) = f_{X|\lambda}(x) = \frac{e^{-\lambda}\lambda^x}{x!}.$$

The parameter λ, though, varies from person to person, so it can be considered the value of a random variable Λ. There are situations where it is reasonable to assume that Λ has a gamma density with parameters β and r:

$$f_\Lambda(\lambda) = \frac{\beta^r}{\Gamma(r)}\lambda^{r-1}e^{-\beta\lambda}.$$

We will make that assumption here.
   From Chapter 3,

$$f_{X,\Lambda}(x, \lambda) = f_{X|\lambda}(x) \cdot f_\Lambda(\lambda),$$

from which it follows that the marginal probability function for X is given by

$$f_X(x) = \int_0^\infty f_{X,\Lambda}(x, \lambda)\, d\lambda = \int_0^\infty f_{X|\lambda}(x) \cdot f_\Lambda(\lambda)\, d\lambda. \qquad (4.11)$$

In this case, Equation 4.11 reduces to

$$
\begin{aligned}
f_X(x) &= \int_0^\infty \frac{\lambda^x}{x!}e^{-\lambda}\frac{\beta^r}{\Gamma(r)}\lambda^{r-1}e^{-\beta\lambda}\, d\lambda \\
&= \frac{1}{x!}\frac{\beta^r}{\Gamma(r)}\int_0^\infty \lambda^{x+r-1}e^{-(\beta+1)\lambda}\, d\lambda \\
&= \frac{1}{x!}\frac{\beta^r}{\Gamma(r)}\frac{\Gamma(x+r)}{(\beta+1)^{x+r}} \\
&= \left(\frac{\beta}{\beta+1}\right)^r\left(\frac{1}{\beta+1}\right)^x\frac{\Gamma(x+r)}{x!\,\Gamma(r)} \\
&= \left(\frac{\beta}{\beta+1}\right)^r\left(\frac{1}{\beta+1}\right)^x\binom{x+r-1}{x}. \qquad (4.12)
\end{aligned}
$$

Setting $p = \beta/(\beta + 1)$ and $q = 1/(\beta + 1)$, we can rewrite Equation 4.12 in a more familiar form:

$$f_X(x) = \binom{x + r - 1}{x} p^r q^x.$$

Thus, if the conditional distribution of X is Poisson, and if its parameter λ is gamma, then the unconditional distribution of X is negative binomial.

Sichel (166) gave the two-handed coordination test to 504 subjects, recording for each the number of errors committed. Table 4.14 shows the results.

TABLE 4.14 Two-handed coordination test

Number of Errors	Observed Frequency	Expected Frequency
0	74	82
1	58	57
2	51	46
3	49	39
4	42	33
5	23	28
6	30	25
7	20	22
8	17	19
9	18	17
10	16	15
11	11	13
12	7	12
13	10	10
14	7	9
15	6	8
16	5	7
17	7	6
18	2	6
19	4	5
20	5	5
21	3	4
22	5	4
23	1	3
24	3	3
25	1	3
26	2	2
27	2	2
28	3	2
29	3	2
30	3	2
31–73	16	13
Total	504	504

The negative binomial with estimated parameters $r = 0.7751$ and $p = 0.0962$ was used to calculate the expected frequencies in column 3. Quite clearly, the model fits very well.

QUESTION 4.6.4   Suppose X_1 and X_2 are independent gamma random variables, X_1 with parameters r and λ, and X_2 with parameters s and λ. Show that $X_1 + X_2$ is a gamma random variable with parameters $r + s$ and λ.

APPENDIX 4.1 A PROPERTY OF THE GAMMA FUNCTION

THEOREM 4.11   Let $r > 0$ and $s > 0$. Then

$$\frac{\Gamma(r)\Gamma(s)}{\Gamma(r+s)} = \int_0^1 u^{r-1}(1-u)^{s-1}\, du.$$

PROOF   By definition,

$$\Gamma(r)\Gamma(s) = \int_0^\infty x^{r-1}e^{-x}\, dx \int_0^\infty y^{s-1}e^{-y}\, dy$$

$$= \int_0^\infty \int_0^\infty x^{r-1}y^{s-1}e^{-(x+y)}\, dx\, dy.$$

Let $u = x/(x+y)$ so that $x = uy/(1-u)$. Then $dx = y\, du/(1-u)^2$ and

$$\Gamma(r)\Gamma(s) = \int_0^\infty \int_0^1 \left(\frac{uy}{1-u}\right)^{r-1} y^{s-1}e^{-(y/(1-u))}\frac{y}{(1-u)^2}\, du\, dy.$$

Now, make the substitution $v = y/(1-u)$, or, equivalently, $y = (1-u)v$. This reduces the double integral to

$$\Gamma(r)\Gamma(s) = \int_0^\infty \int_0^1 (uv)^{r-1}(1-u)^{s-1}v^{s-1}e^{-v}v\, du\, dv$$

$$= \int_0^\infty \int_0^1 u^{r-1}(1-u)^{s-1}v^{r+s-1}e^{-v}\, du\, dv$$

$$= \left(\int_0^\infty v^{r+s-1}e^{-v}\, dv\right)\left[\int_0^1 u^{r-1}(1-u)^{s-1}\, du\right]$$

$$= \Gamma(r+s)\int_0^1 u^{r-1}(1-u)^{s-1}\, du,$$

and the theorem follows.

REVIEW EXERCISES FOR CHAPTER 4

1. Astronomers estimate that as many as 100 billion stars in the Milky Way galaxy may be circled by one or more planets, much as the Earth revolves around the sun. Let p be the probability of intelligent life on any given solar system. How small can p be and still give a 50–50 chance that there is intelligent life on at least one other planet in our galaxy?

2. Suppose that 1% of all items in a supermarket are unmarked. A customer buys ten items and proceeds to check out through the express lane. Estimate the probability that he will be delayed (along with all those behind him in line!) because one or more of his items require a price check. What assumptions are you making?

3. A radioactive source is metered for two hours, during which time the total number of alpha particles counted is 482. What is the probability that during the next minute exactly three particles will be counted? no more than three?

4. A certain young lady is quite popular with her male classmates. She receives, on the average, four phone calls a night. What is the probability that tomorrow night the number of calls she receives will exceed her average by more than one standard deviation?

5. Records were kept during the Second World War of the number of bombs falling in south London and their precise points of impact. The particular part of the city studied was divided up into 576 areas, each of $\frac{1}{4}$ square kilometer. The numbers of areas experiencing x hits, $x = 0, 1, 2, 3, 4, 5$, are listed below (25).

Number of Hits, x	Frequency
0	229
1	211
2	93
3	35
4	7
5	1
	576

Compute the expected frequencies to see how well the Poisson model applies. Use

$$P(X = x) = \frac{e^{-0.93}(0.93)^x}{x!}, \qquad x = 0, 1, 2, \ldots.$$

6. If X and Y are independent random variables with moment-generating functions e^{-4+4e^t} and $(\frac{1}{3} + \frac{2}{3}e^t)^2$, respectively, find $P(X = 2Y)$.

7. In a certain published book of 520 pages, 390 typographical errors occur. What is the probability that one page, selected randomly by the printer as an example of his work, will be free from errors?

8. Suppose X and Y are independent random variables with moment generating functions

$$M_X(t) = e^{-4+4e^t} \quad \text{and} \quad M_Y(t) = e^{-6+6e^t},$$

respectively. Let $Z = X + Y$. Find $f_Z(z)$.

9. A certain chromosome mutation believed to be linked with colorblindness is known to occur, on the average, once in every 10,000 births. If 20,000 babies are born this year in a certain city, what is the probability that at least one will develop colorblindness? What is the "exact" probability model that applies here?

e xam

10. Let X and Y be two independent random variables, each described by a Poisson pdf with $\lambda = 2$. Let $Z = X + Y$. Show that the conditional pdf of X given z is binomial.

11. Traffic records indicate that the number of "fender-benders" along a two-mile through-way during the evening rush hour is Poisson, with the average number of accidents being 0.75. What is the probability that exactly three mishaps occur in the space of two evenings?

12. If the random variable X has a Poisson distribution such that $P(X = 1) = P(X = 2)$, find $P(X = 4)$.

13. Suppose that X is Poisson with parameter λ and Y is Poisson with parameter μ. If X and Y are independent, show that $X + Y$ is Poisson with parameter $\lambda + \mu$.

14. Let Y denote the interval length between consecutive occurrences of an event described by a Poisson distribution. If the Poisson events are occurring at a rate of λ per unit time, show that

$$f_Y(y) = \lambda e^{-\lambda y}, \qquad y > 0.$$

15. Among the most famous meteor showers are the Perseids, which occur each year in early August. In some areas the frequency of visible Perseids can be as high as 40 per hour. Assume the sighting of these meteors is a Poisson event. What is the probability that an observer will have to wait at least five minutes between successive sightings?

16. Suppose that in a certain country, commercial airplane crashes occur at the rate of 2.5 per year. Assuming the frequency of crashes per year to be a Poisson random variable, find the probability that four or more crashes will occur next year. Also, find the probability that the next two crashes will occur within three months of one another.

17. Find the mean and the median for the exponential random variable Y having pdf $f_Y(y) = \lambda e^{-\lambda y}$, $y > 0$. In light of Review Exercise 14, does the relationship between the mean and the median offer any explanation for the familiar adage that "bad things come in threes"? Explain.

exam

18. If X is Poisson with parameter λ and the conditional pdf of Y given x is binomial with parameters x and p, show that the marginal pdf of Y is Poisson with parameter λp.

19. Use the Poisson approximation to calculate the probability that at most one person in 500 will have a birthday on Christmas. Assume there are 365 days in the year.

20. (a) If X is Poisson with parameter λ, show that

$$\frac{f_X(x + 1)}{f_X(x)} = \frac{\lambda}{x + 1}.$$

   (b) Use part (a) to find the largest value for $f_X(x)$, λ being fixed.

21. Flaws in a particular kind of metal sheeting occur at an average rate of one per 10 square feet. What is the probability of two or more flaws in a 5-by-8-foot sheet?

22. Let X denote the number of taxis arriving at a certain station and Y the number of parties wanting a ride. Suppose X and Y are independent Poissons, both with the same parameter. Find the pdf for the excess of users over taxis (positive or negative)—that is, find $f_Z(z)$, where $Z = Y - X$.

23. Let Z be a standard normal random variable. Find
   (a) $P(Z > 2.15)$.   (b) $P(Z < -1.06)$.

(c) $P(0.25 < Z < 1.77)$.   (d) $P(0.25 \leq Z \leq 1.77)$.

(e) $P(-3.06 \leq Z \leq -1.20)$.   (f) $P(Z < 4.31)$.

24. Evaluate the integral $A = \int_0^\infty e^{-4x^2}\, dx$.

25. A political poll of some 200 registered voters shows the Democratic candidate in a gubernatorial race is favored by 110 voters and the Republican candidate by 90, a margin of 10%. Suppose the sentiment for the two candidates in the general population is, in fact, evenly split. Let Y denote the margin percentage in the polls. Use the DeMoivre-Laplace theorem to estimate the probability of Y's being as large as or larger than 0.10.

26. A fair coin is tossed 100 times. Estimate the probability of getting (a) fewer than 45 heads, (b) exactly 46 heads, and (c) between 48 and 53 heads.

27. A random sample of some 747 obituaries published in Salt Lake City newspapers revealed that 46% of the decedents died in the three-month period following their birthday (119). Assess the statistical significance of that finding by estimating the probability that 46% *or more* would die in that particular interval if, in fact, deaths occurred randomly throughout the year. What would you conclude on the basis of your answer?

28. Suppose X is a binomial random variable defined on n independent trials, each having a success probability of $p = \frac{1}{2}$. Find

$$P\left[-1 < \frac{X - n(\frac{1}{2})}{\sqrt{n(\frac{1}{2})(\frac{1}{2})}} < +1\right]$$

for $n = 2, 5, 8, 10, 12$, and 15, and compare the results with the DeMoivre-Laplace limit.

29. Find the points of inflection for an $N(\mu, \sigma^2)$ pdf.

30. Suppose X is an $N(\mu, \sigma^2)$ random variable. Use the properties of moment-generating functions to show that

$$Z = \frac{X - \mu}{\sigma}$$

has an $N(0, 1)$ distribution.

31. Assume that the number of miles a driver gets on a set of radial tires is normally distributed with a mean of 30,000 miles and a standard deviation of 5000 miles. Would the manufacturer of these tires be justified in claiming that 90% of all drivers will get at least 25,000 miles?

32. If $e^{3t + 8t^2}$ is the moment-generating function for the random variable X, find $P(-1 < X < 9)$.

33. A certain type of seed has a probability of 0.8 of germinating. In a package of 100 seeds, what is the probability that at least 75% will germinate?

34. The diameter of the connecting rod in the steering mechanism of a certain foreign sports car must be between 1.480 and 1.500 centimeters, inclusive, to be usable. The distribution of connecting-rod diameters produced by the manufacturing process is normal with a mean of 1.495 cm and a standard deviation of 0.005 cm. What percentage of rods will have to be scrapped?

35. A criminologist has developed a questionnaire for predicting whether a teenager will become a delinquent. Scores on the questionnaire can range from 0 to 100, with higher values reflecting a presumably greater criminal tendency. As a rule of thumb, the criminologist decides to classify a teenager as a potential delinquent if his score exceeds 75. She has already tested the questionnaire on a large sample of teenagers, both delinquent and nondelinquent. Among those considered nondelinquent, scores were normally distributed with a mean of 60 and a standard deviation of 10. Among those considered delinquent, scores were normally distributed with a mean of 80 and a standard deviation of 5.
(a) What proportion of the time will she misclassify a nondelinquent as a delinquent?
(b) What proportion of the time will she misclassify a delinquent as a nondelinquent?

36. The systolic blood pressure of 18-year-old women is normally distributed with a mean of 120 mm Hg and a standard deviation of 12 mm Hg. What is the probability that the blood pressure of a randomly selected 18-year-old woman will be greater than 150? less than 115? between 110 and 130?

37. For any distribution, the *interquartile range* is defined to be the difference between the 75th and the 25th percentiles. Find the interquartile range for the motivation data of Case Study 4.7. What property of a distribution does the magnitude of its interquartile range reflect?

38. A basketball team has a 70% foul-shooting percentage.
(a) Write a formula for the exact probability that out of their next 100 free throws, they will make between 75 and 80, inclusive. Assume the throws are independent.
(b) Using an appropriate approximation, estimate the probability asked for in part (a).

39. A biology professor grades on a curve by assigning C's to all scores from $\mu - \sigma/2$ to $\mu + \sigma/2$; B's, to all scores from $\mu + \sigma/2$ to $\mu + 3\sigma/2$, inclusive; and A's, to all scores exceeding $\mu + 3\sigma/2$. (Grades of D and F are assigned analogously.) If the scores are normally distributed, what percentages of students will receive each grade? What percentages of students will receive each grade if the scores are uniformly distributed from 0 to 100?

40. At a certain Ivy League college, the average freshman score on the verbal part of the SAT is 565, with a standard deviation of 75. If the distribution of scores is normal, what proportion of that school's freshmen have verbal SAT's over 650? under 500?

41. Find the 60th percentile for the verbal SAT scores described in Review Exercise 40. If there are 1000 freshmen at that particular school, what is the standard deviation of the number of students scoring between 570 and 590, inclusive?

42. Let Y be a standard normal random variable. Find an expression for $E(Y^{2k})$ in terms of a gamma function, and then use that result to show that

$$E(Y^{2k}) = (2k - 1)(2k - 3) \cdots 5 \cdot 3 \cdot 1.$$

What does $E(Y^{2k+1})$ equal, for $k = 0, 1, 2, \ldots$?

43. The function

$$f_Y(y) = \frac{1}{\sqrt{2\pi}\,\sigma y} e^{-(1/2)[(\ln y - \ln \lambda)/\sigma]^2}, \qquad y > 0,$$

is called the *lognormal pdf*. Justify its name by showing that if Y has this density, then $X = \ln Y$ is a normal variable with mean $\ln \lambda$ and variance σ^2.

44. In a rush of nostalgia for prehistory, a person decides to gamble with an astragalus bone. He wins if the broad convex side shows, the probability of that event being 0.4. He intends to accuse his opponent of cheating if 30% or fewer of the throws show the broad convex side. How many throws are necessary to make the probability of false accusation less than 0.01?

45. For the motivation data of Case Study 4.7, compute the expected frequency for the "10–14" class, assuming that the appropriate model is a normal curve with a mean of 13.8 and a standard deviation of 6.5.

46. It is estimated that 80% of all 18-year-old girls have weights ranging from 103.5 to 144.5 pounds. Assuming the weight distribution can be adequately approximated by a normal curve, and assuming that 103.5 and 144.5 are equal distances from the average weight, μ, calculate σ.

47. Listed below are the 1971 traffic death rates (per 100 million motor vehicle miles) for each of the 50 states (109).

Ala	6.4	La	7.1	Ohio	4.5
Alaska	8.8	Maine	4.6	Okla	5.0
Ariz	6.2	Mass	3.5	Ore	5.3
Ark	5.6	Md	3.9	Pa	4.1
Cal	4.4	Mich	4.2	RI	3.0
Colo	5.3	Minn	4.6	SC	6.5
Conn	2.8	Miss	5.6	SD	5.4
Del	5.2	Mo	5.6	Tenn	7.1
Fla	5.5	Mont	7.0	Tex	5.2
Ga	6.1	NC	6.2	Utah	5.5
Hawaii	4.7	ND	4.8	Va	4.5
Idaho	7.1	Nebr	4.4	Vt	4.7
Ill	4.3	Nev	8.0	WVa	6.2
Ind	5.1	NH	4.6	Wash	4.3
Iowa	5.9	NJ	3.2	Wisc	4.7
Kans	5.0	NM	8.0	Wy	6.5
Ky	5.6	NY	4.7		

Make a histogram of these observations using as classes "2.0–2.9," "3.0–3.9," and so on. Assuming the distribution of rates can be approximated by a normal distribution, determine the expected frequencies for each of the classes. Let μ and σ have the values 5.3 and 1.3, respectively.

48. The army is developing a new missile and is concerned about its precision with respect to the short or long distance (that is, the x-coordinate of distance). Assume this distance is normally distributed. By observing points of impact, the launchers can adjust the initial trajectory, thereby controlling the mean of the impact distribution. The variance, however, is a problem. It is desired that when aimed properly at least 95% of the misses will fall within an eighth-mile of the target. What is the maximum allowable standard deviation?

49. A professional football team carries three quarterbacks on its traveling squad. Suppose that each quarterback has an 80% chance of completing a game without getting injured. If the team plays a schedule of 12 games, what is the probability that their third-string quarterback makes his first starting appearance in the eleventh game? What is the probability the team has to find a *fourth* quarterback before the season is over? (Assume that the probability of two or more injuries in any one game is zero.)

50. A young couple plan to continue having children until they get their first boy. Suppose the probability of a child's being a boy is $\frac{1}{2}$ and the outcome of each birth is an independent event. What is their expected family size?

51. Show that a geometric random variable, X, is memoryless. That is, prove that
$$P(X = n - 1 + k \,|\, X > n - 1) = P(X = k).$$

52. A somewhat uncoordinated woman is attempting to get a driver's license. Having successfully completed the written exam, she need only pass the road test to achieve her objective. Her abilities in that particular area, though, are somewhat minimal: an unbiased observer would give her no more than a 10% chance of passing. What is the probability she will have to take the test at least seven times in order to get her license? Assume her driving skills remain at the same level, regardless of how many times she fails the test. What is the expected number of times she will have to take the test?

53. A fair die is tossed. What is the probability that the first 5 occurs on the fourth roll?

54. A supermarket chain has a Treasure Hunt game where customers are given a letter (either an A, E, L, S, U, or V) each time they make a purchase. When a customer has collected all six letters, spelling VALUES, he receives a prize. Assuming the letters are given out at random, how many purchases must be made, on the average, to collect a complete set of letters? Hint: Determine the expected number of purchases necessary to get a "second" letter (different than the one received on the first purchase), a "third" letter, and so on.

55. Let $X_1, X_2, \ldots, X_r$ be r independent random variables, each having the geometric pdf with parameter p. Show that $X = X_1 + X_2 + \cdots + X_r$ has pdf
$$f_X(x) = \binom{x - 1}{r - 1} p^x q^{x-r}, \qquad x = r, r + 1, \ldots,$$
and, hence, that $Y = X - r$ has the negative binomial pdf with parameters r and p.

56. A door-to-door encyclopedia salesman is required to document five in-home visits each day. Suppose he has a 30% chance of being invited into any given home, with each home representing an independent trial. If he selects, ahead of time, the addresses of ten households to call on, what is the probability his fifth "success" occurs on the tenth trial? What is the probability he requires fewer than eight addresses to record his fifth success? Suppose he goes to only seven homes: what is the probability he will meet or exceed his quota?

57. The moment-generating function for the negative binomial pdf as given in Definition 4.2 is not the same as the moment-generating function for
$$f_X(x) = \binom{x - 1}{r - 1} p^r q^{x-r}, \qquad x = r, r + 1, \ldots,$$

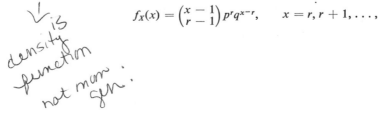

yet both pdf's relate to the occurrence of the rth success in a series of independent Bernoulli trials. Why are the mgf's different?

58. Suppose an underground missile installation has six pads for launching surface-to-air missiles. The missiles are launched one at a time from a single pad until, for whatever reason, the pad malfunctions. Then all missiles are launched from the adjacent pad until *it* malfunctions; then the third pad is used, and so on. Assume each firing on each pad can be considered an independent Bernoulli trial with a success probability of 0.75. What are the chances that the first deployment of the third pad will be to launch the eighth missile? What is the expected number of missiles that will be launched before the fourth pad is needed? The Army plans to stockpile the site with 60 missiles. Is that a reasonable allocation of resources? Explain. Assume that once a pad malfunctions, it cannot be restored to service.

59. If X is a gamma random variable with pdf

$$f_X(x) = \frac{1}{\Gamma(\alpha)\beta^\alpha} x^{\alpha-1} e^{-x/\beta}, \qquad x > 0,$$

then

$$M_X(t) = \left(\frac{1}{1 - \beta t}\right)^\alpha.$$

Also, if Y is an exponential random variable with pdf

$$f_Y(y) = \left(\frac{1}{\lambda}\right) e^{-y/\lambda}, \qquad y > 0,$$

its moment-generating function is given by

$$M_Y(t) = \left(\frac{1}{1 - \lambda t}\right).$$

Using this information, write down the pdf for $Z = Y_1 + Y_2 + Y_3$, where the Y_i's are an independent random sample of size three drawn from an exponential pdf with parameter λ.

60. Prove that the variance of a gamma random variable, as stated in Definition 4.4, is r/λ^2.

61. Use the result of Review Exercise 40 in Chapter 3 to suggest a method for generating random observations from a gamma pdf with parameters r and λ, where r is an integer.

62. Suppose an Antarctic weather station has three electronic wind gauges, an original and two backups. Each gauge's lifetime is exponentially distributed with a mean of 1000 (hours). What is the pdf of Y, the variable measuring the time until the last gauge wears out?

63. Let Z be a standard normal random variable. Show that Z^2 has the gamma pdf with parameters $r = \frac{1}{2}$ and $\lambda = \frac{1}{2}$.

64. The *hazard rate* for a random variable, T, is the likelihood that it will take on a value in the interval $(t, t + dt)$ given that it already exceeds t. In reliability applications, the hazard rate can be interpreted as the likelihood that a piece of equipment, having satisfactorily functioned for t time units, will fail in the next instant. More formally,

$$\text{hazard rate} = h_T(t) = \frac{f_T(t)}{1 - F_T(t)}.$$

Define $H_T(t) = \int_{-\infty}^{t} h_T(t)\, dt$ to be the cumulative hazard rate.

(a)   Express $F_T(t)$ as a function of $H_T(t)$.
(b)   Express $f_T(t)$ as a function of $h_T(t)$ and $H_T(t)$.
(c)   Prove that $\int_{-\infty}^{\infty} h(t)\, dt = \infty$.

65. Find and interpret the hazard rate for an exponential random variable with parameter λ.

66. The two-parameter *Weibull* pdf is defined by

$$f_X(x) = \frac{\beta}{\alpha} x^{\beta-1} e^{-x^\beta/\alpha}, \qquad x > 0, \alpha > 0, \beta > 0.$$

Find the hazard rate for the Weibull distribution. Also, graph $h_X(x)$. Hint: Consider three cases—$\beta < 1$, $\beta = 1$, and $\beta > 1$.

67. If a random variable X has a moment-generating function given by

$$M_X(t) = (1 - 5t)^{-10},$$

find $E(X)$ and Var (X). Also, write down the pdf for X.

68. A mercenary helicopter pilot signs a contract to fly 50 missions to help one of the combatants in an African border war. His services are required, on the average, three times a day. Suppose he has just returned from his 46th mission. Let Y denote the length of time he has to wait until his tour of duty is up. Find the pdf of Y. What is $E(Y)$?

CHAPTER FIVE

Estimation

RONALD AYLMER FISHER (1890–1962)

A towering figure in the development of both applied and mathematical statistics, Fisher took formal training in mathematics and theoretical physics (he graduated from Cambridge in 1912). After a brief career as a teacher, he accepted a post in 1919 as statistician at the Rothamsted Experimental Station. There the day-to-day problems encountered in collecting and interpreting agricultural data led directly to much of his most important work in the theory of estimation and experimental design. Fisher was also a prominent geneticist and devoted considerable time to the development of a quantitative argument that would support Darwin's theory of natural selection. He returned to academia in 1933, succeeding Karl Pearson as the Galton Professor of Eugenics at the University of London. Fisher was knighted in 1952.

5.1 INTRODUCTION

The ability of probability functions to describe, or *model*, experimental data was demonstrated in numerous examples in Chapter 4. In Section 4.2, for example, the Poisson distribution was shown to predict very well the number of Prussian cavalry soldiers kicked to death by their horses. That same family of distributions was also seen to model the number of alpha emissions from a radioactive source as well as the annual number of outbreaks of war, worldwide, from 1500 to 1931. In Section 4.3 another probability function, the normal, was applied to phenomena as diverse as the chest measurements of Scottish soldiers and the performance of 106 randomly selected adults on a motivation test. Still other probability models fitted to data in Chapter 4 included the geometric, the negative binomial, and the gamma. In this chapter we want to scrutinize this relationship between probability models and data more mathematically. Beginning with the material presented here, our focus will shift away from the study of probability and toward the study of statistics.

The application of the methods of probability to the analysis and interpretation of empirical data is known as *statistical inference*. More specifically, statistical inference is the process by which we generalize from a particular sample—that is, from a set of *n* observations—to the theoretical population from which that sample came. Of course, the precise form of the generalization can vary considerably from situation to situation. It may be a single numerical estimate, a *range* of numerical estimates, or even a simple "Yes" or "No."

A few examples may clarify some of these ideas. Imagine the chief programming executive at ABC trying to decide which shows to cancel and which to renew. As input, he may have the day-by-day logs of the programs that are watched by, say, two or three thousand specially selected families, representing various socio-economic, age, and ethnic groups. His problem is to use that sample information to estimate the *total* number of viewers tuned to ABC's programs. (The "population" here is simply the aggregate of all TV viewers.) For a given program, his final inference might be that "21.6% of the potential market is tuned to Program X" (i.e., a single numerical estimate), or it might take the form that "between 18.5% and 23.0% of the market is tuned to Program X" (i.e., a range of estimates).

Comment. Estimation problems of this sort are not uncommon. They represent an area of statistical inference known as *survey sampling*. Political pollsters assessing a candidate's strength among blue-collar workers, consumer researchers testing the marketability of a new laundry detergent, ecologists estimating the number of small-mouth bass in a certain lake, and public health officials trying to determine a community's immunization status against polio—all need to make the same sort of generalization from a sample to a population. ▌

As a second example, suppose a zoologist (for reasons best left unsaid) would like to know whether *Desmodus rotundus*, a small South American vampire bat, prefers blood at room temperature (23°C) or at body temperature (38.5°C) [see (18)]. To get some data, he puts equal numbers of similar bats into two cages. The drinking tubes in cage A are supplied with blood kept at room temperature; those in cage B, with blood kept at body temperature. Suppose that after both groups have had enough time to drink as much as they wanted, he finds that the bats in cage A consumed 3% more blood than the bats in cage B. What should he conclude? Does the 3% excess "prove" that vampire bats prefer blood at room temperature, or is that figure too small to mean anything? Here, one way to phrase the inference would be a simple "Yes" or "No": "Yes, the bats show a preference" or "No, they do not."

These two examples point up the two broad areas into which the subject of statistical inference is traditionally divided: *estimation* and *hypothesis testing*. In the former the inference is numerical; in the latter it becomes a yes or no decision between two conflicting theories. Both areas, as we will see, have wide applicability.

Chapter 5 presents the basic principles of estimation. This will be essentially a two-part endeavor—first it will be necessary to define the mathematical properties a "good" estimator should have; then we will look for procedures that yield estimates having these properties. Hypothesis testing will be taken up in Chapter 6.

5.2 DEFINITIONS

In this section we introduce some of the basic terminology that has come to be associated with statistical estimation. The place to begin is to recall that most probability models—particularly those general enough to be of any practical value—are indexed by (that is, are functions of) one or more constants (see Section 4.1). We call these constants *parameters*. Thus, the Poisson

$$f_Y(y) = \frac{e^{-\lambda}\lambda^y}{y!},$$

is a function of the occurrence rate, λ. The normal,

$$f_Y(y) = \frac{1}{\sqrt{2\pi}\sigma}e^{-(1/2)[(y-\mu)/\sigma]^2},$$

is indexed by *two* parameters, the mean μ and the standard deviation σ. Similarly, the binomial depends on n, the number of trials, and p, the probability of success at any given trial. Each of these distributions, then, is actually a *family* of distributions, where each member of the family has a different parameter value (and a correspondingly different shape) (see Figure 5.1).

The role that parameters play in the inference process is a crucial one. In many situations the family of probability models describing a phenomenon may be known (or at least assumed to be known) but the particular member of the family that *best* describes that phenomenon may be unknown. A criminologist, for example, may have good reason to believe that the distribution of a certain fingerprint abnormality in the general population follows a Poisson distribution, but only

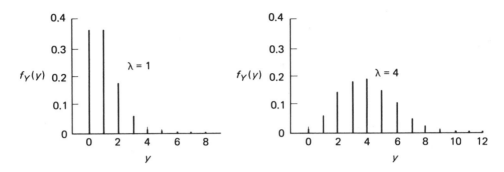

Figure 5.1 Two members of the Poisson family, $f_Y(y) = \dfrac{e^{-\lambda}\lambda^y}{y!}$.

after he does a survey and gets some feeling for the rate at which the abnormality occurs (say, 19.5 times for every 10,000 people) does he know *which* Poisson is the most appropriate model. It is probably fair to say that estimating the unknown parameter (or parameters) of a presumed data model is an intermediate, if not a final, step in almost every inference problem.

By convention, various symbols have come to be associated with the parameters of different distributions: μ and σ with the normal, n and p with the binomial, and so on. However, to facilitate the discussion of parameters and estimates of parameters in general, it will be necessary to adopt a more flexible notation. Accordingly, we will let θ denote an arbitrary parameter. The symbol $f_Y(y; \theta_1, \theta_2, \ldots, \theta_k)$, then, will denote a probability function whose k parameters (known or unknown) are $\theta_1, \theta_2, \ldots, \theta_k$. Most of the problems we will deal with, though, involve only a single parameter. For these, the subscript on θ will be deleted and the pdf written $f_Y(y; \theta)$.

If in a given problem a particular θ_i is unknown, we must estimate it, using the sample data. This will be done through a function known as a *statistic*. By definition, a statistic, $W = h(Y_1, Y_2, \ldots, Y_n)$, is any function of the random variables, $Y_1, Y_2, \ldots, Y_n$, the set of Y_i's being a random sample of size n. Thus, the mean, $Y = (1/n)\sum_{i=1}^{n} Y_i$, the standard deviation, $S = \sqrt{(1/n - 1))\sum_{i=1}^{n}(Y_i - Y)^2}$, and the range, $R = \max_i Y_i - \min_i Y_i$, are all examples of statistics.

Having defined the notions of "parameter" and "statistic," we can now rephrase the main objective of this chapter as outlined in Section 5.1: *we are seeking criteria—and, based on those criteria, procedures—for finding statistics that will serve as "good" estimators of unknown parameters.* Suppose, for example, three lightbulbs are put "on test" and kept lit continuously until they burn out. The resultant data would be the three observed lifetimes—say, 610, 1150, and 1570 hours. If it can be assumed (and there are physical reasons to justify this) that the density function describing the distribution of their lifetimes is of the form $f_Y(y; \lambda) = (1/\lambda)e^{-y/\lambda}$, how should we estimate λ? Should the estimator have the form

$$W = h(Y_1, Y_2, Y_3) = \sum_{i=1}^{3} Y_i \quad \text{or} \quad W = h(Y_1, Y_2, Y_3) = \sum_{i=1}^{3} \frac{3}{Y_i^2}$$

or something entirely different? We will see a little later that according to at least one principle of estimation, W should be $\frac{1}{3} \sum_{i=1}^{3} Y_i$, or, in this case, $3330/3 = 1110$ (recall Example 2.15).

Comment. In Chapter 3 the distinction was made between (1) random variables as generic designations and (2) sample observations as particular realizations of random variables. The same duality exists in connection with estimation. We call the *form* of an expression for estimating an unknown parameter an *estimator*. In the example just described,

$$\frac{1}{3} \sum_{i=1}^{3} Y_i$$

would be an estimator (for λ). However, once the random variables in the estimator are replaced by their sample values [such as $\frac{1}{3}(610 + 1150 + 1570)$], the resulting numerical quantity (1110) is referred to as an *estimate* (for λ). ∎

Basically, there are two formats into which an estimate can be cast: it can be expressed as a point or as an interval. As the name implies, a *point estimate* is a single number that represents, in some sense, our best guess as to the value of the unknown parameter. An *interval estimate*, on the other hand, is a range of numbers generated by a procedure having a high a priori probability (often 0.95 or 0.99) of "including" the unknown parameter. For the lightbulb example, $w = 1110$ would be a point estimate for λ. Without going into the details here (see Review Exercise 32 at the end of Chapter 7), it can be shown that the interval $\left(2 \sum_{i=1}^{3} Y_i/14.4, \, 2 \sum_{i=1}^{3} Y_i/1.24\right)$ has a 95% chance of "containing" λ, in the sense that

$$P\left(\frac{2 \sum_{i=1}^{3} Y_i}{14.4} < \lambda < \frac{2 \sum_{i=1}^{3} Y_i}{1.24}\right) = 0.95.$$

Substituting 3330 for $\sum_{i=1}^{3} Y_i$ gives (462, 5371) as the corresponding interval estimate for λ.

In one important way, interval estimates are superior to point estimates in terms of what they reveal about the data: specifically, the *length* of the interval provides a very useful measure of the estimator's precision. Merely stating, for example, that a point estimate for some θ is, say, $w = 50$, gives no indication of whether the *true* θ is likely to be within just a few units of 50 or whether it could easily be 17, or -35, or $+210$. However, knowing that an interval estimate for θ is (48, 52), we can safely infer that the possibility of the true θ's being something either much less than 48 or much greater than 52 is quite remote.

Because of their simplicity, point estimates will be discussed first. Sections 5.3 through 5.7 present a number of desirable properties to be looked for in an estimator of this type. Section 5.8 then describes two procedures for finding the actual *form* of $W = h(Y_1, Y_2, \ldots, Y_n)$ knowing only $f_Y(y; \theta)$. Interval estimation is introduced in the last two sections of the chapter.

Comment. Applications of these concepts to the normal distribution will be deferred until Chapter 7, where they can be dealt with in a more organized way. Here we will rely primarily on the uniform, the exponential, the Poisson, and the binomial distributions to provide us with examples.   ▌

5.3 PROPERTIES OF POINT ESTIMATORS: UNBIASEDNESS AND EFFICIENCY

Before making any attempt to set down a list of qualities that a good W should possess, it is important to understand one very basic fact about the fundamental nature of estimators: every one, by virtue of its being a function of the sample data, is itself a random variable. This means that its behavior for different random samples will be described by a probability density function. Such a function, in keeping with the notation introduced in Chapter 3, will be denoted $g_W(w)$.

Example 5.1 examines the $g_W(w)$ associated with an estimator for the parameter of a uniform distribution.

EXAMPLE 5.1.   Suppose a sample of size n is drawn from the uniform distribution

$$f_Y(y; \theta) = \begin{cases} 1/\theta, & 0 < y < \theta \\ 0, & \text{elsewhere,} \end{cases}$$

and we decide to estimate θ with the maximum of the y_i's. That is,

$$W = h(Y_1, Y_2, \ldots, Y_n) = \max_i Y_i = Y_{\max} = Y'_n.$$

From Chapter 3, recall that the density function for Y'_i, the ith-order statistic, is

$$f_{Y_i'}(y) = \frac{n!}{(i-1)!\,(n-i)!}[F_Y(y)]^{i-1}[1 - F_Y(y)]^{n-i}f_Y(y),$$

where $F_Y(y)$ is the cdf of Y. For $i = n$ and

$$F_Y(y) = \begin{cases} 0, & y \le 0 \\ y/\theta, & 0 < y < \theta \\ 1, & y \ge \theta, \end{cases}$$

the density function for $W = Y'_n = Y_{\max}$ easily reduces to

$$f_{Y_n'}(y_n) = g_W(w) = \begin{cases} \dfrac{nw^{n-1}}{\theta^n}, & 0 < w < \theta \\ 0, & \text{elsewhere.} \end{cases}$$

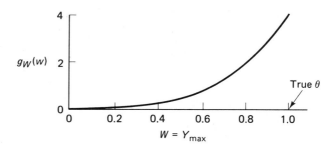

Figure 5.2 Graph of $g_W(w)$.

Figure 5.2 shows $g_W(w)$ for the particular case where $\theta = 1$ and $n = 4$.

QUESTION 5.3.1   For the special case just described—$\theta = 1$ and $n = 4$—find the probability that W will be within $\frac{1}{8}$ of θ.

Realizing that each W has an associated $g_W(w)$ describing its behavior, it is not hard to conceptualize what should be required of a good estimator. As a first condition, it seems reasonable to ask that $g_W(w)$ be somehow "centered" with respect to θ. If it is not, W will tend either to overestimate or underestimate θ, a condition hardly desirable. Figure 5.3(a) shows a W distribution that *does* meet

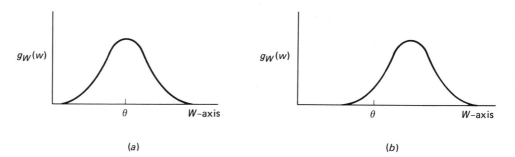

Figure 5.3 Centered and uncentered distributions.

this condition; Figure 5.3(b), one that does not. Of course, we have already seen an example of the latter—using $Y_{\max}$ to estimate the parameter of a uniform distribution. (The notion of W's being centered with respect to θ will be made more precise in the next section when we define *unbiasedness*.)

A second property that a good estimator should possess is *precision*. We will say that an estimator is precise if the dispersion of its distribution is small. In Figure 5.4, both $g_{W_1}(w_1)$ and $g_{W_2}(w_2)$ appear to be centered with respect to the parameter, but clearly the better estimator is the one whose behavior is described by $g_{W_2}(w_2)$ because, of the two, it has a greater chance of being close to the true

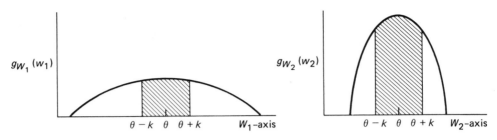

Figure 5.4   Two distributions with different dispersions.

θ. (We will formalize this idea in Section 5.5 by defining what is meant by the *efficiency* of an estimator.)

5.4 UNBIASEDNESS

Although other criteria exist, by far the most common way to decide whether or not an estimator's distribution is suitably "centered" (relative to θ) is by looking at its expected value. If, *on the average*, W is equal to θ, we pronounce the estimator centered (and call it *unbiased*). Definition 5.1 gives the formal statement.

DEFINITION 5.1   Let $Y_1, Y_2, \ldots, Y_n$ be a random sample from $f_Y(y; \theta)$. An estimator $W = h(Y_1, Y_2, \ldots, Y_n)$ is said to be *unbiased* (for θ) if $E(W) = \theta$, for all θ.

EXAMPLE 5.2.   Consider again the problem of estimating the parameter of a uniform distribution with $W = Y_{\max}$. Since $Y_{\max}$ is always less than or equal to θ, it clearly cannot be unbiased. However, we can show that by multiplying it by an appropriate constant, we can transform it into a new statistic that *is* unbiased. From Example 5.1,

$$E(W) = E(Y_{\max}) = \int_0^\theta w g_W(w)\, dw = \int_0^\theta w \frac{n w^{n-1}}{\theta^n}\, dw$$

$$= \frac{n w^{n+1}}{(n+1)\theta^n}\Big|_0^\theta = \left(\frac{n}{n+1}\right)\theta.$$

Suppose, now, a new estimator, W_1, is defined, where

$$W_1 = \left(\frac{n+1}{n}\right) Y_{\max}.$$

Since

$$E(W_1) = \left(\frac{n+1}{n}\right) E(Y_{\max}) = \theta,$$

W_1 is unbiased.

The distribution of W_1 is easily derived. Recall from Chapter 3 that if Y is a continuous random variable with pdf $f_Y(y)$, the pdf of the transformed random variable X, where $X = aY$, is given by

$$g_X(x) = \left(\frac{1}{a}\right) f_Y\left(\frac{x}{a}\right).$$

Applying this result to $W_1 = [(n+1)/n]W$ gives

$$g_{W_1}(w_1) = \frac{n}{n+1} \cdot \frac{n\left[w_1 \Big/ \left(\frac{n+1}{n}\right)\right]^{n-1}}{\theta^n}$$

$$= \frac{\frac{n^{n+1}}{(n+1)^n}(w_1)^{n-1}}{\theta^n}, \qquad \text{for } 0 < w_1 < \left(\frac{n+1}{n}\right)\theta.$$

Figure 5.5 shows $g_{W_1}(w_1)$ for the special case described in Example 5.1: $\theta = 1$ and $n = 4$.

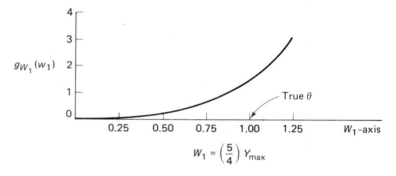

Figure 5.5   Graph of $g_{W_1}(w_1)$.

QUESTION 5.4.1   Find an unbiased estimator for θ based on the *smallest* of n order statistics drawn from a uniform distribution defined over $(0, \theta)$.

QUESTION 5.4.2   Suppose 14, 10, 18, and 21 constitute a random sample of size four drawn from a uniform pdf defined over the interval $(0, \theta)$, where θ is unknown. Find an expression for an unbiased estimator for θ based on Y'_3, the third-order statistic. Evaluate the estimator using the data given. Is it possible that an estimate based on Y'_3 would be clearly inappropriate in a given situation? Explain.

EXAMPLE 5.3.   If the parameter being estimated is, itself, $E(Y)$, then the sample mean will always provide an unbiased estimator. Consider, for example, a random sample of size n taken from the exponential distribution,

$$f_Y(y) = \left(\frac{1}{\lambda}\right)e^{-y/\lambda}, \qquad \text{for } y > 0,$$

where

$$E(Y) = \int_0^\infty y\left(\frac{1}{\lambda}\right)e^{-y/\lambda}\,dy = \lambda.$$

It will follow that

$$W = \bar{Y} = \frac{1}{n} \sum_{i=1}^{n} Y_i$$

will be unbiased for λ. This is an immediate consequence of the distributive property of expected values:

$$E(W) = E(\bar{Y}) = \frac{1}{n} \sum_{i=1}^{n} E(Y_i) = \frac{n\lambda}{n} = \lambda.$$

The derivation of $g_W(w)$ is straightforward if it is remembered that the sum of n independent exponential random variables with parameter λ has the gamma distribution with parameters n and λ (see Review Exercise 59 at the end of Chapter 4). Let

$$Y = \sum_{i=1}^{n} Y_i.$$

If $h_Y(y)$ is the pdf of Y, and if we write W as $W = (1/n)Y = \bar{Y}$, then it follows that

$$g_W(w) = nh_Y(wn)$$

$$= \frac{1}{\Gamma(n)(\lambda/n)^n} w^{n-1} e^{-wn/\lambda}, \qquad 0 < w < \infty.$$

That is, W is gamma with parameters n and λ/n. Figure 5.6 shows $g_W(w)$ for the case where $\lambda = 1$ and $n = 4$.

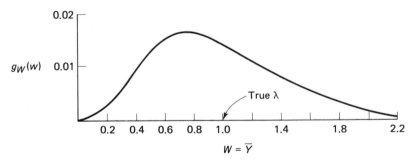

Figure 5.6 Graph of $g_W(w)$.

If a random variable X is distributed according to a Poisson distribution with an occurrence rate of θ, it can be shown (see Review Exercise 14 at the end of Chapter 4) that the distribution of Y, the length of "time" between successive occurrences of X, is exponential with parameter λ, where $\lambda = 1/\theta$. The following example illustrates this duality in a literary context.

Among the many phenomena that can be described by the Poisson is the occurrence of certain words in written material. For example, for a given author the number of times the word "any" appears may follow a Poisson distribution with a parameter of, say, $\lambda = 0.042$ ("any's"/word). If that were true, the number of words between successive "any's" would be approximated by the exponential

$$f_Y(y) = \left(\frac{1}{23.8}\right) e^{-y/23.8}, \qquad y > 0,$$

where $23.8 = 1/0.042$.

The data in Table 5.1 [modified from (76)] represent the number

TABLE 5.1   Preposition gaps

37	135	9	46	210	32	151	78	222	13
85	90	128	488	10	62	112	50	26	6
325	171	29	100	97	191	3	272	56	302
12	67	15	5	157	251	206	30	386	54
310	49	234	2	19	19	185	516	165	226
44	126	110	142	43	25	26	21	72	42

of words between 61 successive occurrences of the Russian preposition к in Pushkin's "The Captain's Daughter." The sum of these 60 observations is 7095, giving a sample mean, $\bar{y}$, of 118.25. There-

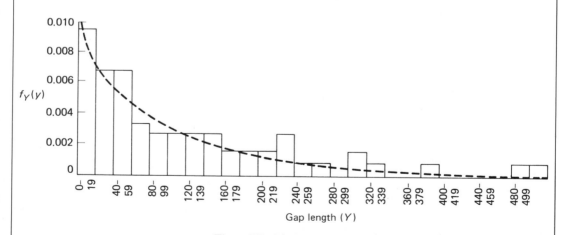

Figure 5.7   The Poisson–exponential relationship.

fore, if the Poisson assumption is true in this case, the particular exponential pdf fitting these data would be

$$f_Y(y) = \left(\frac{1}{118.25}\right) e^{-y/118.25}. \tag{5.1}$$

Figure 5.7 shows the graph of Equation 5.1 superimposed over a histogram of the original data.

EXAMPLE 5.4.   The Poisson distribution,

$$f_Y(y; \lambda) = P(Y = y) = \frac{e^{-\lambda}\lambda^y}{y!}, \qquad y = 0, 1, 2, \ldots,$$

is another probability function for which $E(Y) = \lambda$ (recall Theorem 4.2). This is why the data of Case Studies 4.2, 4.3, and 4.4 were "fitted" using $\bar{y}$ in place of the unknown λ.

QUESTION 5.4.3   Let $Y_1, Y_2, \ldots, Y_n$ be a random sample from any pdf whose mean is θ. Under what conditions is the estimator

$$W = h(Y_1, Y_2, \ldots, Y_n) = \sum_{i=1}^{n} a_i Y_i$$

unbiased?

EXAMPLE 5.5.   If Y is the number of successes in n Bernoulli trials, where each trial has a success probability equal to p, we know that Y has a binomial distribution

$$P(Y = y) = \binom{n}{y} p^y (1 - p)^{n-y}, \qquad y = 0, 1, \ldots, n,$$

and that $E(Y) = np$. Therefore,

$$W = \frac{Y}{n}$$

will be an unbiased estimator for p:

$$E(W) = E\left(\frac{Y}{n}\right) = \frac{1}{n} E(Y) = \frac{np}{n} = p.$$

Real-world applications of the use of Y/n to estimate p are very common. For example, a drug-reaction surveillance program was carried out in nine major hospitals (111). Out of a total of 11,526 patients monitored, it was found that 3240 experienced some form of adverse reaction to their medication. If Y is taken to be the number of patients (out of 11,526) experiencing adverse reactions, it would not be unreasonable to assume that Y is binomial:

$$f_Y(y; p) = P(Y = y) = \binom{11,526}{y} p^y (1 - p)^{11,526-y}, \qquad y = 0, 1, \ldots, 11,526,$$

where p is the probability that a random patient has an adverse reaction. Then, from what we have just derived, an unbiased estimator for p would be $W = Y/11,526$, which reduces in this case to

$$w = \frac{3240}{11,526} = 0.281 \text{ (or 28.1\%)}.$$

QUESTION 5.4.4   Find the distribution of $W = Y/n$.

QUESTION 5.4.5   Any binomial random variable Y can be thought of as the sum of n independent Bernoulli random variables, $Y = Y_1 + Y_2 + \cdots + Y_n$. Show that $W_1 = Y_i$, for any i, is also an unbiased estimator for p. Which estimator, $W = Y/n$ or $W_1 = Y_i$, seems preferable? Why?

Our final example of unbiasedness is an unusual one. We will use the Y/n estimator for the binomial parameter to get a numerical approximation for the value of π. This particular derivation, known as *Buffon's needle problem*, dates back to the eighteenth century.

Comment. Georges-Louis Leclerc, Comte de Buffon (1707–1788), along with Montesquieu, Voltaire, and Rousseau were the four major literary figures of the French Enlightenment. Buffon's monumental work was the 44-volume *Histoire Naturelle, Generale et Particulière*, which established him as one of the foremost theoretical biologists and geologists of the century. A man of eclectic interests, he also studied mathematics extensively, although his contributions in that area were somewhat less grandiose. Probably his best-known work in mathematics today is the "needle" problem described here, although in his own century he was more famous for having authored a French translation of Newton's *Method of Fluxions*. ∎

EXAMPLE 5.6.   Imagine repeatedly dropping a needle 2 inches long onto a grid of horizontal lines, each 2 inches apart (Figure 5.8). We will derive a procedure for estimating π based on the proportion of times the needle crosses one of the lines.

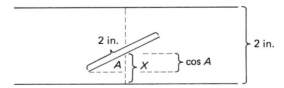

Figure 5.8   Buffon's needle problem.

Let

$X =$ distance from the needle's center to the nearest grid line,

$A =$ angle formed by the needle and the vertical.

Since the needle is falling randomly, both the X distribution and the A distribution are uniform:

$$f_X(x) = 1, \qquad 0 \leq x \leq 1$$

and

$$g_A(a) = \frac{1}{\pi}, \qquad -\frac{\pi}{2} \le a \le \frac{\pi}{2}.$$

Furthermore, *where* the needle falls and the *angle* at which it falls are clearly independent, meaning that $h_{X,A}(x, a)$, the joint pdf of X and A, factors into the product of the corresponding marginals:

$$h_{X,A}(x, a) = f_X(x)g_A(a) = \begin{cases} \frac{1}{\pi}, & 0 \le x \le 1, -\frac{\pi}{2} \le a \le \frac{\pi}{2} \\ 0, & \text{otherwise.} \end{cases}$$

Notice from Figure 5.8 that the needle touches a line if and only if $X \le \cos A$. The probability of this event is the shaded area shown in Figure 5.9. Numerically,

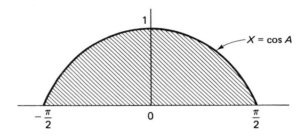

Figure 5.9   Probability of the needle touching a line.

$$P(X \le \cos A) = \int_{-\pi/2}^{\pi/2} \int_0^{\cos a} \left(\frac{1}{\pi}\right) dx\, da = \int_{-\pi/2}^{\pi/2} \left(\frac{\cos a}{\pi}\right) da$$

$$= \frac{2}{\pi}.$$

Suppose, now, the random variable Y is defined to be the number of times (out of n tosses) the needle crosses a line. Clearly, Y is binomial, with p being equal to $2/\pi$:

$$P(Y = y) = \binom{n}{y}\left(\frac{2}{\pi}\right)^y \left(1 - \frac{2}{\pi}\right)^{n-y}.$$

It follows from Example 5.5 that

$$E\left(\frac{Y}{n}\right) = \frac{2}{\pi},$$

implying that for a given value y, an approximation for π would be

$$\pi \doteq \frac{2n}{y}.$$

Comment. This is hardly one of the more efficient ways to estimate π. It can be shown (13) that something of the order of 1,000,000 "trials" would be necessary to be reasonably certain that $2n/Y$ would come within even 0.001 of the true value of π. ∎

Comment. Note that while $2n/y$ is an approximation to π, $2n/Y$ is not an unbiased estimator for π. To convince yourself of that, write out the sum defining $E(2n/Y)$ for n equal to 1. ∎

5.5 EFFICIENCY

Our discussion of unbiasedness has led to three important ideas in connection with point estimates: first, that it makes sense to talk about the expected value of an estimator and that imposing the requirement that $E(W)$ be equal to θ (for all θ) is one way of eliminating systematic bias; second, that even if a given W is *not* unbiased, it can often be easily transformed into a W_1 that is; and, finally, that for any given $f_Y(y; \theta)$, there may be many different W's that satisfy the criterion of unbiasedness. It is this latter observation that we want to examine more fully in this section.

To the practitioner facing the problem of estimating an unknown θ, the lack of uniqueness of unbiased estimators poses an obvious question: *which* unbiased estimator should be used? Are they all equivalent, or are some better than others? For an answer, we need to think back to the *second* property of good estimators that was mentioned at the beginning of Section 5.3: that the dispersion of any estimator should be small, so that the probability of its being close to the true θ will be large. As usual, the concept of dispersion will be translated mathematically into a statement about a variance.

DEFINITION 5.2   Let W_1 and W_2 be two unbiased estimators for θ with variances Var (W_1) and Var (W_2), respectively. We will call W_1 *more efficient than* W_2 if
$$\text{Var}\,(W_1) < \text{Var}\,(W_2).$$
Also, the *relative efficiency* of W_1 with respect to W_2 will be defined as the ratio
$$\text{Var}\,(W_2)/\text{Var}\,(W_1).$$

EXAMPLE 5.7.   As a simple example of two unbiased estimators having very different precisions, consider again the problem of estimating λ in an exponential distribution,
$$f_Y(y) = \left(\frac{1}{\lambda}\right)e^{-y/\lambda}, \qquad y > 0.$$
Since $E(Y) = \lambda$, one unbiased estimator would be $W_1 = \bar{Y}$; another would be $W_2 = Y_1$, where Y_1 is the first observation recorded. We have already seen in Example 5.3 that $g_{W_1}(w_1)$ is a gamma distribution with parameters n and λ/n. Therefore,
$$\text{Var}\,(W_1) = \frac{\lambda^2}{n} \qquad \text{(recall Theorem 4.10)}.$$
Also,
$$g_{W_2}(w_2) = f_Y(y; \lambda) = \left(\frac{1}{\lambda}\right)e^{-y/\lambda}$$

and, from the properties of the exponential,

$$\text{Var}(W_2) = \lambda^2.$$

Clearly, $W_1 = \bar{Y}$ is the better estimator: the variance of its distribution is smaller (by a factor of n) than the variance of W_2.

QUESTION 5.5.1   Show that $W_3 = n \cdot Y_{\min}$ is also an unbiased estimator for λ, and find the relative efficiency of W_1 to W_3 and W_2 to W_3.

EXAMPLE 5.8.   For the problem of estimating the parameter of a uniform distribution, we have already seen that one unbiased estimator is

$$W_1 = \left(\frac{n+1}{n}\right) \cdot Y_{\max}$$

and that

$$g_{W_1}(w_1) = \frac{\frac{n^{n+1}}{(n+1)^n} \cdot (w_1)^{n-1}}{\theta^n}, \qquad 0 < w_1 < \left(\frac{n+1}{n}\right)\theta.$$

By following a similar sort of argument (see Question 5.4.1), we could base a second unbiased estimator for θ on $Y'_1 = Y_{\min}$, the *smallest* of the n order statistics. In particular,

$$f_{Y'_1}(y) = \frac{n}{\theta}\left(1 - \frac{y}{\theta}\right)^{n-1}, \qquad 0 < y < \theta,$$

and

$$E(Y'_1) = \int_0^\theta \frac{ny}{\theta}\left(1 - \frac{y}{\theta}\right)^{n-1} dy = \left(\frac{1}{n+1}\right)\theta.$$

From this latter result, it can be seen that

$$W_2 = (n+1) \cdot Y_{\min}$$

would be a second unbiased estimator for θ. Also, it can easily be verified that the pdf for W_2 is given by

$$g_{W_2}(w_2) = \frac{n}{(n+1)\theta}\left(1 - \frac{w_2}{(n+1)\theta}\right)^{n-1}, \qquad 0 < w_2 < (n+1)\theta.$$

According to Definition 5.2, deciding which estimator is better—W_1 or W_2—requires a comparison of their respective variances. First, consider W_1:

$$E(W_1^2) = \int_0^{[(n+1)/n]\theta} \frac{\frac{n^{n+1}}{(n+1)^n}(w_1)^{n+1}}{\theta^n} dw_1$$

$$= \frac{(n+1)^2}{n(n+2)}\theta^2,$$

in which case

$$\text{Var}(W_1) = E(W_1^2) - [E(W_1)]^2$$

$$= \frac{(n+1)^2}{n(n+2)}\theta^2 - \theta^2$$

$$= \frac{\theta^2}{n(n+2)}.$$

Similarly (see Question 5.5.2),

$$\text{Var}(W_2) = \frac{n\theta^2}{n+2}.$$

Since

$$\frac{\theta^2}{n(n+2)} < \frac{n\theta^2}{n+2} \qquad (\text{for } n > 1),$$

W_1 is, by definition, more efficient than W_2. Actually, the superiority of W_1 is very pronounced—its relative efficiency is n^2:

$$\text{rel. eff. of } W_1 \text{ to } W_2 = \frac{\dfrac{n\theta^2}{n+2}}{\dfrac{\theta^2}{n(n+2)}} = n^2.$$

QUESTION 5.5.2   Carry out the details to verify the formulas for $E(Y'_1)$, $g_{W_2}(w_2)$, and Var (W_2).

CASE STUDY

5.2

On at least one occasion statistical estimation has been used for espionage purposes. During the Second World War, a very simple statistical procedure was developed for estimating German war production. It was based on *serial numbers* (in our terminology, order statistics) and proved to be highly effective.

Every piece of German equipment, whether it was a V-2 rocket, a tank, or just an automobile tire, was stamped with a serial number that indicated the order in which it was manufactured. If the total number of, say, Mark I tanks produced by a certain date was N, each would bear one of the integers from 1 to N. As the war progressed, some of these numbers became known to the Allies—either by the direct capture of a tank or from records seized when a command post was overrun. The problem was to estimate N, the unknown parameter, using only the sample of "captured" serial numbers, $1 \le Y'_1 < Y'_2 < \cdots < Y'_n \le N$.

As a model, it was assumed that the Y''_i's were one of the $\binom{N}{n}$ possible ordered sets of n integers from 1 to N and that each set was equally probable. That is,

$$P(Y'_1 = y'_1 < Y'_2 = y'_2 < \cdots < Y'_n = y'_n) = \binom{N}{n}^{-1}.$$

The procedure originally proposed for estimating N was to add to the largest order statistic the average "gap" in the Y''_i's. Specifically,

$$W_1 = Y'_n + \frac{1}{n-1} \sum_{i>j} (Y'_i - Y'_j - 1),$$

which reduces to

$$W_1 = Y'_n + \frac{Y'_n - Y'_1}{n-1} - 1.$$

Thus, according to this statistic, if five tanks were captured and they bore the numbers 14, 28, 92, 146, and 298, the estimate for the total number of tanks produced would be 368:

$$w_1 = 298 + \frac{284}{4} - 1 = 368.$$

The details will not be gone into here, but it can be shown that W_1 is unbiased and that

$$\text{Var}(W_1) = \frac{n(N-n)(N+1)}{(n-1)(n+1)(n+2)}.$$

Another possible estimator for N, one more in keeping with what was used in Example 5.8 for the continuous case, would be an unbiased function of Y'_n. If

$$W_2 = \left(\frac{n+1}{n}\right) \cdot Y'_n - 1,$$

it can be shown that $E(W_2) = N$. Also,

$$\text{Var}(W_2) = \frac{(N+1)(N-n)}{n(n+2)} \qquad \text{[see (61) for details]}.$$

Both W_1 and W_2 are very good estimators, but the latter is better, the relative efficiency of W_1 to W_2 being slightly less than 1:

$$\frac{\text{Var}(W_2)}{\text{Var}(W_1)} = 1 - \frac{1}{n^2}.$$

As n increases, though, the precision superiority of W_2 diminishes and, for all practical purposes, the two estimators become equivalent.

Comment. When the war was over and the official records of the Speer Ministry impounded, it was found that estimates derived from procedures similar to the one just described were far more accurate than those based on any other source of information. As an example, the serial-number estimate for German tank production in 1942 was 3400, which was very close to the actual figure. The "official" Allied estimate, based on information culled from intelligence reports and espionage activities, was 18,000. Errors of this magnitude were not uncommon. Often the reason for these inflated estimates was the Nazi propaganda machine. Efforts to create the impression that Germany was much stronger than it really was were highly successful. Only the completely objective serial-number procedure remained unaffected! ∎

QUESTION 5.5.3   Suppose an intelligence officer intends to wait until $n = 6$ serial numbers are available and then use W_2 to estimate N. How many observations would he need to have to be able to estimate N with equal precision using W_1? Assume N is large relative to n.

5.6 MINIMUM-VARIANCE ESTIMATORS: THE CRAMER-RAO LOWER BOUND

The concept of relative efficiency has solved one "problem" but, in so doing, raised another. What it provides is a working criterion for choosing between two competing estimators: given W_1 and W_2, both unbiased, we will elect as the "better" estimator the one whose variance is smaller. What it has *not* given us, though, is any reassurance that even the better of W_1 and W_2 is any good. How do we know, for example, that there is not a W_3 somewhere in the class of unbiased estimators whose variance is much smaller than either W_1's or W_2's? The next theorem offers at least a partial answer to that question in the form of a lower bound for the variance of an unbiased estimator.

> **THEOREM 5.1 (Cramer-Rao lower bound)**   Let $Y_1, Y_2, \ldots, Y_n$ be a random sample from $f_Y(y; \theta)$. Let $W = h(Y_1, Y_2, \ldots, Y_n)$ be any unbiased estimator for θ. If $f_Y(y; \theta)$ is such that
>
> $$\frac{\partial^2}{\partial \theta^2} \int f_Y(y; \theta)\, dy = \int \frac{\partial^2}{\partial \theta^2} f_Y(y; \theta)\, dy,$$
>
> then
>
> $$\mathrm{Var}\,(W) \geq \frac{1}{nE\left\{\left[\dfrac{\partial \ln f_Y(y; \theta)}{\partial \theta}\right]^2\right\}} = \frac{1}{-nE\left[\dfrac{\partial^2 \ln f_Y(y; \theta)}{\partial \theta^2}\right]}.$$

PROOF   We will not present a complete statement of the necessary regularity conditions on $f_Y(y; \theta)$. Neither do we give here a proof of the Cramer-Rao lower bound. For that, the interested reader is referred to (31). We will simply establish the equivalence between its two expressions by verifying that

$$E\left\{\left[\frac{\partial \ln f_Y(y; \theta)}{\partial \theta}\right]^2\right\} = -E\left[\frac{\partial^2 \ln f_Y(y; \theta)}{\partial \theta^2}\right].$$

First, let

$$Z = \frac{\partial \ln f_Y(y; \theta)}{\partial \theta} = \frac{\dfrac{\partial f_Y(y; \theta)}{\partial \theta}}{f_Y(y; \theta)}.$$

Applying the formula for differentiating a product gives

$$\frac{\partial Z}{\partial \theta} = \frac{f_Y(y; \theta)\left[\dfrac{\partial^2 f_Y(y; \theta)}{\partial \theta^2}\right] - \left[\dfrac{\partial f_Y(y; \theta)}{\partial \theta}\right]^2}{[f_Y(y; \theta)]^2}$$

$$= \frac{\dfrac{\partial^2 f_Y(y; \theta)}{\partial \theta^2}}{f_Y(y; \theta)} - Z^2.$$

Now, since

$$\int_{-\infty}^{\infty} f_Y(y; \theta)\, dy = 1,$$

and if the differentiation and integration can be interchanged,

$$\frac{\partial^2}{\partial \theta^2} \int_{-\infty}^{\infty} f_Y(y; \theta)\, dy = \int_{-\infty}^{\infty} \frac{\partial^2 f_Y(y; \theta)}{\partial \theta^2}\, dy = 0.$$

Therefore,

$$0 = \int_{-\infty}^{\infty} \frac{\frac{\partial^2 f_Y(y; \theta)}{\partial \theta^2}}{f_Y(y; \theta)} f_Y(y; \theta)\, dy = E\left[\frac{\frac{\partial^2 f_Y(y; \theta)}{\partial \theta^2}}{f_Y(y; \theta)}\right],$$

from which it follows that

$$E(Z^2) = -E\left(\frac{\partial Z}{\partial \theta}\right).$$

EXAMPLE 5.9. Let $Y_1, Y_2, \ldots, Y_n$, be a random sample from the Bernoulli distribution,

$$f_Y(y; p) = p^y (1 - p)^{1-y}, \qquad y = 0, 1; 0 \le p \le 1.$$

It has already been suggested (see Example 5.5) that

$$W = \frac{Y}{n}$$

be used as an (unbiased) estimator of p, where $Y = \sum_{i=1}^{n} Y_i$ = the total number of successes. Of course,

$$\text{Var}(W) = \frac{1}{n^2} \cdot \text{Var}(Y) = \frac{1}{n^2} \cdot np(1 - p) = \frac{p(1 - p)}{n}.$$

To see how W compares, variancewise, with other unbiased estimators, we will determine the Cramer-Rao lower bound for this particular situation. Note that

$$\ln f_Y(y; p) = y \ln p + (1 - y) \ln (1 - p).$$

Therefore,

$$\frac{\partial \ln f_Y(y; p)}{\partial p} = \frac{y}{p} - \frac{1 - y}{1 - p}$$

and

$$\frac{\partial^2 \ln f_Y(y; p)}{\partial p^2} = -\frac{y}{p^2} - \frac{1 - y}{(1 - p)^2}.$$

Taking the expected value of the second derivative gives

$$E\left[\frac{\partial^2 \ln f_Y(y; p)}{\partial p^2}\right] = -\frac{p}{p^2} - \frac{1 - p}{(1 - p)^2} = -\frac{1}{p(1 - p)}.$$

From Theorem 5.1, then, the Cramer-Rao lower bound becomes

$$\frac{1}{-n\left[-\dfrac{1}{p(1 - p)}\right]} = \frac{p(1 - p)}{n}.$$

Notice in this case that the variance of $W = Y/n$ is *equal* to the Cramer-Rao lower bound. This means that W is "optimal" in the sense that among the class of unbiased estimators, none has a smaller variance.

QUESTION 5.6.1   Let $Y_1, Y_2, \ldots, Y_n$ be a random sample from $f_Y(y; \lambda) = (1/\lambda)e^{-y/\lambda}$, $y > 0$. Find the Cramer-Rao lower bound for the variance of an unbiased estimator for λ. Also, compare the variance of $W = \overline{Y}$ to the Cramer-Rao lower bound.

The notion of optimality raised in Example 5.9 motivates the next two definitions and points up the usefulness of Theorem 5.1. Notice that what is defined here to be the *efficiency* of an unbiased estimator is simply its relative efficiency (in the terminology of Section 5.5) when compared to an estimator whose variance achieves the Cramer-Rao lower bound.

DEFINITION 5.3   Let U denote the class of all unbiased estimators, $W = h(Y_1, Y_2, \ldots, Y_n)$, for a given parameter. We say that W^* is a *best* (or *minimum-variance*) estimator if $W^* \in U$ and

$$\text{Var}(W^*) \leq \text{Var}(W), \qquad \text{for all } W \in U.$$

DEFINITION 5.4   An unbiased estimator, $W = h(Y_1, Y_2, \ldots, Y_n)$, is said to be *efficient* if and only if

$$\text{Var}(W) = \frac{1}{nE\left[\dfrac{\partial \ln f_Y(y; \theta)}{\partial \theta}\right]^2}.$$

Also, the *efficiency* of an unbiased W is defined to be the ratio

$$\text{efficiency of } W = \frac{\dfrac{1}{nE\left[\dfrac{\partial \ln f_Y(y; \theta)}{\partial \theta}\right]^2}}{\text{Var}(W)}.$$

Thus, $W = Y/n$ is an *efficient* estimator for the binomial parameter, p. It is also a *best* estimator. Likewise, the sample mean, when used to estimate the parameter in an exponential distribution, is efficient and best (see Question 5.6.1).

Comment. The designations "efficient" and "best" are not synonymous. If the variance of an unbiased estimator is equal to the Cramer-Rao lower bound (as in the case of the binomial and the exponential), then that estimator by definition is a best estimator. The converse, though, is not always true. There are situations for which the variances of *no* unbiased estimators achieve the Cramer-Rao lower bound. Thus, none of them is *efficient* but one (or more) of them would still be termed *best*. ∎

QUESTION 5.6.2   Show that $W = \overline{Y}$ is an efficient estimator for the occurrence rate, λ, in a Poisson distribution,

$$f_Y(y; \lambda) = \frac{e^{-\lambda}\lambda^y}{y!}, \qquad y = 0, 1, 2, \ldots.$$

5.7 FURTHER PROPERTIES OF ESTIMATORS: CONSISTENCY AND SUFFICIENCY

Although the notions of unbiasedness and efficiency lead to the most basic characterizations of point estimates, there are still deeper properties of W and $g_W(w)$ that merit examination. In this section we will briefly consider two of these: the first, *consistency*, is an asymptotic property of an estimator, and the second, *sufficiency*, is a property related to the amount of "information" a given estimator contains.

To understand consistency—or, for that matter, any asymptotic property of an estimator—it is first necessary to think of W as really being W_n, the nth member of an infinite *sequence* of estimators, $W_1, W_2, \ldots, W_n, \ldots$. For example, if r observations were drawn from the exponential pdf, $f_Y(y; \lambda) = (1/\lambda)e^{-y/\lambda}$, we would probably estimate λ with

$$W = W_r = \frac{1}{r} \sum_{i=1}^{r} Y_i,$$

where W_r itself is gamma distributed with parameters r and λ/r (see Example 5.3). But suppose s observations were taken: then the estimator would become

$$W = W_s = \frac{1}{s} \sum_{i=1}^{s} Y_i,$$

with W_s being gamma distributed with parameters s and λ/s. Thus, we see that the *form* of the estimator remains the same (for different sample sizes) but the behavior of $g_{W_n}(w_n)$ may very well change. This suggests that it might prove instructive to examine the limiting behavior of $g_{W_n}(w_n)$ as n gets large. We may find—to W's credit—that $g_{W_n}(w_n)$ has some very desirable properties *in the limit* that it fails to possess for any finite n.

Consistency is one such desirable property of $g_{W_n}(w_n)$. Roughly speaking, an estimator is consistent if, as n gets large, the probability that W_n lies arbitrarily close to the parameter being estimated becomes, itself, arbitrarily close to 1. This has two immediate implications: (1) W_n is *asymptotically unbiased* (see Review Exercise 5 at the end of this chapter) and (2) Var (W_n) converges to 0.

DEFINITION 5.5 An estimator $W_n = h(Y_1, Y_2, \ldots, Y_n)$ is said to be *consistent* (for θ) if it converges stochastically to θ—that is, if for all $\epsilon > 0$ and $\delta > 0$, there exists an $n(\epsilon, \delta)$ such that
$$P(|W_n - \theta| < \epsilon) > 1 - \delta, \qquad \text{for } n > n(\epsilon, \delta).$$

EXAMPLE 5.10. Let $Y_1, Y_2, \ldots, Y_n$ be a random sample from
$$f_Y(y; \theta) = \frac{1}{\theta}, \qquad 0 < y < \theta,$$

and let $W_n = Y_{\max}$. We know that $Y_{\max}$ is biased (for θ), but it might still be consistent. The question is whether or not there is an $n(\epsilon, \delta)$ large enough so that

$$P(|W_n - \theta| < \epsilon) > 1 - \delta, \qquad \text{for } n > n(\epsilon, \delta).$$

Recall from Example 5.1 that

$$g_{W_n}(w_n) = \frac{n(w_n)^{n-1}}{\theta^n}, \qquad 0 < w_n < \theta.$$

Therefore,

$$P(|W_n - \theta| < \epsilon) = P(\theta - \epsilon < W_n < \theta)$$

$$= \int_{\theta-\epsilon}^{\theta} \frac{n(w_n)^{n-1}}{\theta^n} \, dw_n = \frac{w_n^n}{\theta_n} \Big|_{\theta-\epsilon}^{\theta}$$

$$= 1 - \left(\frac{\theta - \epsilon}{\theta}\right)^n.$$

Note that for any $\epsilon > 0$, it is possible to find an n making $[(\theta - \epsilon)/\theta]^n$ as small as desired (in particular, smaller than δ). By Definition 5.5, then, $W_n = Y_{\max}$ is consistent for θ.

Figure 5.10 shows the situation graphically. As n increases, the shape of $g_{W_n}(w_n)$ changes in such a way that it becomes increasingly concentrated within an ϵ-neighborhood of θ.

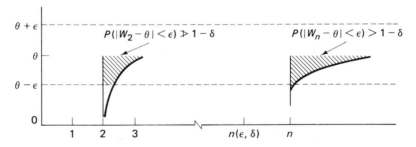

Figure 5.10   The consistency of $Y_{\max}$.

QUESTION 5.7.1   What is the smallest n for which $W_n = Y_{\max}$ has a probability of at least 0.95 of lying within 2% of θ?

EXAMPLE 5.11.   The similarity between the definition of consistency and the statement of Chebyshev's inequality (see Chapter 3) suggests that it might be possible to use the latter to establish the former. As a case in point, consider our usual estimator for the binomial parameter, p:

$$W = Y/n = \text{sample proportion of successes.}$$

Since W is a legitimate random variable and $E(W) = p$, it follows from Chebyshev's inequality that

$$P(|W - p| < \epsilon) > 1 - \frac{\text{Var }(W)}{\epsilon^2}. \tag{5.2}$$

Furthermore, since

$$\text{Var }(W) = \frac{p(1 - p)}{n} \le \frac{1}{4n},$$

we can write Equation 5.2 as

$$P(|W - p| < \epsilon) > 1 - \frac{1}{4n\epsilon^2}.$$

From this latter expression, it can be seen that for $n > (1/4\epsilon^2\delta)$, the probability that W lies in an ϵ-neighborhood of p will be greater than $1 - \delta$. Thus, Y/n is consistent for p.

QUESTION 5.7.2   Is the estimator $W = \bar{Y}$ consistent for the parameter of a Poisson distribution?

QUESTION 5.7.3   Suppose a sample of size n is drawn from an exponential distribution and we choose Y_1, the first observation recorded, to be an estimator for θ. Show that $W = Y_1$ is not consistent.

The next several results deal with the second property of estimators mentioned at the outset of this section, *sufficiency*. For reasons to be outlined shortly, the search for minimum-variance unbiased estimators can always be narrowed to a class of W's that share this particular feature. To develop the idea of sufficiency in any great detail, though, requires considerably more theoretical background than the previous chapters have provided. Consequently, all we can do here is offer a heuristic explanation and present a few simple examples.

If a sample of size n is drawn from some $f_Y(y; \theta)$, we know that the sample space is the totality of n-tuples, $(y_1, y_2, \ldots, y_n)$. It is also true that defining an estimator, W, has the effect of partitioning this sample space into a set of mutually exclusive and exhaustive subsets. For example, if two observations are drawn from a Poisson distribution one possible estimator for the occurrence rate, λ, is

$$W = \bar{Y} = \frac{1}{2} \sum_{i=1}^{2} Y_i.$$

Note that W will equal, say, 3 for any of the following 2-tuples: (0, 6), (1, 5), (2, 4), (3, 3), (4, 2), (5, 1), or (6, 0); likewise, it will equal 2.5 if (0, 5), (1, 4), (2, 3), (3, 2), (4, 1), or (5, 0) is the sample outcome. Each possible outcome, then, can be put into a subset according to the value of W it produces.

Basically, whether or not W is sufficient for λ depends on whether the particular 2-tuple observed—say, (1, 5)—conveys any more information about λ than what is already contained in the statement "$W = 3$." If knowing which partition set an outcome belongs to reveals as much information about the unknown parameter as the particular outcome, itself, then the method of partitioning—that is, W—is said to be *sufficient* (for λ).

The binomial distribution can be used to illustrate this idea in a little more detail. Suppose that in four Bernoulli trials, two successes and two failures—in that order—are observed. That is, the sample outcome is (s, s, f, f) or, in terms of the usual Bernoulli random variable, (1, 1, 0, 0). From this, the conventional estimate for p would be 0.5:

$$w = \frac{\text{number of successes}}{n} = \frac{2}{4} = 0.5.$$

Suppose, now, we look at the conditional distribution of (1, 1, 0, 0) given that $w = 0.5$. By definition,

$$P((1, 1, 0, 0)|w = 0.5) = \frac{P((1, 1, 0, 0) \text{ and } W = 0.5)}{P(W = 0.5)}$$

$$= \frac{P(1, 1, 0, 0)}{P(\text{exactly 2 successes in 4 trials})}$$

$$= \frac{p^2(1 - p)^2}{\binom{4}{2}p^2(1 - p)^2} = \binom{4}{2}^{-1}$$

$$= P((0, 1, 0, 1)|w = 0.5)$$

$$= P((0, 0, 1, 1)|w = 0.5) \text{ and so on.}$$

Notice that these conditional probabilities are independent of p. That is, knowing that the particular outcome is, say, $(1, 1, 0, 0)$ adds nothing to our information about p that we do not already know, having been told that $w = 0.5$. (Of course, it follows that any one-to-one function of W, say nW, $W + 3$, and so on, will enjoy this same property.)

> **DEFINITION 5.6** Let $Y_1, Y_2, \ldots, Y_n$ be a random sample from $f_Y(y; \theta)$. The statistic $W = h(Y_1, Y_2, \ldots, Y_n)$ is said to be *sufficient* for θ if, for all θ and all possible sample points, the conditional pdf of $Y_1, Y_2, \ldots, Y_n$ given w does not depend on θ, either in the function itself or in the function's domain. In the discrete case, W is sufficient if
>
> $$P(Y_1 = y_1, \ldots, Y_n = y_n | w = h(y_1, y_2, \ldots, y_n))$$
>
> does not involve θ.

In establishing that a particular statistic is sufficient, we do not usually use Definition 5.6 directly. Various factorization criteria are available that are much easier to work with. One of them, the Fisher-Neyman criterion, is outlined in Theorem 5.2. Its proof will not be given.

> **THEOREM 5.2 (Fisher–Neyman criterion)** Let $Y_1, Y_2, \ldots, Y_n$ be a random sample from $f_Y(y; \theta)$. Then $W = h(Y_1, Y_2, \ldots, Y_n)$ is a sufficient statistic for θ if and only if the joint density of the Y_i's factors into the product of the pdf of W and a second function that does not depend on θ. That is, W is sufficient if and only if
>
> $$f_Y(y_1; \theta) \cdots f_Y(y_n; \theta) = g_W(h(y_1, \ldots, y_n); \theta) \cdot s(y_1, \ldots, y_n).$$

EXAMPLE 5.12.   It is a simple matter to use the Fisher-Neyman criterion to show that

$$W = Y = \sum_{i=1}^{n} Y_i = \text{total number of successes in } n \text{ trials}$$

is a sufficient statistic for the binomial parameter, p. The joint probability function of the Y_i's is the product of n Bernoullis,

$$f_Y(y_1; p) \cdots f_Y(y_n; p) = p^{\sum_{i=1}^{n} y_i} (1 - p)^{n - \sum_{i=1}^{n} y_i},$$

and Y is binomial with parameters n and p,

$$g_Y(y; p) = \binom{n}{y} p^y (1 - p)^{n-y} = \binom{n}{\sum_{i=1}^{n} y_i} p^{\sum_{i=1}^{n} y_i} (1 - p)^{n - \sum_{i=1}^{n} y_i}.$$

Therefore, by choosing

$$s(y_1, y_2, \ldots, y_n) = \binom{n}{\sum_{i=1}^{n} y_i}^{-1},$$

we have that

$$f_Y(y_1; p) \cdots f_Y(y_n; p) = g_W(w; p) \cdot s(y_1, y_2, \ldots, y_n),$$

proving that $W = Y$ is sufficient for p.

Comment. In any real problem, we would not use Y itself to estimate p but, rather, Y/n, the latter being unbiased. If Y is sufficient for p, though, so is Y/n. ∎

QUESTION 5.7.4   Show that $W = \bar{Y}$ is sufficient for θ in the exponential pdf, $f_Y(y; \theta) = (1/\theta)e^{-y/\theta}$, $0 < y < \infty$.

EXAMPLE 5.13.   As a final example of sufficiency, we will prove that the largest order statistic, Y_{max}, is a sufficient statistic for the parameter of a uniform distribution, $f_Y(y; \theta) = 1/\theta$, $0 < y < \theta$. First of all, note that the joint pdf of the y_i's is simply $(1/\theta)^n$ and recall that

$$g_W(w) = \begin{cases} \dfrac{nw^{n-1}}{\theta^n}, & \text{for } 0 < w < \theta \\ 0, & \text{otherwise,} \end{cases}$$

where $W = Y_{max}$. If we write

$$f_Y(y_1; \theta) \cdots f_Y(y_n; \theta) = \left(\frac{1}{\theta}\right)^n$$

$$= \frac{nw^{n-1}}{\theta^n} \cdot \frac{1}{nw^{n-1}}$$

$$= g_W(w) \cdot s(y_1, y_2, \ldots, y_n),$$

it follows that Y_{max} is a sufficient statistic.

QUESTION 5.7.5   Show that $W = Y_{min}$ is a sufficient statistic for the threshold parameter θ in the exponential pdf

$$f_Y(y; \theta) = e^{-(y-\theta)}, \qquad \theta < y < \infty.$$

In concluding this section, it is appropriate to comment briefly on the way sufficiency relates to two of the other properties discussed earlier—unbiasedness and efficiency. Basically, the importance of sufficiency derives from the fact that the minimum-variance unbiased estimator for any parameter will necessarily be a function of a sufficient statistic. It can be shown that contained in the set, U, of unbiased estimators for a given parameter is a subset, U_S, whose elements are functions of sufficient statistics. Furthermore, U_S has the property that each of its members has a smaller variance than every unbiased estimator in its complement. This means that in looking for a minimum-variance unbiased estimator we need only be concerned with W's that are functions of sufficient statistics. To further simplify matters, in many situations [depending on the nature of $f_Y(y; \theta)$] it can be proved that U_S has only *one* member. Thus, to construct a minimum-variance estimator (in these situations) we need simply find a sufficient statistic and perform whatever transformations are necessary to make it unbiased.

Comment. Explicit consideration of point estimation can be traced back to the time of Galileo, perhaps even earlier. Much of the structure that we now associate with the subject, however, was formulated in the early years of the present century by Sir Ronald A. Fisher. It was Fisher who introduced the notions of unbiasedness, efficiency, and sufficiency. ∎

5.8 ESTIMATING PARAMETERS:
THE METHOD OF MAXIMUM LIKELIHOOD
AND THE METHOD OF MOMENTS

Thus far, our discussion has centered exclusively on the *properties* of estimators—how those estimators were arrived at has not been a concern. Given that an estimator for, say, the Poisson parameter is $W = \bar{Y}$, we have simply tried to establish, after the fact, whether or not it satisfies certain desirable criteria. Is it unbiased? efficient? consistent? sufficient? What is perhaps the more fundamental question, though, still remains: how was W chosen to be $\bar{Y}$ in the first place, as opposed to $\prod_{i=1}^{n} Y_i$, $Y_{\max}$, or any of the other infinitely many possible statistics?

In this section we answer that question. Two different methods of estimation will be described, the *method of maximum likelihood* and the *method of moments*. While these are not the only techniques for finding estimators, they are among the most common. From a theoretical standpoint, the method of maximum likelihood is particularly important, having as it does the property that the estimators it produces are always functions of sufficient statistics. From a practical standpoint, though, it is sometimes difficult to apply. For certain $f_Y(y; \theta)$ the equations that arise in carrying out the method of maximum likelihood are nonlinear and have solutions that can only be approximated. In these situations the method of moments can sometimes be of real value. Although "moment" estimators are not necessarily sufficient, efficient, or even unbiased, they are, nevertheless, "reasonable" and they can often be obtained with a minimum of mathematical difficulty.

The first procedure to be discussed will be the method of maximum likelihood. Let $Y_1, Y_2, \ldots, Y_n$ be a random sample from $f_Y(y; \theta)$ and L its joint pdf:

$$L = f_{Y_1, Y_2, \ldots, Y_n}(y_1, y_2, \ldots, y_n; \theta) = f_Y(y_1; \theta) \cdots f_Y(y_n; \theta)$$

$$= \prod_{i=1}^{n} f_Y(y_i; \theta).$$

Up to now, the joint pdf of a random sample has always been thought of as a function *of the sample*. That is, $L = L(y_1, y_2, \ldots, y_n) = L(y_1, y_2, \ldots, y_n; \theta)$, meaning that θ is fixed and the y_i's are variables. For the purposes of finding estimators, though, it will make sense to reverse the roles of the parameter and the data and think of L as a function of θ—with the y_i's considered fixed. Accordingly, we will write

$$L = L(\theta) = L(\theta; y_1, y_2, \ldots, y_n) = \prod_{i=1}^{n} f_Y(y_i; \theta).$$

> **DEFINITION 5.7** Let $Y_1, Y_2, \ldots, Y_n$ be a random sample from $f_Y(y; \theta)$. If L, the joint pdf of the Y_i's, is thought of as being a function of θ and if the y_i's are thought of as being fixed, then
>
> $$L = L(\theta) = \prod_{i=1}^{n} f_Y(y_i; \theta)$$
>
> is called the *likelihood function*.

A simple numerical example will illustrate how we intend to make use of $L(\theta)$. Suppose three observations—$y_1 = 2.0$, $y_2 = 4.0$, and $y_3 = 3.6$—are drawn from $f_Y(y; \theta) = (1/\theta)e^{-y/\theta}$, $y > 0$. By definition, the corresponding likelihood function is

$$L(\theta) = \prod_{i=1}^{3} \left(\frac{1}{\theta}\right)e^{-y_i/\theta}$$

$$= \left(\frac{1}{\theta^3}\right)e^{-(1/\theta)\sum_{i=1}^{3} y_i}$$

$$= \left(\frac{1}{\theta^3}\right)e^{-9.6/\theta}$$

(see Figure 5.11). Notice that $L(\theta)$ is greatest when $\theta = 3.2$. That is, for the given three observations, the joint pdf is maximized when the parameter is assigned the value 3.2. This amounts to saying that, in terms of the joint pdf, the value $\theta = 3.2$ "best explains" or is "most compatible with" the data. It would not be unreasonable, then, to set W equal to 3.2 (which, in this case, equals $\bar{y}$). Definition 5.8 summarizes the "principle" of using $L(\theta)$ to locate a suitable estimate for θ.

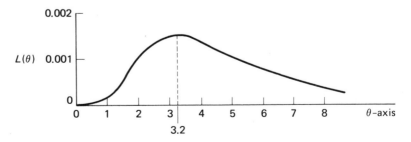

Figure 5.11 Graph of $L(\theta)$.

> **DEFINITION 5.8 (Maximum-likelihood estimation)**   Let $Y_1, Y_2, \ldots,$
> Y_n be a random sample from $f_Y(y; \theta)$ and let $L(\theta)$ be the corresponding likeli-
> hood function. As a general procedure for constructing estimators, we will look
> for the w that maximizes $L(\theta)$. Any value that does—that is, any w for which
>
> $$L(w) \geq L(\theta), \qquad \text{for all } \theta \neq w,$$
>
> is said to be a *maximum-likelihood estimate* (or MLE) for θ. We will sometimes
> write $w = \hat{\theta}$.

In many cases of interest, $L(\theta)$ is differentiable and its maximization is a familiar
calculus problem: the basic procedure is to differentiate $L(\theta)$ with respect to θ,
set the derivative equal to 0, and solve for θ. Provided $L''(\theta) < 0$, the solution will
be the maximum-likelihood estimator. (An exception is noted in Example 5.16.)

Comment. Often it will be easier to maximize $\ln L(\theta)$ rather than $L(\theta)$. Of course,
the monotonicity of the natural logarithm insures that the same w maximizes both
functions.   ∎

Comment. If the presumed data model is a function of k unknown parameters,
then the vector of maximum-likelihood estimates, $(w_1, w_2, \ldots, w_k)$, is gotten by
solving simultaneously the system of equations,

$$\frac{\partial L(\theta_1, \theta_2, \ldots, \theta_k)}{\partial \theta_i} = 0, \qquad i = 1, 2, \ldots, k. \quad ∎$$

EXAMPLE 5.14.   Let $Y_1, Y_2, \ldots, Y_n$ be a random sample from the Poisson distribu-
tion, $f_Y(y; \lambda) = e^{-\lambda}\lambda^y/y!$, $y = 0, 1, 2, \ldots$. The corresponding likelihood function, L,
and its natural log are

$$L(\lambda; y_1, y_2, \ldots, y_n) = \prod_{i=1}^{n} \frac{e^{-\lambda}\lambda^{y_i}}{y_i!} = \frac{e^{-n\lambda}\lambda^{\sum_{i=1}^{n} y_i}}{\prod_{i=1}^{n} y_i!}$$

and

$$\ln L(\lambda; y_1, y_2, \ldots, y_n) = -n\lambda + \left(\sum_{i=1}^{n} y_i\right) \ln \lambda - \ln \left(\prod_{i=1}^{n} y_i!\right).$$

Setting the derivative of $\ln L$ (with respect to λ) equal to 0 gives

$$-n + \frac{1}{\lambda} \sum_{i=1}^{n} y_i = 0. \tag{5.3}$$

Let w denote the value of λ that satisfies Equation 5.3. Then

$$w = \frac{1}{n} \sum_{i=1}^{n} y_i = \bar{y}.$$

Thus, the maximum-likelihood estimate for the parameter of the Poisson is the one
we have been using all along. Note that $w = \bar{y}$ locates the *maximum* of $\ln L$ and not
the minimum, because the second derivative of $\ln L$ is $-(1/\lambda^2) \sum_{i=1}^{n} y_i$, a number always
less than or equal to 0.

QUESTION 5.8.1   Find the maximum-likelihood estimate for the parameter, p, in a binomial distribution. Assume that a random sample $y_1, y_2, \ldots, y_n$ of Bernoulli random variables has been observed. Let $y = y_1 + y_2 + \cdots + y_n$ be the value of the corresponding binomial random variable.

CASE STUDY

5.3

Recall the data of Case Study 4.4, giving the annual number of outbreaks of war from 1500 to 1931 (Table 5.2).

TABLE 5.2   Outbreaks of war

Number of Outbreaks in a Given Calendar Year	Number of Years
0	223
1	142
2	48
3	15
4	4
5+	0
	432

If Y denotes the number of wars started in a given year—and if it is assumed that the distribution of Y can be described by a Poisson—then

$$P(Y = y) = \frac{e^{-\lambda}\lambda^y}{y!}, \qquad y = 0, 1, 2, \ldots.$$

The maximum-likelihood estimate for λ in this case is 0.69:

$$w = \bar{y} = \frac{1}{432}(223 \cdot 0 + 142 \cdot 1 + 48 \cdot 2 + 15 \cdot 3 + 4 \cdot 4)$$

$$= \frac{299}{432} = 0.69 \text{ "new" wars/year.}$$

The specific model being proposed, then, is

$$f_Y(y; \lambda) = f_Y(y; 0.69) = P(Y = y) = \frac{e^{-0.69}(0.69)^y}{y!}, \qquad y = 0, 1, \ldots.$$

As Table 4.6 makes clear, the agreement between the observed war distribution and the Poisson predictions based on a w of 0.69 is remarkably good.

QUESTION 5.8.2   Estimate the probability that for the next *two* years no wars will break out anywhere in the world. What assumptions are you making?

EXAMPLE 5.15.   If a series of independent Bernoulli trials is terminated with the occurrence of the first success, the probability function for K, the length of the series, will be given by the *geometric distribution*,

$$f_K(k; p) = P(K = k) = (1 - p)^{k-1}p, \qquad k = 1, 2, \ldots,$$

where p is the (unknown) probability of success at any trial. If n such series are observed, the data will consist of n lengths, $k_1, k_2, \ldots, k_n$. To get the maximum-likelihood estimator for p, note, first of all, that

$$L(p; k_1, k_2, \ldots, k_n) = L(p) = (1 - p)^{\sum_{i=1}^{n} k_i - n} p^n. \tag{5.4}$$

Taking the ln of Equation 5.4 gives

$$\ln L(p) = \left(\sum_{i=1}^{n} k_i - n\right) \ln (1 - p) + n \ln p,$$

from which it follows that

$$\frac{d}{dp}[\ln L(p)] = \left(\frac{1}{1 - p}\right)\left(n - \sum_{i=1}^{n} k_i\right) + \frac{n}{p}.$$

Let w denote the value of p for which $d/dp\,[\ln L(p)] = 0$. It can be easily seen that

$$w = \frac{n}{\sum_{i=1}^{n} k_i} = \frac{1}{\bar{k}}.$$

Again, notice that

$$\frac{d^2}{dp^2}[\ln L(p)] = \frac{1}{(1 - p)^2}\left(n - \sum_{i=1}^{n} k_i\right) - \frac{1}{p^2} < 0,$$

guaranteeing that the differentiation has, indeed, produced a maximum.

QUESTION 5.8.3   A statistically minded fraternity man keeps records on how many girls he has to ask before one of them says she will be his date for a Saturday football game. His school plays five home games and his five acceptances come on the third, sixth, fourth, second, and ninth girls he asks. Assume that the probability, p, that any girl he asks will accept his invitation is constant from girl to girl. Estimate p.

CASE STUDY

5.4

In Case Study 4.9, the geometric distribution was suggested as a possible model for the number of utterances per song bout for the North American cardinal. Table 5.3 reproduces the original data.

TABLE 5.3   Utterances per bout

Number of Utterances Per Bout	Frequency
1	132
2	52
3	34
4	9
5	7
6	5
7	5
8+	6
	250

If K denotes the length of a bout, what is being proposed is that

$$f_K(k;p) = P(K = k) = (1 - p)^{k-1}p, \qquad k = 1, 2, \ldots.$$

In this context, p is the probability that a given utterance will terminate a bout.

Note from Table 5.3 that the average bout length is 2.09 (if all the 8+'s are set equal to 8):

$$\bar{k} = \tfrac{1}{250}(132 \cdot 1 + 52 \cdot 2 + \cdots + 6 \cdot 8) = \tfrac{522}{250}$$
$$= 2.09.$$

By the result established in Example 5.15, the maximum-likelihood estimator for p is

$$W = \frac{n}{\sum\limits_{i=1}^{n} K_i},$$

or, in this case, $w = 1/2.09 = 0.48$. Table 5.4 shows the corresponding expected frequencies $[= 250 \cdot f_K(k;0.48)]$.

TABLE 5.4   Observed and expected frequencies

Number of Utterances Per Bout	Observed Frequency	Expected Frequency
1	132	120.0
2	52	62.4
3	34	32.4
4	9	16.9
5	7	8.8
6	5	4.6
7	5	2.4
8+	6	2.5
	250	250.0

Comment. Keep in mind that finding the maximum-likelihood estimator in no way validates our choice of $f_K(k; p)$. In this particular situation, the suitability of the geometric model is certainly open to question. Table 5.4 shows there is considerable disagreement between the observed and expected bout-length frequencies for certain values of k. ▮

QUESTION 5.8.4   How might the disagreement between the "observed frequency" and "expected frequency" columns be quantified? Suggest a formula.

EXAMPLE 5.16.   Maximum-likelihood estimators are not always gotten by setting a derivative equal to 0. Consider, for instance, a sample of size n drawn from the uniform distribution over $(0, \theta)$. The likelihood function is

$$L(\theta; y_1, y_2, \ldots, y_n) = \begin{cases} (1/\theta)^n, & \text{for } 0 \leq y_i \leq \theta \\ 0, & \text{otherwise,} \end{cases}$$

and, since each y_i must be less than or equal to θ, the range of θ extends from $y_{\max}$ to infinity. To find the θ that maximizes $L(\theta)$ we need simply graph $(1/\theta)^n$. It is clear by inspection that, being monotonic, $L(\theta)$ is maximized at its left endpoint (Figure 5.12). Thus,

$$w = y_{\max}$$

(recall Example 5.2).

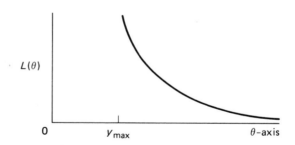

Figure 5.12   Graph of $(1/\theta)^n$.

QUESTION 5.8.5   Find the maximum-likelihood estimator for θ in the pdf $f_Y(y; \theta) = e^{-(y-\theta)}$, $y > \theta$. Assume a random sample of size n has been drawn.

A second procedure for estimating parameters, one that is often more tractable than the method of maximum likelihood in situations where $f_Y(y; \theta_1, \theta_2, \ldots, \theta_k)$

has a complicated form, is the *method of moments*. Using this method, the set of unknown parameters is estimated by equating the theoretical moments of Y to its corresponding sample moments.

Recall that the first k moments of Y, if they exist, are given by

$$\mu_{(j)} = E(Y^j) = \int_{-\infty}^{\infty} y^j f_Y(y; \theta_1, \theta_2, \ldots, \theta_k)\, dy, \qquad j = 1, 2, \ldots, k.$$

This implies that the $\mu_{(j)}$'s will be functions of the θ_j's. That is, we will be able to write

$$\mu_{(1)} = \mu_1(\theta_1, \theta_2, \ldots, \theta_k),$$

$$\mu_{(2)} = \mu_2(\theta_1, \theta_2, \ldots, \theta_k),$$

$$\vdots \qquad \qquad \vdots$$

$$\mu_{(k)} = \mu_k(\theta_1, \theta_2, \ldots, \theta_k).$$

Analogs to the $\mu_{(j)}$'s can be defined for the y_i's. Specifically, we will let the first k *sample* moments be denoted

$$m_{(j)} = \frac{1}{n} \sum_{i=1}^{n} y_i^j, \qquad j = 1, 2, \ldots, k.$$

These, of course, should be approximately equal to the corresponding $\mu_{(j)}$'s if the presumed $f_Y(y; \theta_1, \theta_2, \ldots, \theta_k)$ is a suitable model. It is precisely this observation that motivates the method of moments. Definition 5.9 gives the details.

DEFINITION 5.9 (Moment estimation)   Let $Y_1, Y_2, \ldots, Y_n$ be a random sample from $f_Y(y; \theta_1, \theta_2, \ldots, \theta_k)$. Let $\mu_{(j)}$ and $m_{(j)}, j = 1, 2, \ldots, k$, be the first k theoretical and sample moments, respectively. As a general procedure for estimating the θ_i's, we will solve simultaneously the system of equations,

$$\mu_{(j)} = \mu_j(\theta_1, \theta_2, \ldots, \theta_k) = m_{(j)}, \qquad j = 1, 2, \ldots, k.$$

The solutions, $w_1, w_2, \ldots, w_k$, will be called the method of moments estimates. By convention, w_i is often replaced by the symbol $\hat{\theta}_i$.

Comment. The method of moments was first proposed near the turn of the century by the great British statistician, Karl Pearson. The method of maximum likelihood goes back much further: both Gauss and Daniel Bernoulli made use of the technique—the latter as early as 1777 (127). Fisher, though, in the early years of the

twentieth century, was the first to make a thorough study of the method's properties, and the procedure is often credited to him. ∎

EXAMPLE 5.17.   Let $Y_1, Y_2, \ldots, Y_n$ be a random sample from

$$f_Y(y; \theta) = (\theta + 1)y^\theta, \qquad 0 < y < 1.$$

Suppose we want to find the method of moments estimate for θ. First, a simple integration gives $\mu_{(1)}$:

$$\mu_{(1)} = E(Y) = \int_0^1 y \cdot (\theta + 1)y^\theta \, dy$$

$$= \frac{\theta + 1}{\theta + 2}.$$

Then, setting $\mu_{(1)}$ equal to the first sample moment, we have

$$\frac{\theta + 1}{\theta + 2} = \frac{1}{n} \sum_{i=1}^{n} y_i = \bar{y}. \qquad (5.5)$$

Let w denote the value of θ that satisfies Equation 5.5. Then w is the method-of-moments
estimate:

$$w = \frac{2\bar{y} - 1}{1 - \bar{y}}.$$

QUESTION 5.8.6   Find the maximum-likelihood estimator for θ based on a random sample of size n
taken from this same pdf. Evaluate both estimators for the sample $y_1 = 0.5$, $y_2 = 0.4$,
and $y_3 = 0.9$.

CASE STUDY

5.5   Although hurricanes generally strike only the eastern and southern
coastal regions of the United States, they do occasionally sweep
inland before completely dissipating. The U.S. Weather Bureau confirms that in the period from 1900 to 1969 a total of 36 hurricanes
moved as far as the Appalachians. In Table 5.5 are listed the maximum
24-hour precipitation levels recorded for those 36 storms during the
time they were over the mountains (68).

TABLE 5.5 Maximum 24-hour precipitation recorded
for 36 inland hurricanes (1900–1969)

Year	Name	Location	Maximum Precipitation (inches)
1969	Camille	Tye River, Va.	31.00
1968	Candy	Hickley, N.Y.	2.82
1965	Betsy	Haywood Gap, N.C.	3.98
1960	Brenda	Cairo, N.Y.	4.02
1959	Gracie	Big Meadows, Va.	9.50
1957	Audrey	Russels Point, Ohio	4.50
1955	Connie	Slide Mt., N.Y.	11.40
1954	Hazel	Big Meadows, Va.	10.71
1954	Carol	Eagles Mere, Pa.	6.31
1952	Able	Bloserville 1-N, Pa.	4.95
1949		North Ford #1, N.C.	5.64
1945		Crossnore, N.C.	5.51
1942		Big Meadows, Va.	13.40
1940		Rhodhiss Dam, N.C.	9.72
1939		Caesars Head, S.C.	6.47
1938		Hubbardston, Mass.	10.16
1934		Balcony Falls, Va.	4.21
1933		Peekamoose, N.Y.	11.60
1932		Caesars Head, S.C.	4.75
1932		Rockhouse, N.C.	6.85
1929		Rockhouse, N.C.	6.25
1928		Roanoke, Va.	3.42
1928		Caesars Head, S.C.	11.80
1923		Mohonk Lake, N.Y.	0.80
1923		Wappinger Falls, N.Y.	3.69
1920		Landrum, S.C.	3.10
1916		Altapass, N.C.	22.22
1916		Highlands, N.C.	7.43
1915		Lookout Mt., Tenn.	5.00
1915		Highlands, N.C.	4.58
1912		Norcross, Ga.	4.46
1906		Horse Cove, N.C.	8.00
1902		Sewanee, Tenn.	3.73
1901		Linville, N.C.	3.50
1900		Marrobone, Ky.	6.20
1900		St. Johnsbury, Vt.	0.67

Figure 5.13 shows a histogram of the data : its skewed shape suggests that *Y*, the maximum 24-hour precipitation, might be well approximated by a member of the gamma family,

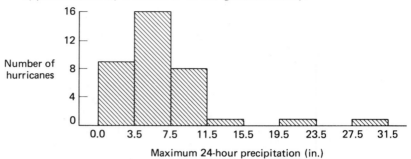

Figure 5.13   Inland hurricane precipitation.

$$f_Y(y; \alpha\ \beta) = \frac{1}{\alpha^\beta \Gamma(\beta)} y^{\beta-1} e^{-y/\alpha}, \qquad y > 0.$$

Here, α and β are the parameters to be estimated. The complexity of $f_Y(y; \alpha, \beta)$, though, makes the method of maximum likelihood unwieldy. As an alternative, we will find the method-of-moments estimates.

Recall from Theorem 4.10 that the first two moments of a gamma distribution are given by

$$\mu_{(1)} = E(Y) = \alpha\beta$$

and

$$\mu_{(2)} = E(Y^2) = \alpha^2 \beta(\beta + 1).$$

For the data of Table 5.13 the first two sample moments are

$$m_{(1)} = \frac{1}{36} \sum_{i=1}^{36} y_i = \frac{1}{36}(262.35)$$

$$= 7.29$$

and

$$m_{(2)} = \frac{1}{36} \sum_{i=1}^{36} y_i^2 = \frac{1}{36}(3081.2177)$$

$$= 85.59.$$

The system of equations that derives from setting the $\mu_{(j)}$'s equal to their corresponding $m_{(j)}$'s reduces to

$$\alpha\beta = 7.29,$$

$$\alpha^2 \beta(\beta + 1) = 85.59.$$

Solving the first for β and substituting the result into the second gives

$$\alpha^2 \left(\frac{7.29}{\alpha}\right)\left(\frac{7.29}{\alpha} + 1\right) = 7.29\alpha + 53.14 = 85.59.$$

Therefore,

$$\hat{\alpha} = \frac{85.59 - 53.14}{7.29} = 4.45$$

and

$$\hat{\beta} = \frac{7.29}{\hat{\alpha}} = 1.64.$$

Figure 5.14 shows the fitted model

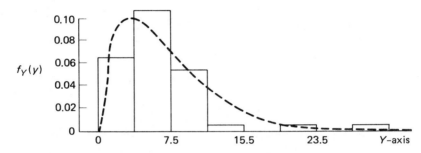

Figure 5.14   Original data with fitted model.

$$f_Y(y; 4.45, 1.64) = \frac{1}{4.45^{1.64}\Gamma(1.64)} y^{0.64} e^{-y/4.45}$$

superimposed over the original data.

Considering the relatively small number of observations in the sample, the agreement is quite good. This would come as no surprise to a meteorologist: the gamma distribution is a very frequently used model for describing precipitation levels.

QUESTION 5.8.7   Find the method-of-moments estimates for μ and σ^2 based on a random sample of size n taken from an $N(\mu, \sigma^2)$ distribution.

5.9 INTERVAL ESTIMATION

It has already been mentioned that all point estimates, no matter how many desirable properties such as unbiasedness and sufficiency they may possess, share the same fundamental weakness: they provide no indication of their inherent precision. We know, for example, that $W = \bar{Y}$ is the "best" estimator for the Poisson parameter. But suppose a sample of size n is taken from a Poisson distribution and it is found that $w = \bar{y} = 3.8$. What exactly does this tell us about the true value of λ? Can we feel reasonably certain, for example, that λ lies somewhere *close* to w—say, in the interval from 3.7 to 3.9? Or, on the other hand, is W so variable that there is a good chance that $|W - \lambda|$ is fairly large?

These are questions any experimenter would be likely to raise; unfortunately, given only the fact that $w = 3.8$, there is no way to answer them. Clearly, what is needed is to combine a point estimate's numerical value with some sort of statement about its variability. This can be done in several ways. Perhaps the most obvious is simply to write the estimate as $w \pm \sqrt{\text{Var}(W)}$, thus letting the magnitude of the standard deviation reflect the precision of W. An even more informative approach, though, is a format known as a *confidence interval*.

We will introduce the notion of a confidence interval with an example. Consider again the problem of estimating the parameter of a uniform distribution with $W = [(n + 1)/n] \cdot Y_{\max}$. We know that W can range from 0 to $[(n + 1)/n]\theta$, and that it does so according to the pdf

$$g_W(w) = \frac{n^{n+1}}{(n + 1)^n} \frac{w^{n-1}}{\theta^n}, \qquad \text{for } 0 \leq w \leq \theta$$

(see Example 5.2). Note that with $g_W(w)$ specified, it is possible to solve two different kinds of problems related to W: (1) we can find the probability that W lies between two specified numbers and (2) we can find the numbers that "contain" W with a specified probability. The latter type will lead us to the definition of a confidence interval.

As an illustration, suppose we want to find two limits, a and b, for which

$$P(a < W < b) = 0.90$$

and

$$P(W \leq a) = P(W \geq b) = 0.05.$$

That is, a and b are the numbers along the W-axis that "cut off" areas of 0.05 in either tail of $g_W(w)$ (Figure 5.15). By inspection, we see that a is the solution of

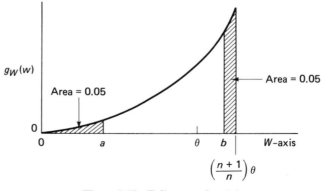

Figure 5.15   Tail areas of $g_W(w)$.

$$\int_0^a \frac{n^{n+1}}{(n+1)^n} \frac{w^{n-1}}{\theta^n} \, dw = 0.05.$$

After integrating,

$$\frac{n^{n+1}}{(n+1)^n} \frac{a^n}{n\theta^n} = 0.05,$$

implying that

$$a = a(\theta) = \sqrt[n]{0.05}\left(\frac{n+1}{n}\right)\theta.$$

Similarly, for the right-hand tail of $g_W(w)$,

$$\int_b^{[(n+1)/n]\theta} \frac{n^{n+1}}{(n+1)^n} \frac{w^{n-1}}{\theta^n} \, dw = 0.05,$$

which gives

$$b = b(\theta) = \sqrt[n]{0.95}\left(\frac{n+1}{n}\right)\theta.$$

Putting these results together gives the two limits we set out to find:

$$P\left[\sqrt[n]{0.05}\left(\frac{n+1}{n}\right)\theta < W < \sqrt[n]{0.95}\left(\frac{n+1}{n}\right)\theta\right] = 0.90.$$

Our real interest, though, will not be in the interval $[a(\theta), b(\theta)]$ but in one of its related forms, where the endpoints do not involve θ. Mathematically, the statement

$$P\left[\sqrt[n]{0.05}\left(\frac{n+1}{n}\right)\theta < W < \sqrt[n]{0.95}\left(\frac{n+1}{n}\right)\theta\right] = 0.90$$

is the same as

$$P\left[\frac{W}{\sqrt[n]{0.95}\left(\frac{n+1}{n}\right)} < \theta < \frac{W}{\sqrt[n]{0.05}\left(\frac{n+1}{n}\right)}\right] = 0.90.$$

This second formulation, though, has a much different interpretation. Imagine that a very large number of samples of size n have been drawn from the uniform distribution over $(0, \theta)$ and for each one, the interval

$$\left(\frac{w}{\sqrt[n]{0.95}\left(\frac{n+1}{n}\right)}, \frac{w}{\sqrt[n]{0.05}\left(\frac{n+1}{n}\right)}\right)$$

is computed. What the second statement says is that 90% of these intervals will include (or contain) the true θ as an interior point. For this reason, the set of estimates

$$\left(\frac{w}{\sqrt[n]{0.95}\left(\frac{n+1}{n}\right)}, \frac{w}{\sqrt[n]{0.05}\left(\frac{n+1}{n}\right)}\right)$$

is called a *90% confidence interval* (for θ).

5.6

Table 5.6 lists 30 random samples of size $n = 5$ drawn from the particular uniform distribution having θ equal to 1. Listed next to

TABLE 5.6   A sampling experiment

Sample	y_{max}	w	90% Interval	Contains θ?
(0.90, 0.90, 0.73, 0.75, 0.54)	0.90	1.08	(0.91, 1.64)	Yes
(0.08, 0.28, 0.53, 0.91, 0.89)	0.91	1.09	(0.92, 1.66)	Yes
(0.77, 0.19, 0.21, 0.51, 0.99)	0.99	1.19	(1.00+, 1.80)	No
(0.33, 0.85, 0.84, 0.56, 0.65)	0.85	1.02	(0.86, 1.55)	Yes
(0.38, 0.37, 0.97, 0.21, 0.73)	0.97	1.16	(0.98, 1.76)	Yes
(0.07, 0.60, 0.83, 0.10, 0.39)	0.83	1.00	(0.84, 1.51)	Yes
(0.59, 0.38, 0.30, 0.65, 0.27)	0.65	0.78	(0.66, 1.18)	Yes
(0.91, 0.68, 0.48, 0.06, 0.10)	0.91	1.09	(0.92, 1.66)	Yes
(0.12, 0.21, 0.19, 0.67, 0.60)	0.67	0.80	(0.67, 1.22)	Yes
(0.53, 0.24, 0.83, 0.16, 0.60)	0.83	1.00	(0.84, 1.51)	Yes
(0.32, 0.51, 0.47, 0.20, 0.66)	0.66	0.79	(0.67, 1.20)	Yes
(0.86, 0.69, 0.93, 0.68, 0.62)	0.93	1.12	(0.94, 1.69)	Yes
(0.93, 0.11, 0.44, 0.17, 0.87)	0.93	1.12	(0.94, 1.69)	Yes
(0.81, 0.01, 0.87, 0.47, 0.95)	0.95	1.14	(0.96, 1.73)	Yes
(0.03, 0.07, 0.06, 0.99, 0.43)	0.99	1.19	(1.00+, 1.80)	No
(0.75, 0.23, 0.94, 0.18, 0.13)	0.94	1.13	(0.95, 1.71)	Yes
(0.19, 0.84, 0.54, 0.42, 0.14)	0.84	1.01	(0.85, 1.53)	Yes
(0.65, 0.38, 0.65, 0.35, 0.07)	0.65	0.78	(0.66, 1.18)	Yes
(0.36, 0.74, 0.38, 0.36, 0.57)	0.74	0.89	(0.75, 1.35)	Yes
(0.13, 0.40, 0.51, 0.50, 0.12)	0.51	0.61	(0.52, 0.93)	No
(0.68, 0.55, 0.05, 0.94, 0.69)	0.94	1.13	(0.95, 1.71)	Yes
(0.51, 0.28, 0.73, 0.10, 0.34)	0.73	0.88	(0.74, 1.33)	Yes
(0.71, 0.56, 0.21, 0.64, 0.85)	0.85	1.02	(0.86, 1.55)	Yes
(0.96, 0.59, 0.05, 0.13, 0.64)	0.96	1.15	(0.97, 1.75)	Yes
(0.88, 0.90, 0.56, 0.49, 0.07)	0.90	1.08	(0.91, 1.64)	Yes
(0.36. 0.62, 0.35, 0.11, 0.91)	0.91	1.09	(0.92, 1.66)	Yes
(0.04, 0.31, 0.86, 0.79, 0.45)	0.86	1.03	(0.87, 1.56)	Yes
(0.58, 0.52, 0.07, 0.27, 0.11)	0.58	0.70	(0.58, 1.06)	Yes
(0.83, 0.91, 0.27, 0.95, 0.20)	0.95	1.14	(0.96, 1.73)	Yes
(0.04, 0.32, 0.28, 0.55, 0.48)	0.55	0.66	(0.56, 1.00+)	Yes

each sample is y_{max}, w, and the corresponding 90% confidence interval,

$$\left(\frac{w}{\sqrt[5]{0.95}\left(\frac{6}{5}\right)}, \frac{w}{\sqrt[5]{0.05}\left(\frac{6}{5}\right)} \right).$$

The last column indicates whether each of the intervals is "correct"— that is, whether it contains as an interior point the value $\theta = 1$. A tally of that column reveals that 27 of the intervals, or 90%, *do* contain the true θ.

QUESTION 5.9.1   Let $Y_1, Y_2, \ldots, Y_n$ be a random sample from the uniform distribution defined over $(0, \theta)$. Find the formula for a 90% confidence interval for θ based on the estimator $W_1 = (n + 1) \cdot Y_{min}$. How would you expect intervals based on W_1 to compare with those given in Table 5.6?

Most applications of confidence intervals involve the parameters of the normal distribution (an exception is noted in Section 5.10). For that reason, the methodology of interval estimation will be developed more fully in Chapter 7. Nevertheless, a few general comments are in order here. First, it should be noted that there is nothing fundamentally special about the figure *90%* as a measure of confidence. We could just as easily have defined a 50%, a 78%, or a 99.9% confidence interval. In practice, though, convention dictates that the confidence "coefficient" almost always be set equal to either 90%, 95%, or 99%. Finally, it must not be forgotten that the "confidence" in a confidence interval refers to the procedure and not to any particular realization of that procedure. *Before* any data are collected, it is true that the (random) interval

$$\left(\frac{W}{\sqrt[n]{0.95}\left(\frac{n+1}{n}\right)}, \frac{W}{\sqrt[n]{0.05}\left(\frac{n+1}{n}\right)} \right)$$

has a 90% chance of containing θ. *After* the sample is drawn, though, and w is computed, the resulting interval is no longer a random variable: all we can say is that that particular interval either contains θ or does not.

Comment. Robert Frost was certainly more familiar with iambic pentameters than he was with estimated parameters, but in 1942 he wrote a couplet that sounded very much like a poet's perception of a confidence interval:

> We dance round in a ring and suppose,
> But the Secret sits in the middle and knows.

[See (97).]   ∎

5.10 CONFIDENCE INTERVALS
FOR THE BINOMIAL PARAMETER

The binomial distribution affords us the opportunity to apply the basic confidence-interval notions of Section 5.9 to a problem of considerable practical importance: given n independent Bernoulli trials, resulting in a total of y successes, construct an interval estimate for the unknown success probability, p. The fact that this is not an uncommon data structure is what makes the problem so important. We begin this section by deriving both approximate and exact interval estimates for p. The section concludes with a discussion of an important related problem—sample-size determination.

First, we recall an approximation—namely, the DeMoivre-Laplace statement that for large n the quantity

$$\frac{\frac{Y}{n} - p}{\sqrt{\frac{p(1-p)}{n}}}$$

behaves like a standard normal. Thus,

$$P\left[-z_{\alpha/2} < \frac{\frac{Y}{n} - p}{\sqrt{\frac{p(1-p)}{n}}} < z_{\alpha/2}\right] \doteq 1 - \alpha, \qquad (5.6)$$

where $P(Z \leq -z_{\alpha/2}) = P(Z \geq +z_{\alpha/2}) = \alpha/2$. As in Section 5.9, the problem is to isolate the unknown parameter in the center of the inequalities; the endpoints will then define an approximate $100(1-\alpha)\%$ confidence interval for p.

Note that the statement inside the brackets in Equation 5.6 is the same as

$$\frac{\left|\frac{Y}{n} - p\right|}{\sqrt{\frac{p(1-p)}{n}}} < z_{\alpha/2},$$

which is equivalent to

$$\left(\frac{Y}{n} - p\right)^2 < z_{\alpha/2}^2 \left[\frac{p(1-p)}{n}\right].$$

Expanded out and written in terms of powers of p, this becomes

$$p^2\left(1 + \frac{z_{\alpha/2}^2}{n}\right) - p\left(\frac{2Y}{n} + \frac{z_{\alpha/2}^2}{n}\right) + \left(\frac{Y}{n}\right)^2 < 0. \qquad (5.7)$$

The left-hand side of this inequality defines a parabola in p; the confidence interval we are looking for lies between its two roots (see Figure 5.16).

The quadratic formula applied to the left-hand side of Inequality 5.7 gives

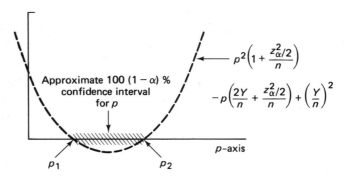

Figure 5.16 Parabola defined by Inequality 5.7.

$$p_1 = \frac{\dfrac{Y}{n} + \dfrac{z_{\alpha/2}^2}{2n} - \dfrac{z_{\alpha/2}}{\sqrt{n}}\sqrt{\left(\dfrac{Y}{n}\right)\left(1 - \dfrac{Y}{n}\right) + \dfrac{z_{\alpha/2}^2}{4n}}}{1 + \dfrac{z_{\alpha/2}^2}{n}}$$

and

$$p_2 = \frac{\dfrac{Y}{n} + \dfrac{z_{\alpha/2}^2}{2n} + \dfrac{z_{\alpha/2}}{\sqrt{n}}\sqrt{\left(\dfrac{Y}{n}\right)\left(1 - \dfrac{Y}{n}\right) + \dfrac{z_{\alpha/2}^2}{4n}}}{1 + \dfrac{z_{\alpha/2}^2}{n}}.$$

The set of p values from p_1 to p_2 certainly qualify as a $100(1 - \alpha)\%$ confidence interval for p. In practice, though, all terms in $z_{\alpha/2}^2/n$ are usually dropped—on the grounds that $z_{\alpha/2}$ will be quite small relative to n. (If $z_{\alpha/2}$ is *not* much smaller than n, the DeMoivre-Laplace approximation of Equation 5.6 should not have been used in the first place.) Omitting the $z_{\alpha/2}^2/n$ terms leaves

$$\left(\frac{y}{n} - z_{\alpha/2}\sqrt{\frac{\dfrac{y}{n}\left(1 - \dfrac{y}{n}\right)}{n}}, \; \frac{y}{n} + z_{\alpha/2}\sqrt{\frac{\dfrac{y}{n}\left(1 - \dfrac{y}{n}\right)}{n}}\right)$$

as the approximate $100(1 - \alpha)\%$ confidence interval for p.

CASE STUDY

5.7

Food-poisoning outbreaks are quite often the result of contaminated salads. In 1967, the New York City Department of Health established the following standards for bacterial counts in salads:

Total plate count: less than 100,000 organisms/gram
Enterococci: less than 1000 organisms/gram

Coliforms: less than 100 organisms/gram
Staphylococci (coagulase positive): 0 organisms/gram

That same year the department routinely examined some 220 tuna salads marketed by various retail and wholesale outlets. A total of 179 were found to be unsatisfactory according to at least one of these criteria (162).

Here the unknown parameter is p, the "true" proportion of contaminated tuna salads in New York City. One way of estimating p would be to construct its, say, 95% confidence interval. Substituting $z_{0.025} = 1.96$ and $y/n = 179/220 = 0.81$ into the expression for a 95% confidence interval for p gives

$$\left(0.81 - 1.96\sqrt{\frac{(0.81)(0.19)}{220}},\ 0.81 + 1.96\sqrt{\frac{(0.81)(0.19)}{220}}\right),$$

which reduces to the set of values from 0.76 to 0.86.

QUESTION 5.10.1   Find the 95% confidence interval for p, using the data of Case Study 5.7 but without ignoring terms in $z_{\alpha/2}^2/n$.

Comment. It *is* possible to find exact confidence intervals for a binomial parameter, but the computations are very tedious. Suppose that y successes are observed in n independent Bernoulli trials. If p_1 and p_2 are the true lower and upper $100(1 - \alpha)\%$ confidence limits for p, then

$$P(p_1 < p < p_2 \,|\, Y = y) = 1 - \alpha.$$

(See Figure 5.17.) Notice that if p_1 were moved further to the left it would be less compatible with the data (that is, with y). The largest p_1 that would *not* be included in the confidence interval is the solution of

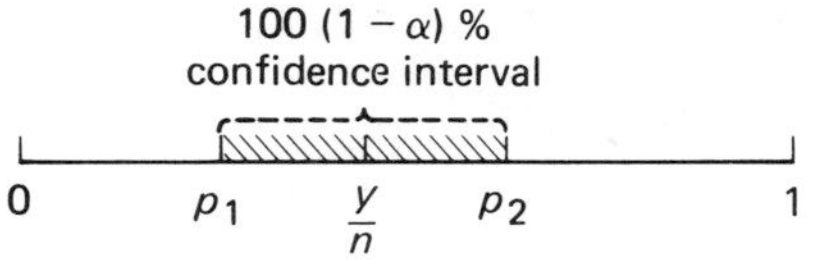

Figure 5.17 Exact confidence interval for binomial parameter.

$$\sum_{j=y}^{n} \binom{n}{j} p_1^j (1 - p_1)^{n-j} = \frac{\alpha}{2}.$$

That is, any value less than or equal to p_1 would not be sufficiently compatible with the observed sample proportion of y/n to warrant its inclusion in the confidence interval. Similarly, the smallest value of p_2 lying outside the interval is the solution of

$$\sum_{j=0}^{y} \binom{n}{j} p_2^j (1 - p_2)^{n-j} = \frac{\alpha}{2}.$$

Tables of (p_1, p_2) values for $1 - \alpha = 0.95$ and 0.99 and for values of n and y from 2 to 100 are readily available (157). If, for example, 33 successes are observed in 42 trials ($y/n = \frac{33}{42} = 0.786$), the exact 95 % confidence interval for p is (0.632, 0.897). [The approximate 95 % confidence intervals are (0.662, 0.910) *without* terms in $z_{\alpha/2}^2/n$ and (0.641, 0.883) *with* terms in $z_{\alpha/2}^2/n$.] ∎

We conclude this section with an experimental "design" question that is directly related to the construction of a confidence interval for p. Suppose a researcher wants to estimate the parameter p in a series of n independent Bernoulli trials, *where n is yet to be determined.* Of course, regardless of n, we know that the point estimator for p will be Y/n. We also know that the standard deviation of Y/n will decrease as n increases. Therefore, as the sample size increases, so does the precision of the estimation procedure. Unfortunately, the cost of any study will also increase with n. We are thus faced with a trade-off—the desire, on one hand, to have as precise an estimator as possible and the necessity, on the other, of keeping costs to a minimum. These two conflicting objectives will often force an experimenter to pose the following sort of question: "What is the smallest n that will 'guarantee' (with a probability of, say, $1 - \alpha$) that Y/n will be within some specified distance, d, of p?" As it turns out, the answer is relatively easy to derive.

Phrasing the question more formally, we are looking for the smallest n such that

$$P\left(\left|\frac{Y}{n} - p\right| < d\right) = 1 - \alpha. \tag{5.8}$$

But note that

$$\left|\frac{Y}{n} - p\right| < d$$

is the same as

$$-d < \frac{Y}{n} - p < d$$

or

$$\frac{-d}{\sqrt{\dfrac{p(1-p)}{n}}} < \frac{\dfrac{Y}{n} - p}{\sqrt{\dfrac{p(1-p)}{n}}} < \frac{d}{\sqrt{\dfrac{p(1-p)}{n}}}.$$

We can approximate Equation 5.8, then, with the expression

$$P\left[\frac{-d}{\sqrt{\dfrac{p(1-p)}{n}}} < Z < \frac{d}{\sqrt{\dfrac{p(1-p)}{n}}}\right] = 1 - \alpha.$$

However, since $P(-z_{\alpha/2} < Z < z_{\alpha/2}) = 1 - \alpha$, it must be true that

$$z_{\alpha/2} = \frac{d}{\sqrt{\dfrac{p(1-p)}{n}}}$$

or, solving for n,

$$n = \frac{z_{\alpha/2}^2 p(1 - p)}{d^2}.$$ (5.9)

Equation 5.9 is not an acceptable final answer, though, because it involves the unknown p. However, since $0 \leq p \leq 1$, the product $p(1 - p)$ will always be less than or equal to $\frac{1}{4}$. Therefore, we can insure that Equation 5.8 is satisfied in even the most "difficult" of situations (when p is actually $\frac{1}{2}$) by choosing as the sample size

$$n = \frac{z_{\alpha/2}^2}{4d^2}.$$ (5.10)

As an application of Equation 5.10, suppose that for the survey described in Case Study 5.7 the inspectors want the *sample* proportion of contaminated tuna salads to be within 0.03 of the *true* proportion—with 95% confidence. That is, they wish to examine the smallest number of salads, n, such that

$$P\left(\left|\frac{Y}{n} - p\right| < 0.03\right) = 0.95.$$

According to Equation 5.10, n should be 1067:

$$n = \frac{(1.96)^2}{4(0.03)^2} = 1067.$$

One final comment about this problem deserves mention. If any prior information about the value of p is available, it may be possible to reduce substantially the necessary sample size by *not* making the $p(1 - p) = \frac{1}{4}$ assumption. For example, suppose a preliminary, small-scale food inspection has indicated that a sizable proportion of tuna salads (at least 75%) are contaminated. We would then determine n by using Equation 5.9 rather than Equation 5.10. The unknown p would be replaced by 0.75:

$$n = \frac{(1.96)^2(0.75)(0.25)}{(0.03)^2} = 800.$$

In this case the minimum sample size was lowered from 1067 to 800, a reduction of 25%.

QUESTION 5.10.2   An attorney for a civil liberties group is interviewing female nurses in a large metropolitan hospital to see whether they feel they have been victims of sexist discrimination on the part of the male doctors they work with. Let p denote the true proportion of nurses who would consider themselves victims of such discrimination. How many nurses must the attorney interview in order to be 60% certain that the sample proportion of nurses who say, "Yes, I have been discriminated against," is within 0.05 of p?

REVIEW EXERCISES FOR CHAPTER 5

1. Epidemiologists at a county hospital located near a nuclear waste disposal plant are monitoring all admissions having a tentative diagnosis of chronic leukemia. The numbers of patients falling into that category for the past twelve months were 2, 1, 0,

0, 1, 1, 1, 4, 1, 0, 2, and 3. Assume the occurrence of such patients (per month) can be described by a Poisson pdf. Estimate the relevant parameter. Is the estimator on which your estimate is based unbiased?

2. Let W_1 and W_2 be any two unbiased estimators for the parameter θ. If $W = aW_1 + bW_2$ is also to be an unbiased estimator for θ, what must be true of a and b?

3. A marketing research firm did two independent surveys of the proportion of viewers in a certain metropolitan area who watched "Charlie's Angels." The first survey revealed that 35 of the 100 persons questioned were tuned to the show. In the second survey, 100 of the 350 individuals contacted watched the show. Give two unbiased estimators for the true proportion of people watching "Charlie's Angels." Make both of your estimators use the information contained in both surveys. Intuitively, what would be a reasonable way to combine the information in the two samples? Explain.

4. Let X have a uniform pdf over the interval $(0, \theta)$. Find an unbiased estimator for θ^2. Hint: Consider a function of X^2.

5. An estimator $W_n = W_n(X_1, X_2, \ldots, X_n)$ for the parameter θ is said to be *asymptotically unbiased* if

$$\lim_{n \to \infty} E(W_n) = \theta.$$

Let $X_1, X_2, \ldots, X_n$ be a random sample from a uniform distribution over $(0, \theta)$. Let $W_n = X'_n$. Is W_n asymptotically unbiased?

6. Let X_1 and X_2 be a random sample of size 2 from an exponential pdf, $f_X(x) = (1/\lambda)e^{-x/\lambda}$, $x > 0$. Define Y to be the *geometric mean* of X_1 and X_2: $Y = \sqrt{X_1 X_2}$. Show that $4Y/\pi$ is an unbiased estimator for λ.

7. A student takes a freshman chemistry exam that the professor claims has always resulted in bell-shaped grade distributions when he has given the test in the past. If the student makes a grade of 66, estimate μ, the class average.

8. If W is an unbiased estimator for θ, is W^2 an unbiased estimator for θ^2?

9. Let Y_i be the ith observation recorded in a sample of size n drawn from a uniform pdf defined over $(0, \theta)$. Is $W = Y_i$ unbiased? If not, construct a W_1 based on W that does have that property. Hint: See Theorem 4.11.

10. We say that W is a *best linear unbiased estimator* for a parameter θ if (1) W is a linear function of the observations, (2) $E(W) = \theta$, for all θ, and (3) $\text{Var}(W) \leq \text{Var}(W^*)$, for all other estimators W^* satisfying conditions (1) and (2). Suppose X is a random variable with mean μ and variance σ^2. Show that $\bar{X} = (1/n)(X_1 + X_2 + \cdots + X_n)$ is the best linear unbiased estimator for μ. Hint: Write as a general linear estimator, $W = a_1 X_1 + \cdots + a_n X_n + c$; show that $c = 0$ and take partial derivatives with respect to the a_i's to show that $a_i = 1/n$.

11. As an alternative to imposing unbiasedness, an estimator's distribution can be "centered" by requiring that its *median* be equal to θ. If it is, W is said to be *median unbiased*. Let $Y_1, Y_2, \ldots, Y_n$ be a random sample from the uniform pdf over $(0, \theta)$. For arbitrary n, is $W_1 = [(n + 1)/n]Y'_n$ median unbiased? Is it *ever* median unbiased?

12. Let Y have the pdf

$$f_Y(n) = \binom{n + r - 1}{n} p^r q^n, \qquad n = 0, 1, \ldots.$$

Show that $W = (r - 1)/(r + Y - 1)$ is an unbiased estimator for p.

13. A secretary spends much of her morning trying to place long distance calls for her supervisor. The people she is trying to reach are hard to find, and she has to make 16 calls before making connections (on the seventeenth call) with the fourth (and last) name on her list. Assuming each attempted call to be an independent trial with the same probability, p, of success, estimate p.

14. Let $Y_1, Y_2, \ldots, Y_n$ be a random sample from a Poisson pdf with parameter λ. Let $W_1 = Y_1$ and $W_2 = \bar{Y}$ be two unbiased estimators for λ. Compute their relative efficiency.

15. It can be shown that if $2n + 1$ random observations are taken from a continuous symmetric pdf with mean μ, and if $f_X(\mu) \neq 0$, then the sample median, X'_{n+1}, is an unbiased estimator for μ, and, as n gets large, the variance of X'_{n+1} converges to

$$\frac{1}{8[f_X(\mu)]^2 n}$$

[see (51)]. Let $X_1, X_2, \ldots, X_{2n+1}$ be a random sample from a uniform pdf over the interval $(0, \theta)$. Then $W_1 = X'_{n+1}$ is an unbiased estimator for $\theta/2$. Construct an unbiased estimator for $\theta/2$ using X'_{2n+1} and compute the relative efficiency of the latter to X'_{n+1}.

16. The parameter for an exponential pdf can be estimated using a linear combination of selected order statistics:

$$W = \text{estimator for } \lambda = c_1 X'_{\alpha_1} + c_2 X'_{\alpha_2} + \cdots + c_k X'_{\alpha_k},$$

where $1 \leq \alpha_1 < \alpha_2 < \cdots < \alpha_k \leq n$, and n is the number of observations available. For various values of n and k, the optimal choices for the c_i's and the α_i's have been determined (95). For example, if $n = 30$, and we wish to estimate λ with a linear combination of *three* order statistics, the α_1, α_2, and α_3 that minimize the variance of the resulting unbiased estimator are 17, 26, and 30, respectively, and the estimator is given by

$$W = 0.450522 X'_{17} + 0.219090 X'_{26} + 0.053584 X'_{30}.$$

Here, the relative efficiency of W to $W^* = \bar{X}$ is 0.91. How many observations would be required to estimate λ using $\bar{X}$, but with the same precision afforded by W?

17. A random sample of size 2 is drawn from an exponential pdf with parameter λ. Let W_1 be the unbiased estimator for λ based on the geometric mean (see Review Exercise 6). Let W_2 be the sample mean. Find the relative efficiency of W_1 to W_2.

18. Find the Cramer-Rao lower bound for estimating the parameter θ in a Cauchy distribution,

$$f_X(x) = \frac{1}{\pi[1 + (x - \theta)^2]}, \qquad -\infty < x < \infty, \; -\infty < \theta < \infty.$$

19. Let Y be any random variable with mean μ and variance σ^2. Let the estimator W be the sample mean of n random observations taken on Y. Prove that $\bar{Y}$ is consistent for μ.

20. Let $W_1, W_2, \ldots, W_n, \ldots$ be any sequence of estimators for θ. Show that

$$E[(W_n - \theta)^2] = \text{Var}(W_n) + [\theta - E(W_n)]^2.$$

21. A sequence of estimators $W_1, W_2, \ldots, W_n, \ldots$ for θ is said to be *squared-error consistent* if

$$\lim_{n \to \infty} E[(W_n - \theta)^2] = 0, \qquad \text{for all } \theta.$$

Show that if W_n is squared-error consistent, then W_n is also consistent in the sense of Definition 5.5.

22. Let $Y_1, Y_2, \ldots, Y_n$ be a random sample from the uniform pdf over the interval $(0, \theta)$. Let $W_n = Y'_n$. Is W_n squared-error consistent?

23. Let $X_1, X_2, \ldots, X_n$ be a random sample from a geometric pdf with parameter p. Show that $W = X_1 + X_2 + \cdots + X_n$ is a sufficient statistic for p.

24. The Fisher-Neyman criterion of Theorem 5.2 is not the only criterion for establishing sufficiency. Another is the *factorization theorem*: let $X_1, X_2, \ldots, X_n$ be a random sample of size n from $f_X(x; \theta)$. Then $W = h(X_1, X_2, \ldots, X_n)$ is sufficient for θ if and only if

$$\prod_{i=1}^{n} f_X(x_i; \theta) = \phi(h(x_1, x_2, \ldots, x_n); \theta) q(x_1, x_2, \ldots, x_n)$$

[see (78) for a proof]. Note that ϕ is dependent on the data only through W, and q does not involve θ. Let $X_1, X_2, \ldots, X_n$ be a random sample from $f_X(x) = \theta x^{\theta-1}$, $0 < x < 1; \theta > 0$. Use the factorization theorem to suggest a sufficient statistic for θ.

25. Suppose $X_1, X_2, \ldots, X_n$ is a random sample from a pdf that can be written in *exponential form*:

$$f_X(x; \theta) = e^{K(x) p(\theta) + S(x) + q(\theta)},$$

where the range of x is independent of θ. Show that

$$W = \sum_{i=1}^{n} K(x_i)$$

is a sufficient statistic for θ.

26. Write the exponential pdf $f_X(x) = (1/\lambda)e^{-x/\lambda}$, $x > 0$, in the form suggested in Review Exercise 25 and deduce a sufficient estimator for λ (given a random sample $X_1, X_2, \ldots, X_n$).

27. Let $X_1, X_2, \ldots, X_n$ be a random sample from a *Pareto* distribution

$$f_X(x) = \frac{\theta}{(1 + x)^{\theta+1}}, \qquad 0 < x < \infty, 0 < \theta < \infty.$$

Write $f_X(x)$ in exponential form and deduce a sufficient statistic for θ. Note: The Pareto distribution finds many economic applications. It models such phenomena as the distribution of incomes above a threshold value and the distribution of the amounts of insurance claims above a threshold value (50).

28. Consider a series of three Bernoulli trials, each with the same unknown success probability p. Let X_i denote the number of successes on the ith trial ($X_i = 0$ or 1). It has already been established that $W = X_1 + X_2 + X_3$ is a sufficient estimator for p; show that $W' = X_1 + 2X_1 + 3X_3$ is not.

29. Let $X_1, X_2, \ldots, X_n$ be a random sample from

$$f_X(x) = \left(\frac{1}{\theta^2}\right) x e^{-x/\theta}, \qquad 0 < x < \infty, 0 < \theta < \infty.$$

Find the MLE for θ.

30. Experience has shown that in a certain state the distribution of voters supporting Democratic candidates for local offices is described quite well by the family of pdf's having the form:

$$f_P(p) = \theta p^{\theta-1}, \qquad 0 < p < 1, 0 < \theta < \infty,$$

where P is the percentage of registrants in a precinct who vote Democratic. Suppose voters in five randomly selected precincts are polled immediately before an election and the proportions intending to vote for the Democratic candidate are 0.45, 0.68, 0.87, 0.36, and 0.54, respectively. Estimate θ.

31. Consider a reliability inspection policy where n items are put "on test" and the first r failure times, $T_1 < T_2 < \cdots < T_r$, are recorded. Assume the pdf describing the response variable is exponential with parameter λ:

$$f_T(t) = \left(\frac{1}{\lambda}\right) e^{-t/\lambda}, \qquad t > 0.$$

Write down the appropriate joint pdf, differentiate it, and show that the MLE for λ is given by

$$\hat{\lambda} = \frac{1}{r} \sum_{j=1}^{r} t_j + (n-r)t_r.$$

32. Maximum-likelihood estimators satisfy an *invariance* property: if $\hat{\theta}$ is the MLE for θ, then $g(\hat{\theta})$ is the MLE of $g(\theta)$. Use that property to find the MLE for the standard deviation of a Poisson distribution. Assume the data to be a random sample of size n.

33. Show by a counterexample that unbiasedness is not preserved under invariance.

34. A commuter's trip home consists of first riding a subway to a bus stop and then taking the bus home. The bus she would like to catch arrives uniformly over the interval (θ_1, θ_2). She would like estimates for both θ_1 and θ_2 so she has some idea of when she should be at the stop (θ_1) and when she is probably too late and will have to wait for the next bus (θ_2). Over an eight-day period, she makes certain to be at the stop early so as to not miss the bus and records the following arrival times: 5:15, 5:21, 5:14, 5:23, 5:29, 5:17, 5:15, and 5:18. Estimate θ_1 and θ_2. Also, give the maximum-likelihood estimate for the mean of the arrival distribution.

35. Prove that the MLE for a parameter θ is a function of a sufficient statistic for that parameter.

36. A criminologist is searching through FBI files to document the frequency of occurrence of a rare double-whorl fingerprint. Among six consecutive sets of 100,000 prints scanned by a computer, the numbers of persons having this particular abnormality were 3, 0, 3, 4, 2, and 1. Assume the occurrence of the double whorl is a Poisson phenomenon. Use the method of moments to estimate λ, its rate of occurrence.

37. Let $X_1, X_2, \ldots, X_n$ be a random sample from a uniform pdf over the interval $(0, \theta)$. Find the method-of-moments estimator for θ. Compare the values of the method-of-moments estimate and the maximum-likelihood estimate for the following random sample of size five: 17, 92, 46, 39, and 56.

38. Use the method of moments to estimate θ in the one-parameter beta distribution,

$$f_X(x) = \frac{\Gamma(\theta+1)}{\Gamma(\theta)\Gamma(1)} x^{\theta-1}, \qquad 0 < x < 1.$$

39. Is there any information in a point estimate that may be "lost" in an interval estimate?

40. Let X have an $N(\mu, \sigma^2)$ pdf. Find the value h having the property that

$$P(\mu - h < X < \mu + h) = 0.50.$$

(In the physical sciences, the quantity h is referred to as the *probable error*.)

41. Construct a 90% confidence interval for p using the ESP data given in Case Study 4.5. What does p represent in this particular setting?

42. Refer to the data given in Review Exercise 27 at the end of Chapter 4. Construct a 50% confidence interval for p, the true probability that Salt Lake City residents' death dates are in the three-month period immediately following their birth dates.

43. Toxoplasmosis is a protozoan-caused disease that can be transmitted to people from cats. It is estimated that as many as one-fourth of all adults in the United States have had a mild case of toxoplasmosis at one time or another. Usually the condition is not serious, but if the disease is contracted by a pregnant woman, a birth defect may be the consequence. To assess the magnitude of the potential dangers posed by house cats, a team of epidemiologists checks 205 cats and finds that 121 have toxoplasmosis antibodies, indicating that they were at one time infected with the disease. Let p denote the true (unknown) proportion of cats who have had toxoplasmosis. Construct a 99% and a 75% confidence interval for p.

44. Give a definition of a one-sided confidence interval for p. Find the two one-sided 95% confidence intervals for p based on the toxoplasmosis data of Review Exercise 43.

45. A baseball player has a .315 batting average based on 200 official at-bats. Let p denote his true probability of hitting safely on a typical trip to the plate. Construct a 95% confidence interval for p.

46. Since graduating from Vanderbilt last semester, Vanessa had sought a full-time position with her alma mater. Unfortunately, her IQ of 130 overqualified her for the kinds of positions that were available. Desperate, and having no other recourse, she stole her personnel file, lowered her IQ by 40 points, and resubmitted her application. She was immediately hired as an Assistant Dean in the College of Arts and Sciences and given a joint appointment as a full professor in the Philosophy Department. Her first project now is to study the feasibility of a proposed Vanderbilt-in-Tibet summer program. She decides to take a random sample of Vanderbilt undergraduates and ask them whether they would be in favor of such a program. More or less arbitrarily, she decides to question enough students so that there will be an 80% chance that her estimate will be within 0.02 of the true proportion of undergraduates in favor of the program. What is the smallest number of students she should interview?

47. A public health survey is to be done in an inner-city area to estimate the proportion of children, ages 0 to 14, having adequate polio immunization. It is desired that the final estimate be within 0.05 of the true proportion with probability 0.98. What is the minimum sample size required?

48. What would be the minimum sample size required to achieve the objectives stated in Review Exercise 47 if it could be assumed that the actual proportion of children adequately immunized against polio was at least 65%?

49. Suppose $n = 3$ independent Bernoulli trials yield $y = 1$ success. Find the exact 95% confidence interval for p. (See the Comment following Case Study 5.7.)

CHAPTER SIX
Hypothesis Testing

PIERRE-SIMON, MARQUIS DE LAPLACE (1749–1827)

As a young man, Laplace went to Paris to seek his fortune as a mathematician, disregarding his father's wishes that he enter the clergy. He soon became a protégé of d'Alembert and at the age of 24 was elected to the Academy of Sciences. Laplace was recognized as one of the leading figures of that group for his work in physics, celestial mechanics, and pure mathematics. He also enjoyed some political prestige, and his friend, Napoleon Bonaparte, made him Minister of the Interior for a brief period. With the restoration of the Bourbon monarchy, Laplace renounced Napoleon for Louis XVIII, who later made him a marquis.

6.1 INTRODUCTION

Inferences, as we have seen, often take the form of numerical estimates, either as single points or as confidence intervals. But not always. In many experimental situations the conclusion to be drawn is *not* numerical and is more aptly phrased as a choice between two conflicting theories, or *hypotheses*. Thus, a court psychiatrist is called upon to pronounce an accused murderer either "sane" or "insane"; the FDA "approves" or "rejects" an application by a pharmaceutical house to get a flu vaccine licensed; a geneticist has to decide whether the inheritance of eye color in *Drosophila melanogaster* "does" or "does not" follow classical Mendelian principles. In this chapter we examine the statistical methodology involved in making decisions of this sort.

The process of dichotomizing the possible conclusions of an experiment and then using the theory of probability to choose between the two alternatives is known as *hypothesis testing*. The two competing hypotheses are denoted H_0, the the *null hypothesis*, and H_1, the *alternative hypothesis*. The former is always a statement of either (1) "no effect" or (2) the status quo. We will see in the next section that H_0 plays a role in hypothesis testing analogous to that of the defendant in a jury trial: it will be assumed that H_0 is true (that is, it will be given the benefit of the doubt) unless the data convince us otherwise.

EXAMPLE 6.1.   Recall the problem discussed in Chapter 1 involving the width-to-length ratios of rectangles sewn by the Shoshoni Indians. At issue was whether or not these particular Indians embraced the same esthetic standard as the Greeks, for whom the "ideal" rectangle had a w/l ratio of approximately 0.618. In this context, the Greek standard becomes the status quo and the null hypothesis would be: "The Shoshonis (also) prefer rectangles whose w/l ratio is 0.618." The alternative hypothesis would be: "The Shoshonis do *not* prefer rectangles whose w/l ratio is 0.618."

EXAMPLE 6.2.   The relationship between pain tolerance and smoking habits was investigated in a recent experiment (160). Subjects belonging to a given age group, race, and sex were divided into two further categories: smokers and nonsmokers. Each subject was given a pain-tolerance test: this was a procedure using two mechanically driven rods positioned on either side of the subject's Achilles tendon. The rods would steadily compress the tendon until the subject found the pain too excessive to bear. This threshold point was then used as a measure of pain tolerance. Here the null hypothesis would be that smoking has *no effect* on a person's pain tolerance. Curiously enough, the data seemed to indicate that among whites, it did: smokers had a much lower pain tolerance than nonsmokers—that is, H_0 was "rejected." For blacks and orientals, though, the corresponding H_0 was "accepted."

To choose between H_0 and H_1 statistically, we will need to translate the two hypotheses into quantitative terms. Usually this is done by specifying (i.e., *assuming*) a probability distribution, $f_Y(y)$, for the response variable and then letting H_0 and H_1 be statements about its parameter values. For example, the hypothesis that the Shoshonis prefer rectangles having a w/l ratio of 0.618 might be rephrased to state that the set of all Shoshoni rectangles generates, say, a *normal* distribution of w/l ratios and that the mean of that distribution is $\mu = 0.618$. Thus, we could write

$$H_0: f_Y(y) \sim N(0.618, \sigma^2),$$

$$H_1: f_Y(y) \sim N(\mu, \sigma^2), \qquad \mu \neq 0.618.$$

In practice, the distributional form of $f_Y(y)$ is seldom included in the statement of H_0 or H_1: the hypotheses are simply written

$$H_0: \mu = 0.618,$$

$$H_1: \mu \neq 0.618.$$

Similarly, in the case of Example 6.2, it might be assumed that the pain-tolerance distributions for smokers and nonsmokers are both normal but with means, μ_S and μ_{NS}, that *may* be different. The two conflicting hypotheses, then, would be abbreviated to

$$H_0: \mu_S = \mu_{NS},$$

$$H_1: \mu_S \neq \mu_{NS}.$$

In the next section we will develop some of the mathematical structure associated with H_0 and H_1. Particularly important will be the rationale that is followed in deciding which of the two hypotheses to accept. The final section introduces the notion of optimality (when is one hypothesis test "better" than another?) and then considers the rather formidable problem of actually finding a procedure for testing a given H_0 against a given H_1.

6.2 THE DECISION RULE

We introduce the basic concepts of hypothesis testing with an example. Suppose a handwriting expert, Mr. Q, claims to be able to tell whether a person is a poor credit risk just by the way he crosses his t's and dots his i's. Being a little skeptical, we ask him to demonstrate his ability experimentally. The "test" is straightforward. We present Mr. Q with 18 folders, collectively representing the handwriting of 36 individuals. One of the two samples in each folder was written by someone deemed unreliable by the local savings and loan. The other sample in each folder was submitted by a citizen of absolutely sterling repute. Mr. Q's task is to decide which of the two samples in each folder is the handwriting of a scoundrel. Of course, if he is only guessing, he has a 50-50 chance of making the correct identification in any given folder. That would make his "expected score" 9. Suppose it turned out, though, that he spotted 12 of the credit risks. What should we conclude? Was he lucky, guessing the additional 3 just by chance? Or does he actually possess the talent he claims?

The obvious probability model to superimpose over this experiment is the binomial, with each folder being thought of as a Bernoulli trial. If p is the probability that Mr. Q correctly identifies the handwriting of the credit risk in a given folder, the decision we have to make reduces to a choice between

H_0: $p = \frac{1}{2}$ (Mr. Q is only guessing),

H_1: $p > \frac{1}{2}$ (Mr. Q's system *does* improve his
                chances of making a correct identification).

Comment. For the hypothesis tests we will consider, H_0 will specify one particular value for the parameter in question; H_1, on the other hand, will allow a *range* of values. If the range of H_1 lies entirely on one side of the H_0 value, the alternative is said to be *one-sided*. If H_1 includes *all* values of the parameter except the one singled out in H_0, it is said to be *two-sided*. Thus, H_1: $p > \frac{1}{2}$ is one-sided; in the Shoshoni example, H_1: $\mu \neq 0.618$ was two-sided.

In a real problem, the decision to make H_1 either one-sided or two-sided is not always clear-cut. As a general rule of thumb, if parameter values in only one "direction" (from the H_0 value) have physical meaning—or if values in only one direction are possible—the alternative should be one-sided. Otherwise, H_1 should be two-sided. In the handwriting example, we are presumably not interested in a system that would prove *worse* than chance. Hence, the one-sided alternative makes sense. For the Shoshoni data, though, it was necessary to make H_1 two-sided—there was no a priori reason for the Shoshoni ratio to be *less* than 0.618 as opposed to *greater* than 0.618. ∎

The decision to accept or reject H_0 will be based on the sample data through an intermediary function known as a *test statistic*. By definition, a test statistic, W, is a random variable that summarizes the information in the Y_i's as it pertains to the parameter in question. In this case, since p is the *true* proportion of correct match-ups, it should not seem unreasonable to set the test statistic equal to the corresponding *sample* proportion. (This will be justified on theoretical grounds in Section 6.4.) Equivalently, we can set W equal to simply the *number* of successes. Then

$$W = Y,$$

where Y is the number of correct identifications (out of 18).

Recall the "principle" of hypothesis testing alluded to in Section 6.1: we will assume at the outset that the null hypothesis is correct, and our position will change only if the data show, *beyond all reasonable doubt*, that H_1 is true. The problem, then, is to define in quantitative terms what reasonable doubt means. This is where W comes in. Notice that the scale of possible W-values—$0, 1, \ldots, 18$—can be thought of as a credibility axis for the null hypothesis (Figure 6.1). For example, if the observed proportion of successes were close to 9 (making $y/n = 0.50$), we

Number of successes (Y) **Figure 6.1** Scale of W-values.

would be obliged to accept H_0, since reasonable doubt that p is not $\frac{1}{2}$ would not have been established. On the other hand, if the observed value of W were considerably to the right of 9, our faith in H_0 would surely falter. Certainly if Mr. Q achieved a perfect score ($W = 18$), we would feel obliged to abandon our initial skepticism.

Reasonable doubt, though, does not have to be as extreme as "18 out of 18." Somewhere *between* W-values of 9 and 18 there is a point, call it Y^*, where, for all practical purposes, the credibility of H_0 ends and reasonable doubt begins. This point is called the *critical value*: it completely determines the decision-making process. For any observed number of successes, y, *greater than or equal to* Y^*, we will reject $H_0: p = \frac{1}{2}$. For any y *less than* Y^*, the null hypothesis will be accepted.

The set of values for the test statistic that require the null hypothesis to be rejected is called the *critical region* and is denoted by the symbol C (Figure 6.2). In set notation,

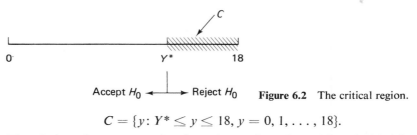

Figure 6.2   The critical region.

$$C = \{y: Y^* \leq y \leq 18, y = 0, 1, \ldots, 18\}.$$

The choice of an exact value for Y^*—so far, all we have decided is that it should be between 9 and 18—is based on probabilistic considerations. Suppose, for example, we set the critical value equal to 11 (Figure 6.3). Under the assumption, then, that H_0 is true, it follows that the probability of W's equaling or exceeding 11 is 0.24:

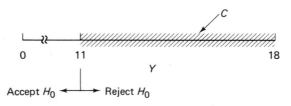

Figure 6.3   Critical value set equal to 11.

$$P(W \geq Y^*) = P(Y \geq 11)$$
$$= \sum_{y=11}^{18} \binom{18}{y}\left(\frac{1}{2}\right)^y\left(1 - \frac{1}{2}\right)^{18-y} = 0.24.$$

(See Figure 6.4.) This implies that if 11 were chosen as the critical value, we would incorrectly reject H_0 almost 25% of the time! For most people, this would not be a satisfactory definition of reasonable doubt. No jury, for example, would convict a defendant knowing they had a 25% chance of being wrong.

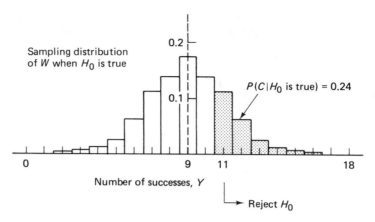

Figure 6.4 Sampling distribution of W ($Y^* = 11$).

Clearly, we must look for a critical value closer to 18. Consider what would happen if Y^* were set equal to 15 (making the sample proportion 0.83). The probability of W's exceeding (or equaling) *this* number—assuming H_0 to be true—is 0.004:

$$P(Y \geq 15) = \sum_{y=15}^{18} \binom{18}{y}\left(\frac{1}{2}\right)^{y}\left(1 - \frac{1}{2}\right)^{18-y} = 0.004.$$

(See Figure 6.5.) Now we may have gone too far to the other extreme. Requiring the sample proportion to be this far out in the tail of the W distribution before we reject H_0 would be like a jury's not convicting a defendant unless the prosecution could produce 10 eyewitnesses.

The decision rule we want is somewhere *between* "Reject H_0 if $y \geq 11$" and "Reject H_0 if $y \geq 15$." Or, what amounts to the same thing, the probability of rejecting H_0 when H_0 is actually true should be something less than 0.24 but

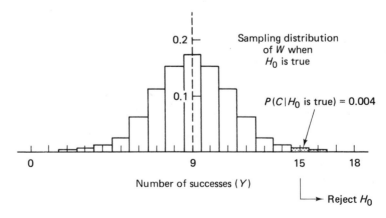

Figure 6.5 Sampling distribution of W ($Y^* = 15$).

greater than 0.004. Obviously, there is no way to "prove" that the optimal probability should be 0.10 or 0.18 or 0.009—or anything else. Over the years, though, a reasonable consensus, from discipline to discipline, has been reached as to how much evidence is enough evidence. *In many situations, researchers define the beginning of reasonable doubt as the value of the test statistic that is equaled or exceeded by chance only 5% of the time (when H_0 is true).*

Thus, according to this criterion, Y^* should be chosen so that

$$P(Y \geq Y^* \mid H_0 \text{ is true}) = 0.05.$$

More specifically, Y^* should satisfy the equation

$$P(Y \geq Y^* \mid H_0 \text{ is true}) = 0.05 = \sum_{y=Y^*}^{18} \binom{18}{y}\left(\frac{1}{2}\right)^y\left(1 - \frac{1}{2}\right)^{18-y}. \qquad (6.1)$$

[Keep in mind that the discreteness of Y may make it impossible to find a Y^* with the property that $P(Y \geq Y^*)$ is *exactly* 0.05.] Here, a little trial and error would show that $Y^* = 13$ is the solution to Equation 6.1. This means that we will reject $H_0: p = \frac{1}{2}$ if and only if Mr. Q makes 13 or more correct identifications (or if $y/18 \geq 13/18 = 0.72$). In this case, we could feel righteously smug that our initial appraisal of Mr. Q's alleged talent was right on target. Having made only 12 correct identifications, Mr. Q. has failed to demonstrate any special ability (relative to our 5% criterion).

Comment. Defining the critical region in terms of the W-value that is exceeded only 5% of the time when H_0 is true is the most common way to quantify reasonable doubt, but there are others. The figure 1% is frequently used, and, to a lesser extent, so are 10% and 0.1%. More will be said about the implications of these "cut-off" points in Section 6.3. ∎

EXAMPLE 6.3.   Recall the Pratt-Woodruff experiment described in Case Study 4.5. At issue was the phenomenon of ESP, with the investigation seeking to gain evidence in support of its existence by looking at the frequency with which subjects could correctly identify one of five different cards. If p denotes the students' probability of guessing any particular card correctly, then the relevant hypotheses to be tested are

$$H_0: p = \tfrac{1}{5} \qquad \text{(students are guessing randomly)}$$

versus

$$H_1: p > \tfrac{1}{5} \qquad \text{(students have ESP)}.$$

The number of trials was $n = 60,000$.

Using the 0.05 criterion, we need to find the Y^* that satisfies

$$P(Y \geq Y^* \mid H_0 \text{ is true}) = 0.05 = \sum_{y=Y^*}^{60,000} \binom{60,000}{y}\left(\frac{1}{5}\right)^y\left(\frac{4}{5}\right)^{60,000-y}.$$

As written, this would be a formidable task, but we can greatly simplify it by invoking the normal approximation to the binomial. We can write

$$P(Y \geq Y^* \mid H_0 \text{ is true}) = P\left[\frac{Y - 60,000(\frac{1}{5})}{\sqrt{60,000(\frac{1}{5})(\frac{4}{5})}} \geq \frac{Y^* - 60,000(\frac{1}{5})}{\sqrt{60,000(\frac{1}{5})(\frac{4}{5})}}\right],$$

and from the DeMoivre-Laplace limit theorem we know that the distribution of

$$\frac{Y - 60,000(\frac{1}{5})}{\sqrt{60,000(\frac{1}{5})(\frac{4}{5})}}$$

is approximated by the standard normal (when H_0 is true). Therefore,

$$P(Y \geq Y^* \,|\, H_0 \text{ is true}) \doteq P\left(Z \geq \frac{Y^* - 12,000}{\sqrt{9600}} \right),$$

and since $P(Z \geq 1.64) = 0.05$, it follows that Y^* is the solution of the equation

$$1.64 = \frac{Y^* - 12,000}{\sqrt{9600}}.$$

This gives 12,161 as the value for Y^*. Since the observed number of correct identifications was 12,489, our conclusion is to reject H_0—the data support the existence of ESP.

Comment. For one-sided alternatives, we have seen that the critical region is confined to one tail of the W-distribution. In situations where H_1 is two-sided, though, there are *two* critical regions, one in each tail of the W distribution. For example, if the hypotheses being tested were

$$H_0: p = \tfrac{1}{2}$$

versus

$$H_1: p \neq \tfrac{1}{2}$$

(and 0.05 were taken as the measure of reasonable doubt), the two critical values would be Y_1^* and Y_2^*, where

$$P(Y \leq Y_1^* \,|\, H_0 \text{ is true}) = 0.025$$

and

$$P(Y \geq Y_2^* \,|\, H_0 \text{ is true}) = 0.025.$$

That is, *half* of the prespecified reasonable doubt probability is put into each tail (Figure 6.6). The decision rule, then, calls for H_0 to be rejected if y is either less than or equal to Y_1^* or greater than or equal to Y_2^*.   ∎

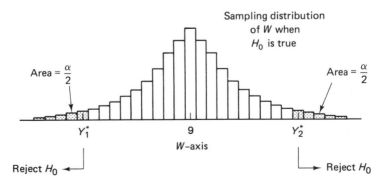

Figure 6.6   Critical region for a two-sided alternative.

As indicated in Example 6.3, the DeMoivre-Laplace limit theorem is frequently used to expedite hypothesis tests for the binomial parameter, p. Theorem 6.1 gives a formal statement of the procedure. The proof is based on the principle introduced in Section 6.4, but the details will not be presented here.

Comment. Theorem 4.3 provides a sufficiently good approximation to the exact distribution of Y/n provided

$$0 < y - 2\sqrt{n\left(\frac{y}{n}\right)\left(1 - \frac{y}{n}\right)} < y + 2\sqrt{n\left(\frac{y}{n}\right)\left(1 - \frac{y}{n}\right)} < n. \qquad (6.2)$$

Since $0 \leq y \leq n$, Equation 6.2 simply says that n and p should be such that y is at least two (estimated) standard deviations away from its minimum and maximum values. ∎

THEOREM 6.1   Suppose a total of y successes are observed in a series of n independent Bernoulli trials and that Inequality 6.2 holds. Let $p = P(\text{success}$ occurs at any given trial) and let α be the probability associated with the definition of reasonable doubt. To test

$$H_0: p = p_0$$

$$\text{versus}$$

$$H_1: p \neq p_0$$

we should reject H_0 if

$$\frac{y - np_0}{\sqrt{np_0(1 - p_0)}} \quad \text{is either} \quad \begin{cases} \leq -z_{\alpha/2} \\ \text{or} \\ \geq +z_{\alpha/2}, \end{cases}$$

where $P(Z \geq z_{\alpha/2}) = P(Z \leq -z_{\alpha/2}) = \alpha/2$. To test

$$H_0: p = p_0$$

$$\text{versus}$$

$$H_1: p > p_0$$

we should reject H_0 if

$$\frac{y - np_0}{\sqrt{np_0(1 - p_0)}} \geq +z_{\alpha}.$$

To test

$$H_0: p = p_0$$

$$\text{versus}$$

$$H_1: p < p_0$$

we should reject H_0 if

$$\frac{y - np_0}{\sqrt{np_0(1 - p_0)}} \leq -z_{\alpha}.$$

QUESTION 6.2.1   Many species of insects are strongly attracted to light at a wavelength of 365 mμ. This is in the near ultraviolet range and is often the wavelength used in outdoor light traps. The phototropic behavior elicited by longer wavelengths, though, is less understood. In a recent experiment (196) the reactions of a particular species of mosquitoes (*Anopheles stephensi*) to these longer wavelengths were studied in a laboratory environment. The experimental apparatus consisted of a metal tube, 56 cm long and 9 cm in diameter. At one end was an incandescent lamp; at the other end, a device that could produce monochromatic light of any desired wavelength. The two sources were adjusted so their intensities at the center of the tube were the same. Mosquitoes were introduced into the tube midway between the two ends. They were exposed simultaneously to both sources of light for $2\frac{1}{2}$ minutes, after which time the two ends of the tube were sealed off and the number of mosquitoes in each arm counted. Between 100 and 200 mosquitoes were placed in the tube for a given "trial"—and the entire procedure was replicated several times. For one series of trials, a total of 958 mosquitoes were exposed to *red* light (690 mμ) coming from one end and "white" light from the other. After $2\frac{1}{2}$ minutes, 642 were found in the red end of the tube. Do the appropriate hypothesis test. Use the 0.01 decision rule. What are you assuming about independence here?

The next two examples show how hypothesis tests can be set up when $f_Y(y)$ is either uniform or exponential. Actually, whatever distribution the Y_i's have, the construction of a test will closely parallel the approach followed for the binomial: first, a sufficient statistic is found for the parameter being tested; then the critical region is defined as the set of W-values least favorable to $H_0: \theta = \theta_0$ (but still admissible under H_1) and having the property that

$$P(W \in C \,|\, H_0 \text{ is true}) = \alpha = \begin{cases} \int_C g_W(w; \theta_0)\, dw \\ \text{or} \\ \sum_C g_W(w; \theta_0), \end{cases} \qquad (6.3)$$

where α is usually set at either 0.05 or 0.01. When Equation 6.3 is solved for the limits of C, the test will be completely specified.

EXAMPLE 6.4.   Suppose countries A and B are at war and A, in an attempt to intimidate B, leaks a high-level communiqué stating that A's armies are deploying a total of 5000 tanks in the field. But B's strategists think that figure might be an exaggeration—so far six of A's tanks have been captured and the highest serial number was 2516. What should B conclude: is A bluffing or not?
  If θ denotes the true tank deployment for country A, what country B wants to test is

$$H_0: \theta = 5000$$

versus

$$H_1: \theta < 5000.$$

We saw in Chapter 5 that a sufficient statistic for θ, assuming the captured serial numbers are uniformly distributed, is

$$W = \frac{n+1}{n} \cdot Y_{\max} = \frac{7}{6} \cdot Y_{\max}.$$

Intuitively, if W is too much *less than* 5000, B should reject H_0.

Comment. Note that we are modeling a discrete random variable with its continuous analog. Serial numbers are necessarily discrete, but the Chapter 5 result giving $g_W(w)$ referred originally to a continuous random variable. However, since the number of serial numbers is quite large, it is not unreasonable to expect the approximation to be more than adequate. ∎

Let 0.05 be retained as the "limit" defining reasonable doubt (Figure 6.7). Then $W*$ is the value for which

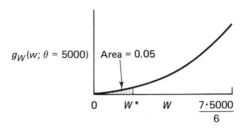

Figure 6.7 Rejection region for tank example.

$$\int_0^{W^*} g_W(w; \theta = 5000) \, dw = \int_0^{W^*} \frac{\frac{6^7}{7^6} w^5}{(5000)^6} \, dw = 0.05$$

(see Example 5.2). Therefore,

$$\frac{6^7 W^{*6}}{7^6 \cdot 6} \cdot \frac{1}{(5000)^6} = 0.05,$$

which gives

$$W^* = \sqrt[6]{0.05} \left[\frac{7(5000)}{6} \right]$$

$$= 3541.$$

Since the observed w is 2935 ($= \frac{7}{6} \cdot 2516$), B should reject H_0 and conclude that A is deploying fewer tanks than the communiqué claims (see Figure 6.8).

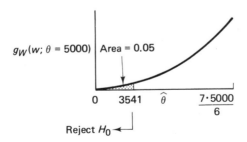

Figure 6.8 Rejection region showing $W*$ computed.

QUESTION 6.2.2   Find the critical region for testing $H_0: \theta = 5000$ against $H_1: \theta < 5000$ if α is set equal to 0.01. Assume, as above, that $n = 6$.

EXAMPLE 6.5. Suppose General Electric markets a new 75-watt lightbulb and guarantees it to last, on the average, 1000 hours. As a check, a consumer group selects four of the bulbs at random, puts them on test, and finds that they last 980, 1005,

992, and 973 hours, respectively, giving an average life of only 987.5 hours. Is this sufficient evidence to invalidate GE's claim?

As we have mentioned before, a simple model for Y, a bulb's lifetime, is the exponential,

$$f_Y(y) = \frac{1}{\lambda} e^{-y/\lambda}, \qquad y > 0.$$

The null and alternative hypotheses would then be written

$$H_0: \lambda = 1000,$$
$$H_1: \lambda < 1000.$$

Recall from Question 5.7.4 that a sufficient statistic for λ is $W = \bar{Y}$; also, from Example 5.3, we know that the density of $\bar{Y}$ is gamma:

$$g_W(w; \lambda = 1000) = \frac{1}{\Gamma(4)(\frac{1000}{4})^4} w^3 e^{-4w/1000}.$$

Again the critical region will be in the left-hand tail of $g_W(w; \lambda = 1000)$. Specifically, we will reject H_0 if $w \leq W^*$, where

$$\int_0^{W^*} \frac{1}{\Gamma(4)(250)^4} w^3 e^{-w/250} \, dw = 0.05.$$

This is not an easy equation to solve, but with tables of the incomplete gamma function (177) it can be shown that

$$W^* = 341.6.$$

Since the observed w is greater than W^*, there is *not* sufficient evidence to reject the manufacturer's claim.

QUESTION 6.2.3   Suppose a consumer group intended to test $H_0: \lambda = 1000$ versus $H_1: \lambda < 1000$ on the basis of *one* observation. Find the corresponding W^* (assuming the 0.05 criterion).

6.3 TYPE I AND TYPE II ERRORS

It is worthwhile at this point to examine the consequences of a decision rule in terms of the *errors* it might lead to. No matter what sort of mathematical facade is put over the decision-making process, there is simply no way to avoid the possibility of making errors—and, what is worse, no way to recognize them even after they occur. Basically, two different kinds of errors can be committed: we can reject H_0 when H_0 is true or we can accept H_0 when H_0 is false. These are called *Type I* and *Type II* errors, respectively. Figure 6.9 shows the four possible "decision"– "state of nature" combinations.

Although there is no way to know, after the fact, whether our decision was incorrect, it *is* possible to compute the *probability* of making an incorrect decision. For example, to test

$$H_0: p = \tfrac{1}{2}$$

versus

$$H_1: p > \tfrac{1}{2}$$

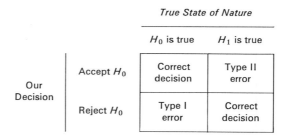

Figure 6.9 The two types of error.

given $n = 18$ observations, we finally decided on a decision rule (see Section 6.2) that called for the rejection of H_0 whenever $y \geq 13$. By definition, then, the probability that we will commit a Type I error is the probability that Y will be greater than or equal to 13 given that $p = \frac{1}{2}$:

$$P(\text{Type I error}) = P(\text{reject } H_0 \mid H_0 \text{ is true})$$
$$= P(Y \geq 13 \mid p = \frac{1}{2})$$
$$= \sum_{y=13}^{18} \binom{18}{y}(\tfrac{1}{2})^y(1 - \tfrac{1}{2})^{18-y}$$
$$= 0.05.$$

This last result, of course, should come as no surprise. The critical value was specifically set equal to 13 so that the probability associated with the critical region *would* be 0.05 (when H_0 was true). This was our definition of reasonable doubt.

The probability of committing a Type I error is referred to as the test's *level of significance* and is denoted α. The concept of level of significance is central to the decision-making process. When the results of any statistical test are summarized, two items of information absolutely *must* be included: (1) whether H_0 was accepted or rejected and (2) the level of significance at which the test was carried out. Thus, we would phrase the conclusion of the handwriting experiment by saying that H_0 *is accepted at the $\alpha = 0.05$ level of significance.*

The purpose in quoting a level of significance is to indicate the extent to which the data support, or fail to support, the null hypothesis. For example, to say that H_0 is rejected at the $\alpha = 0.001$ level of significance is a potentially much stronger statement (in favor of H_1) than to say that H_0 is rejected at the $\alpha = 0.05$ level of significance.

QUESTION 6.3.1   Suppose $H_0 : p = \frac{1}{2}$ is tested against $H_1 : p > \frac{1}{2}$. If H_0 is rejected at the $\alpha = 0.05$ level of significance, will it necessarily be rejected at the $\alpha = 0.01$ level?

Computing the probability of committing a Type II error is complicated by the fact that the alternative hypothesis embraces an entire *range* of parameter values, as opposed to a single point. This means that the assumption that H_1 is true does not uniquely determine the exact sampling distribution for W. It will be necessary, therefore, to condition any Type II error calculation on some specific H_1 parameter value.

As an example, suppose we wanted to find the probability of committing a Type II error in the handwriting experiment *if the true p were 0.7*. By definition,

$$P(\text{Type II error}\,|\,p = 0.7) = P(\text{accept } H_0\,|\,p = 0.7)$$
$$= P(Y \leq 12\,|\,p = 0.7)$$
$$= \sum_{y=0}^{12} \binom{18}{y}(0.7)^y(1 - 0.7)^{18-y}$$
$$= 0.47.$$

This means that if p were actually 0.7 it would go "undetected"—that is, we would erroneously accept $H_0: p = \frac{1}{2}$ — 47% of the time.

The symbol for the probability of a Type II error is β. Figure 6.10 shows the sampling distributions of W when $p = \frac{1}{2}$ and when $p = 0.7$; the areas corresponding to α and β are shaded in.

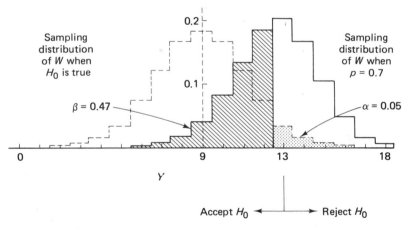

Figure 6.10 Sampling distributions of W, $p = 0.5$ and $p = 0.7$.

Clearly, the magnitude of β is a function of the presumed value for p. If, for example, the handwriting expert could make correct identifications *90%* of the time (Figure 6.11), the probability that our decision rule would lead us to make a Type II error would be 0.006:

$$P(\text{Type II error}\,|\,p = 0.90) = P(\text{accept } H_0\,|\,p = 0.90)$$
$$= P(Y \leq 12\,|\,p = 0.90)$$
$$= \sum_{y=0}^{12} \binom{18}{y}(0.9)^y(1 - 0.9)^{18-y}$$
$$= 0.006.$$

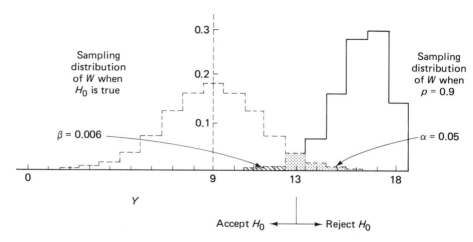

Figure 6.11   Sampling distributions of W, $p = 0.5$ and $p = 0.9$.

Comment. The basic notions of Type I and Type II errors first arose in a quality control context. The pioneering work was done at the Bell Telephone Laboratories: there the terms *producer's risk* and *consumer's risk* were introduced for what we now call α and β. Eventually these ideas were generalized by Neyman and Pearson in the 1930s and evolved into the theory of hypothesis testing as we know it today.

∎

The quantity $1 - \beta$ is often used in place of β to describe the performance characteristics of a test procedure. It follows that if β is the probability of *accepting* H_0 (when H_1 is true), $1 - \beta$ is the probability of *rejecting* H_0 (when H_1 is true). We call $1 - \beta$ the *power* of the test and the set of ordered pairs $(p, 1 - \beta)$ the *power function* of the test. Figure 6.12 shows the power function of the test

$$H_0 \colon p = \tfrac{1}{2}$$

versus

$$H_1 \colon p > \tfrac{1}{2}$$

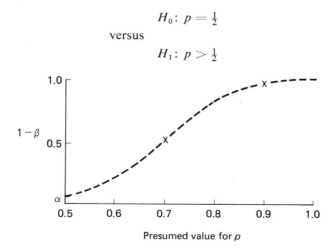

Figure 6.12   Power function.

when the decision rule is "Reject H_0 if $y \geq 13$." [The two X's on the curve represent the points we have just determined, $(0.7, 0.53)$ and $(0.9, 0.994)$.] Notice that when p equals its H_0 value, $1 - \beta = \alpha$. Also, as p gets further and further away from its H_0 value, the power function approaches 1.

Comment. The power function for a one-sided test of the form

$$H_0: p = p_0$$

versus

$$H_1: p < p_0$$

would look like Figure 6.13.

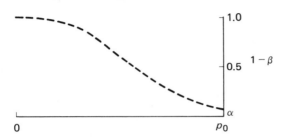

Presumed value for p **Figure 6.13**   Power function.

Figure 6.14 shows the general form of $1 - \beta$ versus p for a two-sided test,

$$H_0: p = p_0$$

versus

$$H_1: p \neq p_0. \quad \blacksquare$$

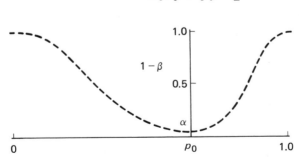

Presumed value for p

Figure 6.14   Power function for two-sided test.

Comment. It may appear from the previous discussion that an experimenter has no control over the probability of committing a Type II error once α is fixed. Such is not the case. The value of β can be manipulated (we would, of course, like it to be as small as possible for a given alternative hypothesis) in two ways:

(1) by choosing a test procedure that has a particularly steep power curve (see Question 6.3.3) and (2) by increasing the sample size. This latter option is especially important and will be explored at considerable length in Chapter 7. A partial solution to the first option, finding an "optimal" test procedure—that is, one that is *most powerful*—is described in Appendix 6.1 on page 257. ▌

QUESTION 6.3.2   Use the DeMoivre-Laplace limit theorem to construct the power function for the ESP study referred to in Example 6.3. Find $1 - \beta$ for p-values of 0.200, 0.201, 0.202, 0.204, and 0.206.

QUESTION 6.3.3   Suppose methods A and B were two different procedures for testing the same H_0 and H_1. (Assume that both A and B used the same sample size and were carried out at the same level of significance.) How could the power functions for A and B be used to tell us which method, if either, was "better?" Is it possible that one method would be better under certain conditions and worse under others? What might the power functions look like then? What would an "ideal" power curve look like?

6.4 A NOTION OF OPTIMALITY: THE GENERALIZED LIKELIHOOD RATIO

In the next several chapters we will be looking closely at some of the particular hypothesis tests that statisticians most often need to use in dealing with real-world problems. All of these have the same conceptual heritage—a very fundamental notion known as the *generalized likelihood ratio*, or GLR. More than just a principle, the generalized likelihood ratio is a working criterion for actually *suggesting* test procedures. In a sense, the GLR does for hypothesis testing what the principle of maximum likelihood does for estimation.

In this section, the generalized-likelihood-ratio principle will be applied to a situation previously discussed—testing hypotheses about the parameter of a uniform distribution. This turns out to be an almost trivial application of the GLR and hardly seems necessary at all, since our intuition has already suggested what seems to be a reasonable answer to the problem (see Example 6.4). Most hypothesis-testing models, though, are not as simplistic or amenable to intuition as the uniform and binomial are, and they require a more formal and systematic analysis based on first principles. This is where the generalized likelihood ratio takes on its special importance.

Let $Y_1, Y_2, \ldots, Y_n$ be a random sample from a uniform distribution defined over $(0, \theta)$, where θ is unknown. Our objective is to test

$$H_0: \theta = \theta_0$$

versus

$$H_1: \theta < \theta_0$$

at a specified level of significance, say, α.

As a starting point, it will be necessary to define two parameter spaces, ω and Ω, where ω is a particular subset of Ω. The former is the set of unknown parameter values admissible under H_0. Here, the only unknown parameter is θ, and the null hypothesis restricts it to a single point. In set notation,

$$\omega = \{\theta : \theta = \theta_0\}.$$

The second parameter space, Ω, is the set of *all* possible values of all unknown parameters. In this case,

$$\Omega = \{\theta : 0 < \theta \leq \theta_0\}.$$

Now, recall the definition of the likelihood function, L, from Definition 5.7. Given a sample from a uniform distribution,

$$L = L(\theta) = \prod_{i=1}^{n} f_Y(y_i; \theta) = \begin{cases} (1/\theta)^n, & \text{for } 0 \leq y_i \leq \theta \\ 0, & \text{otherwise.} \end{cases}$$

For reasons the following will make clear, we will want to maximize $L(\theta)$ twice, once under ω and again under Ω. Since θ can take on only one value—θ_0—under ω,

$$\max_{\omega} L(\theta) = L(\theta_0) = \begin{cases} (1/\theta_0)^n, & \text{for } 0 \leq y_i \leq \theta_0 \\ 0, & \text{otherwise.} \end{cases}$$

Of course, maximizing $L(\theta)$ under Ω—that is, with *no* restrictions—is accomplished by substituting the maximum-likelihood estimator for θ into $L(\theta)$. For the uniform distribution, $W = Y_{\max}$ is the maximum-likelihood estimator (see Example 5.16). Therefore,

$$\max_{\Omega} L(\theta) = \left(\frac{1}{y_{\max}}\right)^n.$$

For notational simplicity, we will let $\max_{\omega} L(\theta_1, \ldots, \theta_k)$ and $\max_{\Omega} L(\theta_1, \ldots, \theta_k)$ be denoted $L(\hat{\omega})$ and $L(\hat{\Omega})$, respectively.

DEFINITION 6.1 Let $Y_1, Y_2, \ldots, Y_n$ be a random sample from $f_Y(y; \theta_1, \ldots, \theta_k)$. The generalized likelihood ratio, λ, is defined to be

$$\lambda = \frac{\max_{\omega} L(\theta_1, \ldots, \theta_k)}{\max_{\Omega} L(\theta_1, \ldots, \theta_k)} = \frac{L(\hat{\omega})}{L(\hat{\Omega})}.$$

For the uniform distribution,

$$\lambda = \frac{(1/\theta_0)^n}{(1/y_{\max})^n} = \left(\frac{y_{\max}}{\theta_0}\right)^n.$$

Note that, in general, λ will always be positive but never greater than 1 (why?). Furthermore, values of the likelihood ratio close to 1 suggest that the data are very compatible with H_0. That is, the observations are "explained" almost as well by the H_0 parameters as by *any* parameters [as measured by $L(\hat{\omega})$ and $L(\hat{\Omega})$]. For these values of λ we should *accept* H_0. Conversely, if $L(\hat{\omega})/L(\hat{\Omega})$ were close to 0, the data would not be very compatible with the parameter values in ω and it would make sense to *reject* H_0. This is the rationale behind the generalized-likelihood-ratio *principle* as stated in the next definition.

> **DEFINITION 6.2** A generalized-likelihood-ratio test (GLRT) is one that rejects H_0 whenever
> $$0 < \lambda \leq \lambda^*,$$
> where λ^* is chosen so that
> $$P(0 < \Lambda \leq \lambda^* \,|\, H_0 \text{ is true}) = \alpha.$$

If we knew the distribution under H_0 of the random variable Λ—say, $g_\Lambda(\lambda \,|\, H_0)$—the critical value λ^* (and, hence, the critical region, C) could be determined by solving the equation

$$\alpha = \int_0^{\lambda^*} g_\Lambda(\lambda \,|\, H_0)\, d\lambda.$$

(See Figure 6.15.) In most situations, though, $g_\Lambda(\lambda \,|\, H_0)$ is *not* known, and it becomes necessary to show that Λ is a monotonic function of some quantity W, where the distribution of W *is* known. Once we have found such a statistic, any test based on W will be equivalent to one based on Λ.

Figure 6.15   Rejection region for GLRT.

Here, a suitable W is easy to find. Note that

$$P(\Lambda \leq \lambda^* \,|\, H_0) = \alpha = P\left[\left(\frac{Y_{\max}}{\theta_0}\right)^n \leq \lambda^* \,|\, H_0\right]$$

$$= P\left(\frac{Y_{\max}}{\theta_0} \leq \sqrt[n]{\lambda^*} \,|\, H_0\right).$$

Let $W = Y_{\max}/\theta_0$ and $W^* = \sqrt[n]{\lambda^*}$. Then

$$P(\Lambda \leq \lambda^* \,|\, H_0) = P(W \leq W^* \,|\, H_0). \tag{6.4}$$

Here the right-hand side of Equation 6.4 can be evaluated from what we already know about the density function for the largest order statistic from a uniform distribution. Let $f_Y(y\,;\theta_0)$ be the density function for $Y_{\max}$. Then

$$g_W(w\,;\theta_0) = \theta_0 f_Y(\theta_0 w),$$

which, from Example 5.1, reduces to

$$\frac{\theta_0 n(\theta_0 w)^{n-1}}{\theta_0^n} = nw^{n-1}, \qquad 0 \le w \le 1.$$

Therefore,

$$P(W \le W^* \,|\, H_0) = \int_0^{W^*} nw^{n-1}\, dw = (W^*)^n = \alpha,$$

implying that the critical value for W is

$$W^* = \sqrt[n]{\alpha}\,.$$

That is, the GLRT calls for H_0 to be rejected if

$$w = \frac{y_{\max}}{\theta_0} \le \sqrt[n]{\alpha}\,.$$

Although written in a slightly different form, this is the same criterion that was decided on in Example 6.4.

Comment. The GLR is applied to other hypothesis-testing situations in a manner very similar to what was described here: first we find $L(\hat{\omega})$ and $L(\hat{\Omega})$, then Λ, and finally W. The algebra involved, though, usually becomes considerably more formidable. For example, in the "normal" model taken up in Chapters 7 and 8, both parameter spaces are two-dimensional and the likelihood function is a product of densities of the form

$$f_Y(y;\mu,\sigma^2) = \frac{1}{\sqrt{2\pi}\,\sigma}\, e^{-(1/2)[(y-\mu)/\sigma]^2}. \quad \blacksquare$$

QUESTION 6.4.1   Let $n_1, n_2, \ldots, n_k$ be a random sample from the geometric probability function

$$f_N(n;p) = pq^{n-1}, \qquad n = 1, 2, \ldots,$$

where $q = 1 - p$. Find $\max_\omega L(p)$, $\max_\Omega L(p)$, and the generalized-likelihood-ratio criterion for testing $H_0 : p = p_0$ versus $H_1 : p \ne p_0$.

APPENDIX 6.1 THE NEYMAN–PEARSON LEMMA

The generalized-likelihood-ratio criterion of Section 6.4 is a modification of an even more basic result in the theory of hypothesis testing, the Neyman-Pearson lemma. While the structure the lemma requires is too restrictive for the result to have much *practical* significance, it does have enormous importance from a conceptual standpoint. Unlike the generalized-likelihood-ratio criterion, which carries with it no guarantee of optimality, a test based on the Neyman-Pearson lemma is "best" in the sense that no other procedure having the same Type I error probability has a lower Type II error probability.

Before giving a formal statement of the Neyman-Pearson lemma, it might be helpful to describe the result in more intuitive terms. Suppose a random sample of size n, $Y_1, Y_2, \ldots, Y_n$, is drawn from $f_Y(y;\theta)$, where θ is unknown but is restricted to be one of only two possible values, either θ_0 or θ_1. Our objective is to test H_0: $\theta = \theta_0$ versus H_1: $\theta = \theta_1$.

Consider any set C, a collection of n-tuples (Figure 6.16) having "size" α—that is, C has the property that

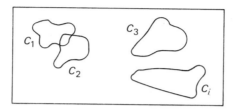

Figure 6.16   Four sets of n-tuples.

$$P[(Y_1, Y_2, \ldots, Y_n) \in C \,|\, H_0 \text{ is true}] = \alpha.$$

It follows that C can be thought of as a critical region for a test of H_0: $\theta = \theta_0$ versus H_1: $\theta = \theta_1$. Specifically, if the observed n-tuple, $(y_1, y_2, \ldots, y_n)$, falls into our preselected C, we will reject H_0. But what if H_1 were true? Then, intuitively, the "best" critical region would be the one having the highest probability of containing $(Y_1, Y_2, \ldots, Y_n)$. Formalizing this notion, we say that C_* is the best critical region of size α if

(1) $P[(Y_1, Y_2, \ldots, Y_n) \in C_* \,|\, H_0 \text{ is true}] = \alpha$

and

(2) $P[(Y_1, Y_2, \ldots, Y_n) \in C_* \,|\, H_1 \text{ is true}] \geq P[(Y_1, Y_2, \ldots, Y_n) \in C \,|\, H_1 \text{ is true}]$

for every other $C \neq C_*$.

The Neyman-Pearson lemma fits into all this by providing us with a working criterion for actually finding C_*. As Theorem 6.2 makes clear, the criterion is based on the likelihood ratio. The first n-tuple we put into C_* will be the one having the smallest value for

$$\frac{L(\theta_0; y_1, y_2, \ldots, y_n)}{L(\theta_1; y_1, y_2, \ldots, y_n)}.$$

This, of course, is the n-tuple most compatible with H_1. Continuing the construction process in this fashion, the next sample point put into C_* will be the one with the *second* smallest value of $L(\theta_0; y_1, y_2, \ldots, y_n)/L(\theta_1; y_1, y_2, \ldots, y_n)$, and so on. C_* will be complete when the sum of the H_0 probabilities associated with its members is α.

THEOREM 6.2 (Neyman-Pearson Lemma)   Let $Y_1, Y_2, \ldots, Y_n$ be a random sample from $f_Y(y; \theta)$, where θ is either θ_0 or θ_1. Let $L(\theta_i; y_1, y_2, \ldots, y_n)$ denote the likelihood function of the observed sample when $\theta = \theta_i$, $i = 0$ or 1. Let k be a positive number. The best critical region, C_*, for testing H_0: $\theta = \theta_0$ versus H_1: $\theta = \theta_1$ is the set of n-tuples, $(y_1, y_2, \ldots, y_n)$ satisfying the following three conditions:

(a) $\dfrac{L(\theta_0 : y_1, y_2, \ldots, y_n)}{L(\theta_1 ; y_1, y_2, \ldots, y_n)} \le k$ for each n-tuple in C_*.

(b) $\dfrac{L(\theta_0 : y_1, y_2, \ldots, y_n)}{L(\theta_1 ; y_1, y_2, \ldots, y_n)} > k$ for each n-tuple not in C_*.

(c) $P[(Y_1, Y_2, \ldots, Y_n) \in C_* \mid H_0 \text{ is true}] = \alpha$.

An example will be helpful at this point. Suppose $Y_1, Y_2, \ldots, Y_{10}$ is a random sample from an $N(\theta, 1)$ distribution, where θ is either 3 or 2, and we want to test

$$H_0: \theta = 3$$

versus

$$H_1: \theta = 2.$$

Set $\alpha = 0.05$. The likelihood ratio is written

$$\frac{L(\theta_0 : y_1, y_2, \ldots, y_{10})}{L(\theta_1 ; y_1, y_2, \ldots, y_{10})} = \frac{(1/\sqrt{2\pi})^{10} e^{-(1/2)\sum_{i=1}^{10}(y_i - 3)^2}}{(1/\sqrt{2\pi})^{10} e^{-(1/2)\sum_{i=1}^{10}(y_i - 2)^2}},$$

which simplifies to

$$e^{\sum_{i=1}^{10} y_i - 25}.$$

According to the Neyman-Pearson lemma, the 10-tuples that should be included in C_* are those for which

$$e^{\sum_{i=1}^{10} y_i - 25} \le k, \tag{6.5}$$

where k is to be chosen so that

$$P(e^{\sum_{i=1}^{10} Y_i - 25} \le k \mid \theta = 3) = 0.05. \tag{6.6}$$

To solve for k in Equation 6.6, we need first to simplify Equation 6.5. Note that the set of y_i's that satisfy

$$e^{\sum_{i=1}^{10} y_i - 25} \le k$$

will be identical to those that satisfy

$$\bar{y} \le \frac{\ln k + 25}{10} = k'.$$

Therefore,

$$P(e^{\sum_{i=1}^{10} Y_i - 25} \le k \mid \theta = 3) = P(\bar{Y} \le k' \mid \theta = 3).$$

But $\bar{Y} \sim N(\theta, \frac{1}{10})$, so that

$$P(\bar{Y} \le k' \mid \theta = 3) = P\left(\frac{\bar{Y} - 3}{1/\sqrt{10}} \le k'' \mid \theta = 3\right) = P(Z \le k'') = 0.05.$$

From tables of the standard normal,

$$P(Z \le -1.64) = 0.05,$$

from which it follows that $k'' = -1.64$ and the best critical region is

$$\frac{\bar{y} - 3}{1/\sqrt{10}} \leq -1.64$$

or, equivalently,

$$\bar{y} \leq 3 - 1.64\left(\frac{1}{\sqrt{10}}\right) = 2.48.$$

This, of course, is the "type" of critical region that makes the most sense intuitively: that is, rejecting H_0 when $\bar{y}$ is too much *less than* 3.

PROOF OF THE NEYMAN-PEARSON LEMMA

Let C be any other critical region of size α (besides C_*) for testing $H_0: \theta = \theta_0$ versus $H_1: \theta = \theta_1$. The lemma will be proved if it can be demonstrated that

$$P[(Y_1, Y_2, \ldots, Y_n) \in C_* | H_1] \geq P[(Y_1, Y_2, \ldots, Y_n) \in C | H_1].$$

To simplify the notation, we will write

$$P[(Y_1, Y_2, \ldots, Y_n) \in C_* | H_1] = \iint \cdots \int_{C_*} L(\theta_1; y_1, \ldots, y_n)\, dy_1 \cdots dy_n$$

as

$$\int_{C_*} L(\theta_1).$$

So, we wish to show that

$$\int_{C_*} L(\theta_1) - \int_C L(\theta_1) \geq 0.$$

It is necessarily the case that

$$C_* = (C_* \cap C) \cup (C_* \cap C^c),$$

from which it follows that

$$\int_{C^*} L(\theta_1) - \int_C L(\theta_1) = \int_{C_* \cap C} L(\theta_1) + \int_{C_* \cap C^c} L(\theta_1) - \int_{C \cap C_*} L(\theta_1) - \int_{C \cap C_*^c} L(\theta_1)$$

$$= \int_{C_* \cap C^c} L(\theta_1) - \int_{C \cap C_*^c} L(\theta_1).$$

But for each n-tuple in C_* (and in $C_* \cap C^c$)

$$\frac{L(\theta_0)}{L(\theta_1)} \leq k.$$

Therefore,

$$\int_{C_* \cap C^c} L(\theta_1) \geq \frac{1}{k} \int_{C_* \cap C^c} L(\theta_0). \tag{6.7}$$

By a similar argument,

$$\int_{C \cap C_*^c} L(\theta_1) \leq \frac{1}{k} \int_{C \cap C_*^c} L(\theta_0). \tag{6.8}$$

Note that Inequalities 6.7 and 6.8 imply that

$$\int_{C_* \cap C^c} L(\theta_1) - \int_{C \cap C_*^c} L(\theta_1) \geq \frac{1}{k} \int_{C_* \cap C^c} L(\theta_0) - \frac{1}{k} \int_{C \cap C_*^c} L(\theta_0).$$

Therefore,

$$\int_{C_*} L(\theta_1) - \int_C L(\theta_1) \geq \frac{1}{k} \left[\int_{C_* \cap C^c} L(\theta_0) - \int_{C \cap C_*^c} L(\theta_0) \right]. \tag{6.9}$$

But

$$\int_{C_* \cap C^c} L(\theta_0) - \int_{C \cap C_*^c} L(\theta_0) = \int_{C_* \cap C^c} L(\theta_0) + \int_{C_* \cap C} L(\theta_0) - \int_{C \cap C_*} L(\theta_0) - \int_{C \cap C_*^c} L(\theta_0)$$

$$= \int_{C_*} L(\theta_0) - \int_C L(\theta_0)$$

$$= \alpha - \alpha$$

$$= 0.$$

The proof is completed by substituting this result into Inequality 6.9.

Comment. A major "weakness" of the Neyman-Pearson lemma is that it requires the null hypothesis to specify completely each parameter in the presumed model. For the kinds of hypothesis tests most often encountered in practice, this is impossible. For instance, a more realistic version of the example just described (see Chapter 7) would assume that μ and σ^2 were both unknown. In that case, H_0 and H_1 would become

$$H_0: \mu = \mu_0 \text{ and } 0 < \sigma^2 < \infty,$$
$$H_1: \mu = \mu_1 \text{ and } 0 < \sigma^2 < \infty$$

and the Neyman-Pearson lemma would no longer apply.   ∎

REVIEW EXERCISES FOR CHAPTER 6

1. There is a theory [see (128)] that the anticipation of a birthday can prolong a person's life. In a recent study set up to examine that notion statistically, it was found that only 8% of 747 people whose obituaries were published in Salt Lake City in 1975 died in the three-month period preceding their birthday (119). Test the appropriate hypothesis at the $\alpha = 0.01$ level of significance.

2. Mr. Cosmo, an astrologer, claims that personality traits are a function of an individual's zodiac sign, and that he has a better than $\frac{1}{12}$ chance (there are 12 zodiac signs) of correctly identifying a person's sign once he knows something about that person's behavior. A skeptic decides to put Mr. Cosmo's claim to a test by asking him to give the zodiac signs for 100 persons. State the decision rule for testing

$$H_0: p = \tfrac{1}{12}$$

versus

$$H_1: p > \tfrac{1}{12}$$

at the $\alpha = 0.05$ level of significance. If Mr. Cosmo guesses 10 of the 100 persons correctly, should we accept or reject H_0?

3. If $H_0: p = p_0$ is rejected in favor of $H_1: p > p_0$ at the $\alpha = 0.01$ level of significance, will it necessarily be rejected at the $\alpha = 0.05$ level of significance? (Recall Question 6.3.1.)

4. A random sample of size $n = 1$ is drawn from a uniform pdf defined over the interval $[0, \theta]$. We decide to test

$$H_0: \theta = 2$$

versus

$$H_1: \theta \neq 2$$

by rejecting H_0 if either $y \leq 0.1$ or $y \geq 1.9$. Find α. Also, find β if the true value of θ is 2.5.

5. Commercial fishermen working certain parts of the Atlantic Ocean sometimes have their efforts hindered by the presence of whales. The problem is to scare away the whales without frightening the fish. In the past, sonar operators have confirmed that 40% of all whales sighted left the area of their own accord, probably to get away from the noise of the boat. Recently, attempts have been made to increase that figure by transmitting underwater the sounds of a killer whale. To date, the technique has been tried on 52 whales: of that number, a total of 24 immediately left the area.
   (a) Set up a null and alternative hypothesis for testing whether broadcasting the cry of a killer whale is significantly more effective in "clearing" a fishing area than doing nothing.
   (b) Test the hypothesis of part (a) using the $\alpha = 0.01$ level of significance.

6. Construct a power curve for the astrological sign experiment described in Review Exercise 2.

7. An urn contains ten chips. An unknown number of the chips are white; the others are red. We wish to test

$$H_0: \text{exactly half the chips are white}$$

versus

$$H_1: \text{more than half the chips are white.}$$

We will draw, without replacement, three chips and reject H_0 if two or more are white. Find α. Also, find β when the urn is (a) 60% white and (b) 70% white.

8. Answer the questions in Review Exercise 7 under the assumption that the three chips are drawn *with* replacement.

9. Suppose the best critical region for a certain hypothesis is of the form

$$\text{Reject } H_0 \text{ if } \sum_{i=1}^{n} y_i \geq k.$$

However, because of the discrete nature of the random variable Y,

$$P\left(\sum_{i=1}^{n} Y_i \geq k_1 \,\middle|\, H_0 \text{ is true}\right) = \alpha_1 > \alpha$$

and

$$P\left(\sum_{i=1}^{n} Y_i \geq k_2 \,\middle|\, H_0 \text{ is true}\right) = \alpha_2 < \alpha,$$

where $k_1 < k_2$. How might a "weighted" decision rule be formulated giving an *exact* Type I error probability of α?

10. Data collected in the past on a certain brand of nickel-cadmium batteries used in 35 mm cameras indicate that the distribution of their lifetimes is exponential with an average life (λ) of one year. That is, if Y denotes a battery's lifetime, then

$$f_Y(y) = e^{-y}, \qquad y > 0.$$

To maintain that standard, two such batteries are sampled at random from a production line and put on an accelerated testing program that simulates normal usage. The quality control inspector wishes to test

$$H_0: \lambda = 1$$

versus

$$H_1: \lambda < 1.$$

If the sum of the two lifetimes is less than or equal to 0.3, the batch from which the sample was chosen will be rejected. Find the corresponding α. Hint: See Example 3.14.

11. Refer to Review Exercise 108 at the end of Chapter 2. Let p denote the probability of the new drug's being successful. Compute α if the hypotheses being tested are $H_0: p = \frac{1}{3}$ versus $H_1: p > \frac{1}{3}$.

12. Suppose that Y_1, Y_2, Y_3, and Y_4, a random sample of size $n = 4$, is taken from a Poisson pdf whose parameter is unknown. Furthermore, suppose that the hypotheses to be tested are

$$H_0: \lambda = 1$$

versus

$$H_1: \lambda > 1.$$

and the decision rule is to be of the form,

$$\text{Reject } H_0 \text{ if } \sum_{i=1}^{n} y_i \geq k.$$

Find the value of k that gives an α approximately equal to 0.05.

13. When the results of statistical hypothesis tests are presented in journals or technical reports, any mention of a level of significance is often deleted—instead, a test's *P value* is quoted. By definition, the P value is the smallest α for which H_0 could be rejected. Find the P value for the whale data described in Review Exercise 5.

14. Use the Neyman-Pearson lemma to derive the form of the best critical region for testing

$$H_0: \lambda = \lambda_0$$

versus

$$H_1: \lambda = \lambda_1 \quad (> \lambda_0)$$

based on a random sample of size n drawn from a Poisson pdf with parameter λ. Will it necessarily be possible to find a critical region for this problem whose *exact* Type I error probability is α?

15. Sal is a pizza inspector for the city health department. Recently he has received a number of complaints directed against a certain pizzeria for allegedly failing to comply with their advertisements. The pizzeria claims that, on the average, each of their large pepperoni pizzas is topped with 2 ounces of pepperoni. The dissatisfied customers feel that the actual amount of pepperoni used is considerably less than that. To settle the matter, Sal decides to do a hypothesis test. First, he assumes that the distribution of pepperoni weights (per pizza) is normal with a mean of μ (ounces) and a standard deviation (σ) of 0.5 ounces. The hypotheses to be tested are

$$H_0: \mu = 2.00$$

versus

$$H_1: \mu < 2.00.$$

Sal decides to take one large pizza at random and weigh the pepperoni it contains. If y, the pepperoni weight, is less than or equal to 1.3 ounces, he will reject H_0.
(a) Compute the α that corresponds to Sal's decision rule.
(b) Compute β if the true average pepperoni weight per pizza is 1.8 ounces.
(c) What should Sal's decision rule be if he wants α to be 0.01?

16. Let $Y_1, Y_2, \ldots, Y_{10}$ be a random sample from an exponential pdf with unknown parameter λ. Find the form of the GLRT for

$$H_0: \lambda = \lambda_0$$

versus

$$H_1: \lambda > \lambda_0.$$

What integral would have to be evaluated to determine the critical value if α were equal to 0.05?

17. Suppose we are testing $H_0: p = p_0$ versus $H_1: p \neq p_0$. There is one way to simultaneously decrease the probabilities of committing both Type I and Type II errors (assuming the decision rule remains fixed): by increasing the sample size. Explain why that should be true.

18. Suppose a single observation is drawn from an $N(\mu, 1)$ distribution for the purpose of testing

$$H_0: \mu = 0$$

versus

$$H_1: \mu = 1.$$

Find the best critical region. Let $\alpha = 0.05$.

19. Suppose a random sample of size 5 is drawn from a uniform pdf

$$f_Y(y; \theta) = \begin{cases} 1/\theta, & 0 < y < \theta \\ 0, & \text{elsewhere.} \end{cases}$$

We wish to test

$$H_0: \theta = 2$$

versus

$$H_1: \theta > 2$$

by rejecting the null hypothesis if $y_{\max} \geq k$. Find the value of k that makes the probability of committing a Type I error equal to 0.05. Also, construct the power curve for the test.

20. Suppose the hypotheses given in Review Exercise 19 were tested using a random sample of size 1, rather than size 5. Construct the corresponding power curve and compare it to the power curve gotten when n equaled 5.

21. Let $Y_1, Y_2, \ldots, Y_{10}$ be a random sample from an $N(\theta, 1)$ pdf. Construct a power curve for the Neyman-Pearson test of

$$H_0: \theta = 3$$

versus

$$H_1: \theta = 2$$

described in Appendix 6.1 on page 257. Set $\alpha = 0.05$.

22. Polygraphs used in criminal investigations typically measure five body functions: (1) thoracic respiration, (2) abdominal respiration, (3) blood pressure and pulse rate, (4) muscular movement and pressure, and (5) galvanic skin response. In principle, the magnitude of these responses when the subject is asked a relevant question ("Did you murder your wife?") indicate whether he is lying or telling the truth. The procedure, of course, as a recent study bore out (80), is not infallible. Seven experienced polygraph examiners were given a set of 40 records—20 were from innocent suspects and 20 from guilty suspects. The subjects had been asked 11 questions, on the basis of which each examiner was to make an overall judgment: "Innocent" or "Guilty." The results are shown below.

		Suspect's True Status	
		Innocent	Guilty
Examiner's	"Innocent"	131	15
Decision	"Guilty"	9	125

What would be the numerical values of α and β in this context? In a judicial setting, should Type I and Type II errors carry equal weight? Explain.

23. Use the Neyman-Pearson lemma to determine the nature of the best critical region for testing $H_0: \theta = \theta_0$ versus $H_1: \theta = \theta_1 \ (< \theta_0)$ if

$$f_X(x) = (1 + \theta)x^{\theta}, \qquad 0 < x < 1.$$

Assume that the data consist of a random sample of size n.

24. A random sample of size 2 is drawn from a uniform pdf defined over the interval $[0, \theta]$. We wish to test

$$H_0: \theta = 2$$

versus

$$H_1: \theta < 2$$

by rejecting H_0 when $y_1 + y_2 \leq k$. Find the value for k that gives a level of significance of 0.05.

25. Suppose the hypotheses of Review Exercise 24 are to be tested with a decision rule of the form, "Reject $H_0: \theta = 2$ if $y_1 y_2 \leq k^*$." Find the value of k^* that gives a level of significance of 0.05. (See Example 3.12.)

26. If Y is a Poisson random variable with parameter λ, it can be shown (see Section 7.4) that

$$\frac{Y - \lambda}{\sqrt{\lambda}}$$

has approximately a standard normal pdf. This suggests that a test of

$$H_0: \lambda = \lambda_0$$

versus

$$H_1: \lambda > \lambda_0$$

can be set up by rejecting H_0 whenever $(y - \lambda_0)/\sqrt{\lambda_0}$ is too large. The following situation is a case in point. On the average, the number of babies born in Cleveland, Ohio, in September is 1472. On January 26, 1977, the city was immobilized by a blizzard. Nine months later, in September of 1977, the number of recorded births was 1718, an increase of 246. If births are assumed to be Poisson events with a September average of 1472, use the normal approximation to test $H_0: \lambda = 1472$ versus $H_1: \lambda > 1472$. Let α be 0.05.

27. Construct a power curve for the "blizzard baby" data of Review Exercise 26.

28. Suppose x successes are observed in a series of n independent Bernoulli trials. Let p denote the probability that a success occurs on any given trial. Use the Neyman-Pearson lemma to derive the form of the best critical region for testing $H_0: p = p_0$ versus $H_1: p = p_1$ $(< p_0)$.

29. Let Y_1 and Y_2 be a random sample of size 2 from the pdf

$$f_Y(y; \theta) = \theta y^{\theta - 1}, \qquad 0 < y < 1.$$

Use the Neyman-Pearson lemma to find the best critical region for testing

$$H_0: \theta = 1$$

versus

$$H_1: \theta = 2.$$

Specify the test for α equal to 0.10.

30. A die is either fair or loaded in such a way that $p_i = P(\text{face 1 appears}) = ki, i = 1, 2, \ldots, 6$. Suggest a reasonable form for a critical region to test

$$H_0: p_i = \tfrac{1}{6}, \qquad i = 1, 2, \ldots, 6$$

versus

$$H_1: p_i = ki, \qquad i = 1, 2, \ldots, 6.$$

(We will return to this problem in Chapter 9.)

31. Let $Y_1, Y_2, \ldots, Y_n$ be a random sample from a pdf depending on a single unknown parameter, θ. Use the Fisher-Neyman criterion (Theorem 5.2) to prove that the Neyman-Pearson critical region for testing $H_0: \theta = \theta_0$ versus $H_1: \theta = \theta_1$ will be a function of the sufficient statistic for θ, if one exists.

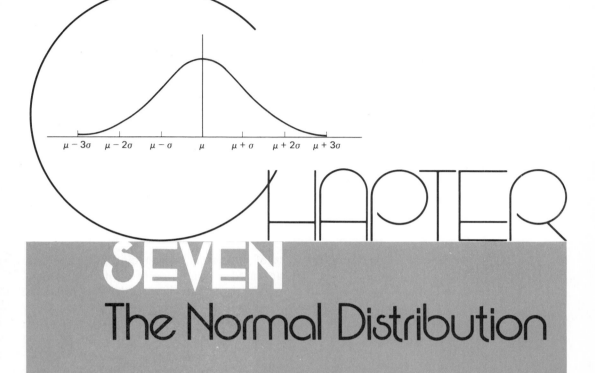

CHAPTER SEVEN
The Normal Distribution

FRANCIS GALTON

I know of scarcely anything so apt to impress the imagination as the wonderful form of cosmic order expressed by the "Law of Frequency of Error" (the normal distribution). The law would have been personified by the Greeks and deified, if they had known of it. It reigns with serenity and in complete self effacement amidst the wildest confusion. The huger the mob, and the greater the anarchy, the more perfect is its sway. It is the supreme law of Unreason.

7.1 INTRODUCTION

Finding probability distributions to describe—and, ultimately, to predict—empirical data is one of the most important contributions a statistician can make to the research scientist. Already we have seen a number of "models" pressed into this sort of service: the exponential described very well the wearout times of radar tubes: the binomial was an obvious choice as a model for the number of correct responses in an ESP experiment; there were physical reasons why the gamma would provide a good fit for rainfall data; and the Poisson was seen to apply to situations as diverse as the number of alpha particles emitted from a radioactive source to the number of major labor strikes occurring weekly in the United Kingdom. But by far the most widely used probability model in statistics is the *normal* distribution,

$$f_Y(y) = \frac{1}{\sqrt{2\pi}\,\sigma} e^{-(1/2)[(y-\mu)/\sigma]^2}, \qquad -\infty < y < \infty. \qquad (7.1)$$

In Chapter 4 we recounted some of the history surrounding this more than 250-year-old function: how it first appeared as a limiting form of the binomial, only to be quickly popularized as a model in its own right by the likes of Gauss, Quetelet, and Galton. Today we recognize that the unique prominence enjoyed by this distribution actually derives from two sources. On the one hand, many real-world phenomena *do* behave according to the bell-shaped pattern that Equation 7.1 prescribes. A second, and more compelling, reason is the *central limit theorem*: the vast majority of inference procedures in common use are based on averages of independent, identically distributed random variables—precisely the sort of quantity, according to the theorem, that tends to be approximated by the normal. The result is that when the sample size, n, is sufficiently large, almost any inference procedure can be reduced to a probability statement about an appropriate normal variable.

In this chapter we examine some of the more basic hypothesis-testing and estimation problems associated with the normal distribution. What is important in this material is the theory, not the applications. Later chapters will investigate problems of far more practical significance. The solutions of those problems, though, will depend on relatively simple extensions of the results proved in the next several sections.

7.2 POINT ESTIMATES FOR μ AND σ²

As we observed in Chapter 4, the normal distribution is a two-parameter family—μ being a measure of location and σ^2 a measure of dispersion. Finding point estimates for these parameters will be our first objective. Because of the normal's relative simplicity, both $\hat{\mu}$ and $\hat{\sigma}^2$ can be easily obtained using the method of maximum likelihood.

Comment. Recall that if Y is a random variable and its density function is Equation 7.1, we will write $Y \sim N(\mu, \sigma^2)$. If $\mu = 0$ and $\sigma^2 = 1$, the variable is said to be a *standard normal* and will be denoted by the letter Z. ∎

> **THEOREM 7.1** Let $Y_1, Y_2, \ldots, Y_n$ be a random sample from an $N(\mu, \sigma^2)$ distribution. The maximum-likelihood estimates for μ and σ^2 are
>
> $$\hat{\mu} = \frac{1}{n} \sum_{i=1}^{n} y_i = \bar{y}$$
>
> and
>
> $$\hat{\sigma}^2 = \frac{1}{n} \sum_{i=1}^{n} (y_i - \bar{y})^2.$$

PROOF   The likelihood function of the y_i's can be written

$$L(\mu, \sigma^2) = \prod_{i=1}^{n} f_Y(y_i; \mu, \sigma^2)$$

$$= \frac{1}{\sqrt{2\pi}\,\sigma} e^{-(1/2)[(y_1-\mu)/\sigma]^2} \cdots \frac{1}{\sqrt{2\pi}\,\sigma} e^{-(1/2)[(y_n-\mu)/\sigma]^2}$$

$$= \left(\frac{1}{2\pi\sigma^2}\right)^{n/2} e^{-(1/2\sigma^2)\sum_{i=1}^{n}(y_i-\mu)^2}.$$

Also,

$$\ln L(\mu, \sigma^2) = -\frac{n}{2} \ln 2\pi\sigma^2 - \frac{1}{2\sigma^2} \sum_{i=1}^{n} (y_i - \mu)^2.$$

Setting the two partial derivatives of $\ln L(\mu, \sigma^2)$ equal to 0 gives

$$\frac{\partial \ln L(\mu, \sigma^2)}{\partial \mu} = \frac{1}{\sigma^2} \sum_{i=1}^{n} (y_i - \mu) = 0 \tag{7.2}$$

and

$$\frac{\partial \ln L(\mu, \sigma^2)}{\partial \sigma^2} = -\frac{n}{2\sigma^2} + \frac{1}{2\sigma^4} \sum_{i=1}^{n} (y_i - \mu)^2 = 0. \tag{7.3}$$

The simultaneous solution of Equations 7.2 and 7.3 gives the statement of the theorem:

$$\hat{\mu} = \frac{1}{n} \sum_{i=1}^{n} y_i = \bar{y}$$

and

$$\hat{\sigma}^2 = \frac{1}{n} \sum_{i=1}^{n} (y_i - \bar{y})^2.$$

The next two theorems list some of the properties of $\hat{\mu}$ and $\hat{\sigma}^2$. All are straightforward applications of the definitions and theorems of Chapter 5.

THEOREM 7.2   Given that $Y_1, Y_2, \ldots, Y_n \sim N(\mu, \sigma^2)$, $\bar{Y}$, the maximum-likelihood estimator for μ, is unbiased, efficient, and consistent. Also, if σ^2 is known, $\bar{Y}$ is sufficient.

PROOF   The unbiasedness of $\bar{Y}$ follows trivially (see Example 5.3). To show that $\bar{Y}$ is efficient, note that

$$\frac{\partial \ln f_Y(v; \mu, \sigma^2)}{\partial \mu} = \frac{y - \mu}{\sigma^2}$$

and

$$\frac{\partial^2 \ln f_Y(v; \mu, \sigma^2)}{\partial \mu^2} = -\frac{1}{\sigma^2}.$$

From Theorem 5.1, then, the Cramer-Rao lower bound for the variance of an unbiased estimator for μ is

$$\frac{1}{-n\left(-\dfrac{1}{\sigma^2}\right)} = \frac{\sigma^2}{n}.$$

But the variance of the maximum-likelihood estimator *does* achieve that bound: Var $(\bar{Y}) = \sigma^2/n$, so the efficiency of $\bar{Y}$ is established.

Chebyshev's inequality can be used to prove that $\bar{Y}$ is consistent. By the statement of the inequality,

$$P(|\bar{Y} - \mu| < \epsilon) > 1 - \frac{\text{Var }(\bar{Y})}{\epsilon^2}$$

or, after substituting for Var $(\bar{Y})$,

$$P(|\bar{Y} - \mu| < \epsilon) > 1 - \frac{\sigma^2}{n\epsilon^2},$$

from which it is clear that $\bar{Y}$ converges stochastically to μ (i.e., is consistent): for any $\delta > 0$, $\bar{Y}$ will be in an ϵ-neighborhood of μ a minimum of $100(1 - \delta)\%$ of the time, provided $n = n(\epsilon, \delta) \geq \sigma^2/\epsilon^2\delta$.

The sufficiency of $\bar{Y}$ follows from the Fisher-Neyman criterion (Theorem 5.2) and the Corollary to Theorem 7.4. The details will be left as an exercise.

THEOREM 7.3   Given that $Y_1, Y_2, \ldots, Y_n \sim N(\mu, \sigma^2)$,

$$\frac{1}{n} \sum_{i=1}^{n} (Y_i - \bar{Y})^2,$$

the maximum-likelihood estimator for σ^2, is biased and consistent. Also, if μ is known, the estimator is sufficient.

PROOF   To establish the consistency and sufficiency of $(1/n) \sum_{i=1}^{n} (Y_i - \bar{Y})^2$ requires a result proved later in this chapter (Theorem 7.7) and will be left as an exercise. To prove that the estimator is biased, note that

$$E\left[\frac{1}{n}\sum_{i=1}^{n}(Y_i - \bar{Y})^2\right] = E\left[\frac{1}{n}\sum_{i=1}^{n}(Y_i^2 - 2Y_i\bar{Y} + \bar{Y}^2)\right]$$

$$= E\left[\frac{1}{n}\left(\sum_{i=1}^{n}Y_i^2 - n\bar{Y}^2\right)\right]$$

$$= \frac{1}{n}\sum_{i=1}^{n}E(Y_i^2) - E(\bar{Y}^2).$$

But

$$E(Y_i^2) = \sigma^2 + \mu^2$$

and

$$E(\bar{Y}^2) = \frac{\sigma^2}{n} + \mu^2. \quad \text{(Why?)}$$

Therefore,

$$E\left[\frac{1}{n}\sum_{i=1}^{n}(Y_i - \bar{Y})^2\right] = \frac{1}{n}(n\sigma^2 + n\mu^2) - \frac{\sigma^2}{n} - \mu^2$$

$$= \left(\frac{n-1}{n}\right)\sigma^2,$$

thus demonstrating that the estimator is biased—specifically, the MLE tends to *underestimate* σ^2.

Comment. In practice, σ^2 is usually estimated not by the $\hat{\sigma}^2$ of Theorem 7.3 but by the *sample variance*, s^2, where

$$s^2 = \left(\frac{n}{n-1}\right)\hat{\sigma}^2 = \frac{1}{n-1}\sum_{i=1}^{n}(y_i - \bar{y})^2.$$

This, of course, is simply the unbiased estimator based on the sufficient statistic:

$$E(S^2) = \left(\frac{n}{n-1}\right)E\left[\frac{1}{n}\sum_{i=1}^{n}(Y_i - \bar{Y})^2\right] = \sigma^2. \quad \blacksquare$$

QUESTION 7.2.1   Show that an equivalent expression for s^2 is

$$s^2 = \frac{n\sum_{i=1}^{n}y_i^2 - \left(\sum_{i=1}^{n}y_i\right)^2}{n(n-1)}.$$

This provides a very convenient computing formula. Accumulating sums and sums of squares is much easier to do, particularly on a calculator, than computing sums of squares of differences. Rounding errors will also tend to be less of a problem with this second formulation.

We conclude this section with an application of the estimators proposed in Theorems 7.2 and 7.3.

CASE STUDY

In the eighth century B.C., the Etruscan civilization was the most advanced in all of Italy. Its art forms and political innovations were destined to leave indelible marks on the entire Western world. Originally located along the western coast between the Arno and Tiber rivers (the region now known as Tuscany), it spread quickly across the Apennines and eventually overran much of Italy. But as quickly as it came, it faded. Militarily it was to prove no match for the burgeoning Roman legions, and by the dawn of Christianity it was all but gone.

No chronicles of the Etruscan empire have ever been found, and to this day its origins remain shrouded in mystery. Were the Etruscans native Italians, or were they immigrants? And if they were immigrants, where did they come from? Much of what *is* known has come from archeological investigations and anthropometric studies—the latter involving the use of body measurements to determine racial characteristics and ethnic origins. The data presented here are an example of one such study.

Research has shown that the maximum head breadth of modern Italian males averages 132.4 mm. Listed in Table 7.1 are the maximum head breadths recorded for 84 male Etruscan skulls uncovered in various archeological digs throughout Italy (7).

TABLE 7.1 Maximum head breadths (mm) of 84 Etruscan males

141	148	132	138	154	142	150
146	155	158	150	140	147	148
144	150	149	145	149	158	143
141	144	144	126	140	144	142
141	140	145	135	147	146	141
136	140	146	142	137	148	154
137	139	143	140	131	143	141
149	148	135	148	152	143	144
141	143	147	146	150	132	142
142	143	153	149	146	149	138
142	149	142	137	134	144	146
147	140	142	140	137	152	145

A graph of these data (Figure 7.1) indicates that the population from which they are a sample may very well be normal. Lacking any other information, we would probably choose $N(\bar{y}, s^2)$ as the particular normal having the best chance of fitting the data. For these 84 observations,

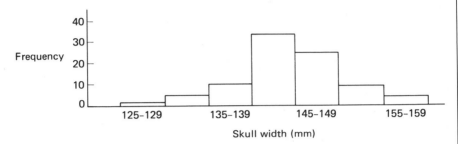

Figure 7.1   Graph of Etruscan head-breadth data.

$$\sum_{i=1}^{84} y_i = 12,077$$

and

$$\sum_{i=1}^{84} y_i^2 = 1,739,315.$$

Therefore,

$$\bar{y} = \frac{12,077}{84} = 143.8,$$

and, using the formula given in Question 7.2.1,

$$s = \sqrt{\frac{84(1,739,315) - (12,077)^2}{84(83)}} = 6.0.$$

Figure 7.2 shows the $N[143.8, (6.0)^2]$ density superimposed over the histogram of Figure 7.1. The fit is clearly quite good.

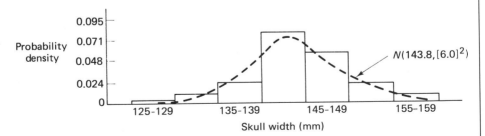

Figure 7.2   Normal curve fitted to histogram.

Comment. From a hypothesis-testing standpoint, it would make sense to test

$$H_0: \mu = 132.4$$

versus

$$H_1: \mu \neq 132.4,$$

where μ is the true average maximum head breadth for Etruscans. Rejecting H_0 could be interpreted as evidence (although evolutionary shifts would have to be taken into account) that the Etruscans and the Italians are of different ethnic origins. ∎

QUESTION 7.2.2   Assuming the model

$$f_Y(y) = \frac{1}{\sqrt{2\pi}\,(6.0)} e^{-(1/2)[(y-143.8)/6.0]^2},$$

compute the expected number of skulls having widths in the 135–139 mm range. (See Case Study 4.6.)

QUESTION 7.2.3   For a random sample of size n, find the method-of-moments estimators for μ and σ^2.

7.3 LINEAR COMBINATIONS OF NORMAL RANDOM VARIABLES

As pointed out in Section 7.1, it is the rule rather than the exception for estimation and hypothesis-testing procedures to be based on *averages*, either of the original data or of some function of the original data. The methodology that has evolved for the normal distribution is no different. We will see in a later section that inferences focusing on μ will be based on $\bar{y}$, while those concerning σ^2 will derive from s^2, which is just an average of $(y_i - \bar{y})^2$ terms. What this implies is that the place to begin our discussion is with the sampling behavior of linear combinations of independent normals.

The main result of this section is Theorem 7.4, which gives the distribution of the sum of two independent normals—and, hence, by induction, of n independent normals. The proof is accomplished by writing the sum as a convolution and then doing the indicated integration. To simplify the algebra, we will consider a special case first; from this and Definition 4.1 the general statement will follow immediately.

Comment. A much easier way to prove Theorem 7.4, using moment-generating functions, is indicated in Question 7.3.1. The lengthier proof given here is still noteworthy, though, because of the manipulative techniques it uses—techniques that remain applicable in situations where moment-generating functions are of no help. ∎

LEMMA   If $Y_1 \sim N(0, \sigma^2)$ and $Y_2 \sim N(0, 1)$, and if Y_1 and Y_2 are independent, $Y_1 + Y_2 \sim N(0, \sigma^2 + 1)$.

PROOF   Written as a convolution, the probability function for $Y_1 + Y_2$ is the integral of the product of two normal densities:

$$f_{Y_1+Y_2}(t) = \int_{-\infty}^{\infty} f_{Y_1}(y) f_{Y_2}(t - y) \, dy$$

$$= \int_{-\infty}^{\infty} \frac{1}{\sqrt{2\pi}\,\sigma} e^{-y^2/2\sigma^2} \frac{1}{\sqrt{2\pi}} e^{-(1/2)(t-y)^2} \, dy$$

$$= \frac{1}{2\pi\sigma} \int_{-\infty}^{\infty} e^{-(1/2)[y^2/\sigma^2 + (t-y)^2]} \, dy.$$

The quantity in the brackets of the exponent can be simplified to give

$$\frac{y^2}{\sigma^2} + y^2 - 2ty + t^2 = \left(\frac{1 + \sigma^2}{\sigma^2}\right)y^2 - 2ty + t^2. \tag{7.4}$$

If we let

$$c = \frac{\sigma}{\sqrt{1 + \sigma^2}},$$

the right-hand side of Equation 7.4 becomes

$$\frac{1}{c^2}y^2 - 2ty + t^2 = \frac{1}{c^2}y^2 - 2ty + c^2t^2 + t^2 - c^2t^2$$

$$= \left(\frac{y}{c} - ct\right)^2 + (1 - c^2)t^2$$

$$= \left(\frac{y - c^2t}{c}\right)^2 + (1 - c^2)t^2.$$

Therefore,

$$f_{Y_1+Y_2}(t) = \frac{1}{2\pi\sigma} \int_{-\infty}^{\infty} e^{-(1/2)[(y-c^2t)/c]^2} e^{-(1/2)(1-c^2)t^2} \, dy$$

$$= \frac{1}{\sqrt{2\pi}\,(\sigma/c)} e^{-(1/2)(1-c^2)t^2} \frac{1}{\sqrt{2\pi}\,c} \int_{-\infty}^{\infty} e^{-(1/2)[(y-c^2t)/c]^2} \, dy.$$

Note that the last factor is 1, being the integral over the entire range of a normal variable with mean c^2t and variance c^2. Furthermore, since

$$1 - c^2 = 1 - \frac{\sigma^2}{1 + \sigma^2} = \frac{1}{1 + \sigma^2}$$

and

$$\frac{\sigma}{c} = \frac{1}{\sqrt{1 + \sigma^2}},$$

the first factor is a normal density with mean 0 and variance $1 + \sigma^2$.

THEOREM 7.4 Let $Y_1 \sim N(\mu_1, \sigma_1^2)$ and $Y_2 \sim N(\mu_2, \sigma_2^2)$. If Y_1 and Y_2 are independent,

(a) The sum, $Y_1 + Y_2$, is also normally distributed.
(b) The mean of the sum is the sum of the original means.
(c) The variance of the sum is the sum of the original variances.

That is,

$$Y_1 + Y_2 \sim N(\mu_1 + \mu_2, \sigma_1^2 + \sigma_2^2).$$

PROOF   We can write

$$Y_1 + Y_2 = \sigma_2 \left(\frac{Y_1 - \mu_1}{\sigma_2} + \frac{Y_2 - \mu_2}{\sigma_2} \right) + \mu_1 + \mu_2.$$

But by the previous lemma,

$$\frac{Y_1 - \mu_1}{\sigma_2} + \frac{Y_2 - \mu_2}{\sigma_2} \sim N\left(0, \frac{\sigma_1^2}{\sigma_2^2} + 1 \right).$$

An application of Theorem 3.3 gives

$$Y_1 + Y_2 \sim N(\mu_1 + \mu_2, \sigma_1^2 + \sigma_2^2).$$

QUESTION 7.3.1   Prove Theorem 7.4 by examining the product of the moment-generating functions for Y_1 and Y_2.

The most important special case of Theorem 7.4 concerns the distribution of the sample mean, which, of course, can be viewed as a linear combination of n identically distributed, independent normals, each having the same mean μ, variance σ^2, and coefficient $1/n$.

COROLLARY If $\bar{Y} = (1/n) \sum_{i=1}^{n} Y_i$ is the sample mean of n independent $N(\mu, \sigma^2)$ random variables, then $\bar{Y} \sim N(\mu, \sigma^2/n)$.

Comment. Because of Theorem 7.2, we already know that $\bar{Y}$ is a "best" estimator—that is, among the class of unbiased estimators for μ, $\bar{Y}$ has the smallest variance. But now with the corollary to Theorem 7.4 we have the *distribution* of $\bar{Y}$, which makes it possible to actually quantify, in a probabilistic sense, $\bar{Y}$'s precision. For example, suppose a single observation is drawn from an $N[100, (5)^2]$ distribution. The probability that Y lies within ± 5 (one standard deviation of Y) of its mean is 0.682:

$$P(95 \le Y \le 105) = P\left(\frac{95 - 100}{5} \le \frac{Y - \mu}{\sigma} \le \frac{105 - 100}{5} \right)$$

$$= P(-1.00 \le Z \le 1.00) = 0.682.$$

If, on the other hand, a random sample of, say, size 4 had been drawn from the same distribution, the probability of the resulting *mean's* falling between 95 and 105 would have increased to 0.955:

$$P(95 \leq \bar{Y} \leq 105) = P\left(\frac{95 - 100}{5/\sqrt{4}} \leq \frac{\bar{Y} - \mu}{\sigma/\sqrt{4}} \leq \frac{105 - 100}{5/\sqrt{4}}\right)$$

$$= P(-2.00 \leq Z \leq 2.00) = 0.955.$$

Table 7.2 lists the probability of $\bar{Y}$'s lying in the interval $\mu - \sigma$ to $\mu + \sigma$ for sample sizes ranging from 1 to 9.

TABLE 7.2   Probabilities for nine sample sizes

n	$P(\mu - \sigma \leq \bar{Y} \leq \mu + \sigma) = P(-\sqrt{n} \leq Z \leq \sqrt{n})$
1	0.682
2	0.841
3	0.916
4	0.955
5	0.975
6	0.986
7	0.992
8	0.995
9	0.997

Comment. Notice above how the denominator of the Z transformation is modified, depending on whether the probability statement is about Y or $\bar{Y}$. The expression $\sigma/\sqrt{n}$ appearing in the latter is often referred to as the *standard error of the mean*. ∎

QUESTION 7.3.2   Use the corollary to Theorem 7.4 and the Fisher-Neyman criterion of Chapter 5 to prove that $\bar{Y}$ is a sufficient statistic for μ.

QUESTION 7.3.3   The IQ's of nine randomly selected persons are recorded. Let $\bar{Y}$ denote their average. Assuming the distribution from which the Y_i's were drawn is normal with a mean of 100 and a standard deviation of 16, what is the probability that $\bar{Y}$ will exceed 103? What is the probability that any arbitrary Y_i will exceed 103? What is the probability that exactly three of the Y_i's will exceed 103?

7.4 THE CENTRAL LIMIT THEOREM

We have now established the normality of a sum of normal random variables. It would be too much to ask that the sum of nonnormal variables also have a normal distribution. Even so, under very general conditions such a sum can be *approximately* normal. The first theorem stating this sort of result was the now familiar Theorem 4.3 of DeMoivre and Laplace. There was a great deal of work on this problem during the nineteenth century, particularly by the Russian school of probabilists [see (104)], and it culminated in a rather general result by A. M. Lyapunov in 1901. Theorem 7.5 gives a version of this result strong enough for most statistical applications. The name *central limit theorem* is due to G. Polya ("Über den zentralen Grenzwertsatz der Wahrscheinlichkeitsrechnung und das Momentenproblem," *Mathematische Zeitschrift*, vol. 8, 1920, pp. 171–181).

THEOREM 7.5 (Central Limit Theorem)   Let $Y_1, Y_2, \ldots$ be an infinite sequence of independent random variables, each with the same distribution. Suppose the mean μ and the variance σ^2 of $f_Y(y)$ are both finite. For any numbers c and d,

$$\lim_{n \to \infty} P\left(c < \frac{Y_1 + \cdots + Y_n - n\mu}{\sqrt{n}\,\sigma} < d\right) = \frac{1}{\sqrt{2\pi}} \int_c^d e^{-(1/2)y^2}\, dy.$$

PROOF

If the Y_i's are Bernoulli random variables, Theorem 7.5 reduces to the DeMoivre-Laplace theorem of Chapter 4. A proof for that special case requiring only modest analytical techniques can be found in (45). More general cases require background more elaborate than we can develop here. However, we can *sketch* a proof by strengthening the hypotheses to assume that the moment-generating functions of the Y_i's exist. Then, not only are the mean and variance finite, but the common moment-generating function, $M(t)$, of the Y_i's has $M^{(r)}(0)$ defined for all $r \geq 0$ and all moments exist.

The use of the hypothesized moment-generating functions is via the following lemma, whose proof is beyond the scope of this text.

LEMMA   Suppose $X, X_1, X_2, \ldots$ are random variables with $\lim_{n \to \infty} M_{X_n}(t)$ $= M_X(t)$ for all t in some interval about 0. Then $\lim_{n \to \infty} F_{X_n}(x) = F_X(x)$ for all real values x.

Thus, it is necessary to establish

$$\lim_{n \to \infty} M_{(Y_1 + \cdots + Y_n - n\mu)/(\sqrt{n}\,\sigma)}(t) = M_Z(t) = e^{(1/2)t^2}$$

where $Z \sim N(0, 1)$.

For notational convenience write

$$\frac{Y_1 + \cdots + Y_n - n\mu}{\sqrt{n}\,\sigma} = \frac{W_1 + \cdots + W_n}{\sqrt{n}},$$

where $W_i = (Y_i - \mu)/\sigma$. Of course, $E(W_i) = 0$ and $\text{Var}(W_i) = 1$. From the first part of Theorem 3.19 it follows that the moment-generating functions for the W_i's exist, since they exist for the Y_i's. Applying both parts of Theorem 3.19, we find that

$$M_{(W_1 + \cdots + W_n)/\sqrt{n}}(t) = \left[M\left(\frac{t}{\sqrt{n}}\right) \right]^n,$$

where $M(t)$ represents the common moment-generating function of W_i.

From the normalization of the W_i's we obtain $M(0) = 1$, $M^{(1)}(0) = E(W_i) = 0$, and $M^{(2)}(0) = \text{Var}(W_i) = 1$. Applying Taylor's theorem with remainder to $M(t)$ gives

$$M(t) = 1 + M^{(1)}(0)t + \tfrac{1}{2}M^{(2)}(r)t^2 = 1 + \tfrac{1}{2}t^2 M^{(2)}(r)$$

for some number r, $|r| < |t|$. Thus

$$\lim_{n \to \infty}\left[M\left(\frac{t}{\sqrt{n}}\right) \right]^n = \lim_{n \to \infty}\left[1 + \frac{t^2}{2n}M^{(2)}(s) \right]^n, \quad |s| < \frac{|t|}{\sqrt{n}}$$

$$= \exp \lim_{n \to \infty} n \ln\left[1 + \frac{t^2}{2n}M^{(2)}(s) \right]$$

$$= \exp \lim_{n \to \infty} \frac{t^2}{2} \cdot M^{(2)}(s) \cdot \frac{\ln\left[1 + \dfrac{t^2}{2n}M^{(2)}(s) \right] - \ln(1)}{\dfrac{t^2}{2n}M^{(2)}(s)}.$$

As remarked previously, the existence of $M(t)$ implies the existence of all its derivatives. In particular $M^{(3)}(t)$ exists, so $M^{(2)}(t)$ is continuous. Hence, $\lim_{t \to 0} M^{(2)}(t) = M^2(0) = 1$. Since $|s| < |t|/\sqrt{n}$, $s \longrightarrow 0$ as $n \longrightarrow \infty$, so

$$\lim_{n \to \infty} M^{(2)}(s) = M^{(2)}(0) = 1.$$

Also, as $n \longrightarrow \infty$, the quantity $(t^2/2n)M^{(2)}(s) \longrightarrow 0 \cdot 1 = 0$, so it plays the role of "Δx" in the definition of the derivative. Hence, we obtain

$$\lim_{n \to \infty}\left[M\left(\frac{t}{\sqrt{n}}\right) \right]^n = \exp \frac{t^2}{2} \cdot 1 \cdot \ln^{(1)}(1) = e^{(1/2)t^2}.$$

Since this last expression is the moment-generating function for a standard normal variable, the theorem is proved.

Our main interest in Theorem 7.5 is to justify the use of the statistical procedures in this chapter for nonnormal variables. Since many of these techniques use the average of independent, identically distributed random variables, the central limit theorem applies: that is, for large n, the sample mean is approximately normal.

This theorem also helps to explain the prevalence of biological and sociological phenomena that are empirically observed to follow a normal distribution. A case in point is inheritance. If a trait is determined by a large number of genes, we may think of the realization of that trait as being the sum of a large number of random variables, each making a small contribution. As a result, the distribution of the trait in the general population is likely to be normal. It is true that these variables need not be independent, but central-limit-type theorems also hold for large classes of dependent variables.

A more concrete application of Theorem 7.5 is the following case study of the dispersal of living organisms.

CASE STUDY

7.2

Suppose we think of a small portion of the earth's surface as planar and equipped with x- and y-axes. At the origin of this system is placed one of some particular kind of organism. This animal, say, wanders off and eventually makes a home and reproduces. The coordinates of its new home can be thought of as a pair of random variables (X'_1, Y'_1). Now, follow some offspring of this animal until it settles down to raise a family at (X'_2, Y'_2). Clearly, the distribution of X'_2 and Y'_2 depends on the values taken on by X'_1 and Y'_1. It seems reasonable, though, to assume that $X_1 = X'_1 - 0$ and $X_2 = X'_2 - X'_1$ are independent and have the same distribution; similarly for $Y_1 = Y'_1 - 0$ and $Y_2 = Y'_2 - Y'_1$.

Assume this process continues for n generations, the result of which is n pairs of random variables, $(X_1, Y_1), (X_2, Y_2), \ldots, (X_n, Y_n)$, where $(X_i, Y_i) = (X'_i - X'_{i-1}, Y'_i - Y'_{i-1})$ represents the x- and y-distances the ith-generation offspring wanders from its $(i-1)$st-generation home. The nature of the X_i's and Y_i's suggests that it would not be inappropriate to assume both sets of variables to be independent. Let $(\tilde{X}_n, \tilde{Y}_n)$ represent the x- and y-distances of the nth generation from its ancestral home—that is, the origin. Then $\tilde{X}_n = X_1 + \cdots + X_n$, $\tilde{Y}_n = Y_1 + \cdots + Y_n$, and, by the central limit theorem, $\tilde{X}_n$ and $\tilde{Y}_n$ are approximately normal if n is large. Furthermore, both variables are independent, with mean 0 and variance $n\sigma^2$, where σ^2 is the variance of each X_i and Y_i.

It will prove to be more useful for our purposes to consider Euclidean distances from the origin, so set

$$R_n = \sqrt{\tilde{X}_n^2 + \tilde{Y}_n^2}.$$

Then

$$F_{R_n}(t) = P(R_n \le t) = P(R_n^2 \le t^2).$$

At this point we preview several results from the next section (Definition 7.1 and Theorem 7.7) and, with the help of the transformation theorem of Section 3.5, assert that

$$f_{R^2_n}(t) = \frac{1}{2}\left(\frac{1}{n\sigma^2}\right)e^{-t/2n\sigma^2}.$$

Therefore,

$$f_{R_n}(t) = 2tf_{R_n^2}(t^2) = \left(\frac{1}{n\sigma^2}\right)te^{-t^2/2n\sigma^2}, \qquad t > 0. \qquad (7.5)$$

From Equation 7.5, it is a simple matter to obtain the proportion of the nth generation to be found outside a circle of radius r:

$$\int_r^\infty \left(\frac{1}{n\sigma^2}\right)te^{-t^2/2n\sigma^2}\,dt = e^{-r^2/2n\sigma^2}.$$

We wish to use this expression to find the radius of the circle that includes essentially the entire population. Let N_n denote the size of the population and let s be the radius of the circle for which the probability of an animal's living outside this circle is $(1/N_n)$. Then $e^{-s^2/2n\sigma^2} = 1/N_n$ or, equivalently, $s = (2n\sigma^2 \ln N_n)^{1/2}$. Assume the population has increased geometrically, so $N_n = e^{kn}$, for some parameter k. It follows that $s = (2n\sigma^2 kn)^{1/2} = \sqrt{2k}\,n\sigma$, and the expected number of the population living outside a circle of radius s is $N_n(1/N_n)$ $= 1$. Thus, a circle of radius s is a theoretical contour containing essentially all of the population. Note: s is a linear function of n (169).

Let us now compare our theoretical arguments to some empirical evidence. Ulbrich (183) plotted the apparent boundaries for the spread of the muskrat, *Ondatra zibethica*, across central Europe during the early years of the present century (see Figure 7.3). If we

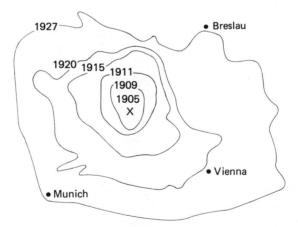

Figure 7.3   Spread of the muskrat population.

consider the contour to be an estimate of the theoretical circle of radius s, then the area it bounds should be an estimate of πs^2. The square root of the area should then be a linear function of n. The five

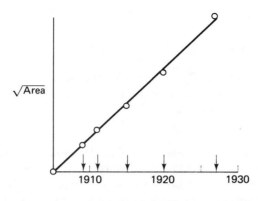

Figure 7.4   Muskrat population dispersion: square root of area occupied versus time.

dates from Ulbrich are plotted against $\sqrt{\text{area}}$ in Figure 7.4. They clearly fit a straight line remarkably well.

QUESTION 7.4.1   Empirical studies have shown that the sum of k independent uniform random variables, when suitably standardized, can be used as an approximation to an $N(0, 1)$ random variable. Furthermore, the approximation is quite good even for k as small as 12. Write down a formula for generating standard normal deviates.

7.5 THE χ^2 DISTRIBUTION;
INFERENCES ABOUT σ^2

Having derived the sampling distribution for the maximum-likelihood estimator for μ, we will now turn our attention to the second parameter of the normal distribution, σ^2. Before testing $H_0: \sigma^2 = \sigma_0^2$, though, or constructing confidence intervals for σ^2, we will need to investigate the properties of sums of the form $\sum_{i=1}^{n} Z_i^2$, where the Z_i's are independent standard normals. Central to these properties is a very important distribution we will be seeing for the first time, the chi square. As later chapters will make abundantly clear, the utility of the chi square extends far beyond the problems being dealt with here.

DEFINITION 7.1   A random variable Y is said to have the *chi square distribution with n degrees of freedom* if

$$f_Y(y) = \frac{1}{2^{n/2}\Gamma(n/2)} y^{(n/2)-1} e^{-y/2}, \qquad y > 0.$$

For notational simplicity, we will write $Y \sim \chi_n^2$.

Comment. The terminology "degrees of freedom" appears in connection with several pdf's introduced in this chapter. The phrase derives from the number of independent squared $N(0, 1)$ variables in sums like the one appearing in Theorem

7.7, such sums being intimately related to the pdf's in question. For our purposes, however, it will be quite sufficient simply to think of "degrees of freedom" as a name given to a parameter. ▮

Note that χ_n^2 is a special case of the gamma density presented in Chapter 4. Specifically, a χ_n^2 variable is identical to a gamma variable (with parameters r and λ) when the parameters of the latter are set equal to $n/2$ and $1/2$, respectively. Figure 7.5 shows how the shape of a χ_n^2 distribution varies with n.

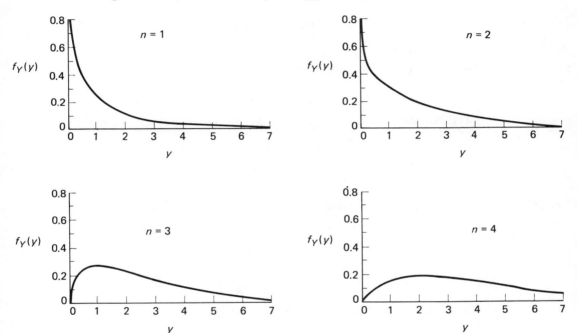

Figure 7.5   Variation of shape of chi square distribution with number of degrees of freedom.

Comment. Table A.3 in the Appendix lists selected percentiles of the χ^2 distribution for degrees of freedom ranging from 1 to 50. For $n > 50$, the percentiles can be easily approximated:

$$\chi_{p,n}^2 \doteq n\left(1 - \frac{2}{9n} + z_{1-p}\sqrt{\frac{2}{9n}}\right)^3,$$

where $\chi_{p,n}^2$ is the unknown $(100p)$th percentile of the χ_n^2 distribution and z_{1-p} is the $(100p)$th percentile of the $N(0, 1)$ [see (38)]. ▮

QUESTION 7.5.1   Show directly—without appealing to the fact that a χ^2 variable is a gamma variable—that $f_Y(y)$ as stated in Definition 7.1 is a true probability density.

QUESTION 7.5.2   Show that $E(\chi_n^2) = n$ and Var $(\chi_n^2) = 2n$.

Like certain other distributions we have encountered—notably, the binomial, Poisson, exponential, and normal—the chi square reproduces itself. That is, the sum of two independent chi squares is itself a chi square.

> **THEOREM 7.6** Let $Y_1 \sim \chi_n^2$ and $Y_2 \sim \chi_m^2$ and let Y_1 and Y_2 be independent. Then
>
> $$Y_1 + Y_2 \sim \chi_{n+m}^2.$$

PROOF   The proof is a special case of Question 4.6.4.

The next theorem shows how the normal and χ^2 distributions are related. This is an important result, one that gives rise to a wide variety of inference procedures.

> **THEOREM 7.7** Let $Z_1, Z_2, \ldots, Z_n$ be n independent standard normal random variables. Then
>
> $$\sum_{i=1}^{n} Z_i^2 \sim \chi_n^2.$$

PROOF   The proof follows by induction. First, suppose $n = 1$. Then

$$F_{Z_1^2}(t) = P(Z_1^2 \leq t) = P(-\sqrt{t} \leq Z_1 \leq \sqrt{t}\,)$$
$$= 2P(0 \leq Z_1 \leq \sqrt{t}\,)$$
$$= \frac{2}{\sqrt{2\pi}} \int_0^{\sqrt{t}} e^{-z^2/2}\, dz.$$

Differentiation gives the density function for Z_1^2:

$$f_{Z_1^2}(t) = F'_{Z_1^2}(t) = \frac{2}{\sqrt{2\pi}} \frac{1}{2\sqrt{t}} e^{-t/2} = \frac{1}{2^{1/2}\Gamma(1/2)} t^{(1/2)-1} e^{-t/2}.$$

Therefore, by inspection, $Z_1^2 \sim \chi_1^2$.

Now, suppose the theorem is true for the first $n-1$ random variables. We can write the sum of the first n Z_i^2's as

$$Z_1^2 + \cdots + Z_n^2 = Z_1^2 + (Z_2^2 + \cdots + Z_n^2).$$

By what we have just proved, $Z_1^2 \sim \chi_1^2$; also, by the induction hypothesis, $\sum_{i=2}^{n} Z_i^2 \sim \chi_{n-1}^2$. The final result then follows from Theorem 7.6.

EXAMPLE 7.1.   Tables have been compiled that list random samples from an $N(0, 1)$ distribution [see, for example, (38)]. By taking independent sets of n of these numbers and adding together their squares, we can simulate the conclusion of Theorem 7.7. Shown in Table 7.3 are the $\sum_{i=1}^{4} Z_i^2$ values computed from 100 independent sets of four $N(0, 1)$ variables.

TABLE 7.3 $\sum\limits_{i=1}^{4} Z_i^2$ values: a sampling experiment

3.472	6.472	8.347	13.025	1.483	7.832
0.772	5.449	1.037	4.744	4.940	2.186
2.920	1.083	3.047	5.627	4.091	1.031
4.532	1.033	3.146	4.004	2.685	4.379
1.510	0.964	1.519	4.668	12.723	2.018
6.018	3.820	4.900	3.300	3.147	5.741
6.613	9.386	4.874	9.775	5.290	6.854
8.992	3.330	2.574	0.611	0.870	1.152
0.738	8.630	5.233	0.579	1.653	1.237
8.484	3.643	2.118	5.813	5.168	4.255
1.079	3.145	4.541	2.052	6.846	0.570
0.476	2.151	0.391	0.758	3.700	2.476
2.680	0.756	3.549	2.694	10.884	1.630
2.392	3.084	2.577	4.354	3.785	2.232
1.348	1.840	6.208	10.938	2.217	1.264
1.330	1.808	1.642	3.434	3.596	4.687
2.650	6.203	1.830	4.865		

As the theorem predicts, these numbers do seem to be described by a χ_4^2 distribution: Figure 7.6 shows a good fit between the function

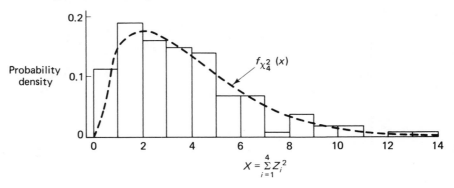

Figure 7.6   Fitting a χ_4^2 distribution.

$$f_{\chi^2_4}(x) = \frac{1}{2^{4/2}\Gamma(\frac{4}{2})}x^{(4/2)-1}e^{-x/2}$$

and the histogram of $\sum\limits_{i=1}^{4} Z_i^2$ values.

QUESTION 7.5.3   Approximate the number of $\sum\limits_{i=1}^{4} Z_i^2$ values expected to fall between 2.0 and 3.0. Assume that 100 sums (of four) are to be recorded.

One final result is necessary before we can derive any inference procedures for σ^2. Theorem 7.8 gives the sampling distribution of the ratio $(n-1)S^2/\sigma^2$, the significance of which will become evident in Theorems 7.9 and 7.10.

THEOREM 7.8 Let $Y_1, Y_2, \ldots, Y_n$ be n independent $N(\mu, \sigma^2)$ random variables. As defined earlier, let

$$S^2 = \frac{1}{n-1} \sum_{i=1}^{n} (Y_i - \bar{Y})^2$$

be their sample variance. Then

$$\frac{(n-1)S^2}{\sigma^2} \sim \chi^2_{n-1}.$$

PROOF   The proof depends on a result given in Appendix 7.2 on page 307 and, in fact, is actually a part of that derivation. We include the following argument for its pedagogical value as an exercise in the use of moment-generating functions.

Note, first of all, that

$$\frac{(n-1)S^2}{\sigma^2} = \sum_{i=1}^{n} \left(\frac{Y_i - \bar{Y}}{\sigma}\right)^2 = \sum_{i=1}^{n} \left[\left(\frac{Y_i - \mu}{\sigma}\right)^2 - \left(\frac{\bar{Y} - \mu}{\sigma}\right)^2\right]$$

$$= \sum_{i=1}^{n} \left(\frac{Y_i - \mu}{\sigma}\right)^2 - \left(\frac{\bar{Y} - \mu}{\sigma/\sqrt{n}}\right)^2$$

or, transposing terms,

$$\frac{(n-1)S^2}{\sigma^2} + \left(\frac{\bar{Y} - \mu}{\sigma/\sqrt{n}}\right)^2 = \sum_{i=1}^{n} \left(\frac{Y_i - \mu}{\sigma}\right)^2.$$

Since $\bar{Y} \sim N(\mu, \sigma^2/n)$, $[(\bar{Y} - \mu)/(\sigma/\sqrt{n})]^2 \sim \chi^2_1$ by Theorem 7.7. By the same result,

$$\sum_{i=1}^{n} \left(\frac{Y_i - \mu}{\sigma}\right)^2 \sim \chi^2_n.$$

Furthermore, it can be shown (see Appendix 7.2 on page 307) that

$$\frac{(n-1)S^2}{\sigma^2} \quad \text{and} \quad \left(\frac{\bar{Y} - \mu}{\sigma/\sqrt{n}}\right)^2$$

are independent. Therefore, the moment-generating function for $\sum_{i=1}^{n} [(Y_i - \mu)/\sigma]^2$ is the product of the moment-generating functions for $(n-1)S^2/\sigma^2$ and $[(\bar{Y} - \mu)/(\sigma/\sqrt{n})]^2$. That is,

$$\left(\frac{1}{1-2t}\right)^{n/2} = E[e^{t(n-1)S^2/\sigma^2}] \cdot \left(\frac{1}{1-2t}\right)^{1/2}, \qquad t < \frac{1}{2}.$$

It follows that

$$E(e^{t(n-1)S^2/\sigma^2}) = \left(\frac{1}{1-2t}\right)^{(n-1)/2}, \qquad t < \frac{1}{2},$$

which implies that

$$\frac{(n-1)S^2}{\sigma^2} \sim \chi^2_{n-1}.$$

With Theorem 7.8 established, it remains a simple matter to invert the variance ratio and construct confidence intervals for σ^2. Theorem 7.9 states the formal result. The proof will be left as an exercise. [Recall that the symbol $\chi^2_{p,n}$ denotes the $(100p)$th percentile of the χ^2_n distribution.]

Comment. Inference procedures for σ^2, whether they be confidence intervals or hypothesis tests, are far less important in real-world applications than those for μ. They are presented here first only because the background results they require have already been developed. The usual inference procedures involving μ require a distribution (the Student t) whose derivation is best delayed until the next section.

■

THEOREM 7.9   Let $Y_1, Y_2, \ldots, Y_n \sim N(\mu, \sigma^2)$. A $100(1-\alpha)\%$ confidence interval for σ^2 is given by

$$\left(\frac{(n-1)s^2}{\chi^2_{1-\alpha/2,\, n-1}},\ \frac{(n-1)s^2}{\chi^2_{\alpha/2,\, n-1}} \right).$$

CASE STUDY

7.3

The chain of events that we call the geological evolution of the earth began hundreds of millions of years ago. Fossils have played a key role in documenting the *relative* times these events occurred, but to establish an *absolute* chronology, scientists rely primarily on radioactive decay. In this example, we look at the variability associated with a dating technique based on a mineral's potassium-argon ratio.

Almost all minerals contain potassium (K) as well as certain of its isotopes, including ^{40}K. But ^{40}K is unstable and decays into isotopes of argon and calcium, ^{40}A and ^{40}Ca. By knowing the rates at which these daughter products are formed, and by measuring the amounts of ^{40}A and ^{40}K present, we can estimate the age of the mineral.

Table 7.4 gives the estimated ages of 19 mineral samples collected from the Black Forest in southeastern Germany (108). Each of the estimates was based on the sample's potassium-argon ratio.

TABLE 7.4 Potassium-argon dates of samples from the Black Forest

Specimen	Estimated Age (millions of years)
1	$y_1 = 249$
2	$y_2 = 254$
3	$y_3 = 243$
4	$y_4 = 268$
5	$y_5 = 253$
6	$y_6 = 269$
7	$y_7 = 287$
8	$y_8 = 241$
9	$y_9 = 273$
10	$y_{10} = 306$
11	$y_{11} = 303$
12	$y_{12} = 280$
13	$y_{13} = 260$
14	$y_{14} = 256$
15	$y_{15} = 278$
16	$y_{16} = 344$
17	$y_{17} = 304$
18	$y_{18} = 283$
19	$y_{19} = 310$

A primary concern in this problem would be the *precision* of the potassium-argon procedure—that is, how close together would repeated age determinations be on the exact same sample? The most direct way of getting at this question would be to construct a confidence interval for σ^2, the procedure's true variance.

For the data given,

$$\sum_{i=1}^{19} y_i = 5261,$$

$$\sum_{i=1}^{19} y_i^2 = 1,469,945,$$

so the sample variance is 733.4:

$$s^2 = \frac{19(1,469,945) - (5261)^2}{19(18)} = 733.4.$$

From Table A.3 in the Appendix we find that

$$P(8.23 < \chi_{18}^2 < 31.53) = 0.95.$$

Therefore, a 95% confidence interval for σ^2 is

$$\left(\frac{(19-1)(733.4)}{31.53}, \ \frac{(19-1)(733.4)}{8.23} \right)$$

$$= (418.7 \text{ million years}^2, 1604.0 \text{ million years}^2).$$

Comment. For the applied statistician, σ is a more meaningful measure of variation than σ^2 (since σ is in the same units as the data). To get a confidence interval for σ, though, we simply take the square root of the confidence limits for σ^2. Here, a 95% confidence interval for σ would be

$$(\sqrt{418.7}, \sqrt{1604.0}) = (20.5 \text{ million years}, 40.0 \text{ million years}). \quad \blacksquare$$

QUESTION 7.5.4   Assume that the *w/l* ratios of Shoshoni rectangles are $N(\mu, \sigma^2)$ random variables (see Case Study 1.2). Construct a 95% confidence interval for σ^2.

We conclude this section by applying the generalized-likelihood-ratio criterion to the problem of testing hypotheses about σ^2. Theorem 7.10 gives an approximate GLRT.

THEOREM 7.10 Let $Y_1, Y_2, \ldots, Y_n \sim N(\mu, \sigma^2)$. To test $H_0: \sigma^2 = \sigma_0^2$ versus $H_1: \sigma^2 \neq \sigma_0^2$ at the α level of significance, reject the null hypothesis if $(n-1)s^2/\sigma_0^2$ is either (a) $\leq \chi^2_{\alpha/2, n-1}$ or (b) $\geq \chi^2_{1-\alpha/2, n-1}$.

PROOF   The two parameter spaces relevant to this problem are

$$\omega = \{(\mu, \sigma^2): -\infty < \mu < \infty, \sigma^2 = \sigma_0^2\}$$

and

$$\Omega = \{(\mu, \sigma^2): -\infty < \mu < \infty, 0 \leq \sigma^2\}.$$

In both, the MLE for μ is $\bar{y}$. In ω, the MLE for σ^2 is simply σ_0^2; in Ω, $\hat{\sigma}^2 = (1/n) \sum_{i=1}^{n} (y_i - \bar{y})^2$ (see Section 7.2). Therefore, the two likelihood functions, maximized over ω and over Ω, are

$$L(\hat{\omega}) = \left(\frac{1}{2\pi\sigma_0}\right)^{n/2} \exp\left[-\frac{1}{2} \sum_{i=1}^{n} \left(\frac{y_i - \bar{y}}{\sigma_0}\right)^2\right]$$

and

$$L(\hat{\Omega}) = \left[\frac{n}{2\pi \sum_{i=1}^{n}(y_i - \bar{y})^2}\right]^{n/2} \exp\left\{-\frac{n}{2} \sum_{i=1}^{n} \left[\frac{y_i - \bar{y}}{\sqrt{\sum_{i=1}^{n}(y_i - \bar{y})^2}}\right]^2\right\}$$

$$= \left[\frac{n}{2\pi \sum_{i=1}^{n}(y_i - \bar{y})^2}\right]^{n/2} e^{-n/2}.$$

It follows that the generalized-likelihood-ratio criterion is given by

$$\lambda = \frac{L(\hat{\omega})}{L(\hat{\Omega})}$$

$$= \left[\frac{\sum_{i=1}^{n}(y_i - \bar{y})^2}{n\sigma_0^2}\right]^{n/2} \exp\left[-\frac{1}{2} \sum_{i=1}^{n} \left(\frac{y_i - \bar{y}}{\sigma_0}\right)^2 + \frac{n}{2}\right]$$

$$= \left(\frac{\hat{\sigma}^2}{\sigma_0^2}\right)^{n/2} e^{-(n/2)(\hat{\sigma}^2/\sigma_0^2) + (n/2)}.$$

Considered as a function of $(\hat{\sigma}^2/\sigma_0^2)$, the numerical behavior of λ is not obvious. It can be shown, however, that λ decreases to 0 as $\hat{\sigma}^2/\sigma_0^2$ becomes either increasingly less than 1 or increasingly greater than 1. Of course, when $\hat{\sigma}^2/\sigma_0^2$ *equals* 1, λ equals 1. Table 7.5 illustrates this pattern for the particular case where $n = 10$.

TABLE 7.5 λ as a function
of $\hat{\sigma}^2/\sigma_0^2$ ($n = 10$)

$\hat{\sigma}^2/\sigma_0^2$	λ
0.25	0.042
0.50	0.381
0.75	0.828
1.00	1.000
1.50	0.623
2.00	0.216
2.50	0.054
3.00	0.011

According to the likelihood-ratio principle, we should reject H_0 for any $\lambda \leq \lambda^*$, where $P(\Lambda \leq \lambda^* \mid H_0) = \alpha$. But as Table 7.5 implies, the shape of $\lambda = \lambda(\hat{\sigma}^2/\sigma_0^2)$ requires that a decision rule of this sort be divided into two critical regions, one for values of $\hat{\sigma}^2/\sigma_0^2$ close to 0 and the other for values of $\hat{\sigma}^2/\sigma_0^2$ much larger than 1. Specifically (see Figure 7.7),

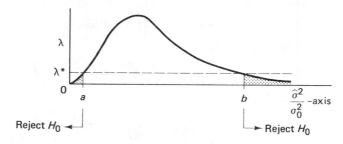

Figure 7.7   Critical region for testing $H_0: \sigma^2 = \sigma_0^2$ versus $H_1: \sigma^2 \neq \sigma_0^2$ (in terms of λ).

$$C = \{(\hat{\sigma}^2/\sigma_0^2): (\hat{\sigma}^2/\sigma_0^2) \leq a \text{ or } (\hat{\sigma}^2/\sigma_0^2) \geq b\}.$$

Comment. At this point it is necessary to make a slight approximation. Just because $P(\Lambda \leq \lambda^* \mid H_0) = \alpha$, it does not follow that

$$P\left[\frac{(1/n) \sum_{i=1}^{n} (Y_i - \bar{Y})^2}{\sigma_0^2} \leq a\right] = \frac{\alpha}{2} = P\left[\frac{(1/n) \sum_{i=1}^{n} (Y_i - \bar{Y})^2}{\sigma_0^2} \geq b\right],$$

and, in fact, the two tails of the critical regions will *not* have exactly the same probability. Nevertheless, the two are numerically close enough so that we will not substantially compromise the likelihood-ratio criterion by setting each one equal to $\alpha/2$. ∎

Note that

$$P\left[\frac{(1/n)\sum\limits_{i=1}^{n}(Y_i - \bar{Y})^2}{\sigma_0^2} \leq a\right] = P\left[\frac{\sum\limits_{i=1}^{n}(Y_i - \bar{Y})^2}{\sigma_0^2} \leq na\right]$$

$$= P\left[\frac{(n-1)S^2}{\sigma_0^2} \leq na\right]$$

$$= P(\chi_{n-1}^2 \leq na)$$

and, similarly,

$$P\left[\frac{(1/n)\sum\limits_{i=1}^{n}(Y_i - \bar{Y})^2}{\sigma_0^2} \geq b\right] = P(\chi_{n-1}^2 \geq nb).$$

Thus, we will choose as critical values $\chi_{\alpha/2,n-1}^2$ and $\chi_{1-\alpha/2,n-1}^2$ and reject H_0 if either

$$\frac{(n-1)s^2}{\sigma_0^2} \leq \chi_{\alpha/2,n-1}^2$$

or

$$\frac{(n-1)s^2}{\sigma_0^2} \geq \chi_{1-\alpha/2,n-1}^2.$$

(See Figure 7.8.)

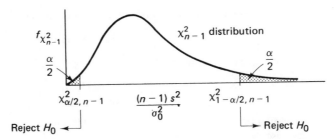

Figure 7.8   Critical region for testing $H_0: \sigma^2 = \sigma_0^2$ versus $H_1: \sigma^2 \neq \sigma_0^2$ (in terms of χ_{n-1}^2).

Comment. One-sided tests for dispersion are set up in a similar fashion. In the case of

$$H_0: \sigma^2 = \sigma_0^2$$

versus

$$H_1: \sigma^2 < \sigma_0^2,$$

H_0 is rejected if

$$\frac{(n-1)s^2}{\sigma_0^2} \leq \chi_{\alpha,n-1}^2.$$

For

$$H_0: \sigma^2 = \sigma_0^2$$

versus

$$H_1: \sigma^2 > \sigma_0^2$$

H_0 is rejected if

$$\frac{(n-1)s^2}{\sigma_0^2} \geq \chi^2_{1-\alpha, n-1}. \quad \blacksquare$$

EXAMPLE 7.2.   Another procedure was used for dating rocks radioactively before the advent of the potassium-argon method described in Case Study 7.3. Based on a mineral's lead content, this earlier method was known to be capable of yielding estimates for this time period that had a standard deviation of 30.4 million years. Recall that the sample standard deviation for the 19 Black Forest rocks (i.e., for the potassium-argon method) was *less*—specifically, it was $\sqrt{733.4}$, or 27.1 million years. Can this be taken as "proof" that the potassium-argon procedure is significantly more precise than its competitor? That is, if σ^2 denotes the (true) variance for the potassium-argon method, we want to test

$$H_0: \sigma^2 = (30.4)^2$$

versus

$$H_1: \sigma^2 < (30.4)^2.$$

Let $\alpha = 0.05$. For a χ^2 variable with 18 degrees of freedom,

$$P(\chi^2_{18} \leq 9.39) = 0.05.$$

Thus, H_0 should be rejected if $(n-1)s^2/\sigma_0^2 \leq 9.39$. But

$$\frac{(19-1)(733.4)}{(30.4)^2} = 14.3,$$

meaning that the potassium-argon method has *not* been shown to be significantly more precise.

QUESTION 7.5.5   The A above middle C is the note given to an orchestra, usually by the oboe, for tuning purposes. Its pitch is defined to be the sound of a tuning fork vibrating at 440 cycles per second (cps). No tuning fork, of course, will always vibrate at *exactly* 440 cps; rather, the pitch, Y, is a random variable. Suppose that Y is normally distributed with $\mu = 440$ and variance σ^2. (Here the parameter σ^2 is a measure of the quality of the tuning fork.) With the standard manufacturing process, $\sigma^2 = 1.1$. A new production technique has just been suggested, however, and its proponents claim it will yield values of σ^2 significantly less than 1.1. To test the claim, six tuning forks are made according to the new procedure. The resulting vibration frequencies are 440.8, 440.3, 439.2, 439.8, 440.6, and 441.1 cps. Do the appropriate hypothesis test. Let $\alpha = 0.05$.

7.6 THE F DISTRIBUTION AND t DISTRIBUTION

There are several important probability distributions intimately related to the normal. The χ^2 density introduced in the previous section is one; in this section we look at two others, the *F distribution* (named after the great British statistician, Sir Ronald A. Fisher) and the *Student t* distribution (named after W. S. Gosset, who published under the pseudonym "Student").[1] Applications of the t and the F will be deferred to later sections. Here we will simply derive their distributions.

> **THEOREM 7.11** Let $U = \sum_{i=1}^{m} X_i^2$ and $V = \sum_{i=1}^{n} Y_i^2$, where $X_i \sim N(0, 1)$, $i = 1, 2, \ldots, m$, and $Y_i \sim N(0, 1), i = 1, 2, \ldots, n$. Let all the X_i's and Y_i's be independent. Define
>
> $$F = \frac{U/m}{V/n}.$$
>
> Then
>
> $$f_F(z) = \frac{\Gamma\left(\dfrac{m+n}{2}\right)}{\Gamma\left(\dfrac{m}{2}\right)\Gamma\left(\dfrac{n}{2}\right)} \frac{m^{m/2} n^{n/2} z^{(m/2)-1}}{(n + mz)^{(m+n)/2}}.$$
>
> The random variable F is said to have an F distribution with m and n degrees of freedom.

PROOF   We will begin by finding the density function for U/V. For notational simplicity, let $c = m/2$ and $d = n/2$. From Theorem 7.7,

$$f_U(u) = \frac{1}{\Gamma(c)2^c} u^{c-1} e^{-u/2}, \qquad u > 0,$$

and

$$f_V(v) = \frac{1}{\Gamma(d)2^d} v^{d-1} e^{-v/2}, \qquad v > 0.$$

For the quotient U/V,

$$F_{U/V}(z) = P(U/V \le z) = P(U/z \le V).$$

(See Figure 7.9.)
Therefore,

$$F_{U/V}(z) = \frac{1}{\Gamma(c)\Gamma(d)2^{c+d}} \int_0^\infty \left(\int_0^{zv} u^{c-1} e^{-u/2} \, du \right) v^{d-1} e^{-v/2} \, dv.$$

[1]Gosset's first gainful employment after leaving Oxford in 1899 was with Messrs. Guinness, a Dublin brewery. Because of company policy, which forbade publication by employees, it was necessary for Gosset to publish his scientific papers under a pen name. The pseudonym he chose was "Student."

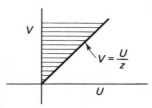

Figure 7.9 Region of integration for $\frac{U}{V} \leq z$.

Thus,

$$f_{U/V}(z) = \frac{1}{\Gamma(c)\Gamma(d)2^{c+d}} \int_0^\infty [(zv)^{c-1}e^{-zv/2}v]v^{d-1}e^{-v/2}\,dv.$$

Note that to obtain the above, we have interchanged the outer integral and d/dz, then used the Fundamental Theorem of Calculus (where does the extra v come from?). Simplification of the above integral gives

$$f_{U/V}(z) = \frac{1}{\Gamma(c)\Gamma(d)2^{c+d}}z^{c-1} \int_0^\infty v^{c+d-1}e^{-[(1+z)/2]v}\,dv.$$

We recognize the integrand as the variable part of a gamma density with $r = c + d$ and $\lambda = (1 + z)/2$, so the integral has value $2^{c+d}\,\Gamma(c + d)/(1 + z)^{c+d}$. Therefore,

$$f_{U/V}(z) = \frac{\Gamma(c + d)}{\Gamma(c)\Gamma(d)} \cdot \frac{z^{c-1}}{(1 + z)^{c+d}}.$$

The statement of the theorem follows immediately from the fact that

$$f_{(n/m)(U/V)}(z) = f_F(z) = \frac{m}{n}f_{U/V}\left(\frac{m}{n}z\right).$$

Comment. The notation $F \sim F_{m,n}$ will denote that the random variable F has an F distribution with m and n degrees of freedom, with m being the degrees of freedom associated with the numerator. Table A.4 in the Appendix gives selected percentiles of $F_{m,n}$ for various m and n. ∎

QUESTION 7.6.1   If $F \sim F_{m,n}$, show that $1/F \sim F_{n,m}$.

QUESTION 7.6.2   Divide the 100 entries in Table 7.2 into 50 groups of 2 and compute the ratio for each group. Compare the upper percentiles of the distribution of 50 quotients with the appropriate upper 0.10, 0.05, and 0.01 percentiles listed in Table A.4.

The next series of definitions and theorems introduce Gosset's best-known contribution to statistical theory, the Student t distribution. This is a critically important result, being the foundation on which confidence intervals for μ and hypothesis tests for μ are based.

DEFINITION 7.2 Let $Z \sim N(0, 1)$, $V^2 \sim \chi_n^2$, and let Z and V^2 be independent. The (Student) t ratio with n degrees of freedom will be defined as the quotient

$$T = \frac{Z}{\sqrt{V^2/n}}.$$

LEMMA $f_T(t)$ is symmetric: $f_T(t) = f_T(-t)$, for all t.

PROOF   Note that

$$-T = \frac{-Z}{\sqrt{V^2/n}},$$

being the ratio of an $N(0, 1)$ to a random variable that is independent of Z, has the same distribution as T. But $-T$ has density $f_T(-t)$ at t. Thus, $f_T(-t) = f_T(t)$.

THEOREM 7.12 The density function for a Student t random variable with n degrees of freedom is

$$f_T(t) = \frac{\Gamma\left(\dfrac{n+1}{2}\right)}{\sqrt{n\pi}\,\Gamma\left(\dfrac{n}{2}\right)\left(1 + \dfrac{t^2}{n}\right)^{(n+1)/2}}, \qquad -\infty < t < \infty.$$

PROOF   Observe that $T^2 = Z^2/(V^2/n) \sim F_{1,n}$. Therefore,

$$f_{T^2}(z) = \frac{n^{n/2}\,\Gamma\left(\dfrac{n+1}{2}\right)}{\Gamma\left(\dfrac{1}{2}\right)\Gamma\left(\dfrac{n}{2}\right)}z^{-1/2}\frac{1}{(n+z)^{(n+1)/2}}.$$

Suppose $t > 0$. By the symmetry of $f_T(t)$,

$$\begin{aligned}
F_T(t) = P(T \leq t) &= \tfrac{1}{2} + P(0 \leq T \leq t) \\
&= \tfrac{1}{2} + \tfrac{1}{2}P(-t \leq T \leq t) \\
&= \tfrac{1}{2} + \tfrac{1}{2}P(0 \leq T^2 \leq t^2) \\
&= \tfrac{1}{2} + \tfrac{1}{2}F_{T^2}(t^2).
\end{aligned}$$

Differentiating $F_T(t)$ gives the stated result:

$$\begin{aligned}
f_T(t) = F_T'(t) &= tF_{T^2}(t^2) \\
&= t\frac{n^{n/2}\,\Gamma\left(\dfrac{n+1}{2}\right)}{\Gamma\left(\dfrac{1}{2}\right)\Gamma\left(\dfrac{n}{2}\right)}(t^2)^{-(1/2)}\frac{1}{(n+t^2)^{(n+1)/2}} \\
&= \frac{\Gamma\left(\dfrac{n+1}{2}\right)}{\sqrt{n\pi}\,\Gamma\left(\dfrac{n}{2}\right)} \cdot \frac{1}{\left[1 + \left(\dfrac{t^2}{n}\right)\right]^{(n+1)/2}}.
\end{aligned}$$

Comment. Over the years, the lower-case t has come to be the accepted symbol for the random variable of Definition 7.2. Here, we will use the lower-case t when the context makes its meaning clear, but in mathematical statements about distributions we will be consistent with random-variable notation and denote the ratio as T. In some instances it is desirable to know how many degrees of freedom are associated with the distribution. When this is the case, we will write t_n (or T_n). ∎

Comment. The t distribution looks very much like the standard normal, although the t is somewhat flatter and has more area in its tails. As n increases, though, the family of t_n distributions converges to the $N(0, 1)$. To illustrate this, Table 7.6 shows the abscissa values cutting off areas of 0.01 and 0.05 in the right-hand tails of Z and t_n for various n between 1 and 100. ∎

TABLE 7.6 Comparison of 95th and 99th percentiles: t_n and Z

	0.05		0.01	
n	t_n	Z	t_n	Z
1	6.314	1.64	31.821	2.33
2	2.920	1.64	6.965	2.33
3	2.353	1.64	4.541	2.33
4	2.132	1.64	3.747	2.33
5	2.015	1.64	3.365	2.33
10	1.812	1.64	2.764	2.33
15	1.753	1.64	2.602	2.33
20	1.725	1.64	2.528	2.33
25	1.708	1.64	2.485	2.33
30	1.697	1.64	2.457	2.33
40	1.684	1.64	2.423	2.33
50	1.676	1.64	2.403	2.33
75	1.666	1.64	2.377	2.33
100	1.660	1.64	2.364	2.33

Comment. Values of $t_{p,n}$, the $100(1 - p)$th percentile of the Student t distribution with n degrees of freedom, are tabulated in Table A.2 in the Appendix. The p values included are 0.20, 0.15, 0.10, 0.05, 0.025, 0.01, and 0.005; n ranges from 1 to 100. For $n > 100$, the Student t and the standard normal are sufficiently similar that percentiles of the normal may be used to approximate the same percentiles of the Student t. ∎

The next theorem gives a particularly important special case of Definition 7.2.

THEOREM 7.13   Let $Y_1, Y_2, \ldots, Y_n \sim N(\mu, \sigma^2)$. Then

$$\frac{\bar{Y} - \mu}{S/\sqrt{n}} \sim T_{n-1}.$$

PROOF   We can rewrite $(\bar{Y} - \mu)/(S/\sqrt{n})$ in the form

$$\frac{\bar{Y} - \mu}{S/\sqrt{n}} = \frac{\dfrac{\bar{Y} - \mu}{\sigma/\sqrt{n}}}{\sqrt{\dfrac{(n-1)S^2}{\sigma^2(n-1)}}}.$$

But

$$\frac{\bar{Y} - \mu}{\sigma/\sqrt{n}} \sim N(0, 1) \quad \text{and} \quad \frac{(n-1)S^2}{\sigma^2} \sim \chi^2_{n-1}.$$

Also, from Appendix 7.2 on page 307,

$$\frac{\bar{Y} - \mu}{\sigma/\sqrt{n}} \quad \text{and} \quad \frac{(n-1)S^2}{\sigma^2}$$

are independent, so the result follows immediately from Definition 7.2.

By inverting the t ratio of Theorem 7.13, we can construct confidence intervals for μ. Theorem 7.14 gives the result. The proof is left as an exercise (see Question 7.6.3).

THEOREM 7.14 Let $Y_1, Y_2, \ldots, Y_n \sim N(\mu, \sigma^2)$. A $100(1 - \alpha)\%$ confidence interval for μ is given by

$$\left(\bar{y} - t_{\alpha/2,n-1}\frac{s}{\sqrt{n}}, \quad \bar{y} + t_{\alpha/2,n-1}\frac{s}{\sqrt{n}}\right).$$

CASE STUDY

7.4

To hunt flying insects, bats emit high-frequency sounds and then listen for their echoes. Until an insect is located, these pulses are emitted at intervals of from 50 to 100 milliseconds. When an insect *is* detected, the pulse-to-pulse interval suddenly decreases—sometimes to as low as 10 milliseconds—thus enabling the bat to pinpoint its prey's position. This raises an interesting question: how far apart are the bat and the insect when the bat first senses that the insect is there? Or, put another way, what is the effective range of a bat's echolocation system?

The technical problems that had to be overcome in measuring the bat-to-insect detection distance were far more complex than the statistical problems involved in analyzing the actual data. The procedure that finally evolved was to put a bat into an 11-by-16-foot room, along with an ample supply of fruit flies, and record the action with two synchronized 16 mm sound-on-film cameras. By examining the two sets of pictures frame by frame, scientists could follow the bat's flight pattern and, at the same time, monitor its pulse frequency. For each insect that was caught, it was therefore possible to estimate the distance between the bat and the insect at the precise moment the bat's pulse-to-pulse interval decreased. Table 7.7 shows the results (63).

TABLE 7.7   Bat-to-insect detection distances for 11 "catches"

Catch Number	Detection Distance (cm)
1	62
2	52
3	68
4	23
5	34
6	45
7	27
8	42
9	83
10	56
11	40

Letting $y_1 = 62$, $y_2 = 52, \ldots, y_{11} = 40$, we have that

$$\sum_{i=1}^{11} y_i = 532 \quad \text{and} \quad \sum_{i=1}^{11} y_i^2 = 29{,}000.$$

Therefore,

$$\bar{y} = \frac{532}{11} = 48.4 \text{ cm}$$

and

$$s = \sqrt{\frac{11(29{,}000) - (532)^2}{11(10)}} = 18.1 \text{ cm.}$$

If the population from which the y_i's are being drawn is normal, the behavior of

$$\frac{\bar{Y} - \mu}{S/\sqrt{n}}$$

will be described by a Student t curve with 10 degrees of freedom. From Table A.2 in the Appendix,

$$P(-2.23 < t_{10} < 2.23) = 0.95.$$

Accordingly, the 95% confidence interval for μ is

$$\left(\bar{y} - 2.23\left(\frac{s}{\sqrt{11}}\right), \bar{y} + 2.23\left(\frac{s}{\sqrt{11}}\right)\right) = \left(48.4 - 2.23\left(\frac{18.1}{\sqrt{11}}\right), 48.4 + 2.23\left(\frac{18.1}{\sqrt{11}}\right)\right)$$

$$= (36.2 \text{ cm}, 60.6 \text{ cm}).$$

Comment. This example is typical of the sort of situation where inferences take the form of confidence intervals rather than hypothesis tests. One of the primary objectives here would be to estimate μ, the true average bat-to-insect detection distance. However, there is no way to phrase such an objective in terms of a hypothesis test, because there is no "standard" value for μ that we can single out to associate with H_0. Presumably, this is the first time measurements such as these have ever been made.   ∎

QUESTION 7.6.3   Prove Theorem 7.14. Hint: Start with the expression

$$P(-t_{\alpha/2, n-1} < T_{n-1} < t_{\alpha/2, n-1}) = 1 - \alpha$$

and make use of the fact that

$$\frac{\bar{Y} - \mu}{S/\sqrt{n}} \sim T_{n-1}.$$

7.7 THE ONE-SAMPLE *t* TEST

Suppose a random sample of size n has been drawn from a distribution presumed to be normal but whose mean and variance are unknown. Often the nature of the data will be such that our immediate concern is the unknown location parameter—that is, we wish to choose between H_0: $\mu = \mu_0$ and H_1: $\mu \neq \mu_0$. An example of this sort of objective was described at some length in Case Study 1.2 and referred to again briefly in Example 6.1. The question is whether or not the Shoshoni Indians had the same aesthetic standards regarding rectangles as did the ancient Greeks. If μ denotes the width-to-length ratio preferred by the Shoshonis, the problem reduces to a choice between H_0: $\mu = 0.618$ and H_1: $\mu \neq 0.618$ where 0.618 is the approximate w/l ratio for golden rectangles.

In deriving a test procedure for H_0: $\mu = \mu_0$ versus H_1: $\mu \neq \mu_0$, once again we appeal to the generalized-likelihood-ratio criterion. Theorem 7.15 gives the decision rule.

THEOREM 7.15   Let $Y_1, Y_2, \ldots, Y_n \sim N(\mu, \sigma^2)$. The GLRT for

$$H_0: \mu = \mu_0$$

versus

$$H_1: \mu \neq \mu_0$$

requires that H_0 be rejected at the α level of significance if

$$\frac{\bar{y} - \mu_0}{s/\sqrt{n}}$$

is either (a) $\leq -t_{\alpha/2, n-1}$ or (b) $\geq +t_{\alpha/2, n-1}$.

PROOF   Since σ^2 is assumed to be unknown, the parameter spaces restricted to H_0 (ω) and $H_0 \cup H_1$ (Ω) are

$$\omega = \{(\mu, \sigma^2): \mu = \mu_0; 0 \leq \sigma^2 < \infty\}$$

and

$$\Omega = \{(\mu, \sigma^2): -\infty < \mu < \infty; 0 \leq \sigma^2 < \infty\}.$$

Without elaborating the details (see Section 7.2 for a very similar problem), it can be readily shown that, under ω,

$$\hat{\mu} = \mu_0 \quad \text{and} \quad \hat{\sigma}^2 = \frac{1}{n} \sum_{i=1}^{n} (y_i - \mu_0)^2;$$

under Ω,

$$\hat{\mu} = \bar{y} \quad \text{and} \quad \hat{\sigma}^2 = \frac{1}{n} \sum_{i=1}^{n} (y_i - \bar{y})^2.$$

Therefore, since

$$L(\mu, \sigma^2) = \left(\frac{1}{\sqrt{2\pi}\,\sigma}\right)^n \exp\left[-\frac{1}{2} \sum_{i=1}^{n} \left(\frac{y_i - \mu}{\sigma}\right)^2\right],$$

direct substitution gives

$$L(\hat{\omega}) = \left[\frac{\sqrt{n}}{\sqrt{2\pi}\,\sqrt{\sum_{i=1}^{n} (y_i - \mu_0)^2}}\right]^n e^{-n/2}$$

$$= \left[\frac{ne^{-1}}{2\pi \sum_{i=1}^{n} (y_i - \mu_0)^2}\right]^{n/2}$$

and

$$L(\hat{\Omega}) = \left[\frac{ne^{-1}}{2\pi \sum\limits_{i=1}^{n} (y_i - \bar{y})^2} \right]^{n/2}.$$

From $L(\hat{\omega})$ and $L(\hat{\Omega})$ we get the likelihood ratio:

$$\lambda = \frac{L(\hat{\omega})}{L(\hat{\Omega})} = \left[\frac{\sum\limits_{i=1}^{n} (y_i - \bar{y})^2}{\sum\limits_{i=1}^{n} (y_i - \mu_0)^2} \right]^{n/2}, \qquad 0 < \lambda \leq 1.$$

As is often the case, it will prove to be more convenient to base a test on a monotonic function of λ, rather than on λ itself. We begin by rewriting the ratio's denominator:

$$\sum_{i=1}^{n} (y_i - \mu_0)^2 = \sum_{i=1}^{n} [(y_i - \bar{y}) + (\bar{y} - \mu_0)]^2$$

$$= \sum_{i=1}^{n} (y_i - \bar{y})^2 + n(\bar{y} - \mu_0)^2.$$

Therefore,

$$\lambda = \left[1 + \frac{n(\bar{y} - \mu_0)^2}{\sum\limits_{i=1}^{n} (y_i - \bar{y})^2} \right]^{-n/2}$$

$$= \left(1 + \frac{t^2}{n-1} \right)^{-n/2}$$

where

$$T = \frac{\bar{Y} - \mu_0}{S/\sqrt{n}} \sim T_{n-1} \qquad \text{(by Theorem 7.13).}$$

Observe that as t^2 increases, λ decreases. This implies that the original GLRT—which, by definition, would have rejected H_0 for any λ that was too small, say, less than λ^*—is equivalent to a test that rejects H_0 whenever t^2 is too large. But since $T \sim T_{n-1}$, "too large" translates numerically into $t_{\alpha/2, n-1}$:

$$0 < \lambda \leq \lambda^* \Longleftrightarrow t^2 \geq (t_{\alpha/2, n-1})^2.$$

But

$$t^2 \geq (t_{\alpha/2, n-1})^2 \Longleftrightarrow t \leq -t_{\alpha/2, n-1} \quad \text{or} \quad t \geq t_{\alpha/2, n-1},$$

and the theorem is proved.

QUESTION 7.7.1   State the GLRT's for testing one-sided alternatives: $H_0: \mu = \mu_0$ versus $H_1: \mu < \mu_0$ and $H_0: \mu = \mu_0$ versus $H_1: \mu > \mu_0$.

As an example of a t test, we will apply the procedure to the Shoshoni rectangle data.

EXAMPLE 7.3.   Whether or not the Shoshoni Indians embrace the golden rectangle as an aesthetic standard reduces to a hypothesis test of

$$H_0: \; \mu = 0.618$$

versus

$$H_1: \; \mu \neq 0.618.$$

If the level of significance is set at $\alpha = 0.05$, the decision rule should be:

Reject H_0 if $\dfrac{\bar{y} - 0.618}{s/\sqrt{20}}$ is either $\begin{cases} \text{(a)} \leq -2.09 \; (= -t_{0.025,19}) \\ \text{(b)} \geq +2.09 \; (= t_{0.025,19}). \end{cases}$

From Table 1.3,

$$\sum_{i=1}^{20} y_i = 13.210 \quad \text{and} \quad \sum_{i=1}^{20} y_i^2 = 8.8878,$$

making

$$\bar{y} = \frac{13.210}{20} = 0.661 \quad \text{and} \quad s = \sqrt{\frac{20(8.8878) - (13.210)^2}{20(19)}} = 0.093.$$

The "observed" t ratio is thus 2.05, and we should accept H_0:

$$\frac{0.661 - 0.618}{0.093/\sqrt{20}} = 2.05.$$

(See Figure 7.10.)

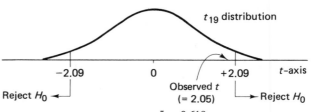

Figure 7.10   Distribution of $\dfrac{\bar{y} - 0.618}{s/\sqrt{20}}$ when H_0 is true.

Comment.   Although the conclusion at the $\alpha = 0.05$ level is to accept H_0, the proximity of the observed t to the upper cutoff value should serve as a warning that the evidence in favor of the null hypothesis was far from overwhelming. In fact, had α been chosen to be 0.10, the critical values would have been ± 1.73, and H_0 would have been rejected.   ∎

QUESTION 7.7.2   Construct a 95% confidence interval for μ. Before carrying out the computations, what do you know about its location relative to the particular value 0.618?

APPENDIX 7.1 POWER CALCULATIONS FOR A ONE-SAMPLE *t* TEST

By definition, the probability of committing a Type I error with the GLRT of Theorem 7.15 is α. But what about a Type II error? If $\mu = \mu_1$, where $\mu_1 \neq \mu_0$, what are the chances that the observed $\bar{y}$ and s will take on values that will "deceive" the decision maker into *accepting* H_0? To answer that, we need the following result, which gives the sampling distribution of the t ratio when H_1, rather than H_0, is true.

THEOREM 7.16   Let $Y_1, Y_2, \ldots, Y_n \sim N(\mu_1, \sigma^2)$. Let

$$T_\Delta = \frac{\bar{Y} - \mu_1 + \Delta}{S/\sqrt{n}} = \frac{\bar{Y} - \mu_0}{S/\sqrt{n}},$$

where $\Delta = \mu_1 - \mu_0$. Then T_Δ is said to be a noncentral T variable with $n - 1$ degrees of freedom and noncentrality parameter γ, where

$$\gamma = \frac{\sqrt{n}\,\Delta}{\sigma}.$$

Also,

$$f_{T_\Delta}(t) = \frac{e^{-(1/2)\gamma^2}\,\Gamma\left(\dfrac{n}{2}\right)}{\sqrt{\pi(n-1)}\,\Gamma\left(\dfrac{n}{2} - \dfrac{1}{2}\right)}\left(1 + \frac{t^2}{n-1}\right)^{-n/2} g(t),$$

where

$$g(t) = \sum_{r=0}^{\infty} \left(\frac{\sqrt{2}\,\gamma t}{\sqrt{n-1}}\right)^r \left[1 + \left(\frac{t^2}{n-1}\right)\right]^{-r/2} \frac{\Gamma[\frac{1}{2}(n+r)]}{r!\,\Gamma(n/2)}.$$

The probability of committing a Type II error, given that $\mu = \mu_1$, is the integral of $f_{T_\Delta}(t)$ between the two critical values defined by the original decision rule:

$P(\text{accept } H_0 \mid H_1 \text{ is true and } \mu = \mu_1)$

$$= P\left[-t_{\alpha/2,\,n-1} < \frac{\bar{Y} - \mu_0}{S/\sqrt{n}} < t_{\alpha/2,\,n-1} \,\Big|\, \mu = \mu_1\right]$$

$$= P\left[-t_{\alpha/2,\,n-1} < \frac{\bar{Y} - \mu_1 + \Delta}{S/\sqrt{n}} < t_{\alpha/2,\,n-1} \,\Big|\, \mu = \mu_1\right]$$

$$= \int_{-t_{\alpha/2,\,n-1}}^{t_{\alpha/2,\,n-1}} f_{T_\Delta}(t)\, dt = \beta.$$

Therefore, the *power* of the test (at $\mu = \mu_1$ and for some specified noncentrality parameter γ) is the sum of two integrals:

$$1 - \beta = \int_{-\infty}^{-t_{\alpha/2,\,n-1}} f_{T_\Delta}(t)\, dt + \int_{t_{\alpha/2,\,n-1}}^{\infty} f_{T_\Delta}(t)\, dt. \tag{7.6}$$

Figure 7.11 (11) gives solutions to Equation 7.6 for a broad range of sample sizes and noncentrality parameters. The ordinate on the graph is simply $1 - \beta$; the abscissa is expressed in terms of ϕ, where $\phi = \gamma/\sqrt{2} = \sqrt{n/2}\,(\Delta/\sigma)$. Making up the body of the table are two sets of curves corresponding to two-sided tests at the $\alpha = 0.01$ and $\alpha = 0.05$ levels of significance. (If H_1 is one-sided, these same two sets of curves give $1 - \beta$ for tests at levels 0.005 and 0.025, respectively.)

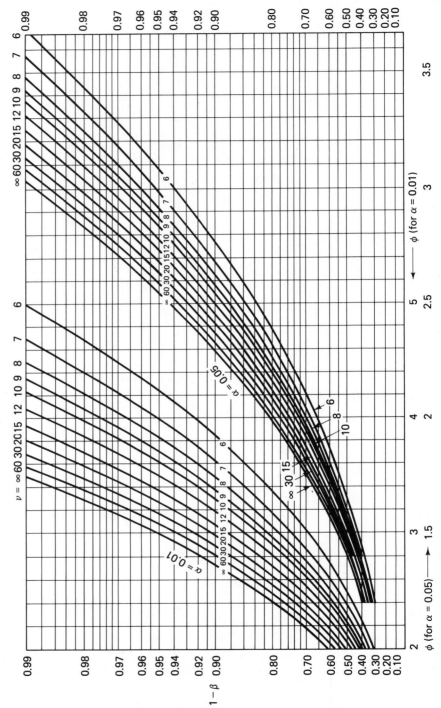

Figure 7.11 Tail areas of the noncentral T.

As an illustration of how Figure 7.11 is used, suppose we wanted to construct the power function for the two-sided, $\alpha = 0.05$ test described in Case Study 1.2—H_0: $\mu = 0.618$ versus H_1: $\mu \neq 0.618$. (Recall from Example 7.3 that the decision rule required that H_0 be rejected if t was either less than or equal to -2.093 or greater than or equal to $+2.093$.) We might begin, for example, by computing the probability that the decision rule will reject H_0 if μ has shifted 0.5 standard deviations (of Y) to the *right* of μ_0. This is equivalent to setting Δ/σ equal to 0.5, in which case $\phi = \sqrt{n/2}\,(\Delta/\sigma) = \sqrt{20/2}\,(0.5) = 1.58$. Note that if an $\alpha = 0.05$ curve corresponding to 19 degrees of freedom had been included in Figure 7.11, its ordinate for an abscissa of 1.58 would be approximately 0.57. (See Figure 7.12.) Thus,

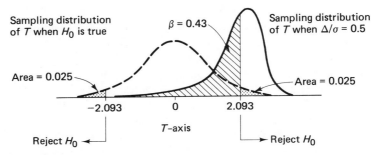

Figure 7.12   Noncentral T distributions ($\Delta/\sigma = 0$ and $\Delta/\sigma = 0.5$).

$$P\left(\text{reject } H_0 \,\Big|\, \frac{\Delta}{\sigma} = 0.5\right) = 1 - \beta$$

$$= \int_{-\infty}^{-2.093} f_{T_\Delta}\left(t;\, \frac{\Delta}{\sigma} = 0.5\right) dt + \int_{2.093}^{\infty} f_{T_\Delta}\left(t;\, \frac{\Delta}{\sigma} = 0.5\right) dt$$

$$\doteq 0.57.$$

Proceeding in this same fashion, suppose that μ has shifted 0.75 standard deviations to the right of μ_0. Then $\phi = \sqrt{20/2}\,(0.75) = 2.37$ and, from Figure 7.11, $1 - \beta = 0.89$. (See Figure 7.13.)

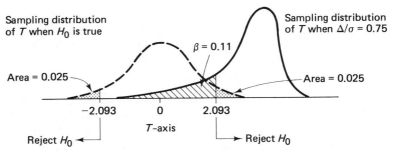

Figure 7.13   Noncentral T distributions ($\Delta/\sigma = 0$ and $\Delta/\sigma = 0.75$).

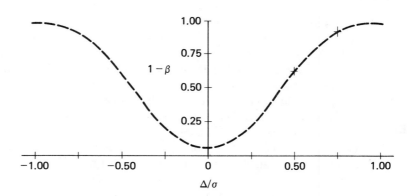

Figure 7.14   Power function for Shoshoni hypothesis test.

Figure 7.14 shows the complete power function for this particular test. The two X's on the curve correspond to Δ/σ values of 0.50 and 0.75. [That the power function is symmetric around $\Delta/\sigma = 0$ follows immediately from the functional representation of $f_{T_\Delta}(t)$ as given in Theorem 7.16.]

In practice, an experimenter is likely to use Figure 7.11 in a way slightly different from the one we have just described. Rather than determine the power function for a test whose decision rule (and sample size) has already been decided upon, he will usually turn the problem around and look for the smallest sample size that guarantees that β will not exceed a certain value for a given Δ/σ.

For example, suppose an experimenter has decided to test $H_0: \mu = \mu_0$ versus $H_1: \mu \neq \mu_0$ at the $\alpha = 0.05$ level of significance. How large a sample should he take? Should n be 10 or 20 or 250—or something even larger? Clearly, this is a fundamental question in the design of every experiment. Unfortunately, it has no simple answer, but with the help of Figure 7.11 we can at least make some ballpark estimates.

Before n can even be estimated, though, the experimenter needs to specify a minimum $1 - \beta$ he is willing to accept for a given amount of shift away from μ_0. For instance, if the true μ is 0.5 standard deviations or more away from μ_0, he may want the test procedure to be capable of rejecting H_0 at least 80% of the time. In symbols, if Δ/σ is greater than or equal to 0.5, $1 - \beta$ should be greater than or equal to 0.80.

Comment. We know, of course, that for any fixed $\Delta/\sigma \neq 0$, $1 - \beta$ approaches 1 as n approaches infinity. The problem is to find the *smallest* (i.e., the most economical) n that will make $1 - \beta$ at least as large as 0.8 when μ shifts 0.5 standard deviations away from μ_0. ∎

Note that if n were 20, ϕ would be $\sqrt{20/2}\,(0.5) = 1.58$. But, from Figure 7.11, $1 - \beta$ for $\phi = 1.58$ and $n = 20$ is approximately 0.56, implying that $n = 20$ is too small. Similarly, if n were 30, ϕ would be $\sqrt{30/2}\,(0.5) = 1.94$, and the corresponding $1 - \beta$ would be 0.75—still too small. But for $n = 40$, $\phi = \sqrt{40/2}\,(0.5) = 2.24$ and $1 - \beta = 0.87$. Therefore, the n we are looking for lies somewhere between 30 and 40.

Comment. Sometimes the greatest benefit to be derived from power calculations is in showing that the objectives of a project are unrealistic. For example, if an experimenter can afford to take only n^* observations and if the calculated $1 - \beta$ against a prespecified (Δ/σ) is much too small, then the project should either be redesigned or not be done at all. ∎

APPENDIX 7.2 SOME DISTRIBUTION RESULTS FOR $\bar{Y}$ AND S^2

THEOREM 7.17   Let $Y_i \sim N(\mu, \sigma^2)$, $i = 1, 2, \ldots, n$, and let the Y_i's be independent. Define

$$\bar{Y} = \frac{1}{n} \sum_{i=1}^{n} Y_i \quad \text{and} \quad S^2 = \frac{1}{n-1} \sum_{i=1}^{n} (Y_i - \bar{Y})^2.$$

Then

(a) $\bar{Y}$ and S^2 are independent, and
(b) $(n-1)S^2/\sigma^2 \sim \chi_{n-1}^2$.

PROOF   The proof of this theorem relies on certain linear algebra techniques as well as a change-of-variables formula for multiple integrals. Definition 7.3 and the lemma that follows review the necessary background results. For further details, see (46) or (209).

DEFINITION 7.3

(a) A matrix A is said to be *orthogonal* if $AA^T = I$.
(b) Let β be any n-dimensional vector over the real numbers. That is, $\beta = (c_1, c_2, \ldots, c_n)$, where each c_j is a real number. The *length* of β will be defined as

$$\| \beta \| = (c_1^2 + \cdots + c_n^2)^{1/2}.$$

(Note that $\| \beta \|^2 = \beta \beta^T$.)

LEMMA

(a) A matrix A is orthogonal if and only if

$$\| A\beta \| = \| \beta \| \quad \text{for each } \beta.$$

(b) If a matrix A is orthogonal, then $\det A = 1$.
(c) Let g be a one-to-one continuous mapping on a subset, D, of n-space. Then

$$\int_{g(D)} f(x_1, \ldots, x_n) \, dx_1 \cdots dx_n = \int_{D} f(g(y_1, \ldots, y_n)) \det J(g) \, dy_1 \cdots dy_n,$$

where $J(g)$ is the Jacobian of the transformation.

Set $X_i = (Y_i - \mu)/\sigma$ for $i = 1, 2, \ldots, n$. Then all the X_i's are $N(0, 1)$. Let A be an $n \times n$ orthogonal matrix whose last row is $(1/\sqrt{n}, 1/\sqrt{n}, \ldots, 1/\sqrt{n})$. Let $\vec{X} = (X_1, \ldots, X_n)^T$ and define $\vec{Z} = (Z_1, Z_2, \ldots, Z_n)^T$ by the transformation $\vec{Z} = A\vec{X}$. (Note that $Z_n = (1/\sqrt{n})X_1 + \cdots + (1/\sqrt{n})X_n = \sqrt{n}\,\bar{X}$.)

For any set D,

$$P(\vec{Z} \in D) = P(A\vec{X} \in D) = P(\vec{X} \in A^{-1}D)$$

$$= \int_{A^{-1}D} f_{X_1,\ldots,X_n}(x_1, \ldots, x_n)\, dx_1 \cdots dx_n$$

$$= \int_D f_{X_1,\ldots,X_n}(g(\vec{z}))\, \det J(g)\, dz_1 \cdots dz_n$$

$$= \int_D f_{X_1,\ldots,X_n}(A^{-1}\vec{z}) \cdot 1 \cdot dz_1 \cdots dz_n,$$

where $g(\vec{z}) = A^{-1}\vec{z}$. But A^{-1} is orthogonal, so, setting $(x_1, \ldots, x_n)^T = A^{-1}z$, we have that

$$x_1^2 + \cdots + x_n^2 = z_1^2 + \cdots + z_n^2.$$

Thus,

$$f_{X_1,\ldots,X_n}(\vec{x}) = (2\pi)^{-n/2} e^{-(1/2)(x_1^2 + \cdots + x_n^2)}$$

$$= (2\pi)^{-n/2} e^{-(1/2)(z_1^2 + \cdots + z_n^2)}.$$

From this we conclude that

$$P(\vec{Z} \in D) = \int_D (2\pi)^{-n/2}\, e^{-(n/2)(z_1^2 + \cdots + z_n^2)}\, dz_1 \cdots dz_n,$$

implying that the Z_j's are independent standard normals.

Finally,

$$\sum_{j=1}^n Z_j^2 = \sum_{j=1}^{n-1} Z_j^2 + n\bar{X}^2 = \sum_{j=1}^n X_j^2 = \sum_{j=1}^n (X_j - \bar{X})^2 + n\bar{X}^2.$$

Therefore,

$$\sum_{j=1}^{n-1} Z_j^2 = \sum_{j=1}^n (X_j - \bar{X})^2,$$

and the conclusion follows for standard normal variables. Also, since $\bar{Y} = \sigma\bar{X} + \mu$ and $\sum_{i=1}^n (Y_i - \bar{Y})^2 = \sigma^2 \sum_{i=1}^n (X_i - \bar{X})^2$, the conclusion follows for $N(\mu, \sigma^2)$ variables.

Comment. As part of the proof presented above, we established a version of *Fisher's lemma*:

Let $X_1, X_2, \ldots, X_n$ be independent standard normal random variables and let A be an orthogonal matrix. Define $(Z_1, \ldots, Z_n)^T = A(X_1, \ldots, X_n)^T$. Then the Z_i's are independent standard normal random variables. ∎

REVIEW EXERCISES FOR CHAPTER 7

1. In the home, the amount of radiation emitted by a color television set does not pose a health problem of any consequence. The same may not be true, though, in department stores, where as many as 15 or 20 sets may be turned on at the same time and in a relatively confined area (recall Example 2.5). The following readings were taken at ten different department stores, each having at least five sets in their display areas. The figure shown for each store is an average radiation level (in milliroentgens per hour) based on readings taken at several different locations within each area (87).

Location	Net mr/hour
1	0.40
2	0.48
3	0.60
4	0.15
5	0.50
6	0.80
7	0.50
8	0.36
9	0.16
10	0.89

(a) Compute the sample mean and the sample standard deviation for these data.
(b) The recommended safety limit for exposure of this sort has been set by the National Council on Radiation Protection at 0.5 mr/hour. Assume that the $\bar{y}$ and s calculated in part (a) are the true μ and σ, respectively, characterizing radiation exposure in department-store display areas. If the distribution of exposures from store to store follows a normal pdf, estimate the proportion of stores giving exposures in excess of the recommended safety limit.

2. For data coming from a normal pdf, the proportion of observations in the interval $(\bar{y} - s, \bar{y} + s)$ is approximately 0.68, and the proportion in the interval $(\bar{y} - 2s, \bar{y} + 2s)$ is approximately 0.95. [These are estimates based on the fact that $P(\mu - \sigma < X < \mu + \sigma) = 0.68$ and $P(\mu - 1.96\sigma < X < \mu + 1.96\sigma) = 0.95$ if $X \sim N(\mu, \sigma^2)$.] Determine the one- and two-standard-deviation intervals for the motivation scores given in Case Study 4.7 and compute the proportion of observations they each contain (see Table 4.8). Recall that for those data, $\bar{y} = 13.8$ and $s = 6.5$.

3. What percentage of normally distributed observations would be expected to lie in the interval $(\bar{y} - s, \bar{y} + 2s)$? "Verify" your answer using the Etruscan skull data given in Case Study 7.1.

4. In Review Exercise 5 at the end of Chapter 5, the notion of an estimator's being *asymptotically unbiased* was introduced. If $Y_1, Y_2, \ldots, Y_n \sim N(\mu, \sigma^2)$, is the sample variance asymptotically unbiased for σ^2?

5. The effects of age on various characteristics of a person's blood are not entirely understood. Whether or not age affects the platelet count was the reason for collecting the data listed on the next page (176). The subjects were 24 female rest-home patients.

In a more typical, nongeriatric population, platelet counts are expected to range from 140 to 440 (thousands per mm³ of blood).

Subject	Count	Subject	Count
1	125	13	180
2	170	14	180
3	250	15	280
4	270	16	240
5	144	17	270
6	184	18	220
7	176	19	110
8	100	20	176
9	220	21	280
10	200	22	176
11	170	23	188
12	160	24	176

Assuming platelet counts to be normally distributed, estimate the percentage of elderly women whose counts would be considered abnormal. Draw a histogram of these data. Does the normality assumption seem tenable?

6. At a certain Southern university, the 1975 freshman class numbered 2670, of whom 72.5% were males. The men averaged 631 on the verbal part of the SAT and 691 on the quantitative part. The women's averages were 623 and 640, respectively. Find the average freshman SAT score at this university, where the average score is defined to be one-half times the sum of the verbal and quantitative scores.

7. Show that the $N(\mu, \sigma^2)$ pdf, where μ is known, can be written in exponential form (see Review Exercise 25 at the end of Chapter 5). Use that fact to suggest a sufficient statistic for σ^2.

8. The cylinders and pistons for a certain internal combustion engine are manufactured by a process that gives a normal distribution of cylinder diameters with a mean of 41.5 cm and a standard deviation of 0.4 cm. Similarly, the distribution of piston diameters is normal with a mean of 40.5 cm and a standard deviation of 0.3 cm. If the piston diameter is greater than the cylinder diameter, the former can be reworked until the two "fit." What proportion of cylinder-piston pairs will need to be reworked?

9. A circuit contains three resistors wired in series. Each is rated at 6 ohms. Suppose, however, that the true resistance of each one is a normally distributed random variable with a mean of 6 ohms and a standard deviation of 0.3 ohm. What is the probability that the combined resistance will exceed 19 ohms? How "precise" would the manufacturing process have to be to make the probability less than 0.005 that the combined resistance of the circuit would exceed 19 ohms?

10. The personnel department of a large corporation gives two aptitude tests to job applicants. One measures verbal ability; the other, quantitative ability. From many years' experience, the company has found that the verbal scores tend to be normally distributed with a mean of 50 and a standard deviation of 10. The quantitative scores are normally distributed with a mean of 100 and a standard deviation of 20, and they appear to be independent of the verbal scores. A composite score, C, is assigned to each applicant, where

$$C = 3(\text{verbal score}) + 2(\text{quantitative score}).$$

If company policy prohibits hiring anyone whose C score is below 375, what percentage of applicants will be summarily rejected?

11. Jasmine commutes to work by first driving her car to a bus stop and then riding the bus the rest of the way. The time it takes her to reach the bus stop is normally distributed with a mean of 15 minutes and a standard deviation of 2 minutes. Once she gets to the bus stop, the time required for the remainder of the trip is normally distributed with a mean of 35 minutes and a standard deviation of 5 minutes. Assume the two times are independent.
 (a) Let W denote the total time taken by Jasmine to get to work. Find $E(W)$ and Var (W).
 (b) If Jasmine is supposed to be at work at 8:00 and she leaves her apartment at 7:05, what is the probability she is late?

12. Prove the Corollary to Theorem 7.4 using moment-generating functions.

13. Two policemen trying to ticket speeders are hiding behind two different stretches of road. Suppose the probability of someone's speeding past the first policeman is 0.1, and the second policeman, 0.3. Also, suppose that a total of 50 cars drive past the first officer and 60 the second. Estimate the probability that at least 25 cars are ticketed.

14. Carry out the details involving Theorem 3.3 to verify the last statement in the proof of Theorem 7.4.

15. Let $Y_1, Y_2, \ldots, Y_n$ be a random sample from an $N(2, 4)$ pdf. How large must n be in order that

$$P(1.9 \le \bar{Y} \le 2.1) \ge 0.99?$$

16. How large a random sample needs to be taken from a normal pdf in order for the probability to be at least 0.90 that the sample mean will be within one standard deviation of μ? Within two standard deviations of μ?

17. An elevator in the athletic dorm at State University has a stated load capacity of 2400 pounds. Suppose ten football players get on at the twentieth floor. If the distribution of weights of football players at State is normally distributed with a mean of 220 pounds and a standard deviation of 20 pounds, what is the probability that there will be ten fewer players at the next practice?

18. Let $Y_1, Y_2, \ldots, Y_9$ be a random sample from an $N(0, 9)$ distribution. On the same axes, draw (a) the original pdf and (b) the pdf for $\bar{Y}$. Find $P(Y_i \ge 1)$ and $P(\bar{Y} \ge 1)$.

19. Let $X_1, X_2, \ldots, X_9$ be an independent random sample from an $N(2, 4)$ pdf and Y_1, Y_2, Y_3, Y_4 an independent random sample from an $N(1, 1)$ pdf. Find $P(\bar{X} \ge \bar{Y})$.

20. A student makes 100 check transactions between receiving two consecutive bank statements. Rather than subtract each amount exactly, he rounds each entry off to the nearest dollar. Let Y_i denote the round-off error associated with the ith transaction.

[It can be assumed that Y_i has a uniform pdf over the interval $(-\frac{1}{2}, \frac{1}{2})$.] Use the central limit theorem to approximate the probability that the student's total accumulated error (either positive or negative) after 100 transactions exceeds \$5. Compare your answer here with the Chebyshev inequality upper bound worked out in Example 3.30.

21. Suppose a fair coin is tossed 200 times. Let $Y_i = 1$ if the ith toss comes up heads and $Y_i = 0$, otherwise, $i = 1, 2, \ldots, 200$.
    (a) Use Chebyshev's inequality to get a lower bound for the probability that Y, the sum of the Y_i's, is within 5 of $E(Y)$.
    (b) Use the DeMoivre-Laplace form of the central limit theorem to approximate $P[E(Y) - 5 \leq Y \leq E(Y) + 5]$.

22. The course record for a certain two-man bobsled race in international competition is 17.10 seconds. Each competing team makes four runs down the course. Their "official time" is the average of those four runs. Suppose the Czechoslovakian team achieves times of 17.1, 17.2, 17.1, and 17.2 seconds, respectively. Assume the timing device is accurate to the nearest tenth of a second—that is, a recorded time of 17.1 seconds means the true time is somewhere between 17.05 and 17.15 seconds. Let

$$Y_i = \text{timing error on the } i\text{th run}$$

$$= i\text{th recorded time} - i\text{th true time}, \qquad i = 1, 2, 3, 4.$$

Assume that Y_i is a uniform random variable. Use the central limit theorem to approximate the probability that the Czechoslovakian team actually set a course record even though their average recorded time was 17.15 seconds. (Assume that the course record is exactly 17.10 seconds.)

23. A certain double-whorl fingerprint pattern is known to be extremely rare, occurring, on the average, in only 1.2 individuals per 100,000. Use the central limit theorem to approximate the probability that in a city whose population is 400,000, there will be at least seven people having fingerprints of that type.

24. Suppose the annual number of earthquakes registering 2.5 or more on the Richter scale and having an epicenter within 40 miles of downtown Memphis follows a Poisson distribution with an average rate of $\lambda = 6.5$ per year. Use the central limit theorem to approximate the probability that there will be nine or more such earthquakes in the coming year. Find the same probability using the Poisson model directly.

25. Suppose 100 dice are tossed. Let Y_i be the number showing on the ith die. Estimate the probability that the sum of the Y_i's exceeds 370.

26. Let Y_1, Y_2, Y_3, and Y_4 be a random sample of size four from the pdf

$$f_Y(y) = 3(1 - y)^2, \qquad 0 < y < 1.$$

Use the central limit theorem to approximate $P(\frac{1}{8} < \bar{Y} < \frac{3}{8})$.

27. The number of students enrolling in a certain statistics course averages 35 for the fall semester. The classroom assigned for the course has 40 seats. Assuming the number of students who sign up for the course to be a Poisson random variable, use the normal

approximation to the Poisson to estimate the probability that the number of students enrolling will exceed the number of available seats.

28. Recently there has been considerable controversy over the possible aftereffects of a nuclear weapons test conducted in Nevada in 1957. Protocol for the test included some 3000 military and civilian "observers." Now, more than 20 years later, eight leukemia cases have been diagnosed among those 3000. The expected number of cases, adjusted for the demographic characteristics of the observers, was three.

    (a)   Use the central limit theorem to assess the significance of the eight cases. Assume the occurrence of leukemia cases follows a Poisson distribution. What would you conclude?

    (b)   Find the "exact" probability asked for in part (a) by using the Poisson pdf directly.

29. The chi square distribution is asymptotically normal as n, the number of degrees of freedom, goes to infinity. (This should come as no surprise, since a χ_n^2 random variable can be written as a sum of n independent squared normals.) Find an expression approximating the pth percentile of a χ_n^2 distribution in terms of the pth percentile for an $N(0, 1)$. Use your formula to approximate the 95th percentile of the χ_{150}^2 distribution.

30. Let $Y_1, Y_2, \ldots, Y_n$ be a random sample from an $N(\mu, \sigma^2)$ pdf. Show that the variance of the sample variance is given by

$$\text{Var}\,(S^2) = \frac{2\sigma^4}{n-1}.$$

31. Given that $Y_1, Y_2, \ldots, Y_n$ is an independent $N(\mu, \sigma^2)$ random sample, prove that the sample variance, S^2, is consistent for σ^2 (see Review Exercise 30).

32. Let $X_1, X_2, \ldots, X_n$ be a random sample of size n from an exponential pdf

$$f_X(x) = \left(\frac{1}{\theta}\right)e^{-x/\theta}, \qquad x > 0, \theta > 0.$$

    (a)   Use moment-generating functions to show that $Y = 2n\bar{X}/\theta$ has a χ_{2n}^2 pdf.

    (b)   Use the result of part (a) to derive a formula for a $100(1-\alpha)\%$ confidence interval for θ. Verify the statement made in Section 5.2 that $(462, 5371)$ is a 95% confidence interval for the exponential parameter given that the sample observations are $x_1 = 610$, $x_2 = 1150$, and $x_3 = 1570$.

    (c)   Why would it be difficult to construct exact confidence intervals for θ without using the transformation suggested in part (a)?

33. Prove Theorem 7.9.

34. Use the results of Review Exercises 29 and 32 to construct a 95% confidence interval for the true average life expectancy of the V805 transmitter tubes discussed in Example 2.15.

35. One of the occupational hazards of being an airplane pilot is the hearing loss that results from being exposed to high noise levels. To document the magnitude of the problem, a team of researchers (94) measured the cockpit noise levels of 18 commercial aircraft. The results (in decibels) are listed below.

Plane	Noise Level (db)	Plane	Noise Level (db)
1	74	10	72
2	77	11	90
3	80	12	87
4	82	13	73
5	82	14	83
6	85	15	86
7	80	16	83
8	75	17	83
9	75	18	80

Assume the distribution of cockpit noise levels from plane to plane is normally distributed. Find a 95% confidence interval for σ^2. Also, find a 95% confidence interval for σ.

36. Let Y be a random variable whose pdf involves k parameters, $\theta_1, \theta_2, \ldots, \theta_k$. Based on a random sample of size n, we wish to test

$$H_0: \theta_1 = \theta_{1_0}, \ldots, \theta_r = \theta_{r_0}, r \leq k$$

versus

$$H_1: H_0 \text{ is false.}$$

Let λ denote the generalized likelihood ratio as given in Definition 6.1. Under quite general conditions, it can be shown that $-2 \ln \Lambda$ has approximately a χ^2 distribution with r degrees of freedom and that H_0 should be rejected if $-2 \ln \lambda \geq \chi^2_{r, 1-\alpha}$. (This is a very useful result in situations where the pdf of the generalized likelihood ratio is difficult to find.) Suppose 100 independent Bernoulli trials result in 60 successes. Test $H_0: p = 0.5$ versus $H_1: p \neq 0.5$ at the $\alpha = 0.05$ level of significance with a critical region based on $-2 \ln \Lambda$.

37. Suppose $X \sim \chi^2_{60}$. Use the central limit theorem to approximate $P(45 < X < 65)$.

38. Potential differences (PD) measured across certain parts of the intestinal tract have proven to be useful in diagnosing gastrointestinal disorders. With that as a precedent, an experiment was set up to see whether PD's could, in a similar way, differentiate diseases involving the esophagus (185). There was some concern, however, about the variability of the procedure. By examining a large number of patients with no diseases of the esophagus, it was found that the variance of their PD's was 9.0. For ten patients with clinically established esophagitis, the proposed procedure gave the following esophageal PD's (in millivolts): 26, 15, 18, 32, 12, 26, 39, 16, 10, and 19. Test H_0: $\sigma^2 = 9.0$ versus $H_1: \sigma^2 \neq 9.0$. Let $\alpha = 0.05$.

39. Use the asymptotic normality of the chi square pdf to suggest a large-sample test of $H_0: \sigma^2 = \sigma_0^2$ versus $H_1: \sigma^2 \neq \sigma_0^2$. (Recall Review Exercise 29.) Apply your procedure to the data given in Review Exercise 38.

40. Confidence intervals are typically two-sided, but the inversion technique of Section 5.9 works just as well to set up *one-sided* confidence intervals. Find the two 95% one-sided confidence intervals for the cockpit-noise-level data of Review Exercise 35.

41. A sheet-metal firm intends to run quality control tests on the shear strength of spot welds. As a guideline for choosing a sample size, they decide to take the smallest number of observations for which there is at least a 95% chance that the ratio of the sample variance to the true variance will be less than 2. Find the minimum n.

42. Staffs from 15 different hospitals participated in a surveillance program to monitor the number of patients experiencing adverse reactions to prescribed medication. The percentages for the 15 hospitals were 5.8, 5.3, 4.5, 3.9, 4.6, 5.4, 7.9, 8.2, 6.9, 5.7, 4.6, 6.3, 8.4, 4.6, and 7.3. Let μ denote the true average percentage represented by these figures. Construct a 95% confidence interval for μ.

43. A random variable Y is said to have a *beta distribution* with parameters r and s if

$$f_Y(y) = \frac{\Gamma(r+s)}{\Gamma(r)\Gamma(s)} y^{r-1}(1-y)^{s-1}, \qquad 0 < y < 1, r > 0, s > 0.$$

Use the fact that $\int_0^1 f_Y(y)\,dy = 1$ to find $E(U/V)$, where U and V are independent χ^2 random variables with m and n degrees of freedom, respectively (recall the proof of Theorem 7.11). Hint: In the integral defining $E(U/V)$, make the substitution $z = w/(1-w)$.

44. Use the expression for $E(U/V)$ derived in Review Exercise 43 to find the expected value of an F distribution.

45. Let Y be a beta random variable with parameters r and s. Find $E(Y)$ and Var (Y).

46. Evaluate:

(a) $\displaystyle\int_0^{\pi/2} \cos^4\theta \sin^6\theta\,d\theta.$ (b) $\displaystyle\int_0^4 x(4-x)^{10}\,dx.$

(See Review Exercise 43.)

47. Great discoveries in science tend to be made by persons who are quite young. Listed below are 12 major scientific breakthroughs from the middle of the sixteenth century to the early part of the twentieth century (200).

Discovery	Discoverer	Date	Age
Earth goes around sun	Copernicus	1543	40
Telescope, basic laws of astronomy	Galileo	1600	34
Principles of motion, gravitation, calculus	Newton	1665	23
Nature of electricity	Franklin	1746	40
Burning is uniting with oxygen	Lavoisier	1774	31
Earth evolved by gradual processes	Lyell	1830	33
Evidence for natural selection controlling evolution	Darwin	1858	49
Field equations for light	Maxwell	1864	33
Radioactivity	Curie	1896	34
Quantum theory	Planck	1901	43
Special theory of relativity, $E = mc^2$	Einstein	1905	26
Mathematical foundations for quantum theory	Schroedinger	1926	39

Let μ denote the true average age at which great scientific discoveries are made.

(a) Construct a 95% confidence interval for μ.

(b) Before constructing a confidence interval for a set of observations extending over a long period, we should be convinced that the x_i's exhibit no biases or trends. If, for example, the age at which scientists made major discoveries decreased from century to century, then the parameter μ would no longer be a constant and the confidence interval would be meaningless. Plot "date" versus "age" for the 12 discoveries given above. Put "Date of discovery" on the abscissa. Does the variability in the ages appear to be random with respect to time?

48. Show that as $n \longrightarrow \infty$, the pdf of a Student t random variable with n degrees of freedom converges to the pdf for an $N(0, 1)$. Hint: To show that the constant term in the pdf for T converges to $1/\sqrt{2\pi}$, use Stirling's formula:

$$\Gamma(n) \sim \sqrt{2\pi/n}\ n^n e^{-n}.$$

49. Use Definition 7.2 to get an expression for the even moments of a Student t random variable with n degrees of freedom. (The odd moments are all zero.) Are all the even moments defined?

50. Let $Y_1, Y_2, \ldots, Y_n$ be a random sample from an $N(\mu, \sigma^2)$ pdf where σ^2 is known. Find an expression for a 95% confidence interval for μ. What is the smallest sample size that will yield an interval having a length less than L?

51. Six readings, 4.3, 5.8, 8.4, 3.7, 5.2, and 5.1, are taken on a response variable known to be normally distributed with an unknown mean, μ, but a known variance of 1.50. Construct a 90% confidence interval for μ. Construct a 50% confidence interval for μ.

52. Theorem 7.14 gives an expression for a $100(1 - \alpha)\%$ confidence interval for μ based on a random sample of size n from an $N(\mu, \sigma^2)$ pdf. Review Exercise 50 gives another. Do you think that both expressions are equally likely to come up in real-world applications? Explain.

53. For random samples from an $N(\mu, \sigma^2)$ pdf, $\bar{Y}$ and S^2 are independent (see Theorem 7.17). It follows that if

$$P\left(-z_{\alpha/2} < \frac{\bar{Y} - \mu}{\sigma/\sqrt{n}} < +z_{\alpha/2}\right) = 1 - \alpha$$

and

$$P\left(\chi^2_{\alpha/2, n-1} < \frac{(n-1)S^2}{\sigma^2} < \chi^2_{1-\alpha/2, n-1}\right) = 1 - \alpha,$$

then

$$P\left[-z_{\alpha/2} < \frac{\bar{Y} - \mu}{\sigma/\sqrt{n}} < +z_{\alpha/2} \quad and \quad \chi^2_{\alpha/2, n-1} < \frac{(n-1)S^2}{\sigma^2} < \chi^2_{1-\alpha/2, n-1}\right]$$
$$= (1 - \alpha)^2.$$

Use the latter expression to set up boundaries in the $\mu\sigma^2$ plane for a *joint* $100(1 - \alpha)^2\%$ confidence "region" for μ and σ^2.

54. Evaluate the integral

$$\int_0^\infty \frac{1}{1 + x^2} dx$$

using the Student t distribution.

55. Construct a 99% confidence interval for the true average department-store radiation exposure level using the data of Review Exercise 1.

56. Use the result of Review Exercise 53 to construct a joint 95% confidence interval for the μ and the σ^2 associated with the cockpit-noise-level data of Review Exercise 35. Graph your answer.

57. The production of a nationally marketed detergent results in certain workers' receiving prolonged exposures to *Bacillus subtilis* enzyme. Nineteen workers were tested to determine the effects of these exposures on various respiratory functions (165). One such function, air-flow rate, is measured by computing the ratio of a person's forced expiratory volume (FEV_1) to his vital capacity (VC). (Vital capacity is the maximum volume of air a person can exhale after taking as deep a breath as possible. FEV_1 is the maximum volume of air a person can exhale in one second.) In persons with no lung dysfunction, the "norm" for the FEV_1/VC ratio is 0.80. It can be assumed that if the enzyme does have an effect, it will be to *reduce* the FEV_1/VC ratio. The data are listed below.

Subject	FEV_1/VC		Subject	FEV_1/VC
RH	0.61		WS	0.78
RB	0.70		RV	0.84
MB	0.63		EN	0.83
DM	0.76		WD	0.82
WB	0.67		FR	0.74
RB	0.72		PD	0.85
BF	0.64		EB	0.73
JT	0.82		PC	0.85
PS	0.88		RW	0.87
RB	0.82			

Do the appropriate hypothesis test. Let $\alpha = 0.025$.

58. A psychologist would like to know whether persons with high IQ's (over 130) are significantly faster at working a certain puzzle than persons with average IQ's. The puzzle requires that five irregularly shaped pieces be fitted together to form a cube. Based on considerable past experience, it is known that persons with average IQ's complete the cube in a mean time of 4.6 minutes. When the psychologist presents the puzzle to four persons with IQ's over 130, she records the following completion times (in minutes): 4.0, 4.2, 4.6, and 4.2. State an appropriate H_0 and H_1 for this problem and carry out the test at the $\alpha = 0.05$ level of significance.

59. A test of a simple hypothesis, $H_0: \theta = \theta_0$, against a simple alternative, $H_1: \theta = \theta_1$, is called a *best test* if the critical region is derived using the Neyman-Pearson lemma as described in Appendix 6.1. Suppose, however, the hypotheses to be tested are $H_0: \theta = \theta_0$ versus $H_1: \theta > \theta_0$. If the *form* of the best critical region for testing $H_0: \theta = \theta_0$ versus $H_1: \theta = \theta_1$ for $\theta_1 > \theta_0$ is the same for *any* value of θ_1 (greater than θ_0), then the test is said to be *uniformly most powerful* (UMP). Let $Y_1, Y_2, \ldots,$ Y_n be a random sample from an $N(\mu, \sigma^2)$ pdf, where σ^2 is known. Find a uniformly most powerful test of $H_0: \mu = \mu_0$ versus $H_1: \mu > \mu_0$. Let $\alpha = 0.05$.

60. An environmentalist group wants to monitor the temperature rise in the water 50 yards downstream from a nuclear reactor. They are concerned that the rise not exceed 3°F. If it does, the ecological balance of the stream will be irreparably damaged. They decide to test

$$H_0: \mu = 3°F$$

versus

$$H_1: \mu > 3°F,$$

where μ is the true average temperature increase in the water. Their plan is to collect 16 water samples and choose between H_0 and H_1 on the basis of the average temperature rise, $\bar{y}$. Similar studies suggest that the standard deviation (σ) of the Y_i's will be 1°F. For a decision rule, they decide to reject H_0 if $\bar{y} \geq 3.5°F$.

(a) Find α.
(b) What are their chances of committing a Type II error if, in fact, the true μ is 3.8°F?

61. Construct a power curve for the hypothesis test of Review Exercise 60.

62. Let $Y_1, Y_2, \ldots, Y_n$ be a random sample from an $N(\mu, \sigma^2)$ pdf where σ^2 is known. Find the GLRT for

$$H_0: \mu = \mu_0$$

versus

$$H_1: \mu \neq \mu_0.$$

Let $\alpha = 0.05$. Compare your result with the decision rule for a one-sample t test.

63. A manufacturer of pipe for laying underground electrical cables is concerned about the pipe's rate of corrosion and whether a special coating may retard that rate. As a way of measuring corrosion, they examine a short length of pipe and record the depth of the maximum pit. Their tests have shown that in a year's time in the particular kind of soil they have to deal with the average depth of the maximum pit in a foot of pipe is 0.0042 inch. The standard deviation is 0.0003 inch. Ten pipes are coated with a new plastic and buried in the same soil. After one year, the following maximum pit depths are recorded: 0.0039, 0.0041, 0.0038, 0.0044, 0.0040, 0.0036, 0.0034, 0.0046, 0.0035, and 0.0036. Can it be concluded at the $\alpha = 0.05$ level that the plastic coating is beneficial?

64. The melting point of a certain alloy is required to be 1000°C to meet engineering specifications. Should the true melting point (μ) deviate by more than 30°C, the production process should be stopped and reset. If the standard deviation of the melting point temperatures is 20°C, and we elect to test $H_0: \mu = 1000°C$ versus $H_1: \mu \neq 1000°C$, what will be the probability of detecting a shift of 30°C if ten alloy samples are examined? Assume $\alpha = 0.05$.

65. A quality control engineer needs to monitor a machine that puts cereal into boxes. In particular, he is concerned that the machine may not be putting enough cereal in the boxes. According to the label, each box contains 16 ounces, so he wants to test (at the $\alpha = 0.05$ level of significance)

$$H_0: \mu = 16$$

versus

$$H_1: \mu < 16,$$

where μ is the true average weight. Also, he wants the test to detect a true average weight as small as 15 ounces 95 % of the time. From past experience with this machine, he knows that the weight, Y, per box is normally distributed with a standard deviation, σ, of 0.9 ounce. What is the smallest sample size that will meet his requirements?

66. The pH of the catalyst in a chemical reaction is to be measured by a standard titration procedure. In order for the reaction to proceed at the desired rate, the catalyst pH must be very close to 6.4. It is known that the titration procedure has an associated standard deviation of 0.03. Suppose the chemist wishes to test

$$H_0: \mu = 6.4$$

versus

$$H_1: \mu \neq 6.4$$

at the $\alpha = 0.01$ level of significance and wants to have at least a 95 % chance of rejecting H_0 if the true pH is outside the interval (6.36, 6.44). How many readings should she take?

67. Consider again the hypothesis test of $H_0: \mu = 0.618$ versus $H_1: \mu \neq 0.618$ as described in Case Study 1.2. Suppose it has not yet been decided how many rectangles should be measured. The experimenters, however, want to test H_0 and H_1 at the $\alpha = 0.05$ level of significance and they want the test to be sufficiently precise so that if $|\mu - 0.618|/\sigma \geq 0.60$, H_0 will be rejected at least 90 % of the time. How large should n be?

68. Use Figure 7.11 to construct the power curve for the FEV_1/VC data given in Review Exercise 57.

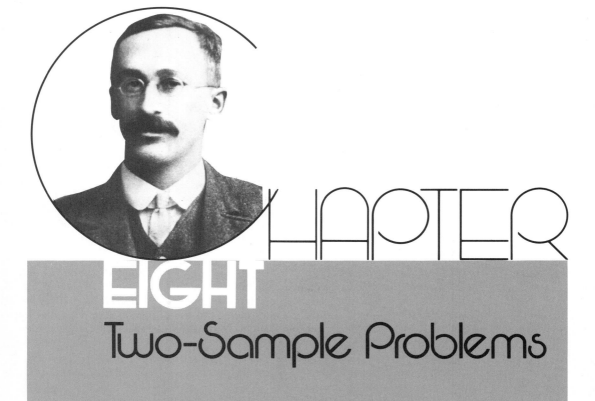

CHAPTER EIGHT

Two-Sample Problems

WILLIAM SEALY GOSSET ("STUDENT") (1876–1937)

After earning an Oxford degree in mathematics and chemistry, Gosset began working in 1899 for Messrs. Guinness, a Dublin brewery. Fluctuations in materials and temperature and the necessarily small-scale experiments inherent in brewing convinced him of the necessity for a new small-sample theory of statistics. Writing under the pseudonym, "Student," he published work with the t ratio that was destined to become a cornerstone of modern statistical methodology. Gosset worked for the Guinness company his entire life.

8.1 INTRODUCTION

The simplicity of the one-sample model makes it the logical starting point for any discussion of statistical inference, but it also limits its applicability to the real world. Very few experiments involve just a single treatment or a single set of conditions. On the contrary, researchers almost invariably design experiments to compare responses to *several* treatments—or, at the very least, to compare a single treatment with a control.

In this chapter, we will examine the simplest of these multitreatment designs, the *two-sample problem*. Structurally, the two-sample problem always falls into one of two different formats: either two (presumably) different treatments are applied to two independent sets of similar subjects or the same treatment is applied to two (presumably) different kinds of subjects. Comparing the effectiveness of germicide A relative to that of germicide B by measuring the zones of inhibition each one produces in two sets of similarly cultured Petri dishes would be an example of the first type. Another would be testing whether monkeys raised by themselves (treatment X) react differently in a stress situation than monkeys raised with siblings (treatment Y). On the other hand, examining the bones of 60-year-old men and 60-year-old women, all lifelong residents of the same city, to see whether both sexes absorb environmental strontium-90 at the same rate would be an example of the second type.

Inference in two-sample problems usually reduces to a comparison of *location* parameters. We might assume, for example, that the population of responses associated with, say, treatment X is normally distributed with mean μ_X and standard deviation σ_X while the Y distribution is normal with mean μ_Y and standard deviation σ_Y. Comparing location parameters, then, reduces to testing $H_0: \mu_X = \mu_Y$. As always, the alternative may be either one-sided, $H_1: \mu_X < \mu_Y$ or $H_1: \mu_X > \mu_Y$, or two-sided, $H_1: \mu_X \neq \mu_Y$. (If the data are binomial, the location parameters are p_X and p_Y, the true "success" probabilities for treatments X and Y, and the null hypothesis takes the form $H_0: p_X = p_Y$.)

Sometimes, although much less frequently, it becomes more relevant to compare the *variabilities* of two treatments, rather than their locations. A food company, for example, trying to decide which of two types of machines to buy for filling cereal boxes would naturally be concerned about the *average* weights of the boxes filled by each type, but they would also want to know something about the *variabilities* of the weights. Obviously, a machine that produced high proportions of "underfills" and "overfills" would be a distinct liability. In a situation of this sort, the appropriate null hypothesis is $H_0: \sigma_X^2 = \sigma_Y^2$.

For comparing the means of two normal populations, the standard procedure is the *two-sample t test*. As described in Section 8.2, this is a relatively straightforward extension of Chapter 7's one-sample t test. For comparing variances, though, it will be necessary to introduce a completely new test—this one based on the F distribution of Section 7.6. The binomial version of the two-sample problem, testing $H_0: p_X = p_Y$, is taken up in Section 8.4.

It was mentioned in connection with one-sample problems that certain inferences, for various reasons, are more aptly phrased in terms of confidence intervals rather than hypothesis tests. The same is true of two-sample problems. In Section 8.5, confidence intervals are constructed for the location *difference* of two populations, $\mu_X - \mu_Y$ (or $p_X - p_Y$), and the variability *quotient*, σ_X^2/σ_Y^2.

8.2 TESTING H_0: $\mu_X = \mu_Y$— THE TWO-SAMPLE t TEST

We will suppose the data for a given experiment consist of two independent random samples, $X_1, X_2, \ldots, X_n$ and $Y_1, Y_2, \ldots, Y_m$, representing either of the models referred to in Section 8.1. Furthermore, the two populations from which the X's and Y's are drawn will be presumed normal. Let μ_X and μ_Y denote their means. Our problem will be to derive a procedure for testing H_0: $\mu_X = \mu_Y$. Of course, to accept H_0 is to accept the equivalence of the two treatments (or the two sets of subjects), at least in terms of the average effects they elicit.

As it turns out, the precise form of the test we are looking for depends on the variances of the X and Y populations. If it can be assumed that σ_X^2 and σ_Y^2 are equal, it is a relatively straightforward task to produce the GLRT for H_0: $\mu_X = \mu_Y$. (This is, in fact, what we will do in Theorem 8.2.) But if the variances of the two populations are *not* equal, the problem becomes much more complex. This second case, known as the Behrens-Fisher problem, is more than 50 years old and remains one of the more famous "unsolved" problems in statistics. What headway investigators *have* made has been confined to approximate solutions [see, for example, Sukhatme (174) or Cochran (28)]. These, however, will not be discussed here; we will restrict our attention to testing H_0: $\mu_X = \mu_Y$ when it can be assumed that $\sigma_X^2 = \sigma_Y^2$.

For the one-sample test that $\mu = \mu_0$, the GLRT was shown to be a function of a special case of the t ratio introduced in Definition 7.2 (recall Theorem 7.13). We begin this section with a theorem that gives still another special case of Definition 7.2. This one will do for the two-sample GLRT what $\sqrt{n}(\bar{Y} - \mu_0)/S$ did for the one-sample GLRT.

THEOREM 8.1   Let $X_1, X_2, \ldots, X_n \sim N(\mu_X, \sigma^2)$ and $Y_1, Y_2, \ldots, Y_m \sim N(\mu_Y, \sigma^2)$ and let the X's and Y's be independent. Let S_X^2 and S_Y^2 be the two sample variances, and S_p^2, the *pooled variance*, where

$$S_p^2 = \frac{(n-1)S_X^2 + (m-1)S_Y^2}{n+m-2} = \frac{\sum_{i=1}^{n}(X_i - \bar{X})^2 + \sum_{i=1}^{m}(Y_i - \bar{Y})^2}{n+m-2}.$$

Then

$$T = \frac{\bar{X} - \bar{Y} - (\mu_X - \mu_Y)}{S_p\sqrt{\frac{1}{n} + \frac{1}{m}}} \sim T_{n+m-2}.$$

PROOF   The method of proof here is very similar to what was used for Theorem 7.13. Note that an equivalent formulation of T would be

$$T = \frac{\dfrac{\bar{X} - \bar{Y} - (\mu_X - \mu_Y)}{\sigma\sqrt{\dfrac{1}{n} + \dfrac{1}{m}}}}{\sqrt{S_p^2/\sigma^2}}$$

$$= \frac{\dfrac{\bar{X} - \bar{Y} - (\mu_X - \mu_Y)}{\sigma\sqrt{\dfrac{1}{n} + \dfrac{1}{m}}}}{\sqrt{\dfrac{1}{n + m - 2}\left[\displaystyle\sum_{i=1}^{n}\left(\dfrac{X_i - \bar{X}}{\sigma}\right)^2 + \sum_{i=1}^{m}\left(\dfrac{Y_i - \bar{Y}}{\sigma}\right)^2\right]}}.$$

But $E(\bar{X} - \bar{Y}) = \mu_X - \mu_Y$ and $\text{Var}(\bar{X} - \bar{Y}) = \sigma^2/n + \sigma^2/m$, so the numerator of the ratio is clearly an $N(0, 1)$. In the denominator,

$$\sum_{i=1}^{n}\left(\frac{X_i - \bar{X}}{\sigma}\right)^2 = \frac{(n - 1)S_X^2}{\sigma^2} \sim \chi_{n-1}^2$$

and

$$\sum_{i=1}^{m}\left(\frac{Y_i - \bar{Y}}{\sigma}\right)^2 = \frac{(m - 1)S_Y^2}{\sigma^2} \sim \chi_{m-1}^2,$$

so that by Theorem 7.6,

$$\sum_{i=1}^{n}\left(\frac{X_i - \bar{X}}{\sigma}\right)^2 + \sum_{i=1}^{m}\left(\frac{Y_i - \bar{Y}}{\sigma}\right)^2 \sim \chi_{n+m-2}^2.$$

Also, from Appendix 7.2 on page 307, it follows that the numerator and denominator of the ratio are independent. By Definition 7.2, then,

$$\frac{\bar{X} - \bar{Y} - (\mu_X - \mu_Y)}{S_p\sqrt{\dfrac{1}{n} + \dfrac{1}{m}}} \sim \frac{N(0, 1)}{\sqrt{\dfrac{\chi_{n+m-2}^2}{n + m - 2}}} \sim T_{n+m-2}.$$

QUESTION 8.2.1   It was easy to verify from first principles that the numerator of the t ratio had a mean of 0 and a variance of 1. From what result, though, does the *normality* of the numerator derive?

Theorem 8.2 gives the GLRT for testing the equality of two normal means against a two-sided alternative. The modifications for one-sided alternatives should be readily apparent. Case Studies 8.1 and 8.2 show how Theorem 8.2 is applied.

THEOREM 8.2 Let $X_1, X_2, \ldots, X_n \sim N(\mu_X, \sigma^2)$ and $Y_1, Y_2, \ldots, Y_m \sim N(\mu_Y, \sigma^2)$ and let the X's and Y's be independent. At the α level of significance, the GLRT for H_0: $\mu_X = \mu_Y$ versus H_1: $\mu_X \neq \mu_Y$ calls for H_0 to be rejected if

$$t = \frac{\bar{x} - \bar{y}}{s_p \sqrt{\dfrac{1}{n} + \dfrac{1}{m}}} \quad \text{is either} \quad \begin{cases} \leq -t_{\alpha/2, n+m-2} \\ \text{or} \\ \geq +t_{\alpha/2, n+m-2}. \end{cases}$$

PROOF   (See Appendix 8.1 on page 344.)

CASE STUDY

8.1

Cases of disputed authorship are not very common but when they do occur, they can be very difficult to resolve. Speculation has persisted for several hundred years that some of Shakespeare's works were written by Sir Francis Bacon. And whether it was Alexander Hamilton or James Madison who wrote certain of the Federalist Papers is still an open question. A similar, though more recent, dispute centers around Mark Twain (17).

In 1861, a series of ten essays appeared in the *New Orleans Daily Crescent*. Signed "Quintus Curtius Snodgrass," the essays purported to chronicle the author's adventures as a member of the Louisiana militia. While historians generally agree that the accounts referred to actually did happen, there seems to be no record of anyone named Quintus Curtius Snodgrass. Adding to the mystery is the fact that the style of the pieces bears unmistakable traces—at least to some critics—of the humor and irony that made Mark Twain so famous.

Most typically, efforts to unravel these sorts of "yes, he did—no, he didn't" controversies rely heavily on literary and historical clues. But not always. There is also a statistical approach to the problem. Studies have shown that authors are remarkably consistent in the extent to which they use words of a certain length. That is, a given author will use roughly the same proportion of, say, three-letter words in something he writes this year as he did in whatever he wrote last year. The same holds true for words of any length. *But*, the proportion of three-letter words that author A consistently uses will very likely be different from the proportion of three-letter words that author B uses. It follows that by comparing the proportions of words of a certain length in essays known to be the work of Mark Twain to the proportions found in the ten Snodgrass essays, we should be able to assess the likelihood of the two authors' being one and the same.

Table 8.1 shows the proportions of three-letter words found in eight Twain essays and in the ten Snodgrass essays. (Each of the Twain works was written at approximately the same time the Snodgrass essays appeared.)

TABLE 8.1   Proportion of three-letter words

Twain	Proportion	QCS	Proportion
Sergeant Fathom letter	0.225	Letter I	0.209
Madame Caprell letter	0.262	Letter II	0.205
Mark Twain letters in		Letter III	0.196
Territorial Enterprise		Letter IV	0.210
First letter	0.217	Letter V	0.202
Second letter	0.240	Letter VI	0.207
Third letter	0.230	Letter VII	0.224
Fourth letter	0.229	Letter VIII	0.223
First *Innocents Abroad* letter		Letter IX	0.220
First half	0.235	Letter X	0.201
Second half	0.217		

If $x_1 = 0.225$, $x_2 = 0.262, \ldots, x_8 = 0.217$, and $y_1 = 0.209$, $y_2 = 0.205, \ldots, y_{10} = 0.201$, then

$$\bar{x} = \frac{1.855}{8} = 0.232 \quad \text{and} \quad \bar{y} = \frac{2.097}{10} = 0.210.$$

To analyze these data, we need to decide what the magnitude of the difference between the sample means, $\bar{x} - \bar{y} = 0.232 - 0.210 = 0.022$, actually tells us. Let μ_X and μ_Y denote the proportions of three-letter words in *all* essays written by Twain and by Snodgrass, respectively. Of course, not having examined the complete works of the two authors, we have no way of evaluating either μ_X or μ_Y, so they become the unknown parameters of the problem. What needs to be decided, then, is whether an observed *sample* difference (in the proportions of three-letter words) as large as 0.022 implies that the two *true* proportions, μ_X and μ_Y, are themselves not the same. Or is 0.022 small enough to still be compatible with the hypothesis that they are? Put more formally, we must choose between

$$H_0: \mu_X = \mu_Y$$

and

$$H_1: \mu_X \neq \mu_Y.$$

Since $\quad \sum_{i=1}^{8} x_i^2 = 0.4316 \quad \text{and} \quad \sum_{i=1}^{10} y_i^2 = 0.4406,$

the two sample variances are

$$s_X^2 = \frac{8(0.4316) - (1.855)^2}{8(7)}$$

$$= 0.0002103$$

and

$$s_Y^2 = \frac{10(0.4406) - (2.097)^2}{10(9)}$$

$$= 0.0000955.$$

Combined, they give a pooled standard deviation of 0.012:

$$s_p = \sqrt{\frac{\sum_{i=1}^{8} (x_i - 0.232)^2 + \sum_{i=1}^{10} (y_i - 0.210)^2}{n + m - 2}}$$

$$= \sqrt{\frac{(n - 1)s_X^2 + (m - 1)s_Y^2}{n + m - 2}}$$

$$= \sqrt{\frac{7(0.0002103) + 9(0.0000955)}{8 + 10 - 2}}$$

$$= \sqrt{0.0001457}$$

$$= 0.012.$$

According to Theorem 8.1, if $H_0 : \mu_X = \mu_Y$ is true, the sampling distribution of

$$T = \frac{\bar{X} - \bar{Y}}{S_p\sqrt{\frac{1}{8} + \frac{1}{10}}}$$

is described by a Student t curve with 16 $(= 8 + 10 - 2)$ degrees of freedom. If we elect to test H_0 versus H_1 at the $\alpha = 0.01$ level of significance, the null hypothesis should be rejected if either

(a) $t \geq t_{\alpha/2, n+m-2} = t_{0.005, 16} = 2.92$, or
(b) $t \leq -t_{0.005, 16} = -2.92$ (see Figure 8.1).

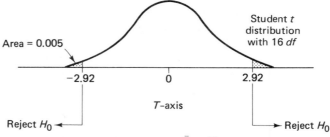

Figure 8.1   Distribution of $\dfrac{\bar{X} - Y}{S_p\sqrt{\frac{1}{8} + \frac{1}{10}}}$ when H_0 is true.

But

$$t = \frac{0.232 - 0.210}{0.012\sqrt{\frac{1}{8} + \frac{1}{10}}}$$

$$= 3.86,$$

a value falling considerably to the right of $t_{0.005, 16}$. Therefore, we *reject* H_0—it would appear that Twain and Snodgrass were not the same person.

Comment. The X_i's and Y_i's in this problem, being proportions, are not normal random variables, so the basic assumption of Theorem 8.2 is not met. Nevertheless, the probabilistic behavior of the t ratio is often only minimally affected by the nonnormality of the two populations being sampled. (More will be said about this property, known as *robustness*, in Chapter 13.) Suffice it to say that for this set of data, the t test provides a very adequate analysis.

QUESTION 8.2.2   Should the alternative hypothesis here be one-sided or two-sided? Explain.

CASE STUDY

8.2 Poverty Point is the name given to a number of widely scattered archaeological sites throughout Louisiana, Mississippi, and Arkansas. These are the remains of a society thought to have flourished during the period from 1700 B.C. to 500 B.C. Among their characteristic artifacts are ornaments that were fashioned out of clay and then baked. Described in this example is a method for "dating" the various Poverty Point sites by using a property of baked clay known as thermoluminescence (83).

When certain substances are heated, they emit light in proportion to the amount of radiation to which they have been exposed. This is thermoluminescence. Furthermore, when these same substances are heated to a high enough temperature (i.e., *annealed*) they "lose" whatever exposure they had previously accumulated. This is what happened to the clay ornaments when they were first baked—over 2000 years ago. Each of them now provides a record of the total cosmic and background radiation it received since the time of its baking.

By calibrating samples of clay to determine how much thermoluminescence is produced for a given amount of incident radiation,
scientists can estimate the age of the artifacts.

Table 8.2 gives the estimated ages of eight clay ornaments, four
each found at two geographically separated Poverty Point sites,
Terral Lewis and Jaketown. The question to be answered is whether
the technologies at these two sites evolved at similar rates, as
measured by when they were capable of making these ornaments.

TABLE 8.2   Thermoluminescent dates (years B.C.)

Terral Lewis Estimates (x_i)	Jaketown Estimates (y_i)
1492	1346
1169	942
883	908
988	858

Suppose μ_X and μ_Y denote the true average thermoluminescent
dates for all Terral Lewis and Jaketown artifacts, respectively. The
hypotheses to be tested are

$$H_0: \ \mu_X = \mu_Y$$

versus

$$H_1: \ \mu_X \neq \mu_Y.$$

(The alternative is two-sided because there is no a priori reason for
anticipating which of the two sets of dates, if either, will be older
than the other.) We will choose $\alpha = 0.05$ as the level of significance.

Since the total sample size is 8, the decision rule calls for H_0 to
be rejected if

$$|t| = \frac{|\bar{x} - \bar{y}|}{s_p\sqrt{\frac{1}{4} + \frac{1}{4}}} \geq t_{\alpha/2, n+m-2} = t_{0.025, 6} = 2.45.$$

But

$$\sum_{i=1}^{4} x_i = 4532, \qquad \sum_{i=1}^{4} x_i^2 = 5{,}348{,}458,$$

and

$$\sum_{i=1}^{4} y_i = 4054, \qquad \sum_{i=1}^{4} y_i^2 = 4{,}259{,}708,$$

so that

$$\bar{x} = \frac{4532}{4} = 1133.0,$$

$$\bar{y} = \frac{4054}{4} = 1013.5,$$

$$s_X^2 = \frac{4(5,348,458) - (4532)^2}{4(3)} = 71,234.0,$$

$$s_Y^2 = \frac{4(4,259,708) - (4054)^2}{4(3)} = 50,326.3,$$

and

$$s_p = \sqrt{\frac{3(71,234.0) + 3(50,326.3)}{4 + 4 - 2}} = 246.5.$$

Therefore,

$$|t| = \frac{|1133.0 - 1013.5|}{246.5\sqrt{\frac{1}{4} + \frac{1}{4}}} = 0.68,$$

and our conclusion is to *accept* H_0: if there was a difference in the rate of technological advancement between Terral Lewis and Jaketown, these data do not show it.

QUESTION 8.2.3   How different would $\bar{x}$ and $\bar{y}$ have to be before we could reject $H_0: \mu_X = \mu_Y$ at the $\alpha = 0.05$ level of significance?

8.3 TESTING H_0: $\sigma_X^2 = \sigma_Y^2$— THE *F* TEST

Although by far the majority of two-sample problems are set up to detect possible shifts in location parameters, situations sometimes arise where it is equally important—perhaps even more important—to compare variability parameters. Two machines on an assembly line, for example, may be producing items whose *average* dimensions (μ_X and μ_Y) of some sort—say, thickness—are not significantly different but whose variabilities (as measured by σ_X^2 and σ_Y^2) are. This becomes a critical piece of information if the increased variability results in an unacceptable proportion of items from one of the machines falling outside the engineering specifications (see Figure 8.2).

In this section we will examine the generalized-likelihood-ratio test of H_0: $\sigma_X^2 = \sigma_Y^2$ versus H_1: $\sigma_X^2 \neq \sigma_Y^2$. The data will consist of two independent random samples of sizes n and m: the first—$X_1, X_2, \ldots, X_n$—is assumed to have come from a normal distribution having mean μ_X and variance σ_X^2; the second—$Y_1, Y_2, \ldots, Y_m$—from a normal distribution having mean μ_Y and variance σ_Y^2. (All four parameters are assumed to be unknown.) Theorem 8.3 gives the test procedure that will be used. The proof will not be given, but it follows the same basic pattern we have seen in other GLRT's; the important step is showing that the likelihood ratio is a monotonic function of the F distribution defined in Theorem 7.11.

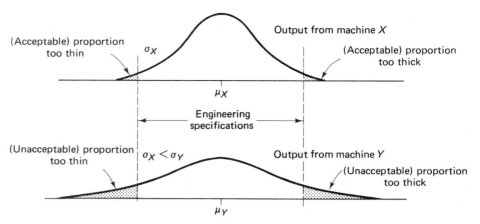

Figure 8.2 Variability of machine outputs.

Comment. Tests of $H_0: \sigma_X^2 = \sigma_Y^2$ arise in another, more routine, context. Recall that the procedure for testing the equality of μ_X and μ_Y depended on whether or not the two population variances were equal. This implies that a test of $H_0: \sigma_X^2 = \sigma_Y^2$ should precede every test of $H_0: \mu_X = \mu_Y$. If the former is accepted, the t test on μ_X and μ_Y is done according to Theorem 8.2; but if $H_0: \sigma_X^2 = \sigma_Y^2$ is rejected, Theorem 8.2 is inappropriate and either the Sukhatme, Cochran, or some other approximation to the two-sample t test must be used. There are, however, some liabilities connected with the test procedure of Theorem 8.3, the nature of which will be discussed in Chapter 13. Suffice it to say that while a preliminary look at $H_0: \sigma_X^2 = \sigma_Y^2$ is certainly desirable before doing a two-sample t test, not all statisticians would endorse the method described here. As it turns out, testing variances is a difficult problem and there is no completely satisfactory solution.   ∎

THEOREM 8.3 Let $X_1, X_2, \ldots, X_n \sim N(\mu_X, \sigma_X^2)$ and $Y_1, Y_2, \ldots, Y_m \sim N(\mu_Y, \sigma_Y^2)$ and let the X's and Y's be independent. An approximate GLRT for

$$H_0: \sigma_X^2 = \sigma_Y^2$$

versus

$$H_1: \sigma_X^2 \neq \sigma_Y^2$$

at the α level of significance calls for H_0 to be rejected if

$$\frac{S_Y^2}{S_X^2} \text{ is either } \begin{cases} \leq F_{\alpha/2,\,m-1,\,n-1} \\ \text{or} \\ \geq F_{1-\alpha/2,\,m-1,\,n-1}. \end{cases}$$

Comment. The GLRT described in Theorem 8.3 is *approximate* for the same sort of reason the GLRT for $H_0: \sigma^2 = \sigma_0^2$ was approximate (see Theorem 7.10). The distribution of the test statistic, S_Y^2/S_X^2, is not symmetric, and the two ranges of variance ratios yielding λ's less than or equal to λ^* (i.e., the left tail and right tail of the critical region) have slightly different areas. For the sake of convenience, though, it is customary to choose the two critical values so that each cuts off the same area, $\alpha/2$. ■

QUESTION 8.3.1   Show that $S_Y^2/S_X^2 \sim F_{m-1,\, n-1}$.

CASE STUDY

8.3   Electroencephalograms are records showing fluctuations of electrical activity in the brain. Among the several different kinds of brain "waves" produced, the dominant ones are usually *alpha* waves. These have a characteristic frequency of anywhere from 8 to 13 cycles per second.

The objective of the experiment described in this example was to see whether sensory deprivation over an extended period of time has any effect on the alpha-wave pattern. The subjects were 20 inmates in a Canadian prison. They were randomly split into two equal-sized groups. Members of one group were placed in solitary confinement; those in the other group were allowed to remain in their own cells. Seven days later, alpha-wave frequencies were measured for all 20 subjects (57), as shown in Table 8.3.

TABLE 8.3 Alpha-wave frequencies (cps)

Nonconfined (x_i)	Solitary Confinement (y_i)
10.7	9.6
10.7	10.4
10.4	9.7
10.9	10.3
10.5	9.2
10.3	9.3
9.6	9.9
11.1	9.5
11.2	9.0
10.4	10.9

Judging from the graph (Figure 8.3), there was an apparent *decrease* in alpha-wave frequency for persons in solitary confine-

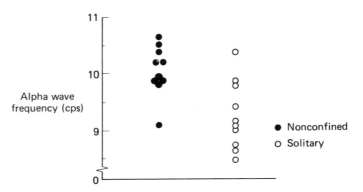

Figure 8.3 Alpha-wave frequencies (cps).

ment. There also appears to have been an *increase* in the variability for that group. We will use the F test to determine whether the observed difference in variability ($s_X^2 = 0.21$ versus $s_Y^2 = 0.36$) is statistically significant.

Let σ_X^2 and σ_Y^2 denote the true variances of alpha-wave frequencies for nonconfined and solitary-confined prisoners, respectively. The hypotheses to be tested are

$$H_0: \sigma_X^2 = \sigma_Y^2$$

versus

$$H_1: \sigma_X^2 \neq \sigma_Y^2.$$

Let $\alpha = 0.05$ be the level of significance. Given that

$$\sum_{i=1}^{10} x_i = 105.8, \qquad \sum_{i=1}^{10} x_i^2 = 1121.26,$$

$$\sum_{i=1}^{10} y_i = 97.8, \qquad \sum_{i=1}^{10} y_i^2 = 959.70,$$

the sample variances become

$$s_X^2 = \frac{10(1121.26) - (105.8)^2}{10(9)} = 0.21$$

and

$$s_Y^2 = \frac{10(959.70) - (97.8)^2}{10(9)} = 0.36.$$

Dividing the sample variances gives an observed F ratio of 1.71:

$$F = \frac{s_Y^2}{s_X^2} = \frac{0.36}{0.21} = 1.71.$$

Both n and m are 10, so we would expect S_Y^2/S_X^2 to behave like an F random variable with 9 and 9 degrees of freedom (assuming H_0: $\sigma_X^2 = \sigma_Y^2$ is true). From Table A.4 in the Appendix, we see that the values cutting off areas of 0.025 in either tail of that distribution are 0.248 and 4.03. (See Figure 8.4.) Our conclusion, then, is to *accept* H_0. (In light of the Comment preceding Theorem 8.3, it would now be appropriate to test $H_0: \mu_X = \mu_Y$ using the two-sample t test of Section 8.2.)

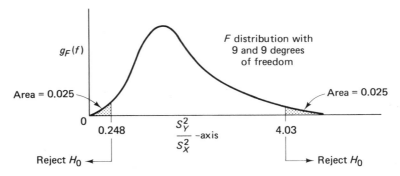

Figure 8.4   Distribution of S_Y^2/S_X^2 when H_0 is true.

QUESTION 8.3.2   Do an F test on the Mark Twain-Quintus Curtius Snodgrass data of Case Study 8.1. Take 0.05 to be the level of significance. Was the method used to test $H_0: \mu_X = \mu_Y$ appropriate?

8.4 BINOMIAL DATA: TESTING $H_0: p_X = p_Y$

Up to this point, the data considered in Chapter 8 have been independent random samples of sizes n and m drawn from two *continuous* distributions—in fact, from two *normal* distributions. Obviously, many other sorts of data might have to be dealt with. The X's and Y's might represent continuous random variables, for example, but have density functions other than the normal. Or they might be *discrete*. In this section we consider the most common example of this latter type: situations where the two sets of data are *binomial*.

Suppose that n Bernoulli trials related to treatment X have resulted in x successes, and m (independent) Bernoulli trials related to treatment Y in y successes. We wish to test whether p_X and p_Y, the *true* probabilities of success for treatment X and treatment Y, are equal:

$$H_0: p_X = p_Y \quad (= p)$$

versus

$$H_1: p_X \neq p_Y.$$

The level of significance will be α.

Here the two parameter spaces are given by

$$\omega = \{(p_X, p_Y): 0 \leq p_X = p_Y \leq 1\}$$

and

$$\Omega = \{(p_X, p_Y): 0 \leq p_X \leq 1, 0 \leq p_Y \leq 1\}.$$

Furthermore, the likelihood function can be written

$$L = p_X^x(1 - p_X)^{n-x} \cdot p_Y^y(1 - p_Y)^{m-y}.$$

Setting the derivative of $\ln L$ with respect to $p \, (= p_X = p_Y)$ equal to 0 and solving for p gives a not too surprising result—namely,

$$\hat{p} = \frac{x + y}{n + m}.$$

That is, the MLE for p under H_0 is the pooled success proportion. Similarly, solving $\partial \ln L/\partial p_X = 0$ and $\partial \ln L/\partial p_Y = 0$ gives the two original sample proportions as the unrestricted MLE's for p_X and p_Y:

$$\hat{p}_X = \frac{x}{n}, \qquad \hat{p}_Y = \frac{y}{m}.$$

Putting $\hat{p}$, $\hat{p}_X$, and $\hat{p}_Y$ back into L gives the generalized likelihood ratio:

$$\lambda = \frac{L(\hat{\omega})}{L(\hat{\Omega})} = \frac{[(x + y)/(n + m)]^{x+y}[1 - (x + y)/(n + m)]^{n+m-x-y}}{(x/n)^x[1 - (x/n)]^{n-x}(y/m)^y[1 - (y/m)]^{m-y}}. \qquad (8.1)$$

Equation 8.1 is such a difficult function to work with that it is necessary to find an approximation to the usual generalized-likelihood-ratio test. There are several available. It can be shown, for example, that $-2 \ln \Lambda$ for this problem has an asymptotic χ^2 distribution with 1 degree of freedom (195). Thus, an approximate two-sided, $\alpha = 0.05$ test is to reject H_0 if $-2 \ln \lambda \geq 3.84$.

Another approach, and the one most often used, is to appeal to the central limit theorem and make the observation that

$$\frac{\dfrac{X}{n} - \dfrac{Y}{m} - E\left(\dfrac{X}{n} - \dfrac{Y}{m}\right)}{\sqrt{\mathrm{Var}\left(\dfrac{X}{n} - \dfrac{Y}{m}\right)}}$$

has an approximate $N(0, 1)$ distribution. Of course, under H_0,

$$E\left(\frac{X}{n} - \frac{Y}{m}\right) = 0$$

and

$$\mathrm{Var}\left(\frac{X}{n} - \frac{Y}{m}\right) = \frac{p(1 - p)}{n} + \frac{p(1 - p)}{m}$$

$$= \frac{(n + m)p(1 - p)}{nm}.$$

If p is now replaced by $(x + y)/(n + m)$, its MLE under ω, we get the statement of Theorem 8.4. Details of the proof will be omitted.

> **THEOREM 8.4** Let x and y denote the numbers of successes observed in two independent sets of n and m Bernoulli trials, respectively. Let p_X and p_Y denote the true success probabilities associated with each set of trials. An approximate GLRT at the α level of significance for
>
> $$H_0: p_X = p_Y$$
>
> versus
>
> $$H_1: p_X \neq p_Y$$
>
> is gotten by rejecting H_0 whenever
>
> $$\frac{\dfrac{x}{n} - \dfrac{y}{m}}{\sqrt{\dfrac{\left(\dfrac{x + y}{n + m}\right)\left(1 - \dfrac{x + y}{n + m}\right)(n + m)}{nm}}} \quad \text{is either} \quad \begin{cases} \leq -z_{\alpha/2} \\ \text{or} \\ \geq +z_{\alpha/2}. \end{cases}$$

The utility of Theorem 8.4 actually extends far beyond the scope we have just described. Any continuous variable can always be dichotomized and "transformed" into a Bernoulli variable. For example, blood pressure can be recorded in terms of "mm Hg," a continuous variable, or as simply "normal" or "abnormal," a Bernoulli variable. The next two case studies illustrate these two sources of binomial data. In the first, the variables begin and end as Bernoullis, while in the second, the initial measurement of "number of nightmares per month" is immediately dichotomized into "often" and "seldom."

CASE STUDY

8.4   Law and order was a key issue for all the presidential contenders in 1968 but particularly so for George Wallace. He and his supporters were the sharpest critics of the courts and the strongest advocates of a renewed commitment to vigorous law enforcement. Whenever one segment of society moralizes to another, though, there is a natural tendency for the accused to question whether the accusers are, themselves, above reproach. "Wallaceites" talked a good game of law and order, but did they practice what they preached? In Nashville, Tennessee, a team of sociologists carried out a rather unorthodox survey that seemed to indicate that maybe they didn't (202).

Before the general election of 1968, the government of Nashville-Davidson County had passed a law requiring all locally operated vehicles to display on their windshield a "Metro sticker" (costing $15). It was far from clear, though, how strictly the law would be enforced. Many motorists thought they could get by without one.

For several days following the sticker deadline (November 1) the investigators made spot checks of various parking lots in and around Nashville to see whether supporters of the various candidates differed significantly in their compliance with the law. Table 8.4 shows some of the results.

TABLE 8.4 Parking-lot surveys of bumper stickers

In Support of[a]	Number of Cars	Number with Stickers
Humphrey	$n_H = 178$	$x_H = 154$
Wallace	$n_W = 361$	$x_W = 270$

[a]A car was assumed to be owned by a Humphrey supporter, for example, if it displayed a Humphrey bumper sticker.

The sample proportions of Humphrey and Wallace supporters obeying the law were $154/178 = 0.865$ and $270/361 = 0.748$, respectively. Presumably, these two figures are unbiased estimates of p_H and p_W, the *true* proportions of Humphrey and Wallace cars bearing a Metro sticker. Can we conclude that p_H and p_W are not equal, as the two sample proportions would indicate, or is the observed difference of 0.117 ($= 0.865 - 0.748$) well within the normal bounds of sampling variability? Put more formally, the hypotheses to be tested are

$$H_0: p_H = p_W \quad (= p)$$

versus

$$H_1: p_H \neq p_W.$$

Let $\alpha = 0.01$.

If H_0 is true, a pooled estimate of p would be the overall sample proportion of cars with Metro stickers. That is,

$$\hat{p} = \frac{154 + 270}{178 + 361} = \frac{424}{539} = 0.787.$$

According to Theorem 8.4, then, the test statistic is equal to 3.12:

$$\frac{0.865 - 0.748}{\sqrt{\dfrac{(0.787)(0.213)(539)}{(178)(361)}}} = 3.12.$$

But $\pm z_{0.005} = \pm 2.58$. Thus, we should reject the null hypothesis: it would appear that the proportion of Wallace supporters breaking the law by not having a Metro sticker was significantly higher than the corresponding proportion for Humphrey supporters.

QUESTION 8.4.1   What other explanation(s), besides the obvious one that Wallace supporters are less law-abiding than Humphrey supporters with respect to local tax ordinances, might account for the observed difference between $\hat{p}_H$ and $\hat{p}_W$?

CASE STUDY

8.5

Over the years, numerous studies have sought to characterize the nightmare sufferer. Out of these has emerged the stereotype of someone with high anxiety, low ego strength, feelings of inadequacy, and poorer-than-average physical health. What is not so well known, though, is whether men fall into this pattern with the same frequency as women. To this end, a recent investigation (77) looked at nightmare frequencies for a sample of 160 men and 192 women. Each subject was asked whether he (or she) experienced nightmares "often" (at least once a month) or "seldom" (less than once a month). The findings are summarized in Table 8.5.

TABLE 8.5 Frequency of nightmares

	Men	Women	Totals
Nightmares often	55	60	115
Nightmares seldom	105	132	237
Totals	160	192	

If p_M and p_W denote the true proportions of men having nightmares often and women having nightmares often, respectively, what we want to test is

$$H_0 : p_M = p_W$$

versus

$$H_1 : p_M \neq p_W.$$

Suppose α is set equal to 0.05. This makes the critical values equal to $\pm z_{0.025}$, or ± 1.96. Substituting the pooled estimate for the "nightmares often" probability, $\frac{115}{352}$, into the expression for the test statistic gives

$$\frac{\frac{55}{160} - \frac{60}{192}}{\sqrt{\frac{(\frac{115}{352})(1 - \frac{115}{352})(352)}{(160)(192)}}} = 0.64.$$

The conclusion, then, is clear: we should accept the null hypothesis—these data provide no convincing evidence that the frequency of nightmares is different for men than for women.

QUESTION 8.4.2   What would the *P* value be for these data (see Review Exercise 13 at the end of Chapter 6)?

8.5 CONFIDENCE INTERVALS FOR THE TWO-SAMPLE PROBLEM

Implicit in Theorems 8.2, 8.3, and 8.4 has been the sampling distribution *when H_0 is true* of $\bar{X} - \bar{Y}$, S_Y^2/S_X^2, and $X/n - Y/m$. More generally, when the null hypothesis is not necessarily true, these test statistics take the form given below. [Note: Statement (1) is included here only for the sake of completeness—it has already appeared as Theorem 8.1. Statement (3) represents a slight modification of the expression that appeared in Theorem 8.4, the difference being that the denominator here is the square root of the unpooled, rather than the pooled, estimator for the variance of $X/n - Y/m$. See (78) for a proof that the asymptotic normality still holds.]

(1)
$$\frac{\bar{X} - \bar{Y} - (\mu_X - \mu_Y)}{S_p \sqrt{\frac{1}{n} + \frac{1}{m}}} \sim T_{n+m-2}.$$

(2)
$$\frac{S_Y^2/\sigma_Y^2}{S_X^2/\sigma_X^2} \sim F_{m-1, n-1}.$$

(3)
$$\frac{\frac{X}{n} - \frac{Y}{m} - (p_X - p_Y)}{\sqrt{\frac{(\frac{X}{n})(1 - \frac{X}{n})}{n} + \frac{(\frac{Y}{m})(1 - \frac{Y}{m})}{m}}} \sim N(0, 1), \text{ approximately.}$$

By inverting these expressions, it is possible to derive $100(1 - \alpha)\%$ confidence intervals for $\mu_X - \mu_Y$, σ_X^2/σ_Y^2, and $p_X - p_Y$. Theorems 8.5, 8.6, and 8.7 state the results. The proofs are very straightforward and will be left as exercises.

> **THEOREM 8.5** Let $X_1, X_2, \ldots, X_n \sim N(\mu_X, \sigma^2)$ and $Y_1, Y_2, \ldots, Y_m \sim N(\mu_Y, \sigma^2)$ and let the X's and Y's be independent. Let s_p be the pooled standard deviation. Then a $100(1 - \alpha)\%$ confidence interval for $\mu_X - \mu_Y$ is given by
>
> $$\left(\bar{x} - \bar{y} - t_{\alpha/2, n+m-2} s_p \sqrt{\frac{1}{n} + \frac{1}{m}}, \; \bar{x} - \bar{y} + t_{\alpha/2, n+m-2} s_p \sqrt{\frac{1}{n} + \frac{1}{m}} \right).$$

CASE STUDY

8.6

Occasionally in forensic medicine, or in the aftermath of a bad accident, identifying the sex of a victim can be a very difficult task. In some of these cases, dental structure provides a useful criterion, since individual teeth will remain in good condition long after other tissues have deteriorated. Furthermore, studies have shown that female teeth and male teeth have different physical and chemical characteristics.

The extent to which X-rays can penetrate tooth enamel, for instance, is different for men than it is for women. Listed in Table 8.6 are "spectropenetration gradients" for eight female teeth and eight male teeth (53). These numbers are measures of the rate of change in the amount of X-ray penetration through a 500-micron section of tooth enamel at a wavelength of 600 mμ as opposed to 400 mμ.

TABLE 8.6 Enamel spectropenetration gradients

Male (x_i)	Female (y_i)
4.9	4.8
5.4	5.3
5.0	3.7
5.5	4.1
5.4	5.6
6.6	4.0
6.3	3.6
4.3	5.0

Let μ_X and μ_Y be the population means of the spectropenetration gradients associated with male teeth and with female teeth, respectively. Note that

$$\sum_{i=1}^{8} x_i = 43.4, \qquad \sum_{i=1}^{8} x_i^2 = 239.32,$$

from which

$$\bar{x} = \frac{43.4}{8} = 5.4$$

and

$$s_X^2 = \frac{8(239.32) - (43.4)^2}{8(7)} = 0.55.$$

Similarly,

$$\sum_{i=1}^{8} y_i = 36.1, \qquad \sum_{i=1}^{8} y_i^2 = 166.95,$$

so that

$$\bar{y} = \frac{36.1}{8} = 4.5$$

and

$$s_Y^2 = \frac{8(166.95) - (36.1)^2}{8(7)} = 0.58.$$

Therefore, the pooled standard deviation is equal to 0.75:

$$s_p = \sqrt{\frac{7(0.55) + 7(0.58)}{8 + 8 - 2}} = \sqrt{0.565} = 0.75.$$

We know that the ratio

$$\frac{\bar{X} - \bar{Y} - (\mu_X - \mu_Y)}{S_p \sqrt{\frac{1}{8} + \frac{1}{8}}}$$

will be approximated by a Student t curve with 14 degrees of freedom. Since $t_{0.025, 14} = 2.14$, the 95% confidence interval for $\mu_X - \mu_Y$ is given by

$$\left(\bar{x} - \bar{y} - 2.14 s_p \sqrt{\tfrac{1}{8} + \tfrac{1}{8}}, \ \bar{x} - \bar{y} + 2.14 s_p \sqrt{\tfrac{1}{8} + \tfrac{1}{8}}\right)$$

$$= [5.4 - 4.5 - 2.14(0.75)\sqrt{0.25}, \ 5.4 - 4.5 + 2.14(0.75)\sqrt{0.25}]$$

$$= (0.1, 1.7).$$

Comment. Here the 95% confidence interval does not include the value 0. This means that had we tested

$$H_0: \mu_X = \mu_Y$$

versus

$$H_1: \mu_X \neq \mu_Y$$

at the $\alpha = 0.05$ level of significance, H_0 would have been rejected. ▌

QUESTION 8.5.1   Refer to the data on thermoluminescence described in Case Study 8.2. Construct a 90% confidence interval for $\mu_X - \mu_Y$.

THEOREM 8.6 Let $X_1, X_2, \ldots, X_n \sim N(\mu_X, \sigma_X^2)$ and $Y_1, Y_2, \ldots, Y_m \sim N(\mu_Y, \sigma_Y^2)$ and let the X's and Y's be independent. A $100(1 - \alpha)\%$ confidence interval for the variance ratio, σ_X^2/σ_Y^2, is given by

$$\left(\frac{s_X^2}{s_Y^2} F_{\alpha/2, m-1, n-1}, \; \frac{s_X^2}{s_Y^2} F_{1-\alpha/2, m-1, n-1} \right).$$

CASE STUDY

8.7

The easiest way to measure the movement, or flow, of a glacier is with a camera. First a set of reference points is marked off at various sites near the glacier's edge. Then these points, along with the glacier, are photographed from an airplane. The problem is this: How long should the time interval be between photographs? If too *short* a period has elapsed, the glacier will not have moved very far and the errors associated with the photographic technique will be relatively large. If too *long* a period has elapsed, parts of the glacier might be deformed by the surrounding terrain, an eventuality that could introduce substantial variability into the point-to-point velocity estimates.

In this example, two sets of flow rates for the Antarctic's Hoseason Glacier are examined (113), one based on photographs taken *three* years apart, the other, *five* years apart. Both sets of data were taken under identical conditions. Also, on the basis of other considerations, it can be assumed that the "true" flow rate for the glacier was constant for the eight years in question. The data are listed in Table 8.7.

TABLE 8.7 Flow rates estimated
for the Hoseason Glacier (meters per day)

Three-Year Span (x_i)	Five-Year Span (y_i)
0.73	0.72
0.76	0.74
0.75	0.74
0.77	0.72
0.73	0.72
0.75	
0.74	

The objective here is to assess the relative variabilities associated with the three- and five-year time periods. One way to do this—assuming the data to be normal—is to construct, say, a 95% confidence interval for the variance ratio. If that interval does not contain the value "1", we infer that the two time periods lead to flow rate estimates of significantly different precision.

From Table 8.7,

$$\sum_{i=1}^{7} x_i = 5.23 \quad \text{and} \quad \sum_{i=1}^{7} x_i^2 = 3.9089,$$

so that

$$s_X^2 = \frac{7(3.9089) - (5.23)^2}{7(6)} = 0.000224.$$

Similarly,

$$\sum_{i=1}^{5} y_i = 3.64 \quad \text{and} \quad \sum_{i=1}^{5} y_i^2 = 2.6504,$$

making

$$s_Y^2 = \frac{5(2.6504) - (3.64)^2}{5(4)} = 0.000120.$$

The two critical values come from Table A.4 in the Appendix:

$$F_{0.025, 4, 6} = 0.109 \quad \text{and} \quad F_{0.975, 4, 6} = 6.23.$$

When all of these quantities are substituted into the statement of Theorem 8.6, we get a 95% confidence interval for σ_X^2/σ_Y^2:

$$\left(\frac{0.000224}{0.000120} 0.109, \; \frac{0.000224}{0.000120} 6.23 \right) = (0.203, 11.629).$$

Thus, although the three-year data had a larger *sample* variance than the five-year data, no conclusions can be drawn about the *true* variances being different, because the ratio $\sigma_X^2/\sigma_Y^2 = 1$ is contained in the confidence interval.

QUESTION 8.5.2   Use the data of Case Study 8.2 to construct a 99% confidence interval for σ_X^2/σ_Y^2.

THEOREM 8.7   Let x and y denote the numbers of successes observed in two independent sets of n and m Bernoulli trials, respectively. If p_X and p_Y denote the true success probabilities, an approximate $100(1 - \alpha)\%$ confidence interval for $p_X - p_Y$ is given by

$$\left(\frac{x}{n} - \frac{y}{m} - z_{\alpha/2}\sqrt{\frac{\left(\frac{x}{n}\right)\left(1 - \frac{x}{n}\right)}{n} + \frac{\left(\frac{y}{m}\right)\left(1 - \frac{y}{m}\right)}{m}}, \right.$$

$$\left. \frac{x}{n} - \frac{y}{m} + z_{\alpha/2}\sqrt{\frac{\left(\frac{x}{n}\right)\left(1 - \frac{x}{n}\right)}{n} + \frac{\left(\frac{y}{m}\right)\left(1 - \frac{y}{m}\right)}{m}}\right).$$

CASE STUDY

8.8

Until almost the end of the nineteenth century the mortality associated with surgical operations—even minor ones—was extremely high. The major problem was infection. The germ theory as a model for disease transmission was still unknown, so there was no concept of sterilization. As a result, many patients died from postoperative complications.

The major breakthrough that was so desperately needed finally came when Joseph Lister, a British physician, began reading about some of the work done by Louis Pasteur. In a series of classic experiments, Pasteur had succeeded in demonstrating the part that yeasts and bacteria play in fermentation. What Lister conjectured was that human infections might have a similar organic origin. To test his theory he began using carbolic acid as an operating-room disinfectant. The data in Table 8.8 show the outcomes of 75 amputations performed by Lister, 35 without the aid of carbolic acid and 40 with it (197).

TABLE 8.8   Mortality rates—Lister's amputations

Carbolic acid used ?

		No	Yes	Totals
Patient	Yes	19	34	53
lived ?	No	16	6	22
	Totals	35	40	

Let p_W (estimated by $\frac{34}{40}$) and $p_{W/O}$ (estimated by $\frac{19}{35}$) denote the true survival probabilities for patients amputated "with" and "without" the use of carbolic acid, respectively. To construct a 95% confidence interval for $p_W - p_{W/O}$ we note that $z_{\alpha/2} = 1.96$; then Theorem 8.7 reduces to

$$\left(\frac{34}{40} - \frac{19}{35} - 1.96 \sqrt{\frac{(\frac{34}{40})(1 - \frac{34}{40})}{40} + \frac{(\frac{19}{35})(1 - \frac{19}{35})}{35}} \, , \right.$$

$$\left. \frac{34}{40} - \frac{19}{35} + 1.96 \sqrt{\frac{(\frac{34}{40})(1 - \frac{34}{40})}{40} + \frac{(\frac{19}{35})(1 - \frac{19}{35})}{35}} \right)$$

$$= (0.31 - 1.96 \sqrt{0.0103}, \, 0.31 + 1.96 \sqrt{0.0103})$$

$$= (0.11, 0.51).$$

Since $p_W - p_{W/O} = 0$ is not included in the interval, we would conclude that the presence or absence of carbolic acid *does* constitute a significant effect. Specifically, patients on whom carbolic acid is used have a better chance of recovery.

Comment. For this analysis to be valid, Lister's decisions regarding when to use carbolic acid and when not to would have to have been made *at random*, uninfluenced in any way by the physical condition of the patient. Obviously, if he chose to use carbolic acid on those patients having a more favorable prognosis to start with, the results are meaningless. At this point, more than 100 years after the fact, whether or not the treatment assignment was done in a completely objective fashion can only be conjectured. ∎

QUESTION 8.5.3   Construct a 70% confidence interval for $p_M - p_W$ in the nightmare-frequency data summarized in Case Study 8.5.

APPENDIX 8.1 A DERIVATION OF THE TWO-SAMPLE t TEST (A PROOF OF THEOREM 8.2)

To begin, we note that both the restricted and unrestricted parameter spaces, ω and Ω, are three-dimensional:

$$\omega = \{(\mu_X, \mu_Y, \sigma): -\infty < \mu_X = \mu_Y < \infty, 0 < \sigma < \infty\}$$

and

$$\Omega = \{(\mu_X, \mu_Y, \sigma): -\infty < \mu_X < \infty, -\infty < \mu_Y < \infty, 0 < \sigma < \infty\}.$$

Since the X's and Y's are independent (and normal),

$$L(\omega) = \prod_{i=1}^{n} f_X(x_i) \prod_{j=1}^{m} f_Y(y_j)$$

$$= \left(\frac{1}{\sqrt{2\pi}\sigma}\right)^{n+m} \exp\left\{-\frac{1}{2\sigma^2}\left[\sum_{i=1}^{n}(x_i - \mu)^2 + \sum_{j=1}^{m}(y_j - \mu)^2\right]\right\}, \qquad (8.2)$$

where $\mu = \mu_X = \mu_Y$. If we take $\ln L(\omega)$ and solve $\partial \ln L(\omega)/\partial\mu = 0$ and $\partial \ln L(\omega)/\partial\sigma^2 = 0$ simultaneously, the solutions will be the restricted maximum-likelihood estimates:

$$\hat{\mu} = \frac{\sum_{i=1}^{n} x_i + \sum_{j=1}^{m} y_j}{n + m} \qquad (8.3)$$

and

$$\hat{\sigma}^2 = \frac{\sum_{i=1}^{n}(x_i - \hat{\mu})^2 + \sum_{j=1}^{m}(y_j - \hat{\mu})^2}{n + m}. \qquad (8.4)$$

Substituting Equations 8.3 and 8.4 into Equation 8.2 gives the numerator of the generalized likelihood ratio:

$$L(\hat{\omega}) = \left(\frac{e^{-1}}{2\pi\hat{\sigma}^2}\right)^{(n+m)/2}.$$

Similarly, the likelihood function unrestricted by the null hypothesis is

$$L(\Omega) = \left(\frac{1}{\sqrt{2\pi}\sigma}\right)^{n+m} \exp\left\{-\frac{1}{2\sigma^2}\left[\sum_{i=1}^{n}(x_i - \mu_X)^2 + \sum_{j=1}^{m}(y_j - \mu_Y)^2\right]\right\}. \qquad (8.5)$$

Here, solving

$$\frac{\partial \ln L(\Omega)}{\partial\mu_X} = 0, \qquad \frac{\partial \ln L(\Omega)}{\partial\mu_Y} = 0, \qquad \frac{\partial \ln L(\Omega)}{\partial\sigma^2} = 0$$

gives

$$\hat{\mu}_X = \bar{x}, \qquad \hat{\mu}_Y = \bar{y},$$

$$\hat{\sigma}_\Omega^2 = \frac{\sum_{i=1}^{n}(x_i - \bar{x})^2 + \sum_{j=1}^{m}(y_j - \bar{y})^2}{n + m}.$$

If these estimates are substituted into Equation 8.5, the maximum value for $L(\Omega)$ simplifies to

$$L(\hat{\Omega}) = (e^{-1}/2\pi\hat{\sigma}_\Omega^2)^{(n+m)/2}.$$

It follows, then, that the generalized likelihood ratio, λ, is equal to

$$\lambda = \frac{L(\hat{\omega})}{L(\hat{\Omega})} = \left(\frac{\hat{\sigma}_\Omega^2}{\hat{\sigma}^2}\right)^{(n+m)/2}$$

or, equivalently,

$$\lambda^{2/(n+m)} = \frac{\sum\limits_{i=1}^{n}(x_i - \bar{x})^2 + \sum\limits_{j=1}^{m}(y_j - \bar{y})^2}{\sum\limits_{i=1}^{n}\left[x_i - \left(\dfrac{n\bar{x} + m\bar{y}}{n+m}\right)\right]^2 + \sum\limits_{j=1}^{m}\left[y_j - \left(\dfrac{n\bar{x} + m\bar{y}}{n+m}\right)\right]^2}.$$

Using the identity

$$\sum_{i=1}^{n}\left(x_i - \frac{n\bar{x} + m\bar{y}}{n+m}\right)^2 = \sum_{i=1}^{n}(x_i - \bar{x})^2 + \frac{m^2 n}{(n+m)^2}(\bar{x} - \bar{y})^2,$$

we can write $\lambda^{2/(n+m)}$ as

$$\lambda^{2/(n+m)} = \frac{\sum\limits_{i=1}^{n}(x_i - \bar{x})^2 + \sum\limits_{j=1}^{m}(y_j - \bar{y})^2}{\sum\limits_{i=1}^{n}(x_i - \bar{x})^2 + \sum\limits_{j=1}^{m}(y_j - \bar{y})^2 + \dfrac{nm}{n+m}(\bar{x} - \bar{y})^2}$$

$$= \frac{1}{1 + \dfrac{(\bar{x} - \bar{y})^2}{\left[\sum\limits_{i=1}^{n}(x_i - \bar{x})^2 + \sum\limits_{j=1}^{m}(y_j - \bar{y})^2\right]\left(\dfrac{1}{n} + \dfrac{1}{m}\right)}}$$

$$= \frac{n + m - 2}{n + m - 2 + \dfrac{(\bar{x} - \bar{y})^2}{s_p^2[(1/n) + (1/m)]}},$$

where s_p^2 is the pooled variance:

$$s_p^2 = \frac{1}{n + m - 2}\left[\sum_{i=1}^{n}(x_i - \bar{x})^2 + \sum_{j=1}^{m}(y_j - \bar{y})^2\right].$$

Therefore, in terms of the observed t ratio, $\lambda^{2/(n+m)}$ simplifies to

$$\lambda^{2/(n+m)} = \frac{n + m - 2}{n + m - 2 + t^2}. \tag{8.6}$$

    At this point the proof is almost complete. The generalized-likelihood-ratio criterion, rejecting H_0: $\mu_X = \mu_Y$ when $0 < \lambda \le \lambda^*$, is clearly equivalent to rejecting the null hypothesis when $0 < \lambda^{2/(n+m)} \le \lambda^{**}$. But both of these, from Equation 8.6, are the same as rejecting H_0 when t^2 is too large. Thus, the decision rule in terms of t^2 is

        Reject H_0: $\mu_X = \mu_Y$ in favor of H_1: $\mu_X \ne \mu_Y$ if $t^2 \ge t^{*^2}$.

Or, phrasing this in still another way, we should reject H_0 if either $t \ge t^*$ or $t \le -t^*$, where

$$P(-t^* < T < t^* \,|\, H_0\text{: } \mu_X = \mu_Y \text{ is true}) = 1 - \alpha.$$

By Theorem 8.1, though, $T \sim T_{n+m-2}$, which makes $\pm t^* = \pm t_{\alpha/2, n+m-2}$, and the theorem is proved.

APPENDIX 8.2 POWER CALCULATIONS
FOR A TWO-SAMPLE t TEST

Power calculations for a two-sample t test proceed along the same lines established for the one-sample t test and described in Appendix 7.1 on page 302. Basically, there are two different questions that can be answered: (1) Given n, m, and α, what is the probability that $H_0: \mu_X = \mu_Y$ will be rejected if, in fact, the X and Y distributions have shifted apart a distance Δ? (2) Given α, what are the smallest values of n and m for which the probability of making a Type II error is no larger than β—for some fixed $\mu_X - \mu_Y$? [In both these questions the location shift $(\Delta = \mu_X - \mu_Y)$ is usually expressed in terms of standard deviations (Δ/σ).] Figure 7.11 can be used to approximate both answers.

As an example of the first situation, imagine testing

$$H_0: \mu_X = \mu_Y$$

versus

$$H_1: \mu_X \neq \mu_Y$$

with $n = 13$, $m = 9$, and $\alpha = 0.01$. We might have reason to ask the following: if μ_X has shifted 1.5 standard deviations to the right of μ_Y $[\Delta/\sigma = (\mu_X - \mu_Y)/\sigma = 1.5]$, what is the probability that H_0 will be rejected? Recall that the abscissa in Figure 7.11 is scaled in terms of ϕ, where

$$\phi = \frac{\Delta}{\sigma_{\hat{\Delta}}}\left(\frac{1}{\sqrt{2}}\right)$$

and $\sigma_{\hat{\Delta}}$ is the standard deviation of the sample estimator for Δ—namely, $\bar{X} - \bar{Y}$. Therefore,

$$\sigma_{\hat{\Delta}} = \sigma\sqrt{\frac{1}{n} + \frac{1}{m}} = \sigma\sqrt{\frac{1}{13} + \frac{1}{9}}$$

and

$$\phi = \frac{\Delta}{\sigma\sqrt{\frac{1}{13} + \frac{1}{9}}}\frac{1}{\sqrt{2}}$$

$$= \frac{\Delta}{\sigma}\frac{1}{\sqrt{\frac{1}{13} + \frac{1}{9}}}\frac{1}{\sqrt{2}}$$

$$= (1.5)\left(\frac{1}{0.434}\right)\left(\frac{1}{1.414}\right)$$

$$= 2.4.$$

With the combined sample size totaling 22, the estimator for σ—S_p—will have 20 degrees of freedom. The probability, then, of rejecting H_0 is gotten by entering Figure 7.11 with a ϕ of 2.4 and reading off $1 - \beta$ from the $\alpha = 0.01$ curve having $\nu = 20$ degrees of freedom:

$$1 - \beta = P\left(\text{reject } H_0 \,\middle|\, \frac{\Delta}{\sigma} = 1.5\right) = 0.71.$$

The second problem, choosing n and m to satisfy requirements imposed on α and β, is likely to be much more relevant to an experimenter than the first procedure. As a numerical illustration of this second problem, suppose it has been decided that the hypotheses to be tested are

$$H_0: \mu_X = \mu_Y$$

versus

$$H_1: \mu_X \neq \mu_Y$$

and that the level of significance should be 0.05. The question is, how large should n and m be?

To simplify matters, we will assume that n and m are to be equal. Finally, as a precision requirement, we will insist that the sample size be large enough to enable the test to reject H_0 at least 80% of the time if $|\mu_X - \mu_Y|/\sigma \geq 1.75$. Accordingly, ϕ reduces to

$$\phi = \frac{\Delta}{\sigma_{\hat{\Delta}}} \times \left(\frac{1}{\sqrt{2}}\right) = \frac{\Delta}{\sigma\sqrt{(2/n)}} \times \frac{1}{\sqrt{2}} = \frac{\sqrt{n}}{2}\frac{\Delta}{\sigma} = \frac{\sqrt{n}}{2} \times 1.75$$

$$= 0.875\sqrt{n}.$$

Now, suppose n were 9. Then $\phi = 0.875\sqrt{9} = 2.625$ and $\nu = 9 + 9 - 2 = 16$. From Figure 7.11, the probability of rejecting H_0 under these circumstances (two samples of size 9 and $\Delta/\sigma = 1.75$) is approximately 0.94. This figure, however, considerably exceeds our power requirement of 0.80, implying that a smaller sample size would be adequate. So, suppose n were 4. Then $\phi = 0.875\sqrt{4} = 1.75$, $\nu = 6$, and $1 - \beta = 0.55$. But now the test would not be precise enough. Table 8.9 lists $1 - \beta$ for sample sizes ranging from $n = 4$ to $n = 9$.

TABLE 8.9   Values of $1 - \beta$ as a function of n

n	ϕ	ν	$1 - \beta$
4	1.75	6	0.55
5	1.96	8	0.67
6	2.14	10	0.78
7	2.31	12	0.85
8	2.47	14	0.90
9	2.62	16	0.94

Notice that $1 - \beta$ exceeds 0.80 for the first time when $n = 7$. This means that the experimenter should take two samples of size 7.

REVIEW EXERCISES FOR CHAPTER 8

1. According to recent speculation, height and life expectancy may be related, with short people enjoying longer lives than their taller counterparts. Death certificates of short and tall baseball players and short and tall boxers do seem to bear out that contention (151). Another source of data on which to check out the theory are former U.S.

presidents. Listed below are the 31 presidents who died of natural causes. They are divided into two categories: short ($< 5'8''$) and tall ($\geq 5'8''$).

Short Presidents			Tall Presidents		
President	Height	Age	President	Height	Age
Madison	5 ft 4 in.	85	W. Harrison	5 ft 8 in.	68
Van Buren	5 ft 6 in.	79	Polk	5 ft 8 in.	53
B. Harrison	5 ft 6 in.	67	Taylor	5 ft 8 in.	65
J. Adams	5 ft 7 in.	90	Grant	5 ft $8\frac{1}{2}$ in.	63
J. Q. Adams	5 ft 7 in.	80	Hayes	5 ft $8\frac{1}{2}$ in.	70
			Truman	5 ft 9 in.	88
			Fillmore	5 ft 9 in.	74
			Pierce	5 ft 10 in.	64
			A. Johnson	5 ft 10 in.	66
			T. Roosevelt	5 ft 10 in.	60
			Coolidge	5 ft 10 in.	60
			Eisenhower	5 ft 10 in.	78
			Cleveland	5 ft 11 in.	71
			Wilson	5 ft 11 in.	67
			Hoover	5 ft 11 in.	90
			Monroe	6 ft	73
			Tyler	6 ft	71
			Buchanan	6 ft	77
			Taft	6 ft	72
			Harding	6 ft	57
			Jackson	6 ft 1 in.	78
			Washington	6 ft 2 in.	67
			Arthur	6 ft 2 in.	56
			F. Roosevelt	6 ft 2 in.	63
			L. Johnson	6 ft 2 in.	64
			Jefferson	6 ft $2\frac{1}{2}$ in.	83

Test the appropriate hypothesis. Let $\alpha = 0.05$.

2. Thrombocytopenia is a condition characterized by a chronically lowered blood platelet count. Among its most effective treatments is a splenectomy—the surgical removal of the patient's spleen. The success of such an operation, however, may be influenced by the patient's spleen weight. Listed below are the spleen weights (in grams) of 14

Spleen weights (grams)

Splenectomy Was Successful		Splenectomy Was Unsuccessful
150	136	70
142	122	110
160	200	85
110	160	90
120	102	210
240	152	
152	280	

persons for whom a splenectomy was ultimately successful and of 5 persons for whom the operation was unsuccessful (123). Let μ_X and μ_Y denote the true average spleen weights of persons for whom the operation would be successful and unsuccessful, respectively. At the $\alpha = 0.05$ level of significance, test $H_0: \mu_X = \mu_Y$ versus $H_1: \mu_X \neq \mu_Y$.

3. A time-study engineer is assigned the problem of comparing two different work sequences in a garment factory for measuring the shear strength of polyester fibers. To collect some data he randomly divides 12 workers into two groups. The first group measures the shear strength using work sequence A, the second group, work sequence B. The data recorded were the 12 completion times.

Completion times (seconds)

Work Sequence A	Work Sequence B
220	247
235	223
214	215
197	219
206	207
214	236

Test whether the difference in average completion times is significantly different. Let $\alpha = 0.05$.

4. Serotonin is a substance found in the blood that may or may not be related to psychiatric disorders. Also, its concentration may or may not be affected by chronic LSD usage. A two-sample experiment was set up to see if any relationship could be established between LSD usage and serotonin formation in an animal population (37). Twenty-six rats were given a daily oral dose of 20 μg of LSD-25 per kilogram of body weight. The LSD was dissolved in 1 ml of water. (This particular amount was thought to be comparable in effect to the dosage a person might take.) A similar procedure was followed with a control group of 25 rats, but their "treatment" consisted of just the water. After 30 days, the animals were sacrificed and the concentrations of serotonin in their brains were measured. The results are summarized below.

Serotonin concentration (nmole/g)

Control Group	LSD Group
$\bar{x} = 2.84$	$\bar{y} = 3.20$
$s_X/\sqrt{n} = 0.06$	$s_Y/\sqrt{m} = 0.15$

Test whether the two average serotonin concentrations are significantly different. Let $\alpha = 0.05$.

5. A businessman has two basic routes he can take to and from work each day. The first involves going by the interstate; the second requires that he drive through town. On the average, it takes him 33 minutes to get to work via the interstate and 35 minutes by going through town. The standard deviations for the two routes are 6 minutes and 5 minutes, respectively. Assume the distributions of times for the two routes are both normally distributed.

(a) What is the probability that on a given day driving through town would be the quicker of his two alternatives?

(b) What is the probability that driving through town each way for an entire work week (ten trips) would yield a lower average time than taking the interstate for the entire week?

6. The use of carpeting in hospitals, while having obvious esthetic merits, raises an obvious question: are carpeted floors sanitary? One way to get at an answer is to compare bacterial levels in carpeted and uncarpeted rooms. Airborne bacteria can be counted by passing room air at a known rate over a growth medium, incubating that medium, and then counting the number of bacterial colonies that form. In one such study done in a Montana hospital (190), room air was pumped over a Petri dish at the rate of 1 cubic foot per minute. This procedure was repeated in 16 patient rooms, 8 carpeted and 8 uncarpeted. The results, expressed in terms of "bacteria per cubic foot of air," are listed in the table below.

Carpeted Rooms	Bacteria/ft³	Uncarpeted Rooms	Bacteria/ft³
#212	11.8	#210	12.1
#216	8.2	#214	8.3
#220	7.1	#215	3.8
#223	13.0	#217	7.2
#225	10.8	#221	12.0
#226	10.1	#222	11.1
#227	14.6	#224	10.1
#228	14.0	#229	13.7

For the carpeted rooms,

$$\sum_{i=1}^{8} x_i = 89.6 \quad \text{and} \quad \sum_{i=1}^{8} x_i^2 = 1053.70.$$

For the uncarpeted rooms,

$$\sum_{i=1}^{8} y_i = 78.3 \quad \text{and} \quad \sum_{i=1}^{8} y_i^2 = 838.49.$$

Test whether carpeting has any effect on the level of airborne bacteria in patient rooms. Let $\alpha = 0.05$.

7. If the null hypothesis in Review Exercise 6 is true, we would expect $\bar{x} - \bar{y}$ to be numerically close to 0. How much less than 0 or greater than 0 does $\bar{x} - \bar{y}$ have to be before we can reject H_0 at the $\alpha = 0.05$ level of significance?

8. Why would it not be reasonable always to define the pooled standard deviation as the square root of the average of s_X^2 and s_Y^2?

9. Verify the identity used in Appendix 8.1 on page 347 to prove Theorem 8.2:

$$\sum_{i=1}^{n}\left(x_i - \frac{n\bar{x} + m\bar{y}}{n + m}\right)^2 = \sum_{i=1}^{n}(x_i - \bar{x})^2 + \frac{m^2 n}{(n + m)^2}(\bar{x} - \bar{y})^2.$$

10. Carry out the details to verify Equations 8.3 and 8.4.

11. Let $X_i \sim N(\mu_X, \sigma^2)$, $i = 1, 2, \ldots, n$ and $Y_j \sim N(\mu_Y, \sigma^2)$, $j = 1, 2, \ldots, m$, where σ^2 is known. Use the generalized-likelihood-ratio criterion to derive a test procedure for $H_0: \mu_X = \mu_Y$ versus $H_1: \mu_X \neq \mu_Y$. Compare your procedure to the two-sample t test of Theorem 8.2.

12. An experiment is to be conducted to determine whether vampire bats prefer blood at room temperature or at body temperature. Equal numbers of bats are to be given access to drinking tubes attached to a supply of blood kept at one of the two temperatures. The response variable will be the amount of blood (in milliliters) that each bat drinks [see (18)]. The experimenter wants to test whether the average amounts of room-temperature and body-temperature blood consumed are equal (against a two-sided alternative that they are not). If α is going to be set at 0.05 and if the experimenter wants to have at least an 85% chance of rejecting H_0 when, in fact, $|\mu_X - \mu_Y|/\sigma \geq 1.50$, what is the minimum number of bats that should be put in each group?

13. Construct a power curve for the bacterial-level hypothesis test described in Review Exercise 6.

14. A drug is tested to determine whether it can lower the blood glucose level of diabetic rats. Six rats are given the drug, while five others are used as controls. The blood glucose levels (in mg/ml) of the treated group were 2.02, 1.71, 2.04, 1.50, 1.83, and 1.64; for the controls, 2.15, 1.92, 1.78, 2.04, and 2.22. Is the drug effective?

15. If $X_1, X_2, \ldots, X_n \sim N(\mu_X, \sigma^2)$ and $Y_1, Y_2, \ldots, Y_m \sim N(\mu_Y, \sigma^2)$, show that the pooled variance, S_p^2, as defined in Theorem 8.1, is an unbiased estimator for σ^2.

16. Among the many approximate solutions to the Behrens-Fisher problem (see p. 322) is one due to Hsu (154), where the statistic

$$\frac{\bar{X} - \bar{Y}}{\sqrt{\dfrac{S_X^2}{n} + \dfrac{S_Y^2}{m}}}$$

is assumed to have a Student t distribution with f degrees of freedom, where $f = \min(n - 1, m - 1)$. The true probability of committing a Type I error with such a test will never exceed the nominal α. A set of data for which the Hsu procedure would be an appropriate analysis is described in the next paragraph.

The effectiveness of charcoal filters was investigated recently with an experiment involving protozoa (191). *Paramecium aurelia* were suspended in a hanging drop inside a smoke chamber. Every 60 seconds, a six-second puff of smoke was drawn through the chamber. The movements of the paramecia were watched through a stereomicroscope. The variable recorded was the length of time from the start of the experiment to when the last paramecium died. Altogether, the experiment was replicated 12 times. Six of those times the smoke came from a nonfilter cigarette, the other six times from a cigarette with a charcoal filter, as shown in the table at the top of page 353. Test whether the average survival times for these two groups are significantly different. Use the Hsu statistic and let $\alpha = 0.05$.

Survival time (minutes)

Nonfilter	Charcoal Filter
8	21
7	37
11	24
8	17
9	19
8	14

17. The pre-med advisor at State University is trying to decide whether he should encourage juniors to sign up for a private review course as a means of preparing for the MCAT's. Among the 15 students who had taken the exam most recently, five had enrolled in the review course and ten had not. The average MCAT scores for the first group were 10, 8, 9, 8, and 11; the average scores for the other 10 were 8, 7, 7, 9, 10, 8, 7, 11, 8, and 8. What should the pre-med advisor conclude?

18. Among the standard personality inventories used by psychologists is the Thematic Apperception Test (TAT). A subject is shown a series of pictures and is asked to make up a story about each one. Interpreted properly, the content of the stories can provide valuable insights into the subject's mental well-being. The data below show the TAT results for 40 women, 20 of whom were the mothers of normal children and 20 the mothers of schizophrenic children. In each case the subject was shown the same set of ten pictures. The figures recorded were the numbers of stories (out of ten) that revealed a *positive* parent-child relationship, one where the mother was clearly capable of interacting with her child in a flexible, open-minded way (192).

TAT scores

Mothers of Normal Children					Mothers of Schizophrenic Children				
8	4	6	3	1	2	1	1	3	2
4	4	6	4	2	7	2	1	3	1
2	1	1	4	3	0	2	4	2	3
3	2	6	3	4	3	0	1	2	2

(a) Test $H_0: \sigma_X^2 = \sigma_Y^2$ versus $H_1: \sigma_X^2 \neq \sigma_Y^2$, where σ_X^2 and σ_Y^2 are the variances of the scores of mothers of normal children and the scores of mothers of schizophrenic children, respectively. Let $\alpha = 0.05$.

(b) If $H_0: \sigma_X^2 = \sigma_Y^2$ is accepted in part (a), test $H_0: \mu_X = \mu_Y$ versus $H_1: \mu_X \neq \mu_Y$. Set α equal to 0.05.

(c) Suppose that σ_X^2 is, in fact, equal to σ_Y^2 but that μ_X is shifted to the right of μ_Y a distance of 1.0 standard deviations. What is the probability that H_0 will be rejected?

19. Show that the generalized likelihood ratio for testing $H_0: \sigma_X^2 = \sigma_Y^2$ versus $H_1: \sigma_X^2 \neq \sigma_Y^2$ as described in Theorem 8.3 is given by

$$\lambda = \frac{L(\hat{\omega})}{L(\hat{\Omega})} = \frac{(m+n)^{(n+m)/2}}{n^{n/2} m^{m/2}} \frac{\left[\sum_{i=1}^{n} (x_i - \bar{x})^2\right]^{n/2} \left[\sum_{j=1}^{m} (y_j - \bar{y})^2\right]^{m/2}}{\left[\sum_{i=1}^{n} (x_i - \bar{x})^2 + \sum_{j=1}^{m} (y_j - \bar{y})^2\right]^{(m+n)/2}}.$$

20. Construct the power curve for the two-sample t test suggested for the TAT data of Review Exercise 18.

21. Let $X_1, X_2, \ldots, X_n \sim N(\mu_X, \sigma_X^2)$ and $Y_1, Y_2, \ldots, Y_m \sim N(\mu_Y, \sigma_Y^2)$, where μ_X and μ_Y are known. Derive the GLRT for

$$H_0: \sigma_X^2 = \sigma_Y^2$$

versus

$$H_1: \sigma_X^2 > \sigma_Y^2.$$

Compare your answer with the procedure given in Theorem 8.3.

22. Raynaud's syndrome is characterized by the sudden impairment of blood circulation in the fingers, a condition which results in discoloration and heat loss. The magnitude of the problem is evidenced in the following data where 20 subjects (10 "normals" and 10 with Raynaud's syndrome) immersed their right forefingers in water kept at 19°C. The heat output (in cal/cm²/minute) of the forefinger was then measured with a calorimeter (102).

Normal Subjects		Subjects with Raynaud's Syndrome	
Patient	Heat Output (cal/cm²/min)	Patient	Heat Output (cal/cm²/min)
W.K.	2.43	R.A.	0.81
M.N.	1.83	R.M.	0.70
S.A.	2.43	F.M.	0.74
Z.K.	2.70	K.A.	0.36
J.H.	1.88	H.M.	0.75
J.G.	1.96	S.M.	0.56
G.K.	1.53	R.M.	0.65
A.S.	2.08	G.E.	0.87
T.E.	1.85	B.W.	0.40
L.F.	2.44	N.E.	0.31

Test that the heat-output variances for normal subjects and those with Raynaud's syndrome are the same. Use a two-sided alternative and the 0.05 level of significance.

23. The phenomenon of handedness has been extensively studied in human populations. The percentages of adults who are right-handed, left-handed, and ambidextrous are well documented. What is not so well known is that a similar phenomenon is present in lower animals. Dogs, for example, can be either right-pawed or left-pawed. Suppose that in a random sample of 200 beagles it is found that 55 are left-pawed and that in a random sample of 200 collies 40 are left-pawed. Can we conclude that the true proportion of collies that are left-pawed is significantly different than the true proportion of beagles that are left-pawed?

24. Compute $-2 \ln \lambda$ (see Equation 8.1) for the bumper-sticker data of Case Study 8.4 and use it to test the equality of p_H and p_W. Let $\alpha = 0.01$.

25. Show how the $-2 \ln \lambda$ decision rule of Review Exercise 24 is "equivalent" to the testing procedure outlined in Theorem 8.4.

26. A utility infielder for a National League club batted .260 last season in 300 trips to the plate. This year he hit .250 in 200 at-bats. The owners are trying to cut his pay for next year on the grounds that his output has deteriorated. The player argues, though, that his performances the last two seasons have not been significantly different, so his salary should not be reduced. Who is right?

27. In a study designed to see whether a controlled diet could retard the process of arteriosclerosis, a total of 846 randomly chosen persons were followed over an eight-year period. Half were instructed to eat only certain foods; the other half could eat whatever they wanted. At the end of eight years, 66 persons in the diet group were found to have died of either myocardial infarction or cerebral infarction, as compared to 93 deaths of a similar nature in the control group (199). Do the appropriate analysis.

28. Find the *P* value (the lowest α at which H_0 would be rejected) for the bumper-sticker data of Case Study 8.4.

29. The possible effects of a mouse's early environment on its aggressiveness later in life were studied in a recent experiment (81). Two groups of mice were tested: one group had been raised by their natural mothers, the other group by "foster" mice. (A foster mouse was a female whose own litter had been removed shortly after birth.) When a mouse was three months old it was placed in a box divided into two compartments by a partition. On the other side of the partition was another mouse, one that had had no previous contact with the "test" mouse. The partition was then removed and the behavior of the mice was observed for the next six minutes. Altogether the experiment was done 307 times. Of the 167 mice raised by their natural mothers, a total of 27 began fighting with the mouse on the other side of the partition. Among the remaining 140 mice, each of which had been raised by a foster mouse, a total of 47 began fighting. Use these data to test whether mice raised by natural mothers and mice raised by foster mothers are equally aggressive. Use the 0.05 level of significance.

30. Prove Theorem 8.5.

31. Using the data of Review Exercise 22, construct a 95% confidence interval for the difference in heat outputs between the normals and those with Raynaud's syndrome.

32. Let $X_1, X_2, \ldots, X_n$ and $Y_1, Y_2, \ldots, Y_m$ be independent random samples from pdf's $N(\mu_X, \sigma_X^2)$ and $N(\mu_Y, \sigma_Y^2)$, respectively. Assume that σ_X^2 and σ_Y^2 are known. Derive an expression for a $100(1 - \alpha)\%$ confidence interval for $\mu_X - \mu_Y$.

33. Construct a 95% confidence interval for σ_X^2/σ_Y^2 using the Mark Twain–Quintus Curtius Snodgrass data of Case Study 8.1.

34. Prove Theorem 8.6.

35. Construct a 99% confidence interval for the difference between the two "fighting" proportions described in Review Exercise 29. What does the location of the interval imply would be the result of doing a hypothesis test on the same data, using the $\alpha = 0.01$ level of significance?

36. Intuitively, why should the denominator of the expression given in Statement 3 at the beginning of Section 8.5 not be the same as the denominator of the test statistic given in Theorem 8.4?

CHAPTER NINE

Goodness-of-Fit Tests

KARL PEARSON (1857–1936)

Called by some the founder of twentieth-century statistics, Pearson received his university education at Cambridge, concentrating on physics, philosophy, and law. He was called to the bar in 1881 but never practiced. In 1911 Pearson resigned his chair of applied mathematics and mechanics at University College, London, and became the first Galton Professor of Eugenics, as was Galton's wish. Together with Weldon, Pearson founded the prestigious journal *Biometrika* and served as its principal editor from 1901 until his death.

9.1 INTRODUCTION

The give and take between the mathematics of probability and the empiricism of statistics should be, by now, a theme comfortably familiar. Time and time again we have seen repeated measurements, no matter what their source, exhibiting a regularity of pattern that can be well approximated by one or more of the handful of probability functions introduced in Chapter 4. Until now, all the inferences resulting from this interfacing have been parameter specific, a fact to which the many hypothesis tests about means, variances, and binomial proportions paraded forth in Chapters 6, 7, and 8 bear ample testimony. Still, there are other situations where the basic *form* of $f_Y(y)$, rather than the value of its parameters, is the most important question at issue. These situations are the focus of Chapter 9.

A geneticist, for example, might want to know whether the inheritance of a certain set of traits follows the same set of ratios as those prescribed by Mendelian theory. The objective of a psychologist, on the other hand, might be to confirm or refute a newly proposed model for cognitive serial learning. Probably the most habitual users of inference procedures directed at the entire $f_Y(y)$, though, are statisticians themselves: as a prelude to doing any sort of hypothesis test or confidence interval, an attempt should be made, sample size permitting, to verify that the data are, indeed, representative of whatever distribution that procedure presumes. Usually, this will mean testing to see whether or not the Y_i's are normal.

In general, any procedure that seeks to determine whether a set of data could reasonably have originated from some given probability distribution, or *class* of probability distributions, is called a *goodness-of-fit* test. The principle behind the particular goodness-of-fit test we will look at is very straightforward: first the observed Y_i's are grouped, more or less arbitrarily, into k classes; then each class's "expected" occupancy is calculated on the basis of the presumed model. If it should happen that the set of observed and expected frequencies show considerable disagreement (as measured by the appropriate statistic), our conclusion will be that the supposed $f_Y(y)$ was incorrect.

In practice, the method has two variants, depending on the specificity of the null hypothesis. Section 9.3 describes the version to use when both the form of the presumed $f_Y(y)$ and the values of all its parameters are given. The more typical situation occurs when $f_Y(y)$ is designated but its parameters need to be estimated; this is taken up in Section 9.4.

A somewhat different application of this same idea is the subject of Section 9.5. There the null hypothesis is one of *independence*: that $f_{X,Y}(x, y) = f_X(x) \cdot f_Y(y)$. Such tests are extremely practical—to the extent that they may be the most often used inference procedure in the applied statistician's repertoire.

9.2 THE MULTINOMIAL DISTRIBUTION

Their diversity notwithstanding, many goodness-of-fit tests are based on essentially the same statistic, one whose asymptotic distribution is a chi square. The underlying structure of that statistic, though, derives from the *multinomial distribution*, a k-variate extension of the familiar binomial. In this section we define the multinomial and state those of its properties that bear directly on the problem of goodness-of-fit testing.

Given a series of n independent Bernoulli trials, each with success probability p, we know that the pdf for Y, the total number of successes, is

$$P(Y = y) = f_Y(y) = \binom{n}{y} p^y(1 - p)^{n-y}, \qquad y = 0, 1, \ldots, n. \qquad (9.1)$$

One of the obvious ways to generalize Equation 9.1 is to consider situations where at each trial k outcomes can occur, rather than just two. This means that Y will be allowed to take on any one of the values $y_1, y_2, \ldots, y_k$ with respective probabilities $p_1, p_2, \ldots, p_k$, the latter satisfying the constraint

$$\sum_{i=1}^{k} p_i = 1.$$

Notice that if n such trials are observed, the resulting distribution of Y-values can be summarized by defining a new set of random variables, $X_1, X_2, \ldots, X_k$, where

$$X_i = \text{the number of times that } Y = y_i, \qquad i = 1, 2, \ldots, k.$$

Of course, $\sum_{i=1}^{k} X_i = n$.

The vector $(X_1, X_2, \ldots, X_k)$ is a discrete multivariate random variable—its joint pdf is the multinomial we are seeking:

$$P(X_1 = x_1, X_2 = x_2, \ldots, X_k = x_k) = f_{X_1, X_2, \ldots, X_k}(x_1, x_2, \ldots, x_k)$$

$$= \frac{n!}{x_1! x_2! \cdots x_k!} p_1^{x_1} p_2^{x_2} \cdots p_k^{x_k}, \qquad (9.2)$$

$$x_i = 0, \ldots, n; i = 1, \ldots, k; \sum_{i=1}^{k} x_i = n.$$

It should be clear that Equation 9.2 can be obtained by appealing to the same arguments that gave rise to the binomial. For example, the combinatorial term in the multinomial is a direct extension of Theorem 2.10: the number of ways to arrange n items, of which x_1 are of one type, x_2 of a second type, $\ldots$, and x_k of a kth type, is $n!/(x_1! x_2! \cdots x_k!)$. When $k = 2$, and the two types are denoted simply as successes and failures, $n!/(x_1! x_2! \cdots x_k!)$ reduces to the familiar

$$\binom{n}{y} = \frac{n!}{y!(n - y)!}.$$

Theorem 9.1 states a not-unexpected relationship between the binomial and the multinomial.

> **THEOREM 9.1** Let the vector $(X_1, X_2, \ldots, X_k)$ be a multinomial random variable with parameters $n, p_1, \ldots, p_k$. Then the marginal pdf of $X_i, i = 1, 2, \ldots, k$, is the binomial with parameters n and p_i.

PROOF   We will verify the theorem for $k = 3$. Let (X, Y, Z) have a *trinomial* distribution with parameters n, p_X, p_Y, and p_Z. What is to be shown is that

$$f_X(x) = \binom{n}{x} p_X^x (1 - p_X)^{n-x}.$$

By definition,

$$f_X(x) = \sum_y \sum_z \frac{n!}{x!\,y!\,z!} p_X^x\, p_Y^y\, p_Z^z,$$

$$y = 0, 1, \ldots, n - x; \; z = 0, 1, \ldots, n - x; \; y + z = n - x;$$

$$= \sum_{y=0}^{n-x} \frac{n!}{x!\,y!\,(n - x - y)!} p_X^x\, p_Y^y (1 - p_X - p_Y)^{n-x-y}. \tag{9.3}$$

Following the procedure of Example 3.21, we will first factor the "answer" out of the right-hand side of Equation 9.3 and then confirm that what remains sums to 1. The first step gives

$$f_X(x) = \frac{n!}{x!\,(n - x)!} p_X^x (1 - p_X)^{n-x} \sum_{y=0}^{n-x} \frac{(n - x)!}{y!\,(n - x - y)!} p_Y^y \frac{(1 - p_X - p_Y)^{n-x-y}}{(1 - p_X)^{n-x}}$$

$$= \frac{n!}{x!\,(n - x)!} p_X^x (1 - p_X)^{n-x} \sum_{y=0}^{n-x} \frac{(n - x)!}{y!\,(n - x - y)!} \left(\frac{p_Y}{1 - p_X}\right)^y \left(1 - \frac{p_Y}{1 - p_X}\right)^{n-x-y}.$$

Then, by inspection, note that the value of the sum appearing in the right-hand side of $f_X(x)$ is, indeed, 1, being the summation over all the values of a binomial pdf whose parameters are $n - x$ and $[p_Y/(1 - p_X)]$. This proves the theorem for $k = 3$.

It follows immediately from Theorem 9.1, in the general case, that $E(X_i) = np_i$ and $\text{Var}(X_i) = np_i(1 - p_i)$. Also it can be shown, although it is not a consequence of the theorem, that the maximum-likelihood estimates for the p_i's are the direct analogs of their binomial counterparts: namely, $\hat{p}_i = x_i/n$ (recall Question 5.8.1).

One final property of the multinomial deserves mention. Any random variable, discrete or continuous, can be "reduced" to a multinomial by partitioning its range into a set of k nonoverlapping intervals. For example, suppose Y is a continuous random variable with pdf $f_Y(y)$ defined over the entire real line. Take as the k intervals the set, $(-\infty, a_1), [a_1, a_2), \ldots, [a_{k-1}, \infty)$, and let

$$p_i = \int_{a_{i-1}}^{a_i} f_Y(y)\, dy; \quad i = 1, 2, \ldots, k; a_0 = -\infty; a_k = \infty. \tag{9.4}$$

Then, if n measurements are taken on Y, and if X_i is the number of Y's falling into the ith interval, the pdf for the vector $(X_1, X_2, \ldots, X_k)$ will be the multinomial with parameters n and the p_i's of Equation 9.4.

QUESTION 9.2.1   An Army enlistment officer categorizes potential recruits by IQ into three classes— class I: < 90, class II: 90–110, and class III: > 110. Given that the IQ distribution of the population from which the recruits are drawn is $N(100, (16)^2)$, compute the probability that of seven enlistees, two will belong to class I, four to class II, and one to class III.

9.3 GOODNESS-OF-FIT TESTS: ALL PARAMETERS KNOWN

The simplest version of a goodness-of-fit test arises when an experimenter is able to specify completely the probability model from which the sample data are alleged to have come. It might be supposed, for example, that the Y_i's are being generated by a Poisson pdf with parameter λ equal to 6.3, or by a normal distribution with $\mu = 500$ and $\sigma = 100$. For cases such as these, the hypotheses to be tested will be written

$$H_0: f_Y(y) = f_0(y)$$

versus

$$H_1: f_Y(y) \neq f_0(y),$$

where $f_Y(y)$ and $f_0(y)$ are the true and the presumed pdf's, respectively. In some situations it will prove more convenient to characterize the model in terms of the probabilities associated with the k nonoverlapping intervals described at the end of Section 9.2. Then the problem takes the form

$$H_0: p_1 = p_{1_0}, p_2 = p_{2_0}, \ldots, p_k = p_{k_0},$$

versus

$$H_1: p_i \neq p_{i_0}, \qquad \text{for at least one } i.$$

The first statistic for testing either of these sets of hypotheses was proposed by Karl Pearson in 1900. Couched in the language of the multinomial, Pearson's method requires that the n observations be grouped into k intervals (or k "classes" if Y is discrete) and that $p_{1_0}, p_{2_0}, \ldots, p_{k_0}$ [or $f_0(y)$] be specified. Theorem 9.2 defines Pearson's statistic, gives its asymptotic distribution, and locates its critical region.

THEOREM 9.2 Let $(X_1, X_2, \ldots, X_k)$ be a multinomial random variable with parameters $n, p_1, p_2, \ldots, p_k$. Then:

(a) The cdf of the random variable

$$\sum_{i=1}^{k} \frac{(X_i - np_i)^2}{np_i}$$

converges to the cdf of the χ^2 distribution with $k - 1$ degrees of freedom. (For approximation purposes when n is finite, it is usually recommended that the k classes be defined so that np_i is greater than or equal to 5, for all i.)

(b) At the α level of significance, $H_0: p_1 = p_{1_0}, \ldots, p_k = p_{k_0}$ is rejected in favor of H_1: at least one $p_i \neq p_{i_0}$ if

$$c = \sum_{i=1}^{k} \frac{(x_i - np_{i_0})^2}{np_{i_0}} \geq \chi^2_{1-\alpha, k-1}.$$

PROOF   A formal proof lies beyond the scope of this text. We will present a heuristic argument for part (a) for the special case $k = 2$. That $c \geq \chi^2_{1-\alpha, k-1}$ is a reasonable critical region is evident by inspection. If agreement between the actual data and the presumed model were perfect, each x_i would equal its corresponding np_{i_0} (recall Theorem 9.1) and c would be 0 (and, of course, H_0 should be accepted). Conversely, as the discrepancies between the observed and expected frequencies proliferate, and c increases, the credibility of H_0 should surely diminish. On intuitive grounds, then, a test rejecting H_0 when c is large is eminently justifiable.

Now, returning to part (a), suppose $k = 2$. Then

$$\begin{aligned}
C &= \frac{(X_1 - np_1)^2}{np_1} + \frac{(X_2 - np_2)^2}{np_2} \\
&= \frac{(X_1 - np_1)^2}{np_1} + \frac{[n - X_1 - n(1 - p_1)]^2}{n(1 - p_1)} \\
&= \frac{(X_1 - np_1)^2 (1 - p_1) + (-X_1 + np_1)^2 p_1}{np_1(1 - p_1)} \\
&= \frac{(X_1 - np_1)^2}{np_1(1 - p_1)}.
\end{aligned}$$

From Theorem 9.1, $E(X_1) = np_1$ and $\text{Var}(X_1) = np_1(1 - p_1)$, the two implying that C can be written

$$C = \left[\frac{X_1 - E(X_1)}{\sqrt{\text{Var}(X_1)}} \right]^2.$$

By Theorem 4.3, then, C is the square of a variable that is asymptotically $N(0, 1)$, and the statement of part (a) follows (for $k = 2$) from Theorem 7.7. [A proof of the general statement can be accomplished by showing that the limit of the moment-generating function for C—as n goes to ∞—is the moment-generating function for a χ^2_{k-1} random variable. See (60) for details.]

Comment. Although Pearson formulated his statistic before any general theories of hypothesis testing had been developed, it has since been shown that C is, in fact, asymptotically equivalent to the generalized-likelihood-ratio test of $H_0 \colon p_1 = p_{1_0}$, $p_2 = p_{2_0}, \ldots, p_k = p_{k_0}$. ∎

QUESTION 9.3.1   Given a set of n measurements categorized into a multinomial vector, $(X_1, X_2, \ldots, X_k)$, write out the generalized likelihood ratio, λ, for testing

$$H_0 \colon p_1 = p_{1_0}, p_2 = p_{2_0}, \ldots, p_k = p_{k_0}$$

versus

$$H_1 \colon \text{at least one } p_i \neq p_{i_0}.$$

CASE STUDY

9.1

Inhabiting many tropical waters is a small (< 1 mm) crustacean, *Ceriodaphnia cornuta*, that occurs in two distinct morphological forms: one has a series of "horns" protruding from its exoskeleton, while the other is more rounded (Figure 9.1). Zaret (207) describes

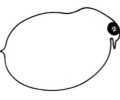

Unhorned   Horned

Figure 9.1 Forms of *C. cornuta*.

an experiment to test whether either variant is more conducive than the other to the survival of the species, in terms of its likelihood of being eaten.

A large number of *C. cornuta* were introduced into a holding tank in a 3-to-1 ratio—three of the unhorned variety were added to every one with horns. Also present in the tank was a natural predator of *C. cornuta*, a small (6-cm) fish, *Melaniris chagresi*. After approximately one hour, long enough for the predator to have completed its feeding, the fish was sacrificed and the contents of its stomach examined. Among the 44 crustacean casualties, the unhorned-to-horned ratio was 40 to 4. Can it be concluded from this that there is a *true* differential predation rate between the two polymorphs?

Here, the two *natural* classes for the response variable are "unhorned" and "horned," and under the null hypothesis that morphology has no effect on survival, it would follow that the probability of either form's being eaten should be proportional to the numbers of each kind available. If $p_1 = P$(unhorned *C. cornuta* is eaten) and $p_2 = P$(horned *C. cornuta* is eaten), the experimenter's objective reduces to a test of

$$H_0: \ p_1 = \tfrac{3}{4}, p_2 = \tfrac{1}{4}$$

versus

$$H_1: \ p_1 \neq \tfrac{3}{4}, p_2 \neq \tfrac{1}{4}.$$

We will let $\alpha = 0.05$.

Since $k = 2$, the behavior of C will be approximated by a χ_1^2 distribution, for which the 0.05 critical value is 3.84 (see Figure 9.2). Substituting the values for the x_i's and p_{i_0}'s into the test statistic gives a c value of 5.93:

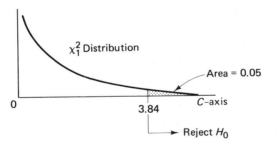

$$c = \frac{[40 - 44(\tfrac{3}{4})]^2}{44(\tfrac{3}{4})} + \frac{[4 - 44(\tfrac{1}{4})]^2}{44(\tfrac{1}{4})}$$

$$= 5.93.$$

Figure 9.2 χ_1^2 distribution.

Our conclusion, then, is to *reject H_0*—it would appear that morphology *does* have an effect on *C. cornuta*'s chances of being eaten, the unhorned variety being significantly tastier!

Comment. The final conclusion reached in this experiment was not exactly what the preceding analysis suggested. Using similar goodness-of-fit arguments, Zaret was able to show that the actual reason for the disparity in predation rates was not the presence or absence of horns but, rather, the enlarged eyespot characteristic of the latter. It is this feature that renders the otherwise nearly transparent *C. cornuta* more visible—and, as a result, more edible. ▮

QUESTION 9.3.2   Verify that the common belief in the propensity of babies to choose an inconvenient hour for birth has some basis in observation. A maternity hospital reported that out of one year's total of 2650 births, 494 occurred between midnight and 4 A.M. (175). Use the goodness-of-fit test to show that these data are not what we should expect if births are assumed to occur uniformly in all time periods. Let $\alpha = 0.05$.

CASE STUDY

9.2

Examples of goodness-of-fit testing are quite common in genetics, where often the easiest way to support a theory about the inheritance of a particular trait is to perform a number of crosses and show (via a χ^2 test) that the appearance of the offspring is, in fact, what was predicted by the theory. The following study is a case in point.

The feathers of a frizzle chicken occur in three variations (or *phenotypes*)—extreme frizzle, mild frizzle, and normal. It has been suggested that two alleles (F and f) control frizzle and that F and f interact according to a phenomenon known as incomplete dominance. What the latter predicts is that progeny whose genetic complement is (F, F) will be extreme frizzles, those having an (F, f) genotype will be mild frizzles, while the (f, f) offspring will be normal.

If two Ff hybrids are crossed, and if the two alleles recombine randomly, the ratio of extreme frizzles to mild frizzles to normals should, of course, be 1 : 2 : 1. Table 9.1 shows the actual phenotypes produced in 93 such crossings (171).

TABLE 9.1   Offspring of Ff crossings

Phenotype	Observed Frequency
Extreme frizzle	23
Mild frizzle	50
Normal	20
	93

Let $p_1 = P$(offspring is extreme frizzle), $p_2 = P$(offspring is mild frizzle), and $p_3 = P$(offspring is normal). Then the hypotheses to be tested can be written

$$H_0: p_1 = \tfrac{1}{4}, p_2 = \tfrac{1}{2}, p_3 = \tfrac{1}{4}$$

versus

$$H_1: \text{at least one } p_i \neq p_{i_0}.$$

If we elect 0.05 to be the level of significance, the critical value is 5.991, the 95th percentile of the χ_2^2 distribution. The value of c, though, is much less than that:

$$c = \frac{[23 - 93(\tfrac{1}{4})]^2}{93(\tfrac{1}{4})} + \frac{[50 - 93(\tfrac{1}{2})]^2}{93(\tfrac{1}{2})} + \frac{[20 - 93(\tfrac{1}{4})]^2}{93(\tfrac{1}{4})}$$

$$= 0.71.$$

Thus, the assumptions that F and f recombine randomly and that the extent of frizzle depends on whether the progeny's genotype is (F, F), (F, f), or (f, f) are, together, supported by the data.

QUESTION 9.3.3   Table 9.2 lists the lengths of the World Series for the 50 years from 1926 to 1975. Test whether these data are compatible with the model that each World Series game is an independent Bernoulli trial with

$$p = P(\text{AL wins}) = P(\text{NL wins}) = \tfrac{1}{2}.$$

TABLE 9.2 World Series lengths (1926–1975)

Number of Games	Number of Years
4	9
5	11
6	8
7	22
	50

(Refer to Review Exercise 2 at the end of Chapter 3 for the appropriate p_{i_0}'s.) Let $\alpha = 0.01$.

9.4 GOODNESS-OF-FIT TESTS: PARAMETERS UNKNOWN

More common than the sort of problems described in Section 9.3 are situations where the experimenter has reason to believe the response variable follows some particular *family* of pdf's—say, the gamma or the normal—but has little or no prior information to suggest what values should be assigned to the model's parameters. In cases such as these, we will carry out the goodness-of-fit test by first estimating all unknown parameters, preferably with the method of maximum likelihood, and then computing c_1, the obvious analog of Pearson's c:

$$c_1 = \sum_{i=1}^{k} \frac{(x_i - n\hat{p}_{i_0})^2}{n\hat{p}_{i_0}}.$$

We pay a price, though, for having to rely on the data to fill in details about the presumed model: each parameter estimated reduces by one the number of degrees of freedom associated with the χ^2 distribution approximating the sampling distribution of C_1—and as the number of degrees of freedom decreases, so does the power of the test. Theorem 9.3, stated here without proof, gives the distribution of the test statistic when the multinomial p_i's are functions of a single unknown parameter, θ.

THEOREM 9.3 Suppose $p_1(\theta), p_2(\theta), \ldots$, and $p_k(\theta)$ are continuously differentiable functions for θ in some interval I, satisfying the following conditions for each $\theta \in I$:

(a) $\sum_{i=1}^{k} p_i(\theta) = 1$.

(b) $p_i(\theta) > \epsilon > 0, 1 \leq i \leq k$.

(c) $p_i'(\theta) \neq 0, 1 \leq i \leq k$.

Then for each n there is a maximum-likelihood estimator, $\hat{\theta}_n$, such that $\hat{\theta}_n$ converges to θ. Furthermore, the cdf of

$$C_1 = \sum_{i=1}^{k} \frac{[X_i - np_i(\hat{\theta}_n)]^2}{np_i(\hat{\theta}_n)}$$

converges to the cdf of a χ^2 distribution with $k - 2$ degrees of freedom.

Comment. In the more general case, where the p_i's are functions of r unknown parameters, the analogous C_1 is asymptotically chi square with $k - 1 - r$ degrees of freedom. ∎

Comment. MLE's are not the only class of estimates for which Theorem 9.3 is true. It holds for any set of $\hat{\theta}$'s that are both asymptotically normal and asymptotically efficient [see (31)]. ∎

EXAMPLE 9.1. Consider again the distribution of the annual number, Y, of outbreaks of war as described in Case Study 4.4. By the discussion on pp. 146–147, the presumption of the Poisson as a model for Y would certainly not be unreasonable here, although there is no obvious, a priori value to substitute for the outbreak *rate*, λ. We do know from Example 5.14, though, that the MLE for the Poisson parameter is the sample mean—in this case, 0.69:

$$\hat{\lambda} = \bar{y} = \frac{0(223) + 1(142) + 2(48) + 3(15) + 4(4)}{432} = 0.69. \qquad (9.5)$$

With Equation 9.5, $f_0(y)$ is completely specified, and the suitability of the Poisson model will be decided by testing

$$H_0: \ Y \sim \text{Poisson}$$
$$\text{versus}$$
$$H_1: \ Y \nsim \text{Poisson}.$$

Note that the "4+" class has an expected frequency of only 2. To comply with the np_{i_0} guideline mentioned in Theorem 9.2 (and also applying to Theorem 9.3), we need to collapse the "3" and "4+" classes into a single category, "3+" (Table 9.3).

TABLE 9.3 Observed and expected war outbreak distributions

Number of Wars, y, Beginning in a Given Year	Observed Frequency (x)	Expected Frequency $[432 \cdot f_0(y)]$
0	223	216.7
1	142	149.5
2	48	51.6
3+	19	14.2
	432	432

With $k = 4$ classes and one estimated parameter, the critical value for C_1 (assuming the 0.05 level of significance) is $\chi^2_{0.95, 2} = 5.991$. But, from Table 9.3,

$$c_1 = \frac{(223 - 216.7)^2}{216.7} + \frac{(142 - 149.5)^2}{149.5} + \frac{(48 - 51.6)^2}{51.6} + \frac{(19 - 14.2)^2}{14.2}$$
$$= 2.43.$$

Our conclusion is to *accept* H_0: for the historical period in question the annual number of outbreaks of war, worldwide, was, indeed, a phenomenon well described by a Poisson pdf.

QUESTION 9.4.1   Do a goodness-of-fit test on the Prussian cavalry data of Case Study 4.2. Let 0.01 be the level of significance.

QUESTION 9.4.2   At the 0.05 level, test the adequacy of the geometric pdf as a model for the bird-call data given in Case Study 4.9.

EXAMPLE 9.2.   Certainly the most frequent distributional assumption made in statistics, whether implicitly or explicitly, is that $Y \sim N(\mu, \sigma^2)$. As we have already seen, this was the starting point for all the t tests, χ^2 tests, and F tests presented in Chapters 7 and 8. In practice, tests for normality are typically based on the two-parameter extension of Theorem 9.3, since both μ and σ^2 will be unknown.

Suppose we wanted to test the normality of the Etruscan-skull data of Case Study 7.1 (as a first step, perhaps, in testing $H_0: \mu = \mu_0$ or $H_0: \sigma = \sigma_0$). Since Y is continuous, it will be necessary to group the observations into a set of nonoverlapping intervals and determine from H_0 the probability associated with each one. Table 9.4 shows one possible partitioning.

TABLE 9.4   Skull-width data

Skull Width (mm)	Observed Frequency, x_i
125–129	1
130–134	4
135–139	10
140–144	33
145–149	24
150–154	9
155–159	3
	84

To compute the expected frequency corresponding to, say, the "140–144" class, we must first determine the probability of a single random observation's falling into that range. Using the continuity correction together with the usual Z transformation, and recalling that $\bar{y} = 143.8$ and $s = 6.0$, we get

$$P(140 < Y < 144) \doteq P(139.5 < Y < 144.5)$$

$$= P\left(\frac{139.5 - 143.8}{6.0} < Z < \frac{144.5 - 143.8}{6.0}\right)$$

$$= P(-0.72 < Z < 0.12) = 0.3120.$$

It follows that the expected frequency for the "140–144" class is 26.2:

$$n\hat{p}_{i_0} = 84(0.3120) = 26.2.$$

Table 9.5 lists the $\hat{p}_{i_0}$'s and $n\hat{p}_{i_0}$'s for the seven classes of Table 9.4.

TABLE 9.5   Observed and expected skull-width distributions

Skull Width	x_i	$\hat{p}_{i_0}$	$84\hat{p}_{i_0}$
≤ 129	1	0.0087	0.7
130–134	4	0.0519	4.4
135–139	10	0.1752	14.7
140–144	33	0.3120	26.2
145–149	24	0.2811	23.6
150–154	9	0.1336	11.2
155+	3	0.0375	3.2
	84	1.000	84.0

Comment. Notice that the first and last classes have been made open-ended. This is a necessary adjustment whenever the model being fit has an infinite range. If this were not done, the $\hat{p}_{i_0}$'s would not sum to 1, a condition required of the multinomial distribution. ▌

Table 9.6 shows the *final* set of classes, the first two and the last two intervals in Table 9.5 having been collapsed to comply with the "$n\hat{p}_{i_0} \geq 5$" condition. As indicated at the bottom of the last column, the calculated value for c_1 is 3.61.

TABLE 9.6   Revised skull-width classes

Skull Width (mm)	x_i	$\hat{p}_{i_0}$	$84\hat{p}_{i_0}$	$(x_i - 84\hat{p}_{i_0})^2/84\hat{p}_{i_0}$
≤ 134	5	0.0606	5.1	0.00
135–139	10	0.1752	14.7	1.50
140–144	33	0.3120	26.2	1.76
145–149	24	0.2811	23.6	0.01
150+	12	0.1711	14.4	0.34
	84	1.0000	84.0	3.61

Since $r = 2$ parameters were estimated and the final calculations were based on $k = 5$ classes, the number of degrees associated with C_1 is $5 - 2 - 1$, or 2. Had we elected to carry out the test at the $\alpha = 0.05$ level of significance, the corresponding critical value would be 5.991 ($= \chi^2_{0.95,2}$), and our conclusion would be to *accept* the normality assumption.

QUESTION 9.4.3   Do a goodness-of-fit test to determine whether an exponential model adequately describes the Pushkin data given in Case Study 5.1. Use the 0.05 level of significance. Define as an initial set of classes, 0–39, 40–79, . . . , 240–279, and 280+. Do whatever combining is necessary to meet the "$n\hat{p}_{i_0} \geq 5$" restriction.

Too often inference is pictured as a static, one-shot process: a confidence interval is constructed or a hypothesis is tested, a conclusion is drawn, and the experimenter moves on to another problem, having disposed of days, maybe even months or years, of data with a simple "yes" or "no." In reality, though, interpreting data is more of a *dynamic* process, with the result of one test serving as input for a second, and *that*, in turn, suggesting still a third, and so on. The next example illustrates this idea by examining the relationship between parameter estimation, goodness-of-fit testing, and model building for a set of data related to the genetic phenomenon of *linkage*.

CASE STUDY

9.3   Corn can be either starchy (S) or sugary (s) and can have a green (G) base leaf or a white (g) one. Suppose the alleles for these two factors occur on separate chromosomes and are hence independent. Then each parent with alleles SsGg produces with equal likelihood gametes of the form (S, G), (S, g), (s, G), and (s, g). If two such hybrid parents are crossed, the phenotypes of the offspring will occur in the proportions suggested by Table 9.7. That is, the probability of

TABLE 9.7 A dihybrid cross

Alleles of First Parent

		SG	Sg	sG	sg
Alleles	SG	(S, G)	(S, G)	(S, G)	(S, G)
of Second	Sg	(S, G)	(S, g)	(S, G)	(S, g)
Parent	sG	(S, G)	(S, G)	(s, G)	(s, G)
	sg	(S, G)	(S, g)	(s, G)	(s, g)

an offspring of type (S, G) is $\frac{9}{16}$; type (S, g), $\frac{3}{16}$; type (s, G), $\frac{3}{16}$; and type (s, g), $\frac{1}{16}$.

Table 9.8 shows the $9:3:3:1$ set of ratios fit to a set of 3839 SsGg × SsGg crossings (21). With no parameters having been esti-

TABLE 9.8 Fitting the $9:3:3:1$ ratio

Phenotype	x_i	p_{i_0}	$3839p_{i_0}$	$(x_i - 3839p_{i_0})^2/3839p_{i_0}$
Starchy green	1997	$\frac{9}{16}$	2159.44	12.22
Starchy white	906	$\frac{3}{16}$	719.81	48.16
Sugary green	904	$\frac{3}{16}$	719.81	47.13
Sugary white	32	$\frac{1}{16}$	239.94	180.21
	3839	1	3839.00	287.72

mated, and given $k = 4$ classes, the critical value for, say, a 0.01 test would be $\chi^2_{0.99,3}$, or 11.341. But the calculated value for c is 287.72, a number clearly indicating the inappropriateness of the initial model.

One possible explanation for the decided lack of fit between the data and the $9:3:3:1$ model is a type of genetic behavior known as *crossing-over*. Suppose a parent has one chromosome with alleles Sg and another with alleles sG. In the process of meiosis, the two members of the chromosome pair move close together, uncoil, and, in some cases, exchange parts—this latter being the phenomenon of crossing-over. It follows that there is a probability, p, of the chromosome pairs Sg and sG changing to SG and sg. Of course, whether or not a cross-over occurs, either chromosome is equally likely to be present in the gamete. The probability, then, of any of the four possible allele combinations being present in the gamete can be determined from Theorem 2.7. For example,

$$P(\text{SG is in gamete}) = P(\text{SG is in gamete}|\text{cross-over}) \cdot P(\text{cross-}$$
$$\text{over}) + P(\text{SG is in gamete}|\text{no cross-over})$$
$$\cdot P(\text{no cross-over})$$
$$= \tfrac{1}{2} \cdot p + 0(1 - p) = \tfrac{1}{2}p.$$

Similarly,

$$P(\text{Sg is in gamete}) = \tfrac{1}{2}(1 - p),$$
$$P(\text{sG is in gamete}) = \tfrac{1}{2}(1 - p),$$

and

$$P(\text{sg is in gamete}) = \tfrac{1}{2}p.$$

There is no reason to assume that p is the same for both parents, so the other parent would admit a similar array of gamete probabilities, but with p replaced by p'. The relative likelihood of a given phenotype can then be inferred from Table 9.7. For example,

$$P[(\text{s, g}) \text{ phenotype}] = P[(\text{s, g}) \text{ allele from first parent } \textit{and}$$
$$(\text{s, g}) \text{ allele from second parent}]$$
$$= \tfrac{1}{2}p \cdot \tfrac{1}{2}p' = \tfrac{1}{4}pp'.$$

Similarly, the probability of an offspring with phenotype (s, G) will be $\tfrac{1}{4}(1 - pp')$, the sum of the probabilities of three different matings:

$$P[(\text{s, G}) \text{ phenotype}] = P[(\text{s, g}) \text{ allele} \times (\text{s, G}) \text{ allele}]$$
$$+ P[(\text{s, G}) \text{ allele} \times (\text{s, G}) \text{ allele}]$$
$$+ P[(\text{s, G}) \text{ allele} \times (\text{s, g}) \text{ allele}]$$
$$= \tfrac{1}{2}p \cdot \tfrac{1}{2}(1 - p') + \tfrac{1}{2}(1 - p) \cdot \tfrac{1}{2}(1 - p')$$
$$+ \tfrac{1}{2}(1 - p) \cdot \tfrac{1}{2}p'$$
$$= \tfrac{1}{4}(1 - pp').$$

By the same reasoning, $P[(\text{S, g}) \text{ phenotype}] = \tfrac{1}{4}(1 - pp')$ and $P[(\text{S, G}) \text{ phenotype}] = \tfrac{1}{4}(2 + pp')$.

Since each phenotype probability depends on the same parameter product, pp' can be replaced by a single parameter θ, where $0 < \theta < 1$. Table 9.9 summarizes the probability distribution for the four phenotypes. Note that if $p = p' = \tfrac{1}{2}$, making $\theta = \tfrac{1}{4}$, the probabilities of Table 9.9 reduce to the $9:3:3:1$ ratios of independent traits. If $p = p' = 1$, and $\theta = 1$, the simple $3:1$ dominant-recessive scheme arises.

Before the data of Table 9.8 can be tested against the theoretical construct of Table 9.9, the unknown θ needs to be estimated. For-

TABLE 9.9 Phenotype probability
distribution

Phenotype	Probability
Starchy, green	$\frac{1}{4}(2 + \theta)$
Starchy, white	$\frac{1}{4}(1 - \theta)$
Sugary, green	$\frac{1}{4}(1 - \theta)$
Sugary, white	$\frac{1}{4}(\theta)$

tunately, its MLE is readily accessible. The likelihood function has
the form,

$$L = [\tfrac{1}{4}(2 + \theta)]^{x_1} \cdot [\tfrac{1}{4}(1 - \theta)]^{x_2} \cdot [\tfrac{1}{4}(1 - \theta)]^{x_3} \cdot [\tfrac{1}{4}(\theta)]^{x_4}.$$

Taking the ln of L and substituting for the x_i's gives

$$\ln L = 1997 \ln (2 + \theta) + (906 + 904) \ln (1 - \theta) + 32 \ln (\theta)$$
$$- 3839 \ln (4).$$

Therefore,

$$\frac{d \ln L}{d\theta} = \frac{1997}{2 + \theta} - \frac{1810}{1 - \theta} + \frac{32}{\theta},$$

which, when set equal to 0, yields the maximum-likelihood estimate
$\hat{\theta} = 0.0357$. Table 9.10 shows the initial data "refit" using the cross-
over model. Notice the improvement between the last column of
Table 9.8 and the last column of Table 9.10.

TABLE 9.10 Fitting the revised model

Phenotype	x_i	$\hat{p}_{i_0}$	$3839\hat{p}_{i_0}$	$(x_i - 3839\hat{p}_{i_0})^2/3839\hat{p}_{i_0}$
Starchy, green	1997	0.5089	1953.67	0.96
Starchy, white	906	0.2411	925.58	0.41
Sugary, green	904	0.2411	925.58	0.50
Sugary, white	32	0.0089	34.17	0.14
	3839	1.0000	3839.00	2.01

Since one parameter was estimated, the χ^2 statistic now has
$4 - 1 - 1 = 2$ degrees of freedom. Also, with the 95th percentile of
χ_2^2 being 5.991, and c_1 equal to 2.01, we *accept* H_0 at the 0.05 level,
thus lending credence both to the notion of crossing-over and to the
probability model we set up to characterize it.

QUESTION 9.4.4   The Japanese morning glory (*Pharbitis nil*) admits a number of recessive traits, two of which are crumpled leaves and variegated leaves. The results of cross-breeding hybrid morning glories bearing a dominant and recessive gene in each trait were the 2176 progeny categorized in Table 9.11 (85). Estimate the recombination parameter, θ, and fit these data using the methodology of Case Study 9.3.

TABLE 9.11   Offspring of cross-bred morning glories

Phenotype	Observed Frequency
Normal	1450
Variegated only	184
Crumpled only	191
Variegated and crumpled	351

The sensitivity of the χ^2 procedure to parameter estimation was illustrated dramatically in Case Study 9.3. When the parameter θ was "estimated" to be $\frac{1}{4}$, yielding the no cross-over $\frac{9}{16}, \frac{3}{16}, \frac{3}{16}, \frac{1}{16}$ distribution, H_0 was resoundingly rejected. Yet with θ set equal to its MLE, the null hypothesis was accepted. A failsafe method of avoiding disagreements of this sort would be to assign the parameters those particular values that minimize the expression for c_1. Not surprisingly, $\hat{\theta}$'s obtained in such a way are known as *minimum χ^2 estimates*. The concluding remarks in this section relate the notion of minimum χ^2 estimation to the method of maximum likelihood and, ultimately, to the statement of Theorem 9.3. (For notational simplicity, we will restrict our attention to the case of a single parameter; the generalization to more than one parameter is straightforward.)

If, for all i, the multinomial p_i's are really $p_i(\theta)$'s, then C_1 is similarly $C_1(\theta)$, and for a fixed sample, $x_1, x_2, \ldots, x_n$,

$$C_1(\theta) = \sum_{i=1}^{k} \frac{[x_i - np_i(\theta)]^2}{np_i(\theta)}.$$

Assuming appropriate differentiability conditions, we can find the minimum χ^2 estimate by setting $C_1'(\theta)$ equal to 0 and solving for θ. To begin,

$$C_1'(\theta) = -\sum_{i=1}^{k} \left\{ \frac{2[x_i - np_i(\theta)]p_i'(\theta)}{p_i(\theta)} + \frac{[x_i - np_i(\theta)]^2 p_i'(\theta)}{np_i^2(\theta)} \right\}. \tag{9.6}$$

In general, setting Equation 9.6 equal to 0 gives a decidedly difficult expression to solve. For large n, though, the second term being added is often negligible, so without unduly compromising matters, we can solve, instead, Equation 9.7:

$$\sum_{i=1}^{k} \frac{[x_i - np_i(\theta)]p_i'(\theta)}{p_i(\theta)} = 0. \tag{9.7}$$

But this is just an old friend in a new suit. To see why, consider the likelihood equation,

$$L(\theta) = [p_1(\theta)]^{x_1}[p_2(\theta)]^{x_2} \cdots [p_k(\theta)]^{x_k}.$$

Taking the ln of $L(\theta)$ and differentiating gives

$$\ln L(\theta) = \sum_{i=1}^{k} x_i \ln p_i(\theta)$$

and

$$\frac{d \ln L(\theta)}{d\theta} = \sum_{i=1}^{k} \frac{x_i p_i'(\theta)}{p_i(\theta)}.$$

But since $\sum_{i=1}^{k} p_i(\theta) = 1$, it follows that $\sum_{i=1}^{k} p_i'(\theta) = 0$, and we can write

$$\sum_{i=1}^{k} \frac{n p_i(\theta) \cdot p_i'(\theta)}{p_i(\theta)} = 0. \qquad (9.8)$$

Now, combine Equation 9.8 with $[d \ln L(\theta)]/d\theta$ and compare the result with Equation 9.7:

$$\frac{d \ln L(\theta)}{d\theta} = \sum_{i=1}^{k} \frac{x_i p_i'(\theta)}{p_i(\theta)} - \sum_{i=1}^{k} \frac{n p_i(\theta) p_i'(\theta)}{p_i(\theta)}$$

$$= \sum_{i=1}^{k} \frac{(x_i - n p_i) p_i'(\theta)}{p_i(\theta)}.$$

It follows that the MLE for θ is essentially the same as the minimum χ^2 estimate. Or, to put it another way, we now have some reassurance that the method of maximum likelihood gives estimates appropriate for goodness-of-fit testing, and that the statement of Theorem 9.3, calling for any unknown θ's to be replaced by their MLE's, is not unreasonable. (In Case Study 9.3 the minimum χ^2 estimate for θ is 0.035785; the MLE, 0.035712. The value of c_1 in the first case is 2.0153; in the second case, 2.0154).

9.5 CONTINGENCY TABLES AND TESTS OF HOMOGENEITY

We have seen that inferences both about parameters and about entire pdf's are two frequent objectives of statistical methodology. A third are inferences about *independence*: does knowing the value of a first random variable provide us with any insight into the probable behavior of a second? Examples of this latter type of problem are commonplace, even outside the scientific community. Perhaps the most familiar are the various studies purportedly linking the occurrence of cancer to such factors as smoking, industrial pollutants, saccharin, cyclamates, and pan-fried hamburgers. Needless to say, questions of independence are by no means confined to health-related issues. Merchants ask if inventory requirements differ in downtown and suburban stores, educators speculate on the relationship between a city's busing efforts and students' performance on standarized exams, and not a few parents have wondered if juvenile delinquency is related to the use of Dr. Spock's baby book.

The notion of independence, of course, has already turned up in a number of different contexts, both probabilistic and statistical. To recast it in an inference setting requires nothing new, conceptually, and we can even use the same notation that was introduced in Chapter 2. Let $A_1, A_2, \ldots, A_r$ be a partition of the sample space S,

$$\bigcup_{i=1}^{r} A_i = S, \qquad A_i \cap A_j = \varnothing, i \neq j.$$

Suppose $B_1, B_2, \ldots, B_c$ is another partition of S. Then the question of whether criteria A and criteria B are independent can be written

$$H_0: \ A_i \text{ and } B_j \text{ are independent for } 1 \leq i \leq r, 1 \leq j \leq c$$

versus

$$H_1: \ A_i \text{ and } B_j \text{ are } not \text{ independent for } 1 \leq i \leq r, 1 \leq j \leq c.$$

The methodology for choosing between H_0 and H_1 follows closely the pattern established by Theorems 9.2 and 9.3. First, we select a random sample, $Y_1, Y_2, \ldots, Y_n$, from S, and then we define the random variable X_{ij} to be the number of observations belonging to the intersection $A_i \cap B_j$. Under H_0,

$$P[Y_k \in (A_i \cap B_j)] = P(Y_k \in A_i) \cdot P(Y_k \in B_j) = p_i q_j, \qquad k = 1, 2, \ldots, n,$$

where $\sum_{i=1}^{r} p_i = 1$ and $\sum_{j=1}^{c} q_j = 1$. Therefore, $E(X_{ij}) = np_i q_j$, and, by analogy with C and C_1, the goodness-of-fit statistic

$$C_2 = \sum_{i=1}^{r} \sum_{j=1}^{c} \frac{(X_{ij} - np_i q_j)^2}{np_i q_j}$$

has an asymptotic χ^2 distribution. Typically, the p_i's and the q_j's will both be estimated, the former with MLE's

$$\hat{p}_i = \frac{1}{n} \sum_{j=1}^{c} x_{ij}$$

and the latter by

$$\hat{q}_j = \frac{1}{n} \sum_{i=1}^{r} x_{ij}.$$

However, since

$$\sum_{i=1}^{r} \hat{p}_i = 1 \quad \text{and} \quad \sum_{j=1}^{c} \hat{q}_j = 1,$$

only $(r - 1) + (c - 1)$ parameters need to be estimated directly. Since the Y's are being categorized into a total of $r \cdot c$ classes, it follows from the generalization of Theorem 9.3 that the number of degrees of freedom associated with C_2 is

$$rc - 1 - (r - 1) - (c - 1) = rc - r - c + 1 = (r - 1)(c - 1).$$

Data for a test of independence are generally presented in tabular form, with rows representing the categories of one criteria and columns the categories of the other. Such displays are called *contingency tables*—a name probably due to Karl Pearson.

Market researchers often gather survey information by telephone. Ideally, the numbers dialed should be random, thus affording every subscriber an equal chance of being included in the sample. In practice, though, doing a survey that way would be very costly because of the sizable number of nonproductive calls resulting from dialing invalid numbers. The obvious remedy is to sample from the telephone directory, but that raises another problem: unlisted numbers would not be included. Nationally, more than 15% of all telephone numbers are unlisted, and the percentage is much higher in certain locales (in Los Angeles, for example, more than 35% of all telephone numbers are unlisted). What this means is that the directory method is unreliable unless the factors being studied are independent of whether or not the telephone is listed.

Suppose a political poll is being taken in California to assess the popularity of a proposition to reduce property taxes. For a directory sample to be unbiased in that context, the proportion of subscribers who own their own homes should not be significantly different for those with listed phones as compared to those with unlisted phones.

The following responses (slightly modified) were obtained in a recent survey by Pacific Bell of 1000 subscribers (139). From the

TABLE 9.12 Pacific Bell subscribers survey

	Listed	Unlisted	
Own	628	146	774
Rent	172	54	226
	800	200	1000

row totals we estimate p_1, the probability of a respondent's being a homeowner, as $\frac{774}{1000}$; p_2, the probability of his being a renter, as $\frac{226}{1000}$. Similarly, q_1, the probability of a respondent's having a listed phone, would be estimated by $\frac{800}{1000}$; and q_2, the probability of his having an unlisted phone, as $\frac{200}{1000}$. Under the null hypothesis asserting independence between home ownership and having a listed phone, the estimated probability of a respondent's being a home-owner with a listed phone is $\hat{p}_1\hat{q}_1$, or $(0.774)(0.8) = 0.6192$. Thus, the expected *number* of such people is $1000(0.6192)$, or 619.2. Table 9.13 shows the similarly calculated expected frequencies (in parentheses) for each of the four data categories.

TABLE 9.13   Expected frequencies

	Listed	Unlisted	
Own	628 (619.2)	146 (154.8)	774
Rent	172 (180.8)	54 (45.2)	226
	800	200	1000

Here, the value of c_2 is 2.77:

$$\frac{(628 - 619.2)^2}{619.2} + \frac{(146 - 154.8)^2}{154.8} + \frac{(172 - 180.8)^2}{180.8} + \frac{(54 - 45.2)^2}{45.2}$$

$$= 2.77.$$

It follows that if we were to test

H_0: ownership and listing status are independent

versus

H_1: ownership and listing status are dependent

at the $\alpha = 0.05$ level, the null hypothesis would be accepted: with $(2 - 1)(2 - 1) = 1$ degree of freedom, the critical χ^2 value is $\chi^2_{0.95,1} = 3.84$. (At the 10% level, though, the null hypothesis would have been rejected.)

Comment. The hazards in drawing inferences from telephone data are well documented. One of the most frequently cited "mistakes" was the *Literary Digest* presidential poll in 1936 that confidently predicted a Landon victory over Roosevelt. ▮

QUESTION 9.5.1   A group of 57 elderly persons visiting a certain clinic on an outpatient basis were categorized as being either "compliers" or "noncompliers" according to whether or not they followed their doctor's orders in taking medication. Each of them was also classified by religious affiliation. Table 9.14 summarizes the responses (186).

TABLE 9.14 Compliance with doctor's orders

	Catholic	Protestant
Compliers	10	15
Noncompliers	7	25

Clearly, the sample proportion of Catholic compliers ($\frac{10}{17}$ = 0.59) is considerably higher than the sample proportion of Protestant compliers ($\frac{15}{40}$ = 0.38). Is this difference statistically significant? At the 0.05 level, test to see whether religion and compliance are independent.

Note: These data were collected to test a theory that the authoritarian nature of the Catholic church induces in its members a greater tolerance for taking orders than would be characteristic of members of Protestant faiths.

CASE STUDY

9.5

The question of whether or not birth order is related to juvenile delinquency was examined in a large-scale study using a high school population. A total of 1154 girls attending public high school were given a questionnaire that measured the degree to which each had exhibited delinquent behavior, in terms of criminal acts, immoral conduct, and so on. After duly tabulating and interpreting the results, the researchers decided to categorize some 111 of the girls as "delinquent." Each girl in the initial sample was also asked to indicate her birth order, as being either (1) the oldest, (2) in between, (3) the youngest, or (4) an only child. Table 9.15 details the delinquency-by-birth-order breakdown (120).

TABLE 9.15 Delinquency and birth order

		Birth Order			
		Oldest	In Between	Youngest	Only Child
Delinquent?	Yes	24	29	35	23
	No	450	312	211	70
		474	341	246	93
	Percent "Yes"	5.1	8.5	14.2	24.7

We wish to test

H_0 : birth order and delinquency are independent

versus

H_1 : birth order and delinquency are dependent.

Let $\alpha = 0.05$. The MLE for p_1, the probability of a girl's showing delinquent behavior, is

$$\hat{p}_1 = \frac{24 + 29 + 35 + 23}{1154} = \frac{111}{1154}.$$

Similarly, the MLE for q_1, the probability of a girl's being an oldest child, is

$$\hat{q}_1 = \frac{24 + 450}{1154} = \frac{474}{1154}.$$

Therefore, under H_0, the expected number of oldest girls exhibiting delinquent behavior would be

$$n\hat{p}_1\hat{q}_1 = 1154 \left(\frac{111}{1154}\right)\left(\frac{474}{1154}\right) = 45.6.$$

Table 9.16 shows the entire set of $n\hat{p}_i\hat{q}_j$ expected values, together with the original x_{ij}'s.

TABLE 9.16

	Oldest	In Between	Youngest	Only Child
Yes	24 (45.6)	29 (32.8)	35 (23.7)	23 (8.9)
No	450 (428.4)	312 (308.2)	211 (222.3)	70 (84.1)

At the 0.05 level, we should reject H_0 if $c_2 \geq \chi^2_{0.95,(2-1)(4-1)}$ = $\chi^2_{0.95,3}$ = 7.81. But

$$c_2 = \frac{(24 - 45.6)^2}{45.6} + \frac{(29 - 32.8)^2}{32.8} + \cdots + \frac{(70 - 84.1)^2}{84.1}$$

$$= 42.5,$$

a number far to the right of the critical value. Thus, we reject H_0 and conclude that birth order and delinquency are dependent. (This is not an unexpected result in light of the considerable differences in the "Percent 'Yes'" figures shown at the bottom of Table 9.15.)

QUESTION 9.5.2   As part of a survey conducted in the state of Washington to characterize drinking habits, a random sample of 208 women were (1) asked whether or not they drank and (2) assigned to one of four socioeconomic strata determined on the basis of occupation, education, and family income ("I" was the highest class; "IV," the lowest). Of the 66 women in class I, 42 drank; of the 56 in class II, 38 drank; of the 57 in class III, 34 drank; and of the 29 in class IV, 6 drank (99). At the 0.01 level of significance, test whether drinking habits and socioeconomic status can be considered independent factors for Washington women. (A similar survey done among Washington *men* revealed the two factors to be independent. Interpret that finding in conjunction with the conclusion drawn from the data for Washington women.)

A variation on the idea of a contingency table occurs when the row or column totals are not random variables but are predetermined and fixed. This would be the situation, for example, if the purpose of an experiment were to determine whether the distribution of blood types varies among tribes of American Indians. Logistically, it would be quite difficult to obtain a random sample from a mixed group of tribes; more likely, the experimenter would sample randomly *within each tribe*. Of course, if that were the protocol followed, the number selected from each tribe would be preselected and not a random variable. What appears to be a test of independence, then, would be more aptly described as a comparison of several finite pdf's. Such problems are called *tests of homogeneity*. Their analysis is exactly the same as a χ^2 test for independence.

CASE STUDY

9.6   Evidence exists suggesting that *concentrated* industries, those where a relatively few firms control a major share of the market, are more profitable than unconcentrated industries. This proposition is somewhat controversial, one point of contention being the lack of a mutually agreed-upon definition of profitability. Since different accounting methods yield different profit figures, a first step in discussing profit versus concentration is to determine whether concentrated and unconcentrated industries tend to use different bookkeeping procedures.

Valuation of inventory is a significant factor in assessing the rate of return in a business. Three basic methods are in common use. One is known as FIFO (first in, first out), where inventories are valued at the latest purchase prices, but the cost of goods sold is carried at the oldest prices. In an inflationary period, this method reports higher profits than its obverse, LIFO (last in, first out). The reader, by now a firm believer in expected values, may prefer the third method, whereby inventories are valued by a weighted average of old and new prices.

Categorizing the second criterion, concentration, is relatively straightforward. The measure used here is the percentage of the total value of shipments in the industry coming from the top eight firms. These percentages will be grouped into three classes: ≥ 90, 50–89, and < 50.

We wish to test the null hypothesis that the use of the three accounting methods is independent of an industry's concentration measure. The sample consisted of 42 firms from industries whose concentration percentage was 90 or more, an additional 66 representing industries in the 50–89 range, and a final 52 selected from industries with concentration ratios less than 50 (67). All 160 firms were then classified according to their accounting method. Table 9.17 shows the original data, together with the expected frequencies (in parentheses).

TABLE 9.17   Accounting methods of 160 firms

	Concentration Class of Firm			
	≥ 90	50–89	< 50	
LIFO	24 (21.0)	25 (33.0)	31 (26.0)	80
FIFO	9 (9.7)	19 (15.3)	9 (12.0)	37
Average Cost	9 (11.3)	22 (17.7)	12 (14.0)	43
	42	66	52	160

Evaluating the usual test statistic for the nine different classes gives

$$c_2 = \frac{(24 - 21.0)^2}{21.0} + \frac{(25 - 33.0)^2}{33.0} + \cdots + \frac{(12 - 14.0)^2}{14.0}$$

$$= 6.82.$$

With the critical value ($\chi^2_{0.95,4}$) equal to 9.49, we can accept H_0 at the 5% level of significance.

QUESTION 9.5.3   Rework the data given in Review Exercise 29 at the end of Chapter 8 as a test for homogeneity. Use the 0.05 level of significance.

REVIEW EXERCISES FOR CHAPTER 9

1. Hogarth is a graduating senior who needs one more math course to complete his major. He would prefer to take either Numerical Analysis or Experimental Design. Unfortunately, those are both popular courses and tend to fill up quickly. While waiting in line to finalize his registration, he is told that each course has only two openings left and that there are seven other senior math majors in front of him. Given the available choices, suppose the probability is $\frac{1}{4}$ that a senior math major will want to take Numerical Analysis, $\frac{1}{3}$ that he will opt for Experimental Design, and $\frac{5}{12}$ that he will look for something else. What is the probability that Hogarth will be able to enroll in either Numerical Analysis or Experimental Design? Hint: Consider the complement.

2. Suppose a loaded die is tossed 12 times, where

$$p_i = P(i \text{ spots appear}) = ki, \qquad i = 1, 2, \ldots, 6.$$

Let X_i, $i = 1, 2, \ldots, 6$, denote the number of times the face with i spots appears. Find the probability that all the X_i's equal 2.

3. Suppose that five randomly selected students take the quantitative portion of the college boards, scores on which are normally distributed with a mean of 500 and a standard deviation of 100. What is the probability that two students earn scores below 400, two get scores between 400 and 650, inclusive, and one scores above 650?

4. Recall the pipeline-missile problem described in Example 2.14. Suppose six missiles are fired at the pipeline. What is the probability that two land within 20 feet to the left of the pipeline and at least three land within 20 feet to the right?

5. Suppose the $1 : 2 : 1$ model proposed for the extreme: mild: normal phenotype ratios in the frizzle fowl data of Case Study 9.2 is correct. Write down a formula giving the probability that in 93 hybrid crossings, 23 of the progeny will be extreme frizzles, 50 will be mild frizzles, and the remaining 20 will be normal. Would the magnitude of this probability itself be of any help in making an inference about the validity of the $1 : 2 : 1$ model? Explain.

6. Let the vector of random variables (X_1, X_2, X_3) have the trinomial pdf with parameters n, p_1, p_2, and $p_3 = 1 - p_1 - p_2$. That is,

$$P(X_1 = x_1, X_2 = x_2, X_3 = x_3) = \frac{n!}{x_1! x_2! x_3!} p_1^{x_1} p_2^{x_2} p_3^{x_3},$$

$$x_i = 0, 1, \ldots, n; \; i = 1, 2, 3; \; x_1 + x_2 + x_3 = n.$$

By definition, the moment-generating function for (X_1, X_2, X_3) is given by

$$M_{X_1, X_2, X_3}(t_1, t_2, t_3) = E(e^{t_1 X_1 + t_2 X_2 + t_3 X_3}).$$

Show that

$$M_{X_1, X_2, X_3}(t_1, t_2, t_3) = (p_1 e^{t_1} + p_2 e^{t_2} + p_3 e^{t_3})^n.$$

7. If $M_{X_1, X_2, X_3}(t_1, t_2, t_3)$ is the moment-generating function for (X_1, X_2, X_3), then $M_{X_1, X_2, X_3}(t_1, 0, 0)$, $M_{X_1, X_2, X_3}(0, t_2, 0)$, and $M_{X_1, X_2, X_3}(0, 0, t_3)$ are the moment-generating functions for the marginal pdf's of X_1, X_2, and X_3, respectively. Use this fact, together with the result of Review Exercise 6, to verify the statement of Theorem 9.1.

8. Let $(x_1, x_2, \ldots, x_k)$ be the vector of sample observations representing a multinomial random variable with parameters $n, p_1, p_2, \ldots,$ and p_k. Find the maximum-likelihood estimates for the p_i's.

9. Four chips are allocated at random to three urns. Let X_i be the number of chips placed in the ith urn. Find $E(X_1^2 X_3)$.

10. Fifty observations are drawn from the pdf

$$f_Y(y) = 6y(1 - y), \qquad 0 < y < 1.$$

Let X_i be the number of observations lying in the interval $((i - 1)/4, i/4)$, $i = 1, 2, 3, 4$.

(a) Write down a formula for $f_{X_1, X_2, X_3, X_4}(10, 15, 15, 10)$.

(b) Find Var (X_3).

11. One hundred samples of size 2 are drawn from an urn containing six red chips and four white chips. (Each selection is made without replacement.) Test the adequacy of the hypergeometric model if zero whites were obtained 35 times; one white, 55 times; and two whites, 10 times. Use the 0.10 decision rule.

12. A sociologist interviews 100 families where the husband and wife, early in their marriage, decided to keep having children until they had their first girl. (All 100 couples did eventually have a female child.) Let X_i denote the total number of children in the ith family. The distribution of the x_i's is shown below.

Number of Children, x_i	Number of Families
1	55
2	19
3	12
4	8
5	3
6	3

Assume that p, the probability that any given child is a girl, is $\frac{1}{2}$. Do a goodness-of-fit test to see whether the distribution of the x_i's can be adequately described by a geometric pdf. Let $\alpha = 0.05$.

13. Consider again the results of the sampling described in Review Exercise 11.

Number of White Chips, x_i	Number of Samples
0	35
1	55
2	10

Suppose, however, that we did not know whether the samples had been drawn with replacement or without replacement. Test whether sampling *with* replacement is a reasonable model. Let $\alpha = 0.05$.

14. Records kept at an eastern racetrack showed the following distribution of winners as a function of their starting-post position. All 144 races were run with a full field of eight horses.

Starting post	1	2	3	4	5	6	7	8
Number of winners	32	21	19	20	16	11	14	11

Test an appropriate goodness-of-fit hypothesis.

15. A coin is tossed 200 times and comes up heads 120 times. Is the coin "fair"? Do this problem two ways—first as a goodness-of-fit test and second with a procedure from Chapter 7. In each case let $\alpha = 0.05$. Are the two procedures equivalent?

16. It was stated in Case Study 4.8 that the mean (μ) and the standard deviation (σ) of pregnancy durations are 266 days and 16 days, respectively. Accepting those as the true parameter values, test whether the additional assumption that pregnancy durations are normally distributed is supported by the following data.

70 Pregnancy Durations
(County General Hospital, 1978)

251	264	234	283	226	244	269	241	276	274
263	243	254	276	241	232	260	248	284	253
265	235	259	279	256	256	254	256	250	269
240	261	263	262	259	230	268	284	259	261
268	268	264	271	263	259	294	259	263	278
267	293	247	244	250	266	286	263	274	253
281	286	266	249	255	233	245	266	265	264

Let $\alpha = 0.10$ be the level of significance. Use "220–229," "230–239," and so on as classes.

17. There have been a number of reports in the medical literature that season of birth and incidence of schizophrenia may be related, with a higher proportion of schizophrenics being born during the early months of the year. A recent study (70) following up on this hypothesis looked at 5139 persons born in England or Wales during the years 1921–1955 who were first admitted during the period 1970–1971 to a psychiatric ward in either England or Wales with a diagnosis of schizophrenia. Of these 5139, 1383 were born in the first quarter of a year. Based on census figures in the two countries, the expected number of persons (out of a random 5139) who would be born in the first quarter is 1292.1. Do an appropriate χ^2 test.

18. According to Table 2.1 the proportion of letters in written English that are vowels (A, E, I, O, U, Y) is 0.4021. Make a tally of all the letters on page 2 of this book, classifying them as either consonants or vowels. Do your data support Dewey's 0.4021 : 0.5979 vowel-to-consonant ratio?

19. Two traits that have been widely studied in tomato plants are *height* ("tall" versus "dwarf") and *leaf type* ("cut" versus "potato"). "Tall" and "cut" are dominant. When a homozygous "tall, cut" is crossed with a "dwarf, potato" the resulting progeny is called a dihybrid. (Its phenotype, of course, will be "tall" and "cut.") When dihybrids are crossed, the phenotypes "tall, cut," "tall, potato," "dwarf, cut," and "dwarf, potato" should appear in a 9 : 3 : 3 : 1 ratio, provided the alleles governing the two traits segregate independently. In one experiment done with these two traits a total of 1611 progeny of dihybrid crosses were categorized by phenotype (171).

Phenotype	Frequency
Tall, cut	926
Tall, potato	288
Dwarf, cut	293
Dwarf, potato	104

Test the appropriateness of the $9:3:3:1$ model. Let $\alpha = 0.01$.

20. Is it a reasonable hypothesis that the random sample of size 40 given below came from the following beta pdf?

$$f_Y(y) = 6y(1 - y), \qquad 0 < y < 1.$$

0.18	0.06	0.27	0.58	0.98
0.55	0.24	0.58	0.97	0.36
0.48	0.11	0.59	0.15	0.53
0.29	0.46	0.21	0.39	0.89
0.34	0.09	0.64	0.52	0.64
0.71	0.56	0.48	0.44	0.40
0.80	0.83	0.02	0.10	0.51
0.43	0.14	0.74	0.75	0.22

Divide the data into five classes. Draw the corresponding histogram and the presumed model on the same graph (refer to the remarks on scaling made in Case Study 4.7). Carry out the hypothesis test at the $\alpha = 0.05$ level.

21. Recall Review Exercise 22 at the end of Chapter 3, where it was claimed that in a certain state the length of time, Y, that a person convicted of grand theft auto actually spends in jail is described by the pdf

$$f_Y(y) = \tfrac{1}{9}y^2, \qquad 0 \le y \le 3.$$

As part of a general investigation into the inequities of judicial sentences, a district attorney reviews the records of 50 persons convicted of grand theft auto. Of those 50, 8 served less than one year in prison, 16 served between one and two years, and 26 between two and three years. Are these data compatible with the presumed $f_Y(y)$? Let $\alpha = 0.05$.

22. During the course of a season, a particular baseball player is involved in 200 games in which he has exactly four official at-bats. His distribution of hits during those games is summarized below.

Number of Hits	Number of Games
0	44
1	86
2	56
3	12
4	2

Test the assumption that X, the number of hits the player gets in four at-bats, behaves like a binomial random variable. Let 0.05 be the level of significance.

23. Is the Poisson distribution a suitable model for the alpha-particle data given in Case Study 4.3? Let $\alpha = 0.05$.

24. A sociologist is studying various aspects of the families of preeminent nineteenth-century scholars. A total of 120 of the scholars in her sample had families consisting of two children. The distribution of boys in this sample is summarized below.

Number of boys: 0   1   2
Number of families: 24 64 32

Can it be concluded that the number of boys in such families is adequately described by a binomial pdf? Let $\alpha = 0.05$.

25. To raise money for a new rectory, the members of a church hold a raffle. A total of n tickets are sold (numbered 1 through n), out of which a total of 50 winners are to be drawn, presumably at random. Listed below are the 50 lucky numbers.

108	110	21	6	44
89	68	50	13	63
84	64	69	92	12
46	78	113	104	105
9	115	58	2	20
19	96	28	72	81
32	75	3	49	86
94	61	35	31	56
17	100	102	117	76
106	112	80	59	73

Set up a goodness-of-fit test that focuses on the randomness of the draw. Use the 0.05 level of significance.

26. Carry out the details for a goodness-of-fit test on the tick data of Case Study 4.11. Use the 0.01 level of significance.

27. As a way of studying the spread of a plant disease known as creeping rot, a field of cabbage plants was divided up into 270 *quadrats*, each quadrat containing the same number of plants. Below are listed the numbers of plants per quadrat showing signs of creeping rot infestation.

Number of Infected Plants/Quadrat	Number of Quadrats
0	38
1	57
2	68
3	47
4	23
5	9
6	10
7	7
8	3
9	4
10	2
11	1
12	1
13+	0

Can the number of plants infected with creeping rot per quadrat be described by a Poisson pdf? What might be a physical reason for the Poisson's not being appropriate in this situation? Which assumption of the Poisson appears to be violated?

28. Do a goodness-of-fit test for normality on the motivation data described in Case Study 4.7.

29. In a study (47) investigating the effect of rubella infections (German measles) on childbirth, a total of 578 pregnancies were classified in retrospect as having been either "normal" or "abnormal," the latter group including abortions, stillbirths, birth defects, and all infant deaths within two years. Altogether, there were 86 abnormal pregnancies. The second variable looked at was *when* the rubella infection occurred— during the first trimester or after the first trimester. It was found that 59 of the 86 abnormal births were among the 202 pregnancies complicated during the first trimester; the remaining 27 were born to mothers who contracted the virus after the first trimester. Is the risk of an abnormal birth dependent on when during the pregnancy the virus is contracted? Let $\alpha = 0.01$.

30. Water witching, the practice of using the movements of a forked twig to locate underground water (or minerals), dates back over 400 years. Its first detailed description appears in Agricola's famous *De re metallica*, published in 1556. That water witching works remains a belief widely held among rural people in Europe and throughout the Americas. [In 1960, the number of "active" water witches in the United States was estimated to be in excess of 20,000 (187).] Reliable evidence either supporting or refuting water witching is hard to come by. Personal accounts of isolated successes or failures tend to be strongly biased by the attitude of the observer. The data below show one of the few authenticated records of *all* the wells dug in a given community.

Results of Well Sinkings at "Witched" and
"Nonwitched" Sites in Fence Lake, New Mexico

	Witched	Nonwitched
Successful	24	25
Unsuccessful	5	7

Test whether or not the "success" of a well was independent of how it was located. Let $\alpha = 0.05$.

31. In the United States, suicide rates tend to be much higher for men than for women— at all ages. That pattern may not extend to all professions, however. Death certificates obtained for the 3637 members of the American Chemical Society who died during the period from April 1948 to July 1967 revealed that 106 of the 3522 male deaths were suicides compared to 13 of the 115 female deaths (101). Do an appropriate test. State H_0 and H_1 explicitly and carry out the test at the $\alpha = 0.01$ level of significance.

32. Analyze the data given in Review Exercise 27 at the end of Chapter 8 with a χ^2 test.

33. A market research study has investigated the relationship between an adult's self-perception and his attitude toward small cars. A total of 299 persons living in a large metropolitan area were surveyed. On the basis of their responses to a questionnaire, each was "assigned" to one of three distinct personality types: (1) cautious conservative, (2) middle-of-the-roader, and (3) confident explorer. At the same time, each was solicited for his overall opinion of small cars. The results are displayed below (86).

		Self-Perception		
		Cautious Conservative	Middle-of-the-Roader	Confident Explorer
Opinion of Small Cars	Favorable	79	58	49
	Neutral	10	8	9
	Unfavorable	10	34	42

Test whether these two traits are independent. Use the 0.01 level of significance.

34. In a recent UFO investigation, a total of 1276 "close encounters" were categorized according to (1) the location of the sighting and (2) the nature of the sighting (84).

	Location of Sighting	
	In Spain	Not in Spain
Nature of Sighting Saucer seen on ground	53	705
Saucer seen hovering	38	412
Miscellaneous	9	59

Test whether "location of sighting" and "nature of sighting" are independent. Let $\alpha = 0.05$. Suppose H_0 is accepted. Could that be considered evidence *in favor of* or *against* the existence of UFO's?

CHAPTER TEN

Regression

FRANCIS GALTON (1822–1911)

Galton had earned a Cambridge mathematics degree and completed two years of medical school when his father died, leaving him with a substantial inheritance. Free to travel, he became an explorer of some note, but when *Origin of the Species* was published in 1859, his interests began to shift from geography to statistics and anthropology (Charles Darwin was his cousin). It was Galton's work on fingerprints that made possible their use in human identification. He was knighted in 1909.

10.1 INTRODUCTION

One of the major objectives of all analytical inquiry is determining the relationships among the various components of the object or system under study. If these relationships are sufficiently understood, there is then a basis for intelligent action through prediction and control.

As an example, consider the formidable problem of relating the incidence of cancer to its many possible causes—diet, genetic makeup, pollution, and cigarette smoking, to name only a few. Or think of the Wall Street financier, trying to predict stock prices from a myriad of potentially important factors involving market indices, economic climate, and corporation data.

This chapter discusses some of the basic techniques for measuring relationships between random variables. To be sure, in some situations many variables are involved, in which case the analysis becomes quite complex. Fortunately, most of the fundamental ideas emerge clearly for the case of only *two* variables, and we shall restrict our attention accordingly.

Section 10.2 introduces a measure of relationship between two random variables; the treatment is much in the spirit of Chapter 3 and, indeed, has only that chapter as a prerequisite. Section 10.3 discusses the historical development of the statistical theory of *linear* relationships. A computational technique for determining the "best" linear fit between two variables is given in Section 10.4. The next two sections develop some inference procedures to answer questions raised in Sections 10.2 and 10.4. Then in Section 10.7 we give two convenient tests of the hypothesis that a pair of random variables are independent. The chapter concludes with some words of wisdom on statistical malpractice.

10.2 COVARIANCE AND CORRELATION

With the exception of the tests for independence described in the latter part of Chapter 9, our statistical efforts thus far have been directed at characterizing (with means, variances, pdf's, and so on) the behavior of either an individual random variable, X, or some function of a sample of random variables, $X_1, X_2, \ldots, X_n$, where the X_i's are independent and identically distributed. In this section we go off in a somewhat different direction and look for a *measure of relationship* between two random variables, X and Y, where X and Y are presumably *dependent* and not necessarily identically distributed. The *covariance of X and Y*, as given in Definition 10.1 is one such measure of relationship.

> **DEFINITION 10.1** Let X and Y be random variables with means μ_X and μ_Y. The *covariance of X and Y*, written Cov (X, Y), is given by
> $$\text{Cov}(X, Y) = E[(X - \mu_X)(Y - \mu_Y)].$$

QUESTION 10.2.1   What sort of *XY*-relationships will give rise to large values of Cov (*X, Y*)? to small values of Cov (*X, Y*)? Sketch the appropriate graphs.

Comment. Note that the concept of covariance generalizes the notion of variance, since Cov $(X, X) = $ Var (X). ▮

A sometimes more convenient form for the covariance is given in Theorem 10.1. The proof is trivial, following directly from the distributive property of the expected value.

THEOREM 10.1   For any random variables X and Y with means μ_X and μ_Y

$$\text{Cov}\,(X, Y) = E(XY) - \mu_X\mu_Y.$$

QUESTION 10.2.2   Show that

$$\text{Cov}\,(aX + b, cY + d) = ac\,\text{Cov}\,(X, Y)$$

for any constants *a, b, c,* and *d*.

The statement of the next theorem should come as no real surprise. If X and Y vary independently, it follows that for a given $x - \mu_X$, $Y - \mu_Y$ will sometimes be positive and other times negative, thus precipitating a certain amount of "canceling" among the set of $(x - \mu_X)(y - \mu_Y)$ values. What the theorem states is that the canceling is, in fact, complete.

THEOREM 10.2   If X and Y are independent,

$$\text{Cov}\,(X, Y) = 0.$$

PROOF   If X and Y are independent, $E(XY) = E(X) \cdot E(Y) = \mu_X\mu_Y$. The statement of the theorem follows immediately, then, from Theorem 10.1.

The converse of Theorem 10.2 is not true, the following example being a case in point.

EXAMPLE 10.1.   Consider the sample space $S = \{(-2, 4), (-1, 1), (0, 0), (1, 1), (2, 4)\}$ with each point equally likely. Define the random variable X to be the first component of the sample point chosen, and Y the second. Therefore, $X(-2, 4) = -2$, $Y(-2, 4) = 4$, and so on. It is a simple matter here to show that X and Y are dependent, yet their covariance is 0. The former is true because

$$\tfrac{1}{5} = P(X = 1, Y = 1) \neq P(X = 1) \cdot P(Y = 1) = \tfrac{1}{5} \cdot \tfrac{2}{5} = \tfrac{2}{25}.$$

To verify the latter, note that $E(XY) = [(-8) + (-1) + 0 + 1 + 8] \cdot \tfrac{1}{5} = 0$, $E(X) = 0$, and $E(Y) = [4 + 1 + 0 + 1 + 4] \cdot \tfrac{1}{5} = 2$. Therefore, Cov $(X, Y) = 0 - 0 \cdot 2 = 0$.

QUESTION 10.2.3   Let U be a random variable uniformly distributed over $[0, 2\pi]$. Define $X = \cos U$ and $Y = \sin U$. Show that X and Y are dependent but that Cov $(X, Y) = 0$.

Theorem 3.13 gave a formula for the variance of a sum of independent random variables. With the notion of covariance, we can generalize that result to include *dependent* random variables.

> **THEOREM 10.3** Let $X_1, X_2, \ldots, X_n$ be any set of random variables. Let $X = X_1 + X_2 + \cdots + X_n$. Then
>
> $$\text{Var}(X) = \sum_{i=1}^{n} \text{Var}(X_i) + 2 \sum_{j<k} \text{Cov}(X_j, X_k).$$

PROOF   The proof is a straightforward exercise in elementary algebra:

$$\text{Var}(X) = E(X^2) - [E(X)]^2$$

$$= E\left[\left(\sum_{i=1}^{n} X_i\right)^2\right] - \left[\sum_{i=1}^{n} E(X_i)\right]^2$$

$$= E\left(\sum_{i=1}^{n} X_i^2 + 2 \sum_{j<k} X_j X_k\right) - \left[\sum_{i=1}^{n} [E(X_i)]^2 + 2 \sum_{j<k} E(X_j)E(X_k)\right]$$

$$= \sum_{i=1}^{n} [E(X_i^2) - [E(X_i)]^2] + 2 \sum_{j<k} [E(X_j X_k) - E(X_j)E(X_k)]$$

$$= \sum_{i=1}^{n} \text{Var}(X_i) + 2 \sum_{j<k} \text{Cov}(X_j, X_k).$$

QUESTION 10.2.4   Let X, Y, and Z be random variables. Show that
$$\text{Cov}(X + Y, Z) = \text{Cov}(X, Z) + \text{Cov}(Y, Z).$$

EXAMPLE 10.2.   We will use Theorem 10.3 to calculate the variance of a hypergeometric random variable. As a physical model, consider an urn containing N chips, r red and w white $(r + w = N)$. A sample of n is to be selected *without* replacement. Let Y denote the number of red chips drawn. From Theorem 2.13,

$$P(Y = y) = \frac{\binom{r}{y}\binom{w}{n - y}}{\binom{N}{n}}, \qquad y = 0, 1, \ldots, \min(r, n). \tag{10.1}$$

Although the moments of Y could be gotten directly from the fundamental definitions, it is more instructive to introduce auxiliary random variables, as we did in the case of a binomial variable. Define

$$Y_i = \begin{cases} 1, & \text{if the } i\text{th chip is red} \\ 0, & \text{otherwise.} \end{cases}$$

Then $Y = Y_1 + Y_2 + \cdots + Y_n$. It is easy to see that $E(Y_i) = r/N$ and $E(Y) = n(r/N) = np$, where $p = r/N$. Thus, $E(Y)$ is the same whether the sampling is done *with* replacement or *without* replacement.

For the variance, though, the mode of sampling makes a difference. The Y_i's are not independent; however, by symmetry, they are identically distributed. Also, any pair $(Y_j, Y_k), j \neq k$, has the same distribution as (Y_1, Y_2), again by symmetry. [There is no reason to suppose it is more (or less) likely to get red chips on the third and fifth draws than on the first and second.] Since $Y_i^2 = Y_i$, $E(Y_i^2) = E(Y_i) = r/N$, and $\mathrm{Var}(Y_i) = r/N - (r/N)^2 = (r/N)(1 - r/N) = p(1 - p)$. For any $j \neq k$, $\mathrm{Cov}(Y_j, Y_k) = \mathrm{Cov}(Y_1, Y_2) = E(Y_1 Y_2) - E(Y_1)E(Y_2)$. But $E(Y_1 Y_2) = 1 \cdot P(Y_1 Y_2 = 1) = 1 \cdot P(\text{first chip is red and second chip is red}) = r/N \cdot (r - 1)/(N - 1)$, so

$$\mathrm{Cov}(Y_j, Y_k) = \frac{r}{N} \cdot \frac{r-1}{N-1} - \frac{r}{N} \cdot \frac{r}{N} = \frac{r}{N}\left(\frac{r-1}{N-1} - \frac{r}{N}\right) = -\frac{r}{N} \cdot \frac{N-r}{N} \cdot \frac{1}{N-1}.$$

Finally, applying Theorem 10.3, we get an expression for $\mathrm{Var}(Y)$:

$$\mathrm{Var}(Y) = \sum_{i=1}^{n} \mathrm{Var}(Y_i) + 2 \sum_{j<k} \mathrm{Cov}(Y_j, Y_k)$$

$$= np(1 - p) - 2\binom{n}{2}p(1 - p) \cdot \frac{1}{N-1}$$

$$= np(1 - p) \cdot \frac{N-n}{N-1}.$$

Recall that the variance for sampling *with* replacement is $np(1 - p)$, so the not-surprising result emerges that it is more efficient to estimate $p = r/N$ by sampling without replacement (particularly when n is relatively large compared to N).

QUESTION 10.2.5   Find $\mathrm{Var}(Y)$ for the hypergeometric pdf of Equation 10.1 *directly*. Hint: Find the *factorial moment*, $E[Y(Y - 1)]$, and write

$$\mathrm{Var}(Y) = E[Y(Y - 1)] + E(Y) - [E(Y)]^2.$$

For historical reasons, and because a measure of relationship should be dimensionless, the covariance is often normalized by dividing out the X and Y standard deviations. In this form, the quantity is called the *correlation coefficient*.

DEFINITION 10.2   Let X and Y be any two random variables. The *correlation coefficient of X and Y* is denoted $\rho(X, Y)$ and is given by

$$\rho(X, Y) = \frac{\mathrm{Cov}(X, Y)}{\sigma_X \sigma_Y} = \mathrm{Cov}(X^*, Y^*),$$

where $X^* = (X - \mu_X)/\sigma_X$ and $Y^* = (Y - \mu_Y)/\sigma_Y$.

It is, of course, immediate that if X and Y are independent, $\rho(X, Y) = 0$. The converse is not true (recall Example 10.1, where $Y = X^2$). However, even though ρ is not a good indicator of functional relationships *in general*, it does recognize *linear* dependence, as Theorem 10.4 demonstrates.

THEOREM 10.4   For any two random variables X and Y,
  (a) $|\rho(X, Y)| \leq 1$.
  (b) $|\rho(X, Y)| = 1$ if and only if $Y = aX + b$ for some constants a and b (except possibly on a set of probability zero).

PROOF   Following the notation of Definition 10.2, let X^* and Y^* denote the normalized transforms of X and Y. Then

$$0 \leq \text{Var}\,(X^* \pm Y^*) = \text{Var}\,(X^*) \pm 2\,\text{Cov}\,(X^*, Y^*) + \text{Var}\,(Y^*)$$
$$= 1 \pm 2\rho(X, Y) + 1$$
$$= 2[1 \pm \rho(X, Y)].$$

But $1 \pm \rho(X, Y) \geq 0$ implies $|\rho(X, Y)| \leq 1$, and the first part of the theorem is proved.

Next, suppose $\rho(X, Y) = 1$. Then $\text{Var}\,(X^* - Y^*) = 0$; however, a random variable with zero variance is constant, except possibly on a set of probability 0. From the constancy of $X^* - Y^*$, it readily follows that Y is a linear function of X. The case for $\rho(X, Y) = -1$ is similar.

The converse of (b) is left as an exercise.

QUESTION 10.2.6   Suppose X and Y have the joint pdf,

$$f_{X,Y}(x, y) = x + y, \qquad 0 < x < 1, 0 < y < 1.$$

Find $\rho(X, Y)$.

We conclude this section with an estimation problem whose full significance will become apparent in Section 10.3. Suppose the outcomes of an experiment are n measurements made on (X, Y)—that is, $(X_1, Y_1), (X_2, Y_2), \ldots, (X_n, Y_n)$. How should we then *estimate* $\rho(X, Y)$?

Since the correlation coefficient can be written in terms of various theoretical moments,

$$\rho(X, Y) = \frac{E(XY) - E(X)E(Y)}{\sqrt{\text{Var}\,(X)}\,\sqrt{\text{Var}\,(Y)}},$$

it would seem reasonable to estimate each component of $\rho(X, Y)$ with its corresponding *sample* moment. That is, we will take $\bar{X}$ and $\bar{Y}$ to be estimators of $E(X)$ and $E(Y)$, replace $E(XY)$ with

$$\frac{1}{n}\sum_{i=1}^{n} X_i Y_i,$$

and substitute

$$\frac{1}{n}\sum_{i=1}^{n} (X_i - \bar{X})^2 \quad \text{and} \quad \frac{1}{n}\sum_{i=1}^{n} (Y_i - \bar{Y})^2$$

for $\text{Var}\,(X)$ and $\text{Var}\,(Y)$. Putting these together gives R, the *sample correlation coefficient*:

$$R = \frac{\dfrac{1}{n}\sum_{i=1}^{n} X_i Y_i - \bar{X}\bar{Y}}{\sqrt{\dfrac{1}{n}\sum_{i=1}^{n} (X_i - \bar{X})^2}\,\sqrt{\dfrac{1}{n}\sum_{i=1}^{n} (Y_i - \bar{Y})^2}}. \tag{10.2}$$

For computational purposes, it is better to express R in a slightly different way:

$$R = \frac{n \sum_{i=1}^{n} X_i Y_i - \left(\sum_{i=1}^{n} X_i \right) \left(\sum_{i=1}^{n} Y_i \right)}{\sqrt{n \sum_{i=1}^{n} X_i^2 - \left(\sum_{i=1}^{n} X_i \right)^2} \sqrt{n \sum_{i=1}^{n} Y_i^2 - \left(\sum_{i=1}^{n} Y_i \right)^2}}. \tag{10.3}$$

QUESTION 10.2.7   Derive Equation 10.3 from Equation 10.2.

CASE STUDY

10.1

On Wall Street, thousands of highly trained analysts use everything from crystal balls to multimillion-dollar computers to predict the price of stocks, bonds, and other financial commodities. While the actual number of variables involved in these predictions is enormous, the kind of reasoning that goes on can be illustrated by looking at just a few simple indicators of general stock market behavior. Two of the most familiar of these are the Dow Jones average and Standard and Poor's stock price index. Since both of these purport to measure the same quantity, "the state of the market," they should be expected to be highly correlated.

The Dow Jones Composite Average is derived from the daily quotations of some 65 stocks listed on the New York Stock Exchange. The Standard and Poor's "500" Composite Price Index is an average of daily prices from a large sample of stocks, where the index is expressed relative to base values calculated in 1941–1943. Table 10.1 gives for each of these the week's end closing average or index for the months of January, February, and March 1978. Also quoted are the prices for one stock (Eastman Kodak Common) that is used in determining the Dow Jones average and one stock (Eaton Corporation Common) that is not (32). (Our "expectation" would be that the correlation will be highest for the Standard and Poor's index, next highest for the Eastman Kodak prices, and lowest for the Eaton Corporation figures).

Let Standard and Poor's index be the X-variable and Dow Jones the Y. Then

TABLE 10.1   End of week closing price, January–March 1978

Week	Dow Jones Composite	Standard and Poor's Composite	Eastman Kodak Common	Eaton Common
1	276.61	91.62	49.75	36.00
2	271.26	89.69	48.50	33.87
3	272.47	89.89	48.87	34.75
4	268.35	88.58	45.75	33.75
5	271.44	89.62	45.12	34.12
6	272.35	90.08	45.25	34.87
7	263.86	87.96	43.37	34.37
8	265.07	88.49	43.75	34.00
9	262.15	87.45	41.62	33.50
10	265.38	88.88	43.50	33.87
11	269.41	90.20	43.25	34.00
12	267.01	89.36	42.12	34.25
13	266.94	89.21	42.25	34.75

$$\sum_{i=1}^{13} x_i = 1161.03, \qquad \sum_{i=1}^{13} x_i^2 = 103{,}705.57,$$

$$\sum_{i=1}^{13} y_i = 3492.3, \qquad \sum_{i=1}^{13} y_i^2 = 938{,}367.32,$$

$$\sum_{i=1}^{13} x_i y_i = 311{,}946.5,$$

and, from Equation 10.3,

$$r = \frac{13(311{,}946.5) - (1161.03)(3492.3)}{\sqrt{13(103{,}705.57) - (1161.03)^2}\,\sqrt{13(938{,}367.32) - (3492.3)^2}}$$

$$= 0.93.$$

Similarly, the sample correlation coefficient between Eastman Kodak Common and the Dow Jones average comes to 0.83, and between Eaton Corporation and Dow Jones, 0.69.

Comment. Despite the statement of Theorem 10.4, the interpretation of r is far from obvious. What does it mean to say that one pair of variables has a sample correlation coefficient of, say, 0.93, and another, 0.69? How much stronger is one relationship than the other? Does it follow that either or both sets of variables are dependent? We will defer such questions to later sections. Here, we are simply illustrating a computing formula.   ∎

QUESTION 10.2.8   When two closely related species are crossed, the progeny will tend to have physical traits that lie somewhere "between" the corresponding traits of the parents. But what about behavioral traits? Are they similarly "mixed"? And does a hybrid whose physical traits favor one particular parent tend to have behavioral patterns that resemble those of the same parent? One attempt at answering these questions was an experiment done with mallard and pintail ducks. A total of 11 males were studied; all were second-generation crosses. A rating scale was devised that measured the extent to which the plumage of each of the ducks resembled the plumage of the first generation's parents. A score of 0 indicated that the hybrid had the same appearance (phenotype) as a pure mallard; a score of 20 meant that the hybrid looked like a pintail. Similarly, certain behavioral traits were quantified and a second scale was constructed that ranged from 0 (completely mallardlike) to 15 (completely pintaillike). Table 10.2

TABLE 10.2   Behavioral and plumage indices
for 11 mallard × pintail hybrids

Male	Plumage Index (x)	Behavioral Index (y)
R	7	3
S	13	10
D	14	11
F	6	5
W	14	15
K	15	15
U	4	7
O	8	10
V	7	4
J	9	9
L	14	11

shows the results (163). Plot these data and compute their sample correlation coefficient.

10.3 SIMPLE LINEAR REGRESSION— A HISTORICAL APPROACH

Our objective, as stated at the outset of this chapter, is to examine the relationship between two random variables. What we have accomplished in the way of preliminaries is the definition of the correlation coefficient, a quantity related to the notion of independence. We now need to look at the question of functional relationships in more detail and, in particular, at the role that $\rho(X, Y)$ plays in their interpretation.

First, a basic question: given two presumably dependent random variables, how should we go about studying their joint behavior? One approach would be to consider the conditional pdf of Y given x as a function of x—that is, make a graph of $f_{Y|x}(y)$ versus x.

EXAMPLE 10.3.   Let X and Y have the joint density of Example 3.8:

$$f_{X,Y}(x, y) = y^2 e^{-y(x+1)}, \qquad x \geq 0, y \geq 0.$$

The marginal pdf for X has already been derived:

$$f_X(x) = \frac{2}{(x + 1)^3}.$$

Therefore, by Definition 3.8 the conditional pdf of Y given x is

$$f_{Y|x}(y) = \frac{y^2 e^{-y}(x + 1)}{\dfrac{2}{(x + 1)^3}} = \frac{1}{2}(x + 1)^3 y^2 e^{-y(x+1)}.$$

Figure 10.1 shows a sketch of $f_{Y|x}(y)$ versus x for the particular x-values 1, 3, and 5.

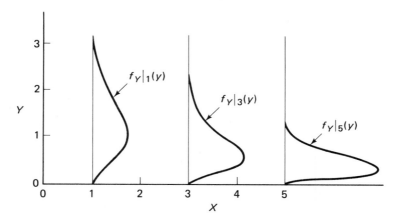

Figure 10.1   Sketch of $f_{Y|x}(y)$ versus x.

QUESTION 10.3.1   Suppose X and Y are two discrete random variables with joint pdf

$$f_{X,Y}(x, y) = \frac{x + y}{15}, \qquad x = 1, 2, 3; y = 0, 1$$

$$= 0, \qquad \text{elsewhere.}$$

Graph $f_{Y|x}(y)$ versus x for $x = 1, 2,$ and 3.

While graphs of $f_{Y|x}(y)$ versus x *do*, indeed, show the relationship between X and Y, they go about it in a rather cumbersome way, as Example 10.3 should have made clear. A more tractable solution would be to replace $f_{Y|x}(y)$ with one of its numerical descriptors—say, $E(Y|x)$—and graph $E(Y|x)$ versus x (see Review Exercise 65 at the end of Chapter 3).

EXAMPLE 10.4.   For the conditional pdf of Example 10.3,

$$E(Y|x) = \int_0^\infty y \cdot \frac{1}{2}(x + 1)^3 y^2 e^{-y(x+1)} \, dy$$

$$= \frac{1}{2(x+1)} \int_0^\infty u^3 e^{-u} \, du,$$

the latter expression resulting from the substitution $u = y(x + 1)$. But

$$\int_0^\infty u^3 e^{-u} \, du = \Gamma(4) = 3! = 6,$$

so the conditional expectation reduces to

$$E(Y|x) = \frac{3}{x+1}.$$

(See Figure 10.2.)

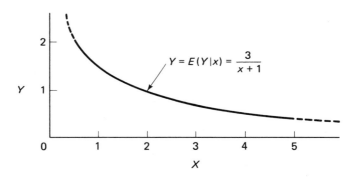

Figure 10.2   A graph of $E(Y|x)$ versus x.

QUESTION 10.3.2   Let X and Y have the joint pdf

$$f_{X,Y}(x, y) = \begin{cases} x + y, & 0 < x < 1, 0 < y < 1 \\ 0, & \text{elsewhere.} \end{cases}$$

Plot $E(Y|x)$ versus x. Is it linear?

In more general terminology, the search for a relationship between $E(Y|x)$ and x is known as the *regression problem* (and the function $h(x) = E(Y|x)$ is referred to as the *regression curve of Y on X*). Experimentally, problems of this sort arise when a researcher records a set of (x_i, y_i) pairs, plots the points on a scatter diagram (Figure 10.3), and then asks for an equation that adequately

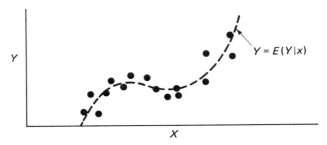

Figure 10.3   Scatter diagram.

describes the functional relationship between the two variables. What the equation is seeking to approximate, of course, is $E(Y|x)$ versus x.

For the remainder of this chapter we will confine our attention to one particular regression problem, the special case where $E(Y|x)$ is a *linear* function of x: $E(Y|x) = a + bx$. Although this is admittedly a very simple data model, it nevertheless occurs quite often in the real world. Furthermore, a number of phenomena whose relationships are not initially linear can be suitably transformed in such a way that the results for linear relationships remain valid. We present an example of this sort in the next section. Finally, many relationships that are not linear in their entirety are still, to a good approximation, linear over some interval of interest. The linear model, then, while simple, is not unrealistic.

Francis Galton, the renowned British biologist and scientist, perhaps more than any other person was responsible for launching *regression analysis* as a worthwhile field of statistical inquiry. Galton was a redoubtable data analyst, whose keen insight enabled him to intuit much of the basic mathematical structure that we now associate with the regression problem. One of his more famous endeavors was an examination of the relationship between parents' heights and their adult children's heights. Table 10.3 shows one of his sets of data (56). Galton used as a measure of parent height the average of the paternal height and 1.08 times the maternal height; this quantity was called the *mid-parent height*.

From cross-tabulations of this sort, Galton made a number of critically important discoveries bearing on the general problem of regression. If X denotes a mid-parent height and Y a child's height, then:

1. The marginal pdf's of both X and Y are normal.
2. $E(Y|x)$ is a linear function of x (see Figure 10.4).
3. Var $(Y|x)$ is constant in x (this property has the marvelous name, *homoscedasticity*).

Sir Francis performed still one other feat of empirical legerdemain on Table 10.3. He replaced any four adjacent figures in the table by their average, which he placed

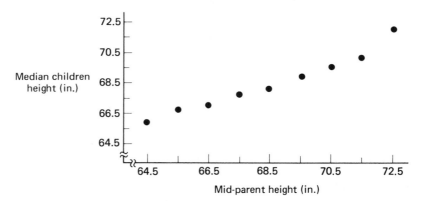

Figure 10.4 Children's heights versus parents' heights.

TABLE 10.3 Number of adult children of various statures born of 205 mid-parents of various statures (All female heights have been multiplied by 1.08)

Height of the Mid-Parents in Inches	Heights of the Adult Children														Total Number of		Medians or Values of M
	Below	62.2	63.2	64.2	65.2	66.2	67.2	68.2	69.2	70.2	71.2	72.2	73.2	Above	Adult Children	Mid-Parents	
Above 72.5	—	—	—	—	—	—	—	—	—	—	1	3	—	—	4	5	—
72.5	—	—	—	—	—	1	2	1	2	7	2	4	—	—	19	6	72.2
71.5	—	—	—	1	3	4	3	5	10	4	9	2	2	—	43	11	69.9
70.5	1	—	1	1	3	12	18	14	7	4	3	3	—	—	68	22	69.5
69.5	—	1	16	4	17	27	20	33	25	20	11	4	5	—	183	41	68.9
68.5	1	—	7	11	16	25	31	34	48	21	18	4	3	—	219	49	68.2
67.5	—	3	5	14	15	36	38	28	38	19	11	—	—	—	211	33	67.6
66.5	—	3	3	5	2	17	17	14	13	4	—	—	—	—	78	20	67.2
65.5	1	—	9	5	7	11	11	7	7	5	2	1	—	—	66	12	66.7
64.5	1	1	4	4	1	5	5	—	2	—	—	—	—	—	23	5	65.8
Below	1	—	2	4	1	2	2	1	1	—	—	—	—	—	14	1	—
Totals	5	7	32	59	48	117	138	120	167	99	64	41	17	14	928	205	
Medians	—	—	66.3	67.8	67.9	67.7	67.9	68.3	68.5	69.0	69.0	70.0	—	—			

at the center of the four figures. Taking this average as the value of the joint density function at that point, he proceeded to draw contour lines—smooth curves connecting points of like value. From these he drew a truly remarkable conclusion:

4. The set of points in the plane for which $f_{X,Y}(x, y) = c$ is an ellipse for each constant $c > 0$. Furthermore, these ellipses are concentric.[1] [See Figure 10.5, reproduced from (126). Also, see (206) for a fuller discussion of the smoothing procedure.]

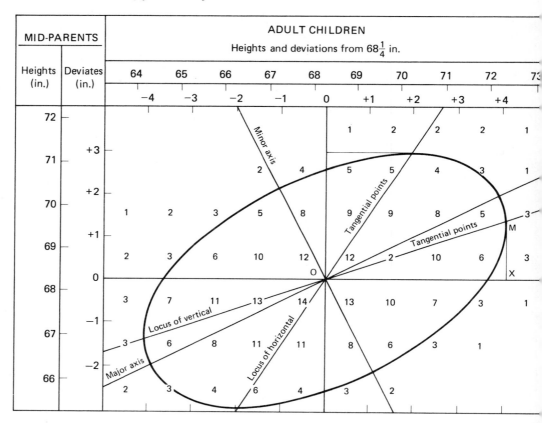

Figure 10.5 Galton's contour lines.

Galton (55) gave the following recollection of the birth of property 4:

At length, one morning, while waiting at a roadside station near Ramsgate for a train, and poring over the diagram in my notebook, it struck me that the lines of equal frequency ran in concentric ellipses. The cases were too few for my

[1]Of this discovery, Pearson said, "That Galton should have evolved all this from his observations is to my mind one of the most noteworthy scientific discoveries arising from pure analysis of observations" (126).

certainty, but my eye, being accustomed to such things, satisfied me that I was approaching the solution. More careful drawings strongly corroborated the first impression.

Galton was curious as to the nature of the probability density functions that would possess these four properties, and he posed the question to J. D. H. Dickson, Tutor at St. Peter's College, Cambridge. Dickson readily supplied the answer, which we will consider in Section 10.6.

To conclude this section, we give a theorem showing the prominent role the correlation coefficient plays in linear regression.

THEOREM 10.5 If X and Y are random variables such that $E(Y|x) = a + bx$ for all values x taken on by X, then

$$E(Y|x) = \mu_Y + \frac{\rho(X, Y)\sigma_Y}{\sigma_X}(x - \mu_X).$$

PROOF The proof for continuous variables is given; for the discrete case, replace the integrals by summations.

First, we note that

$$\int_{-\infty}^{\infty} E(Y|x)f_X(x)\,dx = \mu_Y. \tag{10.4}$$

To see this, write the left-hand side as

$$\int_{-\infty}^{\infty} E(Y|x)f_X(x)\,dx = \int_{-\infty}^{\infty} \left[\int_{-\infty}^{\infty} y\,\frac{f_{X,Y}(x, y)}{f_X(x)}\,dy \right] \cdot f_X(x)\,dx$$

$$= \int_{-\infty}^{\infty} y \left[\int_{-\infty}^{\infty} f_{X,Y}(x, y)\,dx \right] dy,$$

with the latter expression arising from a change in the order of integration. But

$$\int_{-\infty}^{\infty} y \left[\int_{-\infty}^{\infty} f_{X,Y}(x, y)\,dx \right] dy = \int_{-\infty}^{\infty} y f_Y(y)\,dy = \mu_Y.$$

(The result in Equation 10.4 is not in the least surprising; think of what it says intuitively.)

Next, we multiply the equation $E(Y|x) = a + bx$ by $f_X(x)$ and integrate (with respect to x):

$$\int_{-\infty}^{\infty} E(Y|x)f_X(x)\,dx = \int_{-\infty}^{\infty} a f_X(x)\,dx + \int_{-\infty}^{\infty} bx f_X(x)\,dx,$$

in which case

$$\mu_Y = a + b\mu_X \quad \text{or} \quad a = \mu_Y - b\mu_X.$$

Therefore,

$$E(Y|x) = \mu_Y + b(x - \mu_X).$$

Now, transpose μ_Y and multiply by $(x - \mu_X)f_X(x)$:

$$[E(Y|x) - \mu_Y](x - \mu_X)f_X(x) = b(x - \mu_X)^2 f_X(x).$$

An equivalent expression (the reader should verify this) is

$$\int_{-\infty}^{\infty} (y - \mu_Y)(x - \mu_X)f_{X,Y}(x, y)\, dy = b(x - \mu_X)^2 f_X(x). \qquad (10.5)$$

If x is integrated out, Equation 10.5 becomes

$$\text{Cov}\,(X, Y) = b\sigma_X^2$$

or

$$b = \frac{\text{Cov}\,(X, Y)}{\sigma_X^2} = \frac{\rho(X, Y)\sigma_Y}{\sigma_X},$$

thus completing the proof.

If x is considered to be the independent variable and $E(Y\,|\,x)$ the dependent variable, then

$$E(Y\,|\,x) = \mu_Y + \frac{\rho(X, Y)\sigma_Y}{\sigma_X}(x - \mu_X) \qquad (10.6)$$

is the equation of a straight line through the point (μ_X, μ_Y). Equations of this sort are known as *regression lines*.

Comment. The term "regression line" derives from a consequence of Equation 10.6. Suppose we make the simplifying assumption that $\mu_X = \mu_Y = \mu$ and $\sigma_X = \sigma_Y$. Then Equation 10.6 becomes

$$E(Y\,|\,x) - \mu = \rho(X, Y)(x - \mu).$$

But recall that $|\rho(X, Y)| \le 1$—and, in this case, $0 < \rho(X, Y) < 1$. Here the positive sign of $\rho(X, Y)$ tells us that, on the average, tall parents have tall children. However, $\rho(X, Y) < 1$ means (again, *on the average*) that the children's heights are closer to the mean than the parents': a group of parents with mid-height, say, one inch above μ will have children with average height above μ but by less than one inch. Galton called this phenomenon "regression to mediocrity." ▌

10.4 THE METHOD OF LEAST SQUARES

The discussion of any statistical topic generally divides into two parts. First, there is the examination of appropriate models and theories; then comes the application of those results to data. In the regression context, the same duality arises. Suppose there is assumed a linear relationship between X and $E(Y\,|\,x)$—that is, $E(Y\,|\,x) = a + bx$. Given a set of data, what, then, is the "best" choice for the coefficients a and b, or, in the terminology of Theorem 10.5, for

$$\mu_Y - \frac{\rho(X, Y)\sigma_Y}{\sigma_X}\mu_X \quad \text{and} \quad \frac{\rho(X, Y)\sigma_Y}{\sigma_X}?$$

Comment. The technique we explore here seems to have been published first by Legendre in 1805, but in 1809 Gauss claimed he had already derived it as early as 1795. ∎

We will phrase this question in purely geometrical terms, and in more generality than is absolutely necessary for our purposes. The reader may recognize that, in other contexts, what is being posed would be called an interpolation problem.

PROBLEM   Given points (x_1, y_1), (x_2, y_2), . . . , (x_n, y_n) in the plane and a positive integer m, find the polynomial of degree m whose graph is "closest" to the given points.

Note that by dropping any reference to probabilistic notions, we subsume a variety of problems. For instance, both the x_i's and y_i's may be values of random variables. An example would be a study set up to investigate the clinical course of nephritis by taking a random sample of nephritis patients and measuring, for each one, blood pressure (x_i) and heart weight (y_i). On the other hand, it may be better, medically, to preselect certain critical blood pressures, choose patients with those specific x-values, and then measure the corresponding heart weights. If that were the protocol followed, only Y would be a random variable. Still, the differences in the nature of X and Y notwithstanding, the equation we are looking for would be the same.

Comment. The price paid for not imposing any assumptions on X and Y is the lack of any statistical measure of the quality of the estimators we will ultimately derive. Nevertheless, the technique is extremely useful. ∎

Suppose the desired polynomial $p(x)$ is written

$$\sum_{k=0}^{m} a_k x^k,$$

where $a_0, a_1, \ldots, a_m$ are to be determined. The *method of least squares* chooses as "solutions" those coefficients minimizing the sum of the squares of the vertical distances from the data points to the presumed polynomial. That is, the polynomial we will term "best" is the one whose coefficients minimize the function L, where

$$L = \sum_{i=1}^{n} [y_i - p(x_i)]^2.$$

Here, we will treat only the linear case, where $p(x) = a + bx$. The procedure for higher-order polynomials is identical, although the computations become much more tedious.

THEOREM 10.6 Given n points $(x_1, y_1), \dots, (x_n, y_n)$, the straight line $y = a + bx$ minimizing

$$L = \sum_{i=1}^{n} [y_i - (a + bx_i)]^2$$

has slope

$$b = \frac{n \sum_{i=1}^{n} x_i y_i - \left(\sum_{i=1}^{n} x_i\right)\left(\sum_{i=1}^{n} y_i\right)}{n \left(\sum_{i=1}^{n} x_i^2\right) - \left(\sum_{i=1}^{n} x_i\right)^2}$$

and y-intercept

$$a = \frac{\sum_{i=1}^{n} y_i - b \sum_{i=1}^{n} x_i}{n}.$$

PROOF

The proof is accomplished by the usual device of taking the partial derivatives of L with respect to a and with respect to b, setting the resulting expressions equal to 0, and solving. By the first step, we get

$$\frac{\partial L}{\partial a} = \sum_{i=1}^{n} (-2)[y_i - (a + bx_i)]$$

and

$$\frac{\partial L}{\partial b} = \sum_{i=1}^{n} (-2)x_i[y_i - (a + bx_i)].$$

Now, set the right-hand sides of $\partial L/\partial a$ and $\partial L/\partial b$ equal to 0 and simplify. This gives

$$na + \left(\sum_{i=1}^{n} x_i\right)b = \sum_{i=1}^{n} y_i$$

and

$$\left(\sum_{i=1}^{n} x_i\right)a + \left(\sum_{i=1}^{n} x_i^2\right)b = \sum_{i=1}^{n} x_i y_i.$$

An application of Cramer's rule gives the solution for b stated in the theorem. The expression for a follows immediately.

Comment. Consider the random variables associated with the least-squares estimates,

$$B = \frac{n \sum_{i=1}^{n} X_i Y_i - \left(\sum_{i=1}^{n} X_i\right)\left(\sum_{i=1}^{n} Y_i\right)}{n \sum_{i=1}^{n} X_i^2 - \left(\sum_{i=1}^{n} X_i\right)^2}$$

$$= \frac{\frac{1}{n} \sum_{i=1}^{n} X_i Y_i - \bar{X}\bar{Y}}{\frac{1}{n} \sum_{i=1}^{n} (X_i - \bar{X})^2}$$

and

$$A = \bar{Y} - B\bar{X}.$$

Compare these to the coefficients appearing in the statement of Theorem 10.5. For example,

$$B = R \cdot \frac{\sqrt{\dfrac{1}{n} \sum\limits_{i=1}^{n} (Y_i - \bar{Y})^2}}{\sqrt{\dfrac{1}{n} \sum\limits_{i=1}^{n} (X_i - \bar{X})^2}} = R \cdot \frac{\sqrt{\dfrac{1}{n-1} \sum\limits_{i=1}^{n} (Y_i - \bar{Y})^2}}{\sqrt{\dfrac{1}{n-1} \sum\limits_{i=1}^{n} (X_i - \bar{X})^2}} = R \cdot \frac{S_Y}{S_X}.$$

Thus, even though the least-squares line is derived without regard to distributions, it still gives rise to comfortably familiar estimators. ∎

CASE STUDY

10.2

A manufacturer of air conditioning units was having assembly problems due to the failure of a connecting rod to meet finished-weight specifications. Too many rods were being completely finished, then rejected as overweight. To reduce this cost, the manufacturer sought to estimate the relationship between the weight of the finished rod and that of the rough casting. Castings likely to produce rods that were too heavy could then be discarded before undergoing the final tooling process.

To estimate the $E(Y|x)$-versus-x relationship, a total of 25 (x_i, y_i) pairs were measured, x_i being the weight of the ith rough casting and y_i the weight of the finished rod (137). From these data (Table 10.4) the sums needed to evaluate a and b are easily obtained:

$$\sum_{i=1}^{25} x_i = 66.075, \qquad \sum_{i=1}^{25} x_i^2 = 174.673,$$

$$\sum_{i=1}^{25} y_i = 50.12, \qquad \sum_{i=1}^{25} y_i^2 = 100.499,$$

$$\sum_{i=1}^{25} x_i y_i = 132.491.$$

Therefore,

$$b = \frac{25(132.491) - (66.075)(50.12)}{25(174.673) - (66.075)^2} = 0.648$$

and

$$a = \frac{50.12 - 0.648(66.075)}{25} = 0.292,$$

making the least-squares line

$$y = 0.292 + 0.648x. \qquad (10.7)$$

TABLE 10.4 Rough and finished rod weights

Rod Number	Rough Weight	Finished Weight	Rod Number	Rough Weight	Finished Weight
1	2.745	2.080	14	2.635	1.990
2	2.700	2.045	15	2.630	1.990
3	2.690	2.050	16	2.625	1.995
4	2.680	2.005	17	2.625	1.985
5	2.675	2.035	18	2.620	1.970
6	2.670	2.035	19	2.615	1.985
7	2.665	2.020	20	2.615	1.990
8	2.660	2.005	21	2.615	1.995
9	2.655	2.010	22	2.610	1.990
10	2.655	2.000	23	2.590	1.975
11	2.650	2.000	24	2.590	1.995
12	2.650	2.005	25	2.565	1.955
13	2.645	2.015			

Figure 10.6 is a graph of Equation 10.7 superimposed over the data of Table 10.4.

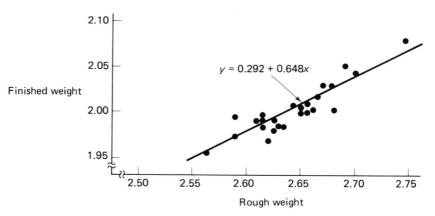

Figure 10.6 Regression line for rough and finished rod weights.

It follows that if the weight of a rough casting is, say, 2.71, the estimate of its finished weight would be 2.05 (see Figure 10.6):

$$\text{estimated weight} = E(Y \mid 2.71)$$
$$= 0.292 + 0.648(2.71)$$
$$= 2.05.$$

If an expected value of 2.05 is judged to be too heavy, then the rough casting should be discarded.

QUESTION 10.4.1   Crickets make their chirping sound by sliding one wing cover very rapidly back and forth over the other. Biologists have long been aware that there is a linear relationship between *temperature* and the frequency with which a cricket chirps, although the slope and *y*-intercept of the relationship varies from species to species. Listed in Table 10.5 are 15 frequency-temperature observations recorded for the striped ground

TABLE 10.5 Measurements taken on temperature
and chirping frequency relationship

Observation Number	Chirps Per Second (x_i)	Temperature, °F (y_i)
1	20.0	88.6
2	16.0	71.6
3	19.8	93.3
4	18.4	84.3
5	17.1	80.6
6	15.5	75.2
7	14.7	69.7
8	17.1	82.0
9	15.4	69.4
10	16.2	83.3
11	15.0	79.6
12	17.2	82.6
13	16.0	80.6
14	17.0	83.5
15	14.4	76.3

cricket, *Nemobius fasciatus fasciatus* (130). Plot these data and find the equation of the least-squares line, $y = a + bx$. Suppose a cricket of this species is observed to chirp 18 times per second. What would be the estimated temperature?

The next case study illustrates why the linear model is not so restrictive as it might initially appear. Transformations similar to the one described in the study can be used to induce linearity in quite a variety of "nonlinear" models. Some of these will be brought up in the exercises.

CASE STUDY

10.3   The used car market is one of keen interest to most Americans, whether as buyers or sellers. Statistically, the market gives rise to any number of regression-type problems. For example, it would certainly be useful to know how the value of a car declines as a function of its age. What kind of function, though, would provide a suitable model?

A first guess might be that the relationship is linear. We will use the least-squares technique to find the best linear model for the data given in Table 10.6, where the second column lists the 1957 national average prices of used cars one to ten years old (24).

TABLE 10.6   1957 national average used car prices

Age	Average Price	Linear Model	Exponential Model
1	$2651	$2116	$2609
2	1943	1861	1938
3	1494	1606	1439
4	1087	1351	1068
5	765	1096	793
6	538	841	589
7	484	585	437
8	290	330	325
9	226	75	241
10	204	−180	179

If age denotes the x-variable and price the y-variable,

$$\sum_{i=1}^{10} x_i = 55, \qquad \sum_{i=1}^{10} x_i^2 = 385,$$

$$\sum_{i=1}^{10} y_i = 9862, \qquad \sum_{i=1}^{10} y_i^2 = 15,502,372,$$

$$\sum_{i=1}^{10} x_i y_i = 32,202.$$

The values $a = 2371.47$ and $b = -255.14$ then follow from an application of Theorem 10.6.

Column 3 of Table 10.6 gives the expected price for each age as modeled by the linear function, $y = 2371.47 - 255.14x$. The magnitude of the discrepancies between the observed and predicted values would seem to imply that, here, the linear model is not adequate.

A plot of the first two columns of Table 10.6 suggests that an exponential function—say, $y = ae^{bx}$—might be a better choice (see Figure 10.7). While the exponential does not immediately fall within the purview of Theorem 10.6, we can make it do so by taking logs. That is, if $y = ae^{bx}$, then

$$\ln y = \ln a + bx, \tag{10.8}$$

which says that $\ln y$ is linear with x. To fit the data to Equation 10.8, then, we simply need to replace the y_i's with their corresponding

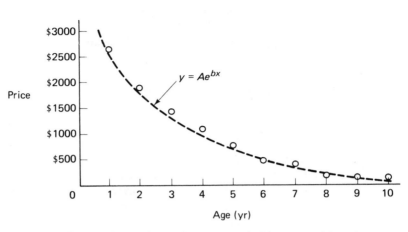

Figure 10.7   Used car prices compared with exponential model.

In y_i's and then apply Theorem 10.6 as usual. This will give ln a = 8.16 and $b = -0.298$. Therefore, $a = e^{8.16} = 3498.19$, making the least-squares model

$$y = 3498.19 \cdot e^{-0.298x}. \tag{10.9}$$

Column 4 shows the predicted prices using Equation 10.9. Now the agreement is quite good. (The graph drawn in Figure 10.7, although labeled $y = ae^{bx}$, is, in fact, $y = 3498.19e^{-0.298x}$.)

QUESTION 10.4.2   One of the factors thought to contribute to the incidence of skin cancer is the ultraviolet radiation coming from the sun. It is well known that the amount of UV radiation a person receives is a function of the shielding thickness of the Earth's ozone layer, which, in turn, depends on the person's latitude. Listed in Table 10.7 are the malignant

TABLE 10.7   Melanoma rates

Location Number	Degrees North Latitude	Melanoma Rate (per 100,000)
1	32.8	9.0
2	33.9	5.9
3	34.1	6.6
4	37.9	5.8
5	40.0	5.5
6	40.8	3.0
7	41.7	3.4
8	42.2	3.1
9	45.0	3.8

skin cancer (melanoma) rates for white males determined for nine areas throughout the United States during the three-year period from 1969 to 1971. The location of each area is given in "degrees north latitude" (44). Fit these data with an exponential model. Let x denote "degrees north latitude" and y "melanoma rate." Plot the data and sketch in the regression line.

10.5 AN INTRODUCTION TO THE LINEAR MODEL

In the geometric view of the previous section, distributions for the x- and y-variables were ignored. The remainder of this chapter examines approximately linear relationships where one or both of the measurements do have an assigned density. This section is devoted to the former—where x is not a random variable but Y is. Examples of this situation are quite common. Recall, for instance, the proposed nephritis study mentioned at the beginning of Section 10.4: certain critical blood pressures (the x-variable) were to be preselected; then patients having those particular blood pressures would have their heart weights (the Y *random* variable) measured. In general, our objective here is to consider the sorts of assumptions that might be imposed on the random variable Y and then explore the various kinds of inference procedures that will inevitably follow.

One model that is often used assumes Y to have the form

$$Y = a + bx + \epsilon, \quad \epsilon \sim N(0, \sigma^2),$$

where the quantity ϵ represents "random error." Of course, this definition of Y is equivalent to requiring that

$$Y \sim N(a + bx, \sigma^2).$$

The use of ϵ simply emphasizes the idea of Y's differing from its mean by an error term.

Comment. A more general form of the preceding model is

$$Y = \sum_{j=1}^{k} b_j f_j(x) + \epsilon, \quad \epsilon \sim N(0, \sigma^2), \tag{10.10}$$

for suitably chosen $f_j(x)$. Even though the $f_j(x)$ may be nonlinear, Equation 10.10 is still referred to as a *linear model* because

$$E(Y) = \sum_{j=1}^{k} b_j f_j(x)$$

is a linear function of the parameters $b_1, b_2, \ldots, b_k$. A further direction of generalization is the extension to more than one variable—that is, letting Y depend on the *set* of nonrandom variables $x_1, x_2, \ldots, x_m$. A comprehensive treatment of these more general models can be found in (62) or (158), to mention but two of a large number of books on the subject. ∎

Placing a distribution on Y does not change our basic interest in estimating the parameters a and b. However, it is now possible to ask for maximum-likelihood estimators, find the distribution of those estimators, and use the resulting pdf's to set up inference procedures. (Compare the statement of Theorem 10.7 to the least-squares estimators given in Theorem 10.6.)

THEOREM 10.7 Let $Y_1, \ldots, Y_n$ be independent random variables, $Y_i \sim N(a + bx_i, \sigma^2)$, $i = 1, \ldots, n$, with parameters a, b, and σ^2 all being unknown. The corresponding maximum-likelihood estimators are given by

$$\hat{A} = \bar{Y} - \hat{B}\bar{x},$$

$$\hat{B} = \frac{n \sum_{i=1}^{n} x_i Y_i - \left(\sum_{i=1}^{n} x_i\right)\left(\sum_{i=1}^{n} Y_i\right)}{n \sum_{i=1}^{n} x_i^2 - \left(\sum_{i=1}^{n} x_i\right)^2} = \frac{\sum_{i=1}^{n} (x_i - \bar{x})(Y_i - \bar{Y})}{\sum_{i=1}^{n} (x_i - \bar{x})^2},$$

and

$$\widehat{\sigma^2} = \frac{1}{n} \sum_{i=1}^{n} (Y_i - \hat{Y}_i)^2,$$

where $\hat{Y}_i = \hat{A} + \hat{B}x_i$, $i = 1, \ldots, n$.

PROOF   Since $Y_i \sim N(a + bx_i, \sigma^2)$, the likelihood function is

$$L = \frac{1}{(2\pi\sigma^2)^{n/2}} \exp\left[-\frac{1}{2\sigma^2} \sum_{i=1}^{n} (y_i - a - bx_i)^2\right]$$

or

$$-2 \ln L = n \ln (2\pi) + n \ln (\sigma^2) + \frac{1}{\sigma^2} \sum_{i=1}^{n} (y_i - a - bx_i).$$

As frequently happens, the maximum of L occurs when the partials of $\ln L$ (or $-2 \ln L$) with respect to a, b, and σ^2 all vanish. Here, the relevant equations are

$$\frac{2}{\sigma^2} \sum_{i=1}^{n} (y_i - a - bx_i)(-1) = 0,$$

$$\frac{2}{\sigma^2} \sum_{i=1}^{n} (y_i - a - bx_i)(-x_i) = 0,$$

$$\frac{n}{\sigma^2} - \frac{1}{(\sigma^2)^2} \sum_{i=1}^{n} (y_i - a - bx_i)^2 = 0.$$

The first two equations depend only on a and b, and the resulting solutions for $\hat{A}$ and $\hat{B}$ have the same form as those found in the proof of Theorem 10.6. Substituting the solutions for the first two equations into the third gives the claimed $\widehat{\sigma^2}$.

QUESTION 10.5.1   Show that an equivalent form for the MLE for B is

$$\hat{B} = \frac{\sum_{i=1}^{n} (x_i - \bar{x})Y_i}{\sum_{i=1}^{n} (x_i - \bar{x})^2}.$$

QUESTION 10.5.2   Prove the following two identities. The second is especially useful for computational purposes.

$$n\widehat{\sigma^2} = \sum_{i=1}^n (Y_i - \bar{Y})^2 - \hat{B} \sum_{i=1}^n (x_i - \bar{x})(Y_i - \bar{Y})$$

$$= \sum_{i=1}^n Y_i^2 - \frac{\left(\sum_{i=1}^n Y_i\right)^2}{n} - \hat{B}\left[\sum_{i=1}^n x_i Y_i - \frac{\left(\sum_{i=1}^n x_i\right)\left(\sum_{i=1}^n Y_i\right)}{n}\right].$$

The next two theorems establish the distributions of $\hat{A}$, $\hat{B}$, and $\widehat{\sigma^2}$ as well as the relationship among the three.

THEOREM 10.8

(a) $\hat{A}$ and $\hat{B}$ are both normally distributed.

(b) $\hat{A}$ and $\hat{B}$ are both unbiased: $E(\hat{A}) = a$ and $E(\hat{B}) = b$.

(c) $\mathrm{Var}\,(\hat{B}) = \dfrac{\sigma^2}{\displaystyle\sum_{i=1}^n (x_i - \bar{x})^2}.$

(d) $\mathrm{Var}\,(\hat{A}) = \dfrac{\sigma^2 \displaystyle\sum_{i=1}^n x_i^2}{n \displaystyle\sum_{i=1}^n (x_i - \bar{x})^2}.$

PROOF   (See Review Exercise 48.)

Comment. For σ^2 known, the random variables

$$\frac{(\hat{B} - b)\sqrt{\sum_{i=1}^n (x_i - \bar{x})^2}}{\sigma} \quad \text{and} \quad \frac{(\hat{A} - a)\sqrt{n \sum_{i=1}^n (x_i - \bar{x})^2}}{\sigma\sqrt{\sum_{i=1}^n x_i^2}}$$

are $N(0, 1)$. Confidence intervals and hypothesis tests for a and b are straightforward (see Review Exercise 51). ∎

In applications of the linear model to real data, σ^2 is usually not known, creating a situation similar to that in Chapter 7, where the Student t—rather than the standard normal—emerges as the distribution of interest. Theorem 10.9 is analogous to Theorem 7.17 in providing the necessary distribution and independence arguments to justify using the Student t. A sketch of the derivation appears in Appendix 10.1 on page 431. (Note the striking similarities between the proof of this result and the proof of Theorem 7.17.)

THEOREM 10.9

(a) The random variables $\hat{B}$, $\bar{Y}$, and $\widehat{\sigma^2}$ are mutually independent.

(b) $\dfrac{n\widehat{\sigma^2}}{\sigma^2} \sim \chi^2_{n-2}$.

Comment. We might have suspected the second part of Theorem 10.9 by recalling the thrust of Chapter 9. The random variable $\widehat{\sigma^2}$ is, in a sense, a measure of goodness-of-fit, and since two parameters have been estimated (a and b), two degrees of freedom are lost, leaving a total of $n - 2$. ∎

Comment. In general, $\hat{A}$ and $\hat{B}$ are not independent (of course, they are when $\bar{x} = 0$, because then $\hat{A} = \bar{Y}$). However, since $\hat{A}$ is a function of $\bar{Y}$ and $\hat{B}$, it and $\hat{B}$ (or any functions of them) are independent of $\widehat{\sigma^2}$. ∎

COROLLARY $[n/(n - 2)]\widehat{\sigma^2}$ is an unbiased estimator of σ^2.

PROOF   The proof follows readily from the basic properties of the expected value:

$$E\left(\frac{n}{n-2}\widehat{\sigma^2}\right) = \frac{\sigma^2}{n-2}E\left(\frac{n\widehat{\sigma^2}}{\sigma^2}\right) = \frac{\sigma^2}{n-2} \cdot (n - 2) = \sigma^2.$$

The next-to-last equality is a result of the fact that $n\widehat{\sigma^2}/\sigma^2 \sim \chi^2_{n-2}$, and the expected value of a chi-square random variable with $n - 2$ degrees of freedom is $n - 2$ (see Question 7.5.2).

We are now in a position to derive the statistics needed for inference procedures on a and b. Hypothesis tests and confidence intervals for b occur more frequently; the interested reader can investigate the procedures for a in the Review Exercises.

THEOREM 10.10   The random variable

$$T = \frac{\sqrt{n - 2}(\hat{B} - b)\sqrt{\sum_{i=1}^{n}(x_i - \bar{x})^2}}{\sqrt{n\widehat{\sigma^2}}}$$

has a Student t distribution with $n - 2$ degrees of freedom.

PROOF   The statistic T is the quotient of the standard normal

$$\frac{\hat{B} - b}{\sqrt{\dfrac{\sigma^2}{\sum_{i=1}^{n}(x_i - \bar{x})^2}}}$$

by the square root of the χ^2_{n-2} variable, $n\widehat{\sigma^2}/\sigma^2$, divided by its degrees of freedom, $n - 2$. Since Theorem 10.9 establishes independence of numerator and denominator, the conclusion is immediate from Definition 7.2.

THEOREM 10.11   A $100(1 - \alpha)\%$ confidence interval for b is given by

$$\left(\hat{b} - t_{\alpha/2,n-2} \sqrt{\frac{n\widehat{\sigma^2}}{(n-2)\sum_{i=1}^{n}(x_i - \bar{x})^2}}, \; \hat{b} + t_{\alpha/2,n-2} \sqrt{\frac{n\widehat{\sigma^2}}{(n-2)\sum_{i=1}^{n}(x_i - \bar{x})^2}} \right).$$

PROOF   (See Review Exercise 52.)

CASE STUDY

10.4

An airline's freight revenues should be a function of the amount of freight carried. If the function is approximately linear in the usual form, $y = a + bx$, we would interpret a to be the fixed or overhead costs and b the marginal revenue—that is, b gives the additional revenue resulting from a unit increase in the amount of cargo consigned. Table 10.8 gives the amount of freight carried (measured in

TABLE 10.8   Airline freight and revenues

Airline	Freight Ton-Miles (millions), x_i	Freight Revenues (millions), y_i
Pan American	860	188
Flying Tiger	681	120
United	645	135
American	529	114
TWA	475	98
Seaboard	359	53
Northwest	246	52
Eastern	207	56
Delta	176	56
Continental	144	29

ton-miles) and the corresponding revenues earned by the ten largest airline freight companies in 1973 (92). For these data,

$$\sum_{i=1}^{10} x_i = 4{,}322, \qquad \sum_{i=1}^{10} y_i = 901,$$

$$\sum_{i=1}^{10} x_i^2 = 2{,}408{,}810, \qquad \sum_{i=1}^{10} y_i^2 = 103{,}195,$$

$$\sum_{i=1}^{10} x_i y_i = 494{,}774.$$

Given these values, we calculate $\hat{a} = 5.902760$ and $\hat{b} = 0.194811$. A further computation yields

$$n\widehat{\sigma^2} = 1489.2624 \quad \text{and} \quad \sum_{i=1}^{10} (x_i - \bar{x})^2 = 540841.6.$$

Interpreting $\hat{b}$ as marginal revenue tells an airline that an increase of 1 million freight ton-miles will result in an additional \$194,811 in revenue. If an airline is willing to accept the validity of the linear model, it might be interested in a confidence interval for b. According to Theorem 10.11 (and noting that $t_{0.025, 8} = 2.306$), a 95% confidence interval for b extends from 0.152 to 0.238:

$$\left(0.194811 - (2.306)\sqrt{\frac{1489.2624}{(8)(540841.6)}}, \right.$$

$$\left. 0.194811 + (2.306)\sqrt{\frac{1489.2624}{8(540841.6)}} \right)$$

$$= (0.194811 - 0.042782, \, 0.194811 + 0.042782) \doteq (0.152, 0.238).$$

QUESTION 10.5.3   Find a 95% confidence interval for the slope of the chirping frequency–temperature data given in Question 10.4.1.

It should be noted that the parameter b in a linear model replaces the correlation coefficient as a measure of linear dependence: to reject the null hypothesis H_0: $b = 0$ is to claim a statistically significant dependence of Y on x. The test procedure is derived from the T statistic of Theorem 10.10. We will state the result but omit our usual GLRT justification.

THEOREM 10.12 To test $H_0 : b = 0$ versus $H_1 : b \neq 0$ at the α level of significance, reject the null hypothesis if

$$\frac{\sqrt{n-2}\,\hat{b}\sqrt{\sum_{i=1}^{n}(x_i - \bar{x})^2}}{\sqrt{n\widehat{\sigma^2}}}$$

is either (1) $\leq -t_{\alpha/2, n-2}$ or (2) $\geq t_{\alpha/2, n-2}$.

Comment. The usual modifications for one-sided tests also apply here. In some applications it would be clear that either $b = 0$ or $b < 0$. An example of this occurs if x = amount of a commodity offered for sale and Y = unit price of the commodity. The next example discusses a situation where it would be appropriate to test $H_0: b = 0$ versus $H_1: b > 0$. ∎

EXAMPLE 10.5.   Case Study 1.4 suggested a possible linear relationship between radioactive contamination (the x-variable) and cancer mortality (the y-variable). Figure 1.2 offered strong visual evidence for such a dependence. As always, however, we must be careful not to mistake random fluctuations for a real phenomenon. Since no reputable medical experts feel that radiation exposure of this sort would ever *decrease* cancer mortality, it makes sense to test $H_0: b = 0$ versus $H_1: b > 0$. The data from Table 1.5 yield the following sums:

$$\sum_{i=1}^{9} x_i = 41.56, \qquad \sum_{i=1}^{9} y_i = 1{,}416.1,$$

$$\sum_{i=1}^{9} x_i^2 = 289.4222, \qquad \sum_{i=1}^{9} y_i^2 = 232{,}498.97,$$

$$\sum_{i=1}^{9} x_i y_i = 7439.37.$$

We then calculate $\sum_{i=1}^{9} (x_i - \bar{x})^2 = 97.507356$, $\hat{b} = 9.231456$, and $9\widehat{\sigma^2} = 1373.9464$. Thus the statistic T (with $b = 0$) has the value

$$\frac{\sqrt{7}\,(9.231456)\sqrt{97.507356}}{\sqrt{1373.9464}} = 6.51.$$

We should reject the null hypothesis only if the test statistic exceeds $t_{\alpha, n-2}$. Here, at the 5% level, $t_{0.05, 7} = 1.895$. Our conclusion, then, is clear-cut: reject H_0.

QUESTION 10.5.4   Insect flight ability can be measured in a laboratory by attaching the insect to a nearly frictionless rotating arm by means of a very thin wire. The "tethered" insect then flies in circles until exhausted. The nonstop distance flown can easily be calculated from the number of revolutions made by the arm. Shown below are measurements of this sort made on *Culex tarsalis* mosquitoes of four different ages. The response variable is the average (tethered) distance flown until exhaustion for 40 females of the species (147).

Age (weeks), x_i	Distance Flown (thousands of meters), y_i
1	12.6
2	11.6
3	6.8
4	9.2

Fit a straight line to these data and test that the slope is 0. Use a two-sided alternative and the 0.05 level of significance.

As useful as inferences about a and b are inferences about $E(Y|x) = ax + b$, the mean of Y for a fixed value of x. For example, if x represents the quantity of goods manufactured and Y the cost for producing that amount, we can well imagine a manager's interest in the average cost for a run of a certain size. A reasonable choice of a point estimator for $E(Y|x)$ is $\hat{Y} = \hat{A} + \hat{B}x$, and it proves to be unbiased:

$$E(\hat{Y}) = E(\hat{A} + \hat{B}x) = E(\hat{A}) + E(\hat{B})x = a + bx.$$

The last equality, of course, is a consequence of the unbiasedness of $\hat{A}$ and $\hat{B}$.

Before we can use $\hat{Y}$ in any kind of confidence interval, we must also know its variance. Fortunately, the necessary computation is relatively straightforward:

$$\begin{aligned}
\text{Var}(\hat{Y}) &= \text{Var}(\hat{A} + \hat{B}x) = \text{Var}(\bar{Y} - \hat{B}\bar{x} + \hat{B}x) \\
&= \text{Var}[\bar{Y} + \hat{B}(x - \bar{x})] \\
&= \text{Var}(\bar{Y}) + (x - \bar{x})^2 \text{Var}(\hat{B}) \qquad \text{(Why?)} \\
&= \frac{1}{n}\sigma^2 + \frac{(x - \bar{x})^2}{\sum_{i=1}^{n}(x_i - \bar{x})^2}\sigma^2 \\
&= \sigma^2\left[\frac{1}{n} + \frac{(x - \bar{x})^2}{\sum_{i=1}^{n}(x_i - \bar{x})^2}\right].
\end{aligned}$$

Out of this last expression comes the not-surprising result that the variance of the estimated average response increases as the value of x becomes more extreme. That is, we are better able to predict the behavior of $E(Y|x)$ for an x-value close to $\bar{x}$ than we are for x-values that are either very small or very large (relative to $\bar{x}$).

QUESTION 10.5.5   Prove that the variance of $\hat{Y}$ can also be written

$$\text{Var}(\hat{Y}) = \frac{\sigma^2 \sum_{i=1}^{n}(x_i - x)^2}{n\sum_{i=1}^{n}(x_i - \bar{x})^2}.$$

By Theorem 10.9, $\hat{Y}$ is independent of $\widehat{n\sigma^2}$, so the statistic

$$\frac{\dfrac{\hat{Y} - (a + bx)}{\sigma\sqrt{\dfrac{1}{n} + \dfrac{(x - \bar{x})^2}{\sum_{i=1}^{n}(x_i - \bar{x})^2}}}}{\dfrac{1}{\sigma}\dfrac{1}{\sqrt{n-2}}\sqrt{\widehat{n\sigma^2}}} = \frac{\sqrt{n-2}[\hat{Y} - (a + bx)]}{\sqrt{\widehat{n\sigma^2}\left[\dfrac{1}{n} + \dfrac{(x - \bar{x})^2}{\sum_{i=1}^{n}(x_i - \bar{x})^2}\right]}}$$

has a Student t distribution with $n - 2$ degrees of freedom. Knowing that, we can readily obtain an expression for a confidence interval for $E(Y|x)$.

THEOREM 10.13 A $100(1 - \alpha)\%$ confidence interval for $E(Y|x) = ax + b$, the expected value of Y at the fixed value x, is given by

$$(\hat{y} - w, \hat{y} + w),$$

where

$$w = t_{\alpha/2, n-2} \frac{1}{\sqrt{n-2}} \sqrt{\widehat{n\sigma^2} \left[\frac{1}{n} + \frac{(x - \bar{x})^2}{\sum\limits_{i=1}^{n} (x_i - \bar{x})^2} \right]}.$$

A variation of the preceding theme is the so-called "prediction problem," where we wish to estimate the particular outcome of a random variable, rather than the parameters of its density or its conditional moments. Recall Case Study 10.4 and imagine a certain airline wanting to predict the revenues it would accrue by carrying a given amount of freight—say, x. To make such an inference we will need to consider both the original random sample, $(x_1, Y_1), \ldots, (x_n, Y_n)$, and an additional observation, (x, Y), with Y being independent of the Y_i's. A *prediction interval* is one that contains Y with a desired probability.

The appropriate statistic is again $\hat{Y}$, but now $\hat{Y} - Y$ will be the quantity of interest, rather than $\hat{Y} - (ax + b)$. First, note that

$$E(\hat{Y} - Y) = E(\hat{Y}) - E(Y) = (ax + b) - (ax + b) = 0.$$

Also,

$$\mathrm{Var}\,(\hat{Y} - Y) = \mathrm{Var}\,(\hat{Y}) + \mathrm{Var}\,(Y)$$

$$= \sigma^2 \left[\frac{1}{n} + \frac{(x - \bar{x})^2}{\sum\limits_{i=1}^{n} (x_i - \bar{x})^2} \right] + \sigma^2$$

$$= \sigma^2 \left[1 + \frac{1}{n} + \frac{(x - \bar{x})^2}{\sum\limits_{i=1}^{n} (x_i - \bar{x})^2} \right].$$

It becomes a simple matter, now, to recast the confidence-interval expression for $E(Y|x)$ into a prediction interval for Y.

THEOREM 10.14 A $100(1 - \alpha)\%$ prediction interval for Y at a fixed value x is given by

$$(\hat{y} - w, \hat{y} + w),$$

where

$$w = t_{\alpha/2, n-2} \frac{1}{\sqrt{n-2}} \sqrt{\widehat{n\sigma^2} \left[1 + \frac{1}{n} + \frac{(x - \bar{x})^2}{\sum\limits_{i=1}^{n} (x_i - \bar{x})^2} \right]}.$$

Comment. The only difference in the confidence interval for $E(Y|x)$ and the prediction interval for Y at the same probability level is that the prediction interval has an extra summand of 1 under the radical. This lengthens the interval in the latter case, which is hardly an unexpected modification, since there is more variability in Y than in $E(Y|x)$. ∎

EXAMPLE 10.6.   In Case Study 10.4, let us suppose American Airlines is considering increasing its capacity for freight to 600 million freight ton-miles. It would not be unreasonable for them to want some intelligent estimate of their possible revenues. A 95% prediction interval would be one way to phrase what they want to know.

The relevant x-value is 600, so the corresponding estimated revenue is $\hat{y} = \hat{a} + 600\hat{b} = 5.902760 + 600(0.194811) = 122.78936$ (millions of dollars). This latter number is the center of the prediction interval. The interval's radius, or half-length, is given by

$$(2.306)\frac{1}{\sqrt{8}}\sqrt{(1489.2624)\left[1 + \frac{1}{10} + \frac{(600 - 432.2)^2}{540841.6}\right]} = 33.770515.$$

Thus, they could anticipate revenues somewhere between \$89.02 million and \$156.56 million:

$$(122.78936 - 33.770515, 122.78936 + 33.770515) = (89.02, 156.56).$$

QUESTION 10.5.6   Construct a 95% confidence interval for $E(Y|600)$ using the airline data of Case Study 10.4. Suppose a Senate subcommittee were investigating allegations of profit gouging in the airline freight industry. Would the committee members be more interested in a 95% confidence interval for $E(Y|600)$ or a 95% prediction interval for Y given that $x = 600$? Explain.

10.6 THE BIVARIATE NORMAL DENSITY

As already mentioned, Galton, having empirically discovered properties 1 through 4 of Section 10.3, enlisted the aid of J. D. H. Dickson in finding a joint density function with these properties. Dickson did so—and with alacrity—the desired distribution being well known and its derivation not particularly difficult. In fact, Pearson (126) commented, "Why Galton did not at once write down the equation to his surface . . . has always been a puzzle to me," and then reasoned, "The explanation of Galton's action possibly lies in the fact that Galton was very modest and throughout his life underrated his own mathematical powers." Taking encouragement from this last remark, let *us* try to derive the appropriate density function $f_{X,Y}(x, y)$.

The first clue to consider is property 4, the elliptical nature of the contour lines. For simplicity, suppose we assume these ellipses are circles centered at the origin. Then $f_{X,Y}(x, y)$ has the form $g(x^2 - 2vxy + y^2)$. The notation of the cross-term constant as $-2v$ is suggested by the recognition that completing the square is often a useful tool when dealing with quadratics. Since the marginal distributions are to be normal, an exponential form seems reasonable—for example,

$$f_{X,Y}(x, y) = Ke^{-(1/2)c(x^2 - 2vxy + y^2)}, \tag{10.11}$$

where $c > 0$ is arbitrary and K must insure that the integral of $f_{X,Y}(x, y)$ is 1.

The integral can be calculated by completing the square in y and writing

$$f_{X,Y}(x, y) = Ke^{-(1/2)c(1-v^2)x^2} \cdot e^{-(1/2)c(y-vx)^2}.$$

Since the first exponent must be negative, we need to insist that $|v| \leq 1$. Then,

$$\int_{-\infty}^{\infty} \int_{-\infty}^{\infty} e^{-(1/2)c(1-v^2)x^2} \cdot e^{-(1/2)c(y-vx)^2} \, dy \, dx$$

$$= \int_{-\infty}^{\infty} e^{-(1/2)c(1-v^2)x^2} \left(\int_{-\infty}^{\infty} e^{-(1/2)c(y-vx)^2} \, dy \right) dx$$

$$= \int_{-\infty}^{\infty} e^{-(1/2)c(1-v^2)x^2} \left(\frac{\sqrt{2\pi}}{\sqrt{c}} \right) dx$$

$$= \frac{\sqrt{2\pi}}{\sqrt{c}} \frac{\sqrt{2\pi}}{\sqrt{c}\sqrt{1-v^2}}$$

$$= \frac{2\pi}{c\sqrt{1-v^2}}.$$

It follows that we should take

$$K = \frac{c\sqrt{1-v^2}}{2\pi}.$$

The constant c can be any positive value, but a convenient form proves to be $c = 1/(1-v^2)$. Substitution of these values for K and c into Equation 10.11 gives

$$f_{X,Y}(x, y) = \frac{1}{2\pi\sqrt{1-v^2}} e^{-(1/2)[1/(1-v^2)](x^2-2vxy+y^2)}$$

$$= \frac{1}{2\pi\sqrt{1-v^2}} e^{-x^2} \cdot e^{-(1/2)[1/(1-v^2)](y-vx)^2}. \qquad (10.12)$$

Recall that our choice of the form of $f_{X,Y}(x, y)$ was predicated on a wish for the marginal pdf's to be normal. A simple integration shows that to be the case:

$$f_X(x) = \int_{-\infty}^{\infty} f_{X,Y}(x, y) \, dy$$

$$= \frac{1}{2\pi\sqrt{1-v^2}} e^{-(1/2)x^2} \int_{-\infty}^{\infty} e^{-(1/2)[1/(1-v^2)](y-vx)^2} \, dx$$

$$= \frac{1}{2\pi\sqrt{1-v^2}} e^{-(1/2)x^2} \cdot \sqrt{2\pi}\sqrt{1-v^2}$$

$$= \frac{1}{\sqrt{2\pi}} e^{-(1/2)x^2}.$$

Since $f_{X,Y}(x, y)$ is symmetric in x and y, $f_Y(y)$ is also $N(0, 1)$.

Next we show that the constant v appearing in Equations 10.11 and 10.12 is really $\rho(X, Y)$. Since $E(X) = E(Y) = 0$,

$$\rho(X, Y) = E(XY) = \int_{-\infty}^{\infty} \int_{-\infty}^{\infty} xy f_{X,Y}(x, y)\, dx\, dy$$

$$= \frac{1}{\sqrt{2\pi}} \int_{-\infty}^{\infty} x e^{-(1/2)x^2} \left(\frac{1}{\sqrt{2\pi}\sqrt{1-v^2}} \int_{-\infty}^{\infty} y e^{-(1/2)[1/(1-v^2)](y-vx)^2}\, dy \right) dx$$

$$= \frac{1}{\sqrt{2\pi}} \int_{-\infty}^{\infty} x e^{-(1/2)x^2} \cdot vx\, dx \qquad \text{(Why?)}$$

$$= v \cdot \frac{1}{\sqrt{2\pi}} \int_{-\infty}^{\infty} x^2 e^{-(1/2)x^2}\, dx = v \cdot \text{Var}\,(X) = v.$$

[To avoid unnecessary notation, we will write $\rho(X, Y)$ as simply ρ.]

There is a two-variable version of Theorem 3.3 allowing a generalization of Equation 10.12 to the case where x is replaced by $(x - \mu_X)/\sigma_X$ and y by $(y - \mu_Y)/\sigma_Y$. To accomplish this, the original density must be multiplied by the derivative of both the X-transformation and the Y-transformation—that is, by $(1/\sigma_X\sigma_Y)$. Definition 10.3 gives the resulting extension of Equation 10.12.

DEFINITION 10.3 Let X and Y be random variables with joint pdf

$$f_{X,Y}(x, y) = \frac{1}{2\pi\sigma_X\sigma_Y\sqrt{1-\rho^2}}$$

$$\cdot \exp\left\{ -\frac{1}{2}\left(\frac{1}{1-\rho^2}\right)\left[\frac{(x-\mu_X)^2}{\sigma_X^2} - 2\rho\frac{x-\mu_X}{\sigma_X}\cdot\frac{y-\mu_Y}{\sigma_Y} + \frac{(y-\mu_Y)^2}{\sigma_Y^2} \right] \right\}$$

for all x and y. Then X and Y are said to have the *bivariate normal distribution*.

Comment. Knowledge of the bivariate normal dates back to the beginning of the nineteenth century: Laplace, for example, wrote about it as early as 1810. Until the time of Galton, though, its applications were primarily confined to problems associated with the theory of astronomical errors. ▮

Comment. There is a natural generalization from the bivariate normal to the *multivariate* normal, the latter having the form

$$f_{\vec{x}}(\vec{x}) = \frac{1}{(2\pi)^{n/2}\sqrt{|A^{-1}|}} \exp\left[-\frac{1}{2}(\vec{x} - \vec{\mu})^T A(\vec{x} - \vec{\mu}) \right],$$

where $\vec{x}$ and $\vec{\mu}$ are n-dimensional column vectors, A is an $n \times n$ symmetric positive-definite matrix, and T denotes the transpose operation. For further details, see (31) or (62). ▮

Theorem 10.15 summarizes what has already been established for the bivariate normal and confirms that it possesses the four properties discovered by Galton.

> **THEOREM 10.15** Suppose X and Y are random variables having the bivariate normal distribution given in Definition 10.3. Then
>
> (a) $X \sim N(\mu_X, \sigma_X^2)$ and $Y \sim N(\mu_Y, \sigma_Y^2)$.
> (b) $\rho(X, Y) = v = \rho$.
> (c) $E(Y|x) = \mu_Y + \dfrac{\rho\sigma_Y}{\sigma_X}(x - \mu_X)$.
> (d) $\text{Var}\,(Y|x) = (1 - \rho^2)\sigma_Y^2$.

PROOF   We have already established (a) and (b). Properties (c) and (d) will be examined for the special case $\mu_X = \mu_Y = 0$ and $\sigma_X = \sigma_Y = 1$. The extension to arbitrary μ_X, μ_Y, σ_X, and σ_Y is straightforward.

First, note that

$$f_{Y|x}(y) = \frac{f_{X,Y}(x, y)}{f_X(x)}$$

$$= \frac{\dfrac{1}{2\pi\sqrt{1 - \rho^2}}\, e^{-(1/2)x^2} e^{-(1/2)[1/(1-\rho^2)](y-\rho x)^2}}{\dfrac{1}{\sqrt{2\pi}}\, e^{-(1/2)x^2}}$$

$$= \frac{1}{\sqrt{2\pi}\,\sqrt{1 - \rho^2}}\, e^{-(1/2)[1/(1-\rho^2)](y-\rho x)^2}. \tag{10.13}$$

But, by inspection, Equation 10.13 is the pdf for an $N(\rho x, 1 - \rho^2)$ random variable. Therefore, $E(Y|x) = \rho x$ and $\text{Var}\,(Y|x) = 1 - \rho^2$. Replacing Y by $(Y - \mu_Y)/\sigma_Y$ and x by $(x - \mu_X)/\sigma_X$ gives the desired results.

QUESTION 10.6.1   Suppose X and Y have a bivariate normal pdf with $\mu_X = 3$, $\mu_Y = 6$, $\sigma_X^2 = 4$, $\sigma_Y^2 = 10$, and $\rho = \frac{1}{2}$. Find $P(5 < Y < 6\frac{1}{2})$ and $P(5 < Y < 6\frac{1}{2}|x = 2)$.

The reader may share the exuberance of Sir Francis, who, upon receiving Dickson's response (basically Theorem 10.15), wrote (56):

> The problem may not be difficult to an accomplished mathematician, but I certainly never felt such a glow of loyalty and respect towards the sovereignty and wide sway of mathematical analysis as when his answer arrived, confirming by purely mathematical reasoning my various and laborious statistical conclusions with far more minuteness than I had dared to hope, because the data ran somewhat roughly and I had to smooth them with tender caution.

Comment. For bivariate normal densities, $\rho(X, Y) = 0$ implies that X and Y are independent, a result not true in general (recall Example 10.1). ∎

Comment. It is tempting to conjecture that any continuous joint pdf with marginals that are normal is, itself, bivariate normal. Such is not the case. For example, let *u* be any odd continuous function defined on the real line, vanishing outside the interval $[-1, 1]$, and with $|u| < (2\pi e)^{-1/2}$. Then define

$$f_{X,Y}(x, y) = \frac{1}{2\pi} e^{-(1/2)(x^2 + y^2)} + u(x)u(y). \quad \blacksquare$$

10.7 TESTING H_0: $\rho(X, Y) = 0$ WHEN X AND Y ARE BIVARIATE NORMAL

Situations arise where it is desirable to test the independence of two random variables. We are not at this point able to formulate such a test in any general context, but for the special case where X and Y are bivariate normal, independence is equivalent to $\rho(X, Y)$'s being 0, and the latter can be tested without too much difficulty.

The sample correlation coefficient, R, seems a likely candidate for a test statistic. It was a prima facie choice of estimator, and it also arose in the least-squares determination of regression coefficients. Furthermore, as the next theorem indicates, R is a maximum-likelihood estimator.

THEOREM 10.16 Given that $f_{X,Y}(x, y)$ is a bivariate normal pdf, the maximum-likelihood estimators for μ_X, μ_Y, σ_X^2, σ_Y^2, and $\rho(X, Y)$, assuming that all are unknown, are $\bar{X}$, $\bar{Y}$, $(1/n) \sum_{i=1}^{n} (X_i - \bar{X})^2$, $(1/n) \sum_{i=1}^{n} (Y_i - \bar{Y})^2$, and R, respectively.

PROOF   The proof is accomplished by the usual method of taking partial derivatives of the likelihood function with respect to each of the five parameters, setting those equal to 0, and solving simultaneously. The details are best left to the energetic reader.

Although we have already made a case for using R to test H_0: $\rho(X, Y) = 0$, it turns out to be more convenient to use a function of R as a test statistic. The derivation of the appropriate density is beyond the scope of this text, so we content ourselves with a statement of the result and an example.

THEOREM 10.17 Under the null hypothesis that $\rho(X, Y) = 0$, the statistic

$$\frac{\sqrt{n-2}\,R}{\sqrt{1-R^2}}$$

has a Student t distribution with $n - 2$ degrees of freedom.

CASE STUDY

10.5

Not long ago, Alvin Toffler wrote a best-seller called *Future Shock* in which he speculated about the cultural, medical, and moral effects of change. One question that was raised involved the extent to which change should be considered a health hazard—if, indeed, it should be considered a health hazard at all. A partial answer was provided in a recent study done on a group of patients hospitalized for various chronic illnesses, some serious and some not. Each patient was asked to fill out a Schedule of Recent Experience (SRE) questionnaire. The questions related to 42 different life-change situations (a new job, another child, and so forth). A composite score was computed that summarized the average degree of change each patient had experienced in the preceding two years. Higher values on the SRE questionnaire reflected greater life-style changes. At the same time, each patient's health condition (the one he was being hospitalized for) was graded according to the Seriousness of Illness Rating Scale (SIRS). Values on this scale ranged from a low of 21 for dandruff to a high of 1020 for cancer.

The patients in the sample had a total of 17 different conditions that were considered chronic. In some cases more than one patient had the same condition. Table 10.9 and Figure 10.8 show the SIRS values associated with the various illnesses as well as the average SRE scores recorded for the patients who had each one (203).

The scatter diagram for Table 10.9 suggests a possible relationship between SRE and SIRS, but the linearity is not overwhelming. Consequently, before fitting a least-squares line through these 17 points it would be a good idea to test formally the null hypothesis that X and Y are independent.

Let ρ denote the true correlation coefficient between X and Y. The hypotheses to be tested, then, are

$$H_0: \ \rho = 0$$

versus

$$H_1: \ \rho \neq 0.$$

To accept H_0 is to conclude that X and Y are independent (assuming the (x_i, y_i) values in Table 10.9 are a random sample from a bivariate normal distribution).

TABLE 10.9 Life-style changes (SRE) and health evaluations (SIRS)

Admitting Diagnosis	Average SRE	SIRS
Dandruff	26	21
Varicose veins	130	173
Psoriasis	317	174
Eczema	231	204
Anemia	325	312
Hyperthyroidism	816	393
Gallstones	563	454
Arthritis	312	468
Peptic ulcer	603	500
High blood pressure	405	520
Diabetes	599	621
Emphysema	357	636
Alcoholism	688	688
Cirrhosis	443	733
Schizophrenia	609	776
Heart failure	772	824
Cancer	777	1020

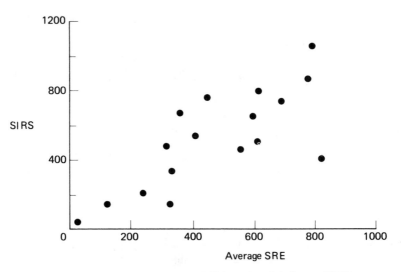

Figure 10.8 Health problems (SIRS) and social change (SRE).

We will reject the null hypothesis only if the sample correlation coefficient, *r*, is either too much less than 0 or too much greater than

0. Let α be 0.05. Then the pdf of the test statistic is the Student t distribution with $n - 2 = 15$ degrees of freedom, and H_0 should be rejected if

$$\frac{\sqrt{n-2}\,r}{\sqrt{1-r^2}}$$

is either (1) less than or equal to -2.13 or (2) greater than or equal to $+2.13$ (see Figure 10.9).

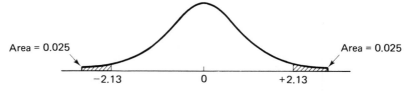

Area = 0.025   Area = 0.025

-2.13 0 $+2.13$

Figure 10.9   Student t distribution with 15 degrees of freedom.

Since

$$\sum_{i=1}^{17} x_i = 7973, \qquad \sum_{i=1}^{17} x_i^2 = 4{,}611{,}291,$$

$$\sum_{i=1}^{17} y_i = 8517, \qquad \sum_{i=1}^{17} y_i^2 = 5{,}421{,}917,$$

$$\sum_{i=1}^{17} x_i y_i = 4{,}759{,}470,$$

it follows that

$$r = \frac{17(4{,}759{,}470) - (7973)(8517)}{\sqrt{17(4{,}611{,}291) - (7973)^2}\,\sqrt{17(5{,}421{,}917) - (8517)^2}}$$

$$= 0.76.$$

Therefore,

$$\frac{\sqrt{n-2}\,r}{\sqrt{1-r^2}} = \frac{\sqrt{15}\,(0.76)}{\sqrt{1-(0.76)^2}} = 4.53.$$

Since the observed t ratio exceeds the upper critical value ($+2.13$), our conclusion is to *reject* the null hypothesis.

QUESTION 10.7.1   Test whether X and $\ln Y$ are independent for the data given in Question 10.4.2. Let X denote "degrees north latitude" and Y "melanoma rate (per 100,000)". Let $\alpha = 0.05$.

Comment. An alternate approach to testing H_0: $\rho = 0$ was given by Fisher (48). He discovered that the statistic

$$\frac{1}{2}\ln\frac{1+R}{1-R}$$

is approximately normal with mean $\frac{1}{2} \ln [(1 + \rho)/(1 - \rho)]$ and variance approximately $1/(n - 3)$. This approximation makes it feasible to determine the power of the test—a difficult task when using $\sqrt{n - 2}\,R/\sqrt{1 - R^2}$. ∎

10.8 EPILOGUE AND CAVEAT

We have seen the usefulness of the correlation coefficient and its intimate relationship with the problem of linear regression. Regression and correlation have emerged, essentially unbidden, in a number of natural phenomena. Yet no discussion of the merits of these concepts would be complete without the warning that *correlation does not imply causality.* An excellent amplification of this dictum can be found in George Bernard Shaw's *The Doctor's Dilemma* (164):

> [C]omparisons which are really comparisons between two social classes with different standards of nutrition and education are palmed off as comparisons between the results of a certain medical treatment and its neglect. Thus it is easy to prove that the wearing of tall hats and the carrying of umbrellas enlarges the chest, prolongs life, and confers comparative immunity from disease; for the statistics show that the classes which use these articles are bigger, healthier, and live longer than the class which never dreams of possessing such things. It does not take much perspicacity to see that what really makes this difference is not the tall hat and the umbrella, but the wealth and nourishment of which they are evidence, and that a gold watch or membership of a club in Pall Mall might be proved in the same way to have the like sovereign virtues. A university degree, a daily bath, the owning of thirty pairs of trousers, a knowledge of Wagner's music, a pew in church, anything, in short, that implies more means and better nurture than the mass of laborers enjoy, can be statistically palmed off as a magic-spell conferring all sorts of privileges.

In 1926, Yule (205) gave one of the first cohesive mathematical treatments of meaningless correlations (the technical term is *spurious* correlations). He began his exposition with the presentation of data showing a very high correlation between Church of England marriages and mortality (see Figure 10.10). He then commented:

> Now I suppose it is possible, given a little ingenuity and goodwill, to rationalize very nearly anything. And I can imagine some enthusiast arguing that the fall in the proportion of Church of England marriages is simply due to the Spread of Scientific Thinking since 1866, and the fall in mortality is also clearly to be ascribed to the Progress of Science; hence both variables are largely or mainly influenced by a common factor and consequently ought to be highly correlated. But most people would, I think, agree with me that the correlation is simply sheer nonsense; that it has no meaning whatever; that it is absurd to suppose that the two variables in question are in any sort of way, however indirect, causally related to one another.

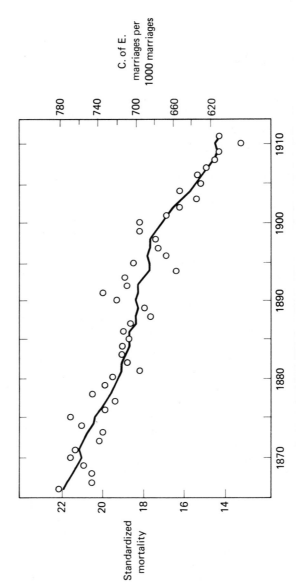

Figure 10.10 Correlation between standardized mortality per 1000 persons in England and Wales (circles) and the proportion of Church of England marriages per 1000 of all marriages (line), 1866–1911 ($r =$ 0.9512).

And yet, if we apply the ordinary test of significance in the ordinary way, the result suggests that the correlation is certainly "significant"—that it lies far outside the probable limits of fluctuations of sampling.

The reader should not suppose the problem of correlation and causality has disappeared; it seems an intrinsic part of discovery and understanding. Only recently a study announced a high correlation between coffee drinking and coronary heart disease. Some newspaper reports of the study portrayed the fragrant essence of the roasted berries of *Coffea arabica* as a menace to public health. Later papers [see (75)] suggested that the risk comes from such factors as cigarette smoking and sucrose intake, both highly correlated with coffee drinking.

There are misuses associated also with the theory of regression. Too often complex problems have been forced into the Procrustean bed of the linear model. This is particularly obvious in the use of linear extrapolation, which recalls Mark Twain's oft-quoted satire in *Life on the Mississippi* (27): "In the space of 176 years the Lower Mississippi has shortened itself 252 miles. That is an average of a trifle more than one mile and a third per year. Therefore any calm person, who is not blind or idiotic, can see that in 742 years from now the lower Mississippi will be one mile and three quarters long, and Cairo, Ill., and New Orleans will have joined their streets together."

It is possible that more blunders have been committed using regression and correlation than with other statistical tools. Be that as it may, the caveat applies to the body of statistical techniques. Statistical tools are powerful indicators of causality and relationship, but they are only that—indicators.

APPENDIX 10.1 A PROOF OF THEOREM 10.9

The strategy for the proof is to express $n\widehat{\sigma^2}$ in terms of the squares of normal random variables and then apply Fisher's lemma (see Appendix 7.2 on page 308). The random variables to be used are $(\hat{B} - b)$, $W_i = Y_i - a - bx_i$, $i = 1, 2, \ldots, n$, and

$$\bar{W} = \frac{1}{n} \sum_{i=1}^{n} W_i = \bar{Y} - a - b\bar{x}.$$

Let us first note that $\hat{B} - b$ can be written as a linear combination of the W_i's. The demonstration requires just a little algebra, beginning with the statement of Question 10.5.1:

$$\hat{B} - b = \frac{\sum_{i=1}^{n} (x_i - \bar{x})Y_i}{\sum_{i=1}^{n} (x_i - \bar{x})^2} - b$$

$$= \frac{\sum_{i=1}^{n} (x_i - \bar{x})Y_i - b \sum_{i=1}^{n} (x_i - \bar{x})^2}{\sum_{i=1}^{n} (x_i - \bar{x})^2}$$

$$= \frac{\sum_{i=1}^{n}(x_i - \bar{x})Y_i - a\sum_{i=1}^{n}(x_i - \bar{x}) - b\sum_{i=1}^{n}(x_i - \bar{x})x_i}{\sum_{i=1}^{n}(x_i - \bar{x})^2}$$

$$= \frac{\sum_{i=1}^{n}(x_i - \bar{x})(Y_i - a - bx_i)}{\sum_{i=1}^{n}(x_i - \bar{x})^2} = \frac{\sum_{i=1}^{n}(x_i - \bar{x})W_i}{\sum_{i=1}^{n}(x_i - \bar{x})^2}.$$

Further,

$$Y_i - \hat{Y}_i = Y_i - \hat{A} - \hat{B}x_i = Y_i - \bar{Y} + \hat{B}\bar{x} - \hat{B}x_i$$
$$= (Y_i - a - bx_i) - (\bar{Y} - a - b\bar{x}) - (x_i - \bar{x})(\hat{B} - b)$$
$$= (W_i - \bar{W}) - (x_i - \bar{x})(\hat{B} - b).$$

Thus,

$$n\widehat{\sigma^2} = \sum_{i=1}^{n}(Y_i - \hat{Y}_i)^2 = \sum_{i=1}^{n}[(W_i - \bar{W}) - (x_i - \bar{x})(\hat{B} - b)]^2$$

$$= \sum_{i=1}^{n}(W_i - \bar{W})^2 + \sum_{i=1}^{n}(x_i - \bar{x})^2(\hat{B} - b)^2 - 2\sum_{i=1}^{n}(W_i - \bar{W})(x_i - \bar{x})(\hat{B} - b)$$

$$= \sum_{i=1}^{n}(W_i - \bar{W})^2 + \sum_{i=1}^{n}(x_i - \bar{x})^2(\hat{B} - b)^2 - 2\sum_{i=1}^{n}W_i(x_i - \bar{x})(\hat{B} - b).$$

This last equality follows since $\sum_{i=1}^{n}\bar{W}(x_i - \bar{x})(\hat{B} - b) = 0$. Now we use the expression for $(\hat{B} - b)$ established at the beginning of the proof to obtain the equality

$$-2\sum_{i=1}^{n}W_i(x_i - \bar{x})(\hat{B} - b) = -2(\hat{B} - b)\sum_{i=1}^{n}(x_i - \bar{x})W_i$$

$$= -2(\hat{B} - b)(\hat{B} - b)\sum_{i=1}^{n}(x_i - \bar{x})^2$$

$$= -2\sum_{i=1}^{n}(x_i - \bar{x})^2(\hat{B} - b)^2.$$

Substituting this last version of the cross-product term into the expression for $n\widehat{\sigma^2}$ gives

$$n\widehat{\sigma^2} = \sum_{i=1}^{n}(W_i - \bar{W})^2 - \sum_{i=1}^{n}(x_i - \bar{x})^2(\hat{B} - b)^2$$

$$= \sum_{i=1}^{n}W_i^2 - n\bar{W}^2 - \sum_{i=1}^{n}(x_i - \bar{x})^2(\hat{B} - b)^2.$$

Now, choose an $n \times n$ orthogonal matrix, M, whose first two rows are

$$\frac{x_1 - \bar{x}}{\sqrt{\sum_{i=1}^{n}(x_i - \bar{x})^2}} \quad \cdots \quad \frac{x_n - \bar{x}}{\sqrt{\sum_{i=1}^{n}(x_i - \bar{x})^2}}$$

and

$$\frac{1}{\sqrt{n}} \quad \cdots \quad \frac{1}{\sqrt{n}}.$$

Define the random variables $Z_1, \ldots, Z_n$ through the transformation

$$\begin{pmatrix} Z_1 \\ \cdot \\ \cdot \\ \cdot \\ Z_n \end{pmatrix} = M \begin{pmatrix} W_1 \\ \cdot \\ \cdot \\ \cdot \\ W_n \end{pmatrix}.$$

By Fisher's lemma, the Z_i's are independent and

$$\sum_{i=1}^{n} Z_i^2 = \sum_{i=1}^{n} W_i^2.$$

Also, $Z_1^2 = \sum_{i=1}^{n} (x_i - \bar{x})^2 (\hat{B} - b)^2$ and $Z_2^2 = n\bar{W}^2$, so

$$\widehat{n\sigma^2} = \sum_{i=1}^{n} W_i^2 - Z_1^2 - Z_2^2 = \sum_{i=3}^{n} Z_i^2.$$

From this follows the independence of $\widehat{n\sigma^2}$, $\hat{B}$, and $\bar{Y}$. Finally, observe that

$$\frac{\widehat{n\sigma^2}}{\sigma^2} = \sum_{i=3}^{n} \left(\frac{Z_i}{\sigma}\right)^2,$$

and the latter has a chi-square distribution with $n - 2$ degrees of freedom.

REVIEW EXERCISES FOR CHAPTER 10

1. Suppose two dice are thrown. Let X be the number showing on the first die and let Y be the larger of the two numbers showing. Find Cov (X, Y).

2. Let X and Y have the joint pdf,

$$f_{X,Y}(x, y) = \begin{cases} \dfrac{x + 2y}{22}, & \text{for } (x, y) = \{(1, 1), (1, 3), (2, 1), (2, 3)\} \\ 0, & \text{elsewhere.} \end{cases}$$

Find Cov (X, Y) and $\rho(X, Y)$.

3. If $M_{X_1, X_2, \ldots, X_k}(t_1, t_2, \ldots, t_k) = E(e^{t_1 X_1 + t_2 X_2 + \cdots + t_k X_k})$ is the joint moment-generating function for the multivariate random variable $(X_1, X_2, \ldots, X_k)$, then

$$E(X_i X_j) = \frac{\partial^2}{\partial t_i \partial t_j} M_{X_1, \ldots, X_k}(t_1, \ldots, t_k) \Big|_{(0, \ldots, 0)}.$$

Let (X_1, X_2, X_3) have the trinomial pdf (see Section 9.2). Use the moment-generating function for the trinomial (see Review Exercise 6 at the end of Chapter 9) to show that

$$\text{Cov } (X_i, X_j) = -np_i p_j.$$

Does it seem "reasonable" for the covariance here to be negative? Explain.

4. Find the covariance for X and Y defined by the joint pdf

$$f_{X,Y}(x, y) = \begin{cases} 8xy, & 0 \leq x \leq y \leq 1 \\ 0, & \text{otherwise.} \end{cases}$$

Also, find $\rho(X, Y)$.

5. Prove that $\rho(a + bX, c + dY) = \rho(X, Y)$ for constants $a, b, c,$ and d. Note that this result allows for a change of scale to one convenient for computation.

6. Suppose X and Y are discrete random variables with the following joint pdf:

(x, y)	$f_{X,Y}(x, y)$
$(1, 2)$	$\frac{1}{2}$
$(1, 3)$	$\frac{1}{4}$
$(2, 1)$	$\frac{1}{8}$
$(2, 4)$	$\frac{1}{8}$

Find the correlation coefficient between X and Y.

7. Suppose Var $(X + Y) = $ Var $(X) + $ Var (Y) and Var $(Y + Z) = $ Var $(Y) + $ Var (Z). Show that Var $(X + Z) = $ Var $(X) + $ Var (Z) if and only if Cov $(X + Y, Y + Z) = $ Var (Y).

8. Let X and Y be random variables with

$$f_{X,Y}(x, y) = \begin{cases} 1, & -y < x < y, 0 < y < 1 \\ 0, & \text{elsewhere.} \end{cases}$$

Show that Cov $(X, Y) = 0$ but that X and Y are dependent.

9. For random variables X and Y, show that

$$\text{Cov } (X + Y, X - Y) = \text{Var } (X) - \text{Var } (Y).$$

10. Suppose Cov $(X, Y) = 0$. Prove that

$$\rho(X + Y, X - Y) = \frac{\text{Var } (X) - \text{Var } (Y)}{\text{Var } (X) + \text{Var } (Y)}.$$

11. Let $X_1, X_2, \ldots, X_n$ be a set of random variables. Define

$$U = \sum_{i=1}^{n} a_i X_i \quad \text{and} \quad V = \sum_{i=1}^{n} b_i X_i.$$

Show that

$$\text{Cov } (U, V) = \sum_{i=1}^{n} a_i b_i \text{ Var } (X_i) + \sum\sum_{i<j} (a_i b_j + a_j b_i) \text{ Cov } (X_i, X_j).$$

12. Let $X_1, X_2, \ldots, X_n$ be independent, identically distributed random variables, each having variance σ^2. Use the result of Review Exercise 11 to show that

$$\text{Cov } (X_i - \bar{X}, \bar{X}) = 0$$

for any i.

13. Let the random variable X take on the values $1, 2, \ldots, n$, each with probability $1/n$. Define Y to be X^2. Find $\rho(X, Y)$ and $\lim_{n \to \infty} \rho(X, Y)$.

14. Suppose the random variables $X_1, X_2,$ and X_3 satisfy the equation $a_1 X_1 + a_2 X_2 + a_3 X_3 = 0$ for constants $a_1, a_2,$ and a_3. Give an expression for Cov (X_i, X_j), $i \neq j$, in terms of the a_i's and the variances of the X_i's, $i = 1, 2, 3$.

15. Let X and Y be defined such that

$$f_{X,Y}(x, y) = \begin{cases} \frac{1}{3}(x + y), & 0 < x < 1, 0 < y < 2 \\ 0, & \text{elsewhere.} \end{cases}$$

Find the expected value and the variance of W, where $W = 4X + 3Y$.

16. There is a theory espoused by some baseball fans that the number of home runs a team hits is markedly affected by the altitude of the club's home park, the rationale being that the air is thinner at the higher altitudes and the balls should go further. Below are the altitudes of the 12 American League ballparks and the number of home runs each of those teams hit during the 1972 season (179).

Club	Altitude (ft)	Number of Home Runs
Cleveland	660	138
Milwaukee	635	81
Detroit	585	135
New York	55	90
Boston	21	120
Baltimore	20	84
Minnesota	815	106
Kansas City	750	57
Chicago	595	109
Texas	435	74
California	340	61
Oakland	25	120

Compute the sample correlation coefficient for these data.

17. The table below gives the mean temperature for 20 successive days and the average butterfat content in the milk of ten cows sampled on those days (136).

Date	Temperature	Percent Butterfat
Apr. 3	64	4.65
Apr. 4	65	4.58
Apr. 5	65	4.67
Apr. 6	64	4.60
Apr. 7	61	4.83
Apr. 8	55	4.55
Apr. 9	39	5.14
Apr. 10	41	4.71
Apr. 11	46	4.69
Apr. 12	59	4.65
Apr. 13	56	4.36
Apr. 14	56	4.82
Apr. 15	62	4.65
Apr. 16	37	4.66
Apr. 17	37	4.95
Apr. 18	45	4.60
Apr. 19	57	4.68
Apr. 20	58	4.65
Apr. 21	60	4.60
Apr. 22	55	4.46

Graph these data and compute their sample correlation coefficient.

18. By late 1971, all cigarette packs had to be labeled with the words, "Warning: The Surgeon General Has Determined That Smoking Is Dangerous to Your Health." The case against smoking rested heavily on statistical, rather than laboratory, evidence. Extensive surveys of smokers and nonsmokers had revealed the former to have much higher risks of dying from a variety of causes, including heart disease. The table below shows (1) the annual cigarette consumption and (2) the mortality rate due to coronary heart disease (CHD) for 21 countries (116).

Year	Country	Cigarette Consumption Per Adult Per Year	CHD Mortality Per 100,000 (Ages 35–64)
1962	United States	3900	256.9
1962	Canada	3350	211.6
1962	Australia	3220	238.1
1962	New Zealand	3220	211.8
1963	United Kingdom	2790	194.1
1962	Switzerland	2780	124.5
1962	Ireland	2770	187.3
1962	Iceland	2290	110.5
1962	Finland	2160	233.1
1963	West Germany	1890	150.3
1962	Netherlands	1810	124.7
1962	Greece	1800	41.2
1962	Austria	1770	182.1
1962	Belgium	1700	118.1
1962	Mexico	1680	31.9
1963	Italy	1510	114.3
1961	Denmark	1500	144.9
1962	France	1410	59.7
1962	Sweden	1270	126.9
1961	Spain	1200	43.9
1962	Norway	1090	136.3

Graph these data and compute their correlation coefficient.

19. Let X and Y be jointly distributed according to the pdf

$$f_{X,Y}(x, y) = \begin{cases} 24xy, & x \geq 0, y \geq 0, x + y \leq 1 \\ 0, & \text{elsewhere.} \end{cases}$$

Sketch a graph showing $f_{Y|x}(y)$ for $x = \frac{1}{4}, \frac{1}{2}$, and $\frac{3}{4}$. Also, draw $Y = E(Y|x)$ on the same graph.

20. Let X and Y have the pdf

$$f_{X,Y}(x, y) = \begin{cases} 2, & 0 < x < y, 0 < y < 1 \\ 0, & \text{elsewhere.} \end{cases}$$

Graph the function $Y = E(Y|x)$.

21. Let X and Y have the trinomial pdf

$$f_{X,Y}(x, y) = \frac{n!}{x!\, y!\, (n - x - y)!} p_X^x p_Y^y (1 - p_X - p_Y)^{n-x-y}$$

(see Section 9.2). Find the regression curve of Y on X.

22. Let X and Y be discrete random variables. The *conditional variance of Y given that $X = x$* is defined to be

$$\text{Var}\,(Y|x) = \sum_{\text{all } y} [y - E(Y|x)]^2 f_{Y|x}(y).$$

(An analogous statement holds when X and Y are continuous.) Show that an equivalent way of expressing $\text{Var}\,(Y|x)$ is

$$\text{Var}\,(Y|x) = E(Y^2|x) - [E(Y|x)]^2.$$

23. Find $\text{Var}\,(Y|x)$ for $x = 1$ and $x = 2$ for the probability model given in Review Exercise 6.

24. Let X and Y have the joint pdf,

$$f_{X,Y}(x, y) = \begin{cases} 2, & 0 \leq x < y < 1 \\ 0, & \text{elsewhere.} \end{cases}$$

Find $\text{Var}\,(Y|x)$.

25. Suppose the regression of Y on X is linear, as is the regression of X on Y. Verify that $\rho(X, Y)$ is the geometric mean of the slopes of the two regression lines. (Recall that the geometric mean of two numbers a and b is $\sqrt{ab}$.)

26. Suppose X and Y have the joint pdf given in Review Exercise 4. Graph $Y = E(Y|x)$. Also, find $\text{Var}\,(Y|x)$ and $P(Y > \frac{3}{4}\,|\,X = \frac{1}{2})$.

27. Suppose two fair dice are tossed. Let X and Y denote the numbers appearing on the first and second die, respectively. Let $Z = X + Y$. Find $E(X|Z = 5)$.

28. Find the regression line of Y on X, where

$$f_{X,Y}(x, y) = \frac{y}{(x + 1)^4} e^{-y/(x+1)}, \qquad 0 < x, 0 < y.$$

Verify that the coefficients agree with the statement of Theorem 10.5.

29. Let X and Y have the bivariate pdf

$$f_{X,Y}(x, y) = 24xy, \qquad 0 < x, 0 < y, x + y < 1.$$

Find the regression line of Y on X.

30. In astronomy, Hubble's law states that there is a linear relationship between the distance of a galaxy from the Earth and the velocity at which that galaxy is moving away from the Earth. Listed below are "distance-velocity" measurements made on 11 different galaxies (26).

Cluster	Distance (millions of light-years)	Velocity (thousands of miles/sec)
Virgo	22	.75
Pegasus	68	2.4
Perseus	108	3.2
Coma Berenices	137	4.7
Ursa Major No. 1	255	9.3
Leo	315	12.0
Corona Borealis	390	13.4
Gemini	405	14.4
Boötes	685	24.5
Ursa Major No. 2	700	26.0
Hydra	1100	38.0

Plot the data and find the equation of the linear relationship. Let "distance" be the "x" variable.

31. Verify that the coefficients $\hat{a}$ and $\hat{b}$ of the least-squares straight line are solutions of the matrix equation

$$\begin{pmatrix} n & \sum x_i \\ \sum x_i & \sum x_i^2 \end{pmatrix}\begin{pmatrix} \hat{a} \\ \hat{b} \end{pmatrix} = \begin{pmatrix} \sum y_i \\ \sum x_i y_i \end{pmatrix}.$$

32. The aging of whisky in charred oak barrels brings about a number of chemical changes that enhance its taste and darken its color. Shown below is the change in a whiskey's proof as a function of the number of years it is stored (156).

Age (years), x	Proof, y
0	104.6
0.5	104.1
1	104.4
2	105.0
3	106.0
4	106.8
5	107.7
6	108.7
7	110.6
8	112.1

(Note: The proof initially decreases because of dilution by moisture in the staves of the barrels.) Graph these data and draw in the least-squares line.

33. Set up (but do not solve) the equations necessary for determining the least-squares estimates for a trigonometric regression,

$$y = a + bx + c \cdot \sin x + d \cdot \cos x.$$

Assume that a random sample $(x_1, y_1), \ldots, (x_n, y_n)$ has been observed.

34. Certain nonlinear regression curves can be made linear by means of a suitable transformation. Recall Case Study 10.3, where the exponential model, $y = ae^{bx}$, was "linearized" by our taking logs of both sides to give

$$\ln y = \ln a + bx.$$

In the latter form, "x" is linear with "$\ln y$." Or, put another way, a graph of x versus y on semilog paper would be a straight line. (Transformations of this sort are extremely useful, because they enable us to use Theorem 10.6 to get estimates for the model's unknown coefficients.) Reexpress the following regression curves in linear form.

(a) $y = ax^b$ (b) $y = a + b/x.$

(c) $y = \dfrac{1}{a + bx}.$ (d) $y = \dfrac{x}{a + bx}.$

(e) $y = \dfrac{c}{1 + e^{a - bx}}.$

35. Find the coefficients for the least-squares fit of the equation $y = ax + (b/x)$ to n points in the plane.

36. Over the years, many efforts have been made to demonstrate that the human brain is appreciably different in structure than the brains of lower-order primates. In point of fact, such differences in gross anatomy are disconcertingly difficult to discern. Listed below are the average areas of the striate cortex (x) and the prestriate cortex (y) found for humans and for three species of chimpanzees (125).

	Areas	
Primate	Striate Cortex (mm²), x	Prestriate Cortex (mm²), y
Homo	2613	7838
Pongo	1876	2864
Cercopithecus	933	1334
Galago	78.9	40.8

Fit these data with an equation of the form

$$y = ax^b.$$

Hint: First make a suitable model transformation (see Review Exercise 34). Graph the data and sketch in the least-squares curve. Is the relationship between striate cortex area and prestriate cortex area in humans similar to the corresponding relationship in chimpanzees?

37. Listed below are the average weights of varsity football players at the University of Texas for selected years from the turn of the century to the middle 1960s. [Data slightly modified from (105).]

Year	Average Weight (lb)
1905	164
1919	163
1932	181
1945	192
1955	194
1965	199

Graph these data and fit them with a straight line. If the linear model were to remain valid for this phenomenon in the years ahead (a highly unlikely state of affairs!), what would the average University of Texas football player weigh in the year 3000? What does this tell you about (1) extrapolating linear models and/or (2) scheduling the University of Texas to play your school on Homecoming in the year 3000?

38. To better understand the plight of people returning to professions after periods of inactivity, a survey was conducted at some 67 randomly selected hospitals throughout the United States: the administrators were asked about their willingness to hire medical technologists who had been away from the field for a certain number of years. The results are summarized on the next page (142).

Years of Inactivity	Percent of Hospitals Willing to Hire
$\frac{1}{2}$	100
$1\frac{1}{2}$	94
4	75
8	44
13	28
18	17

Graph these data and fit them with a straight line.

39. Let a and b be the least-squares coefficients for the set of points $(x_1, y_1), (x_2, y_2)$, $\ldots, (x_n, y_n)$. What are the corresponding coefficients for the points $(x_1, y_1 - c), \ldots,$ $(x_n, y_n - c)$?

40. Mistletoe is a plant that grows parasitically in the upper branches of large trees. Left unchecked, it can seriously stunt a tree's growth. Recently an experiment was done to test a theory that older trees are less susceptible to mistletoe growth than younger trees. A number of shoots were cut from 3-, 4-, 9-, 15-, and 40-year-old Ponderosa pines. These were then side-grafted to 3-year-old nursery stock and planted in a preserve. Each tree was "inoculated" with mistletoe seeds. Five years later, a count was made of the number of mistletoe plants in each stand of trees. (A stand consisted of approximately ten trees; there were three stands of each of the four youngest age groups and two stands of the oldest.) The results are shown below (145).

	Age of Trees (years), x				
	3	4	9	15	40
Number of	28	10	15	6	1
Mistletoe	33	36	22	14	1
Plants, y	22	24	10	9	

(a) Graph the original data. Also, graph x versus $\ln y$.
(b) Fit the data with an exponential model, $y = ae^{bx}$. Sketch in the least-squares curve on the graph of the original data.

41. Prominent among the many useful "growth" models in statistics is the *logistic equation*

$$y = \frac{L}{1 + e^{a-bx}},$$

where L is a growth limit, a is a constant, and b is the rate of increase. One of the applications found for the logistic equation is in the area of strategic arms competition. Listed at the top of the next page is the size of the American Intercontinental Ballistic Missile (ICBM) force from 1960 through 1969 (9). Linearize the logistic model and fit it to the ICBM data (see Review Exercise 34). Let $L = 1054$. Make up a table showing the actual and the predicted sizes of the ICBM force for all the years from 1960 through 1969.

Year	Number of ICBM's	Year	Number of ICBM's
1960	18	1965	854
1961	63	1966	904
1962	294	1967	1054
1963	424	1968	1054
1964	834	1969	1054

42. On September 13, 1942, newspapers around the country printed the following chart released by President Roosevelt, who was trying to encourage citizens to conserve rubber by driving slower. The life of an automobile tire driven at 40 mph was arbitrarily set at 100, and the corresponding tire lives at other speeds are shown on the graph (5).

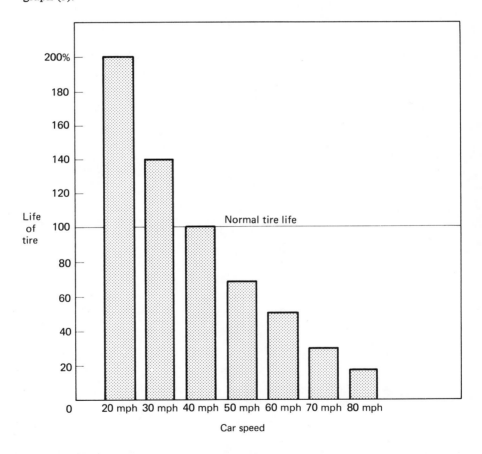

Find an exponential function, $y = ae^{bx}$, relating speed (x) to tire life (y) by calculating the regression coefficients for $\ln y = \ln a + bx$.

43. Refer to the tire-life data of Review Exercise 42. Suppose we arbitrarily decide that tire life at 20 mph will be twice that at 40 mph—that is, for x equal to 20, y should equal 200. Find values for a and b by solving simultaneously the equations

$$100 = ae^{40b} \quad \text{and} \quad 200 = ae^{20b}.$$

Compute the expected tire lives as a function of speed driven for both this model and the least-squares model of Review Exercise 42.

44. Radioactive gold (^{195}Au-aurothiomalate) has an affinity for inflamed tissue and is sometimes used as a tracer to help diagnose arthritis. The following data came from an experiment investigating the length of time, and in what concentrations, ^{195}Au-aurothiomalate is retained in a person's serum (58). Listed below are serum gold concentrations found in ten blood samples taken from patients given 50 mg of auro-thiomalate. Readings were made at various times, ranging from one to seven days after injection. In each case the retention is expressed in terms of the patient's day-zero serum gold concentration.

Days After Injection	Serum Gold (percent of day-zero conc.)
1	94.5
1	86.4
2	71
2	80.5
2	81.4
3	67.4
5	49.3
6	46.8
6	42.3
7	36.6

(a) Graph these data.
(b) The kinetics of radioactive decay would suggest that the retention versus time relationship be exponential. Fit these data with a model of the form, $y = ae^{bx}$.
(c) Estimate the effective half-life of ^{195}Au-aurothiomalate. That is, how long does it take for half the gold to disappear from a person's serum?

45. Suppose we wish to fit a least-squares *parabola*, $y = \hat{a} + \hat{b}x + \hat{c}x^2$, to a set of n points, $(x_1, y_1), (x_2, y_2), \ldots, (x_n, y_n)$. Show that the coefficients $\hat{a}$, $\hat{b}$, and $\hat{c}$ solve the matrix equation

$$\begin{pmatrix} n & \sum x_i & \sum x_i^2 \\ \sum x_i & \sum x_i^2 & \sum x_i^3 \\ \sum x_i^2 & \sum x_i^3 & \sum x_i^4 \end{pmatrix} \begin{pmatrix} \hat{a} \\ \hat{b} \\ \hat{c} \end{pmatrix} = \begin{pmatrix} \sum y_i \\ \sum x_i y_i \\ \sum x_i^2 y_i \end{pmatrix}.$$

46. Derive the matrix equation giving the coefficients of the least-squares polynomial of degree k.

47. Suggestibility as a psychological term refers to a person's acceptance of external influences. The following data were obtained in a study designed to investigate the relationship between suggestibility and chronological age. The subjects were students at a kindergarten-through-twelfth-grade private school. Each of the experiment's several parts involved showing the subject a slide and "intimidating" him to various extents into agreeing with some statement that was incorrect. For example, a slide might depict two geometrical figures of equal area and the subject would be asked whether the figures had (1) different areas or (2) the same area. The examiner, however, would preface the slide with a comment to the effect that most students are, in fact, able to distinguish a difference in area. The more suggestible students, then, will tend to answer (1), even though their initial response may have been (2). Eventually, each age group from 6 to 18 was given several such tests and ultimately assigned a composite suggestibility score that was expressed as a standardized normal deviate. Large positive scores represented the greatest amount of suggestibility; large negative scores, the least amount. The results are shown in Figure 10.11 (107). Use the result given in Review Exercise 45 to fit a parabola to these data.

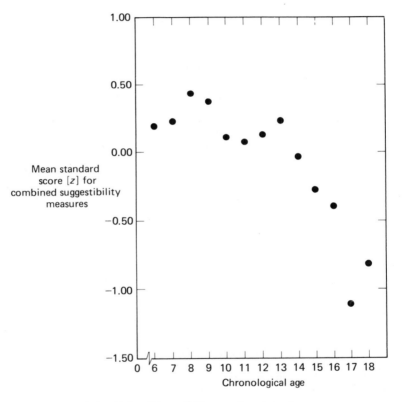

Figure 10.11   Suggestibility as a function of age.

48. Let (x_i, Y_i), $i = 1, 2, \ldots, n$, be a random sample where $Y_i \sim N(a + bx_i, \sigma^2)$.
 (a)   Prove that $E(\bar{Y}) = a + b\bar{x}$.
 (b)   Prove that $E(\hat{B}) = b$. Hint: Write $\hat{B}$ in the form given in Question 10.5.1.
 (c)   Prove that

$$\text{Var}(\hat{B}) = \frac{\sigma^2}{\sum\limits_{i=1}^{n}(x_i - \bar{x})^2}.$$

Hint: Write $\hat{B}$ in the form given in Question 10.5.1.

(d)   Prove that $E(\hat{A}) = a$. Hint: Use (a) and (b).

(e)   Prove that

$$\text{Cov}\,(\hat{B},\,Y_j) = \frac{x_j - \bar{x}}{\sum\limits_{i=1}^{n}(x_i - \bar{x})^2}\,\sigma^2.$$

(f)   Prove that $\text{Cov}\,(\hat{B},\,\bar{Y}) = 0$.

(g)   Prove that

$$\text{Var}\,(\hat{A}) = \frac{\sigma^2\sum\limits_{i=1}^{n}x_i^2}{n\sum\limits_{i=1}^{n}(x_i - \bar{x})^2}.$$

Hint: Use (c) and (f).

49. We wish to estimate the coefficients of the linear model. As usual, n x-values are to be selected. Suppose the values for x must lie in the interval $[-5, 5]$ and, for convenience, n is even. How should the x's be chosen so as to minimize Var $(\hat{B})$?

50. Suppose the x-values referred to in Review Exercise 49 were evenly spaced over the interval $[-5, 5]$. What would be the relative efficiency of estimating b with that sampling scheme as opposed to the optimal strategy?

51. Suppose in Case Study 10.4 that σ were known to be 13.64.
    (a)   Use the $N(0, 1)$ statistic

$$\frac{(\hat{B} - b)}{\sigma}\sqrt{\sum_{i=1}^{n}(x_i - \bar{x})^2}$$

to construct a 95% confidence interval for b.

    (b)   Use the $N(0, 1)$ statistic

$$\frac{(\hat{A} - a)}{\sigma}\sqrt{\frac{n\sum\limits_{i=1}^{n}(x_i - \bar{x})^2}{\sum\limits_{i=1}^{n}x_i^2}}$$

to construct a 95% confidence interval for a.

52. Prove Theorem 10.11.

53. Let $(x_1, Y_1), \ldots, (x_n, Y_n)$, with $Y_i \sim N(a + bx_i, \sigma^2)$, be a random sample and $(u_1, V_1), \ldots, (u_m, V_m)$, with $V_i \sim N(c + du_i, \sigma^2)$, a second random sample independent of the first. Let $\hat{\sigma}_1^2$ be the estimator of σ^2 from the first sample and $\hat{\sigma}_2^2$ the estimator from the second. Find a statistic to test $H_0: b = d$ versus $H_1: b \ne d$.

54. For the six airlines with the largest 1973 passenger revenues, the table at the top of the next page (92) gives the number of passengers carried (x) and the corresponding passenger revenues (Y). Since passenger revenues depend heavily on routes, class of travel, and distances, there could fail to be significance between x and Y. Examine this possibility by testing $H_0: b = 0$ versus $H_1: b > 0$ at the 0.05 level.

Airline	Number of Passengers	Passenger Revenues
United	30,250	1,654
Eastern	26,201	1,130
Delta	24,604	1,021
American	21,163	1,296
TWA	14,148	1,054
Pan American	10,409	1,049

55. Refer to the smoking-CHD mortality data of Review Exercise 18.
   (a)   If x denotes a country's cigarette consumption and y its CHD mortality rate, fit a straight line to the data of the form $y = a + bx$.
   (b)   Test H_0: $b = 0$ versus H_1: $b > 0$. (Why should the alternative be one-sided here?) Let $\alpha = 0.05$.
   (c)   Construct a 95% confidence interval for b.
   (d)   Construct a 95% confidence interval for the expected CHD mortality rate in a country where the cigarette consumption is 2500 per adult per year.

56. Determining small quantities of calcium in the presence of magnesium is a difficult problem for the analytical chemist. Direct precipitation is not feasible. One of the procedures proposed involves the use of alcohol as a solvent. The data below present the results of applying the alcohol method to ten mixtures containing known quantities of CaO. The second column gives, in each instance, the amount of CaO recovered (72).

CaO Present (mg), x	CaO Recovered (mg), y
4.0	3.7
8.0	7.8
12.5	12.1
16.0	15.6
20.0	19.8
25.0	24.5
31.0	31.1
36.0	35.5
40.0	39.4
40.0	39.5

Perhaps the most immediate question to be answered is whether the discrepancies between what was present and what was found are (1) random or (2) intrinsic to the procedure. One way to examine those options is by expressing the amount recovered (y) as a linear function of the amount present (x)—that is, $y = a + bx$—and then testing the null hypotheses that $a = 0$ and $b = 1$.
   (a)   At the 5% level, test H_0: $a = 0$ versus H_1: $a \neq 0$.
   (b)   At the 5% level, test H_0: $b = 1$ versus H_1: $b \neq 1$. Use the statistic of Theorem 10.10.

57. A second method for recovering calcium from magnesium (see Review Exercise 56) has been found that requires fewer precipitations than the alcohol method. In a preliminary study of this second technique, the following data were obtained.

CaO Present (mg), x	CaO Recovered (mg), y
4.0	3.9
8.0	8.1
12.5	12.4
16.0	16.0
20.0	19.8
25.0	25.0
31.0	31.1
36.0	35.8
40.0	40.1
40.0	40.1

Fit these data with a regression equation of the form $y = c + dx$. At the 5% level of significance, test H_0: $b = d$ versus H_1: $b \neq d$. Use the statistic from Review Exercise 53. If H_0 is accepted, it seems clear the new method should be adopted because of its simplicity. What if H_0 is rejected?

58. Set up the test statistic that would be used in making inferences about the y-intercept, a, in the linear model, $y = a + bx$. (Assume that σ^2 is unknown.) What distribution would the statistic have?

59. Refer to the tethered-flight data of Question 10.5.4. Construct a 95% confidence interval for b and a 90% confidence interval for the expected nonstop distance that can be flown by mosquitoes $2\frac{1}{2}$ weeks old. Also, plot the least-squares line on the original data and draw in, as a function of x, the 90% confidence limits for $E(Y|x)$. Finally, draw in, again as a function of x, the 90% prediction limits for Y.

60. The figures below give the monthly sales and the monthly sales expenses for a small manufacturing firm in Norman, Oklahoma (114).

Month	Sales ($ thousands)	Sales Expenses ($ thousands)
4/78	187.1	25.4
5/78	179.5	22.8
6/78	157.0	20.6
7/78	197.0	21.8
8/78	239.4	32.4
9/78	217.8	24.4
10/78	227.1	29.3
11/78	233.4	27.9
12/78	242.0	27.8
1/79	251.9	34.2
2/79	190.0	29.2
3/79	295.8	30.0

(a)   Find the least-squares straight line describing sales expenses (y) as a function of sales (x).
(b)   Test $H_0: b = 0$ versus $H_1: b > 0$ at the 0.05 level of significance.
(c)   Find a 95% confidence interval for b.
(d)   Suppose the company expects monthly sales to become essentially constant at $200,000 per month. Give a 95% confidence interval for the average sales expenses.
(e)   The company is considering what the costs would be if sales were boosted to $300,000 in a certain month. Find the corresponding 95% prediction interval for the sales expenses.
(f)   Use the statistic from Review Exercise 58 to find a 95% confidence interval for the y-intercept, a. Note that, here, a represents the fixed or overhead costs for sales. The large width of a's interval emphasizes the uncertainty in estimating for values of x that are numerically distant from the x-values actually sampled.

61. One of the obvious applications of regression analysis is the situation where two variables, x and y, are related, but one of them—say, y—is difficult to measure. We can finesse such a problem by simply measuring the x-variable and estimating y via the regression function. It may be extremely difficult, for example, to measure the volume of an irregularly shaped object but very simple to weigh it. The table below shows the weight in kilograms and the volume in cubic decimeters of 18 children between the ages of 5 and 8 (16).

Weight, x	Volume, y	Weight, x	Volume, y
17.1	16.7	15.8	15.2
10.5	10.4	15.1	14.8
13.8	13.5	12.1	11.9
15.7	15.7	18.4	18.3
11.9	11.6	17.1	16.7
10.4	10.2	16.7	16.6
15.0	14.5	16.5	15.9
16.0	15.8	15.1	15.1
17.8	17.6	15.1	14.5

(a)   Graph the data and find the least-squares line, $y = \hat{a} + \hat{b}x$.
(b)   Construct a 95% confidence interval for $E(Y|14.0)$.
(c)   Construct a 95% prediction interval for the volume of a child weighing 14.0 kilograms.

62. Show that

$$\sum_{i=1}^{n} (Y_i - \bar{Y})^2 = \sum_{i=1}^{n} (Y_i - \hat{Y}_i)^2 + \sum_{i=1}^{n} (\hat{Y}_i - \bar{Y})^2$$

for any set of points (x_i, Y_i), $i = 1, 2, \ldots, n$.

63. The expression

$$\frac{\sum_{i=1}^{n} (\hat{Y}_i - \bar{Y})^2}{\sum_{i=1}^{n} (Y_i - \bar{Y})^2}$$

is called the *coefficient of determination*. Express the coefficient of determination as a function of the sample correlation coefficient (ignore the fact that x is not a random variable here). What does the coefficient of determination represent (see Review Exercise 62)?

64. Suppose X and Y have a bivariate normal distribution with Var $(X) = $ Var (Y).
   (a) Show that X and $Y - \rho X$ are independent.
   (b) Show that $X + Y$ and $X - Y$ are independent. Hint: See Review Exercise 9.

65. Suppose X and Y have a bivariate normal distribution.
   (a) Prove that $cX + dY$ has a normal distribution for any constants c and d.
   (b) Find $E(cX + dY)$ and Var $(cX + dY)$ in terms of $\mu_X, \mu_Y, \sigma_X, \sigma_Y,$ and $\rho(X, Y)$.

66. Suppose the random variables X and Y have a bivariate normal pdf with $\mu_X = 56$, $\mu_Y = 11$, $\sigma_X^2 = 1.2$, $\sigma_Y^2 = 2.6$, and $\rho = 0.6$. Compute $P(10 < Y < 10.5 \,|\, x = 55)$. Suppose $n = 4$ values were to be observed with x fixed at 55. Find $P(10.5 < \bar{Y} < 11 \,|\, x = 55)$.

67. If the joint pdf of the random variables X and Y is

$$f_{X,Y}(x, y) = k \cdot e^{-(2/3)[(1/4)x^2 - (1/2)xy + y^2]},$$

find $E(X)$, $E(Y)$, Var (X), Var (Y), $\rho(X, Y)$, and k.

68. Give conditions on $a > 0$, $b > 0$, and u so that

$$f_{X,Y}(x, y) = k \cdot e^{-(ax^2 - 2uxy + by^2)}$$

is the bivariate normal density of random variables X and Y each having expected value zero. Also, find Var (X), Var (Y), and $\rho(X, Y)$.

69. Refer to the baseball data of Review Exercise 16. Test the independence of altitude and frequency of home runs. Let $\alpha = 0.05$.

70. Test $H_0: \rho(X, Y) = 0$ versus $H_1: \rho(X, Y) > 0$ for the smoking–CHD mortality rate data of Review Exercise 18. Let $\alpha = 0.05$. Does your answer agree with part (b) of Review Exercise 55?

71. Use the statistic

$$\frac{1}{2} \ln \frac{1 + R}{1 - R}$$

to get an expression for an approximate $100(1 - \alpha)\%$ confidence interval for ρ. Use your result on the data of Question 10.2.8 to construct a 90% confidence interval for the correlation coefficient between the behavioral index and the plumage index of mallard-pintail hybrids.

CHAPTER ELEVEN

The Analysis of Variance

RONALD A. FISHER

No aphorism is more frequently repeated in connection with field trials, than that we must ask Nature few questions or, ideally, one question, at a time. The writer is convinced that this view is wholly mistaken. Nature, he suggests, will best respond to a logical and carefully thought out questionnaire; indeed, if we ask her a single question, she will often refuse to answer until some other topic has been discussed.

11.1 INTRODUCTION

In this chapter we take up an important extension of the two-sample location problem introduced in Chapter 8. The *completely randomized, one-factor design* is a conceptually similar k-sample location problem, but one that requires a substantially different sort of analysis than its prototype. Here, the appropriate test statistic turns out to be a ratio of variance estimates, the sampling behavior of which is described by an F distribution rather than a Student t (recall Theorem 8.2). The name attached to this procedure, in deference to the form of its test statistic, is the *analysis of variance* (or "ANOVA," for short). A very flexible method, the analysis of variance finds many other applications, a particularly important one being the *randomized block design*, which is the subject of Chapter 12.

Comment. Credit for much of the early development of the analysis of variance goes to Sir Ronald A. Fisher. Shortly after the end of World War I, Fisher resigned a public school teaching position that he was none too happy with and accepted a post at the Rothamsted Statistical Laboratory, a facility heavily involved in agricultural research. There he suddenly found himself entangled in problems where differences in the response variable (crop yields, for example) were constantly in danger of being obscured by the high level of uncontrollable heterogeneity in the experimental environment (different soil qualities, drainage gradients, and so on). Quickly seeing that traditional techniques were hopelessly inadequate under these conditions, Fisher set out to look for alternatives and in just a few years succeeded in fashioning an entirely new statistical methodology, a panoply of data-collecting principles and mathematical tools that is today known as *experimental design*. The centerpiece of Fisher's creation—what makes it all work—is the analysis of variance. ∎

Suppose an experimenter wishes to compare the average effects elicited by k different levels of some given factor, where k is greater than or equal to 2. The factor, for example, might be "stop-smoking" therapies and the levels, three specific methods. Or the factor might be crowdedness as it relates to aggression in captive monkeys, with the levels being five different monkey-per-square-foot densities in five separate enclosures. Still another example might be an engineering study comparing the effectiveness of four kinds of catalytic converters in reducing the concentrations of harmful emissions in automobile exhaust. Whatever the circumstances, data from a completely randomized, one-factor design will consist of k independent random samples of sizes $n_1, n_2, \ldots,$ and n_k, the total sample size being denoted $n \left(= \sum_{j=1}^{k} n_j \right)$. We will let Y_{ij} represent the ith observation recorded for the jth level. Table 11.1 shows some additional notation.

The dot notation of Table 11.1 is standard in analysis-of-variance problems. The presence of a dot in lieu of a subscript indicates that that particular subscript has been summed over. Thus, the response total for the jth sample is written

TABLE 11.1 Analysis-of-variance notation

Factor Level

1	2	...	k
Y_{11}	Y_{12}		Y_{1k}
Y_{21}	Y_{22}		
.	.	...	.
.	.		.
.	.		.
Y_{n_11}	Y_{n_22}		Y_{n_kk}

	1	2	...	k
Sample sizes:	n_1	n_2	...	n_k
Sample totals:	$T_{.1}$	$T_{.2}$		$T_{.k}$
Sample means:	$\bar{Y}_{.1}$	$\bar{Y}_{.2}$		$\bar{Y}_{.k}$
True means:	μ_1	μ_2	...	μ_k

$$T_{.j} = \sum_{i=1}^{n_j} Y_{ij}$$

and the corresponding sample mean,

$$\bar{Y}_{.j} = \frac{1}{n_j} \sum_{i=1}^{n_j} Y_{ij} = \frac{T_{.j}}{n_j}.$$

By the same convention, $T_{..}$ and $\bar{Y}_{..}$ will denote the overall total and overall mean, respectively:

$$T_{..} = \sum_{j=1}^{k} \sum_{i=1}^{n_j} Y_{ij} = \sum_{j=1}^{k} T_{.j},$$

$$\bar{Y}_{..} = \frac{1}{n} \sum_{j=1}^{k} \sum_{i=1}^{n_j} Y_{ij} = \frac{1}{n} \sum_{j=1}^{k} n_j \bar{Y}_{.j} = \frac{1}{n} \sum_{j=1}^{k} T_{.j}.$$

Appearing at the bottom of Table 11.1 are a set of "true means," $\mu_1, \mu_2, \ldots, \mu_k$. Each μ_j is an unknown location parameter reflecting the true average response characteristic of level j. Depending on the physical circumstances of the problem, our objective will be either to estimate the μ_j's or test their equality. The latter is perhaps the more common, and the test takes the form

$$H_0: \mu_1 = \mu_2 = \cdots = \mu_k$$

versus

$$H_1: \text{not all the } \mu_j\text{'s are equal.}$$

In the next several sections we will propose a variance-ratio statistic for testing H_0, investigate its sampling behavior under both H_0 and H_1, and introduce a set of computing formulas to simplify its evaluation. We will also explore the possibility of testing more specific *subhypotheses* about the μ_j's—for instance, $H_0^1: \mu_2 = \mu_4$ or $H_0^2: (\mu_2 + \mu_5)/2 = \mu_3$.

11.2 THE *F* TEST

To derive a procedure for testing $H_0: \mu_1 = \mu_2 = \cdots = \mu_k$ we could once again invoke the generalized-likelihood-ratio criterion, compute $\lambda = L(\hat{\omega})/L(\hat{\Omega})$, and begin the search for a monotonic function of λ having a known distribution. But since we have already seen several examples of formal GLRT calculations in Chapters 7 and 8, the benefits of doing still another would be minimal. Instead, we will work backward: the test statistic will be stated at the outset and the "derivation" will be confined to an investigation of some of its properties.

The data structure for a completely randomized, one-factor design was outlined in Section 11.1. To that basic setup we now add a *distribution* assumption—that the Y_{ij}'s are independent and normally distributed with mean μ_j, $j = 1, 2, \ldots, k$, and variance σ^2 (constant for all j). With our previous notation, this could have been written

$$Y_{ij} \sim N(\mu_j, \sigma^2), \tag{11.1}$$

but in analysis-of-variance problems—as in regression problems—distribution assumptions are more typically expressed in terms of *model equations*. In these equations the response variable is represented as the sum of one or more fixed components and one or more random components. Here, one possible model equation would be

$$Y_{ij} = \mu_j + \epsilon_{i(j)}, \tag{11.2}$$

where $\epsilon_{i(j)}$ denotes the "noise" associated with Y_{ij}—that is, the amount by which Y_{ij} differs from its expected value. Of course, from Equation 11.1 it follows that $\epsilon_{i(j)} \sim N(0, \sigma^2)$.

While Equation 11.2 is a perfectly acceptable way to represent the data for a completely randomized, one-factor design, another parameterization is more commonly used. Equations 11.3 and 11.4 define μ, the overall average effect of the k levels and τ_j, the *differential effect* of level j (relative to μ):

$$\tau_j = \mu_j - \mu \tag{11.3}$$

and

$$\mu = \frac{1}{n} \sum_{j=1}^{k} n_j \mu_j. \tag{11.4}$$

In terms of μ and τ_j,

$$Y_{ij} = \mu + \tau_j + \epsilon_{i(j)}. \tag{11.5}$$

The equivalence of Equations 11.2 and 11.5 should be obvious: testing $H_0: \mu_1 = \mu_2 = \cdots = \mu_k$ is the same as testing $H_0: \tau_1 = \tau_2 = \cdots = \tau_k = 0$.

EXAMPLE 11.1. Suppose a completely randomized, one-factor design involves three levels whose true means are $\mu_1 = 2$, $\mu_2 = 7$, and $\mu_3 = 5$. Furthermore, suppose four observations are taken from level 1 and two each from levels 2 and 3 (see Table 11.2). With the second parameterization, the model equation becomes

$$Y_{ij} = \mu + \tau_j + \epsilon_{i(j)}, \qquad i = 1, 2, \ldots, n_j; j = 1, 2, 3,$$

TABLE 11.2

Sample 1 ($\mu_1 = 2$)	Sample 2 ($\mu_2 = 7$)	Sample 3 ($\mu_3 = 5$)
Y_{11}	Y_{12}	Y_{13}
Y_{21}	Y_{22}	Y_{23}
Y_{31}		
Y_{41}		

and the values of the parameters would be

$$\mu = \tfrac{1}{8}[4(2) + 2(7) + 2(5)] = 4.0,$$
$$\tau_1 = 2 - 4.0 = -2.0,$$
$$\tau_2 = 7 - 4.0 = +3.0,$$
$$\tau_3 = 5 - 4.0 = +1.0.$$

QUESTION 11.2.1   Suppose the particular value recorded for Y_{11} was 3.5. What would be the corresponding value for $\epsilon_{1(1)}$? Which, if either, of the two sums

$$\sum_{i=1}^{n_j} (Y_{ij} - \bar{Y}_{.j}) \quad \text{or} \quad \sum_{i=1}^{n_j} (Y_{ij} - \mu_j)$$

is identically zero? What is the expected value of each sum?

QUESTION 11.2.2   (a) Show that, in general,

$$\sum_{j=1}^{k} n_j \tau_j = 0.$$

(b) Verify that the equality in (a) holds for the data of Example 11.1.

Comment. Keep in mind that in any real problem neither μ nor the τ_j's are ever known; however, they can be estimated using the method of least squares. In this context the least-squares function, L, of Section 10.4 becomes

$$L = L(\mu, \tau_1, \tau_2, \ldots, \tau_k) = \sum_{j=1}^{k} \sum_{i=1}^{n_j} (y_{ij} - \mu - \tau_j)^2.$$

Taking derivatives of L with respect to the unknown parameters, and setting those derivatives equal to 0, gives a set of $k + 1$ equations:

$$\frac{\partial L}{\partial \mu} = n\mu + \sum_{j=1}^{k} n_j \tau_j - \sum_{j=1}^{k} \sum_{i=1}^{n_j} y_{ij} = 0,$$

$$\frac{\partial L}{\partial \tau_j} = n_j \mu + n_j \tau_j - \sum_{i=1}^{n_j} y_{ij} = 0, \qquad j = 1, 2, \ldots, k. \tag{11.6}$$

Unique solutions to Equations 11.6 can be obtained only by adding a $(k + 2)$nd equation, the side condition that

$$\sum_{j=1}^{k} n_j \tau_j = 0$$

(recall Question 11.2.2). The resulting estimates, which should come as no surprise, are

$$\hat{\mu} = \bar{y}_{..} \quad \text{and} \quad \hat{\tau}_j = \bar{y}_{.j} - \bar{y}_{..}, \qquad j = 1, 2, \ldots, k. \qquad (11.7)$$

∎

QUESTION 11.2.3   Verify Equations 11.6 and 11.7.

Our development of the analysis of the variance for testing $H_0 : \tau_1 = \tau_2 = \cdots = \tau_k = 0$ will proceed along three lines:

   I. Definition of the test statistic.
  II. Derivation of the distribution of the test statistic under H_0.
 III. Proof that the test statistic is sensitive to departures from H_0.

Taken together, the conclusions of parts I, II, and III will demonstrate the feasibility of using variance ratios to test the equality of means. The important question of the test statistic's optimality, though—specifically, its relationship to the generalized-likelihood-ratio criterion—will be deferred to the exercises.

Part I

We begin with an identity:

$$Y_{ij} = \bar{Y}_{..} + (\bar{Y}_{.j} - \bar{Y}_{..}) + (Y_{ij} - \bar{Y}_{.j})$$

or, equivalently,

$$(Y_{ij} - \bar{Y}_{..}) = (\bar{Y}_{.j} - \bar{Y}_{..}) + (Y_{ij} - \bar{Y}_{.j}), \qquad (11.8)$$

which should be recognized as simply Equation 11.5, with the parameters μ and τ_j having been replaced by their least-squares estimators and $\epsilon_{i(j)}$ by $Y_{ij} - \bar{Y}_{.j}$. Since Equation 11.8 holds for *any* i and j, it must be true that

$$\sum_{j=1}^{k} \sum_{i=1}^{n_j} (Y_{ij} - \bar{Y}_{..})^2 = \sum_{j=1}^{k} \sum_{i=1}^{n_j} [(\bar{Y}_{.j} - \bar{Y}_{..}) + (Y_{ij} - \bar{Y}_{.j})]^2. \qquad (11.9)$$

Expanding the right-hand side of Equation 11.9 gives

$$\sum_{j=1}^{k} \sum_{i=1}^{n_j} (\bar{Y}_{.j} - \bar{Y}_{..})^2 + \sum_{j=1}^{k} \sum_{i=1}^{n_j} (Y_{ij} - \bar{Y}_{.j})^2$$

since the cross-product term vanishes:

$$\sum_{j=1}^{k} \sum_{i=1}^{n_j} (\bar{Y}_{.j} - \bar{Y}_{..})(Y_{ij} - \bar{Y}_{.j}) = \sum_{j=1}^{k} (\bar{Y}_{.j} - \bar{Y}_{..}) \sum_{i=1}^{n_j} (Y_{ij} - \bar{Y}_{.j})$$

$$= \sum_{j=1}^{k} (\bar{Y}_{.j} - \bar{Y}_{..})(0) = 0.$$

Therefore,

$$\sum_{j=1}^{k} \sum_{i=1}^{n_j} (Y_{ij} - \bar{Y}_{..})^2 = \sum_{j=1}^{k} \sum_{i=1}^{n_j} (\bar{Y}_{.j} - \bar{Y}_{..})^2 + \sum_{j=1}^{k} \sum_{i=1}^{n_j} (Y_{ij} - \bar{Y}_{.j})^2$$

or, more conveniently,

$$Q = Q_1 + Q_2. \tag{11.10}$$

It would be well to pause at this point and reflect on what these algebraic manipulations have accomplished. First, consider the term on the left-hand side of Equation 11.10. We will call

$$Q = \sum_{j=1}^{k} \sum_{i=1}^{n_j} (Y_{ij} - \bar{Y}_{..})^2$$

the *total sum of squares* and denote it SS_{total}. What Q measures is the total variability in the data—that is, it quantifies the extent to which the Y_{ij}'s are not all equal. (If they *were* all the same, each would equal $\bar{Y}_{..}$ and Q would be 0.)

On the right-hand side of Equation 11.10,

$$Q_1 = \sum_{j=1}^{k} \sum_{i=1}^{n_j} (\bar{Y}_{.j} - \bar{Y}_{..})^2$$

is an estimate of the extent to which the τ_j's are not all equal: accordingly, Q_1 will be called the *treatment sum of squares* and abbreviated $SS_{\text{treatments}}$.

Finally,

$$Q_2 = \sum_{j=1}^{k} \sum_{i=1}^{n_j} (Y_{ij} - \bar{Y}_{.j})^2$$

measures the variability *within* each level. Deviations of Y_{ij} from $\bar{Y}_{.j}$ estimate *experimental error*, the combined effect on the response variable of factors other than level j. Written SS_{error}, Q_2 is termed the *error sum of squares*.

What Equation 11.10 has accomplished, then, is a *partitioning* of the overall variability among the Y_{ij}'s (as measured by Q) into two components—the first, Q_1, measures that portion of the total variability due to (or "caused by") the fact that the τ_j's are not all 0; the second, Q_2, measures the variability due to factors other than the one being controlled. Conceptually, Q_1 and Q_2 play roles analogous to the numerator and denominator in the two-sample t statistic: Q_2 (like $S_p \sqrt{(1/n) + (1/m)}$) reflects the inherent variability in the data and serves as a yardstick against which the magnitude of the treatment effect, as measured by Q_1, can be compared. (In the two-sample problem, of course, the magnitude of the treatment effect is gauged by $\bar{X} - \bar{Y}$.)

It might seem reasonable at this point to define the quotient Q_1/Q_2 to be our test statistic, but for reasons part II will make clear, both the numerator and denominator first have to be *scaled*—Q_1 by $k - 1$ and Q_2 by $n - k$. The proposed statistic, then, or *F ratio*, for testing

$$H_0: \tau_1 = \tau_2 = \cdots = \tau_k$$

versus

$$H_1: \text{not all the } \tau_j\text{'s} = 0$$

will be

$$F = \frac{Q_1/(k - 1)}{Q_2/(n - k)} = \frac{SS_{\text{treatments}}/(k - 1)}{SS_{\text{error}}/(n - k)}.$$

Part II

In part II we will find the distributions of Q_1 and Q_2 and, from them, of the F ratio, under the null hypothesis that the population means are all equal. First, consider Q_2. Let S_j^2 denote the sample variance for the jth level. In the notation of Table 11.1,

$$S_j^2 = \frac{1}{n_j - 1} \sum_{i=1}^{n_j} (Y_{ij} - \bar{Y}_{.j})^2.$$

It follows by inspection that S_j^2 and Q_2 are directly related:

$$Q_2 = \sum_{j=1}^{k} (n_j - 1)S_j^2. \tag{11.11}$$

Of course, for each j,

$$\frac{(n_j - 1)S_j^2}{\sigma^2} \sim \chi_{n_j - 1}^2$$

(see Theorem 7.8). Furthermore, the k variance ratios are all independent, since the k samples themselves are independent; therefore, by the additive property of chi square variables as stated in Theorem 7.6,

$$\frac{Q_2}{\sigma^2} \sim \chi_{n-k}^2,$$

where, as before, $n = \sum_{j=1}^{k} n_j$.

Finding the distribution of Q_1/σ^2 presents a few more problems than Q_2/σ^2 did, but the basic approach is much the same. We begin by writing Q_1 as the difference of two sums of squares:

$$Q_1 = \sum_{j=1}^{k} \sum_{i=1}^{n_j} (\bar{Y}_{.j} - \bar{Y}_{..})^2 = \sum_{j=1}^{k} n_j(\bar{Y}_{.j} - \bar{Y}_{..})^2$$

$$= \sum_{j=1}^{k} n_j[(\bar{Y}_{.j} - \mu) - (\bar{Y}_{..} - \mu)]^2$$

$$= \sum_{j=1}^{k} n_j[(\bar{Y}_{.j} - \mu)^2 + (\bar{Y}_{..} - \mu)^2 - 2(\bar{Y}_{.j} - \mu)(\bar{Y}_{..} - \mu)]$$

$$= \sum_{j=1}^{k} n_j(\bar{Y}_{.j} - \mu)^2 - n(\bar{Y}_{..} - \mu)^2.$$

Therefore,

$$\frac{Q_1}{\sigma^2} = \sum_{j=1}^{k} \left(\frac{\bar{Y}_{.j} - \mu}{\sigma/\sqrt{n_j}}\right)^2 - \left(\frac{\bar{Y}_{..} - \mu}{\sigma/\sqrt{n}}\right)^2.$$

Under the assumption that H_0 is true, both $E(\bar{Y}_{.j})$ and $E(\bar{Y}_{..})$ equal μ, implying that

$$\frac{\bar{Y}_{.j} - \mu}{\sigma/\sqrt{n_j}} \sim N(0, 1), \qquad \text{for all } j,$$

and

$$\frac{\bar{Y}_{..} - \mu}{\sigma/\sqrt{n}} \sim N(0, 1).$$

Now, by an application of Fisher's lemma, similar to that in Appendices 7.2 on page 308 and 10.1 on page 431, it can be shown that $Q_1/\sigma^2 \sim \chi^2_{k-1}$ and that Q_1 and $\bar{Y}_{..}$ are independent. The details of the argument will be omitted.

The random variables Q_1 and Q_2 are also independent. Since $\bar{Y}_{..}$ is a function of the $\bar{Y}_{.j}$'s, Q_1 is a function of those same variables. Furthermore, Q_2 is a weighted sum of the sample variances, $S_1^2, S_2^2, \ldots, S_k^2$, where each S_i^2 is independent of $\bar{Y}_{.j}$, $i \neq j$, the two having been derived from independent random samples. Finally, recall that S_j^2 is independent of $\bar{Y}_{.j}$ by Theorem 7.17. From these facts the independence of Q_1 and Q_2 readily follows.

The upshot of these distribution and independence arguments is that, by Theorem 7.11, the proposed F ratio has an F distribution (when H_0 is true). That is,

$$F = \frac{\dfrac{Q_1}{k-1}}{\dfrac{Q_2}{n-k}} = \frac{\dfrac{Q_1}{(k-1)\sigma^2}}{\dfrac{Q_2}{(n-k)\sigma^2}} \sim \frac{\dfrac{\chi^2_{k-1}}{k-1}}{\dfrac{\chi^2_{n-k}}{n-k}} \sim F_{k-1,\,n-k}.$$

Part III

It remains to be demonstrated that the proposed F ratio will be sensitive to departures from H_0—that is, that the probability of F's lying in the critical region (which has yet to be defined) given that H_1 is true will be greater than α, the probability of its lying in the critical region when H_0 is true. To demonstrate this, we need to compute the expected values of $Q_1/(k-1)$ and $Q_2/(n-k)$ when H_1 is true. The denominator of the F ratio will be considered first. From Equation 11.11,

$$E(Q_2) = \sum_{j=1}^{k} (n_j - 1)E(S_j^2).$$

But for any j, *and regardless of whether H_0 or H_1 is true*, S_j^2 is an unbiased estimator for σ^2. Thus,

$$E(Q_2) = (n-k)\sigma^2$$

or, equivalently,

$$E\left(\frac{Q_2}{n-k}\right) = \sigma^2. \tag{11.12}$$

The derivation of $E[Q_1/(k-1)]$ is somewhat more involved. First, we simplify the expression for $E(Q_1)$ from a double sum to a single sum and collect terms:

$$E(Q_1) = E\left[\sum_{j=1}^{k}\sum_{i=1}^{n_j}(\bar{Y}_{.j} - \bar{Y}_{..})^2\right] = E\left[\sum_{j=1}^{k} n_j(\bar{Y}_{.j} - \bar{Y}_{..})^2\right]$$

$$= E\left[\sum_{j=1}^{k} n_j(\bar{Y}_{.j}^2 - 2\bar{Y}_{.j}\bar{Y}_{..} + \bar{Y}_{..}^2)\right]$$

$$= E\left[\sum_{j=1}^{k} n_j\bar{Y}_{.j}^2 - n\bar{Y}_{..}^2\right]. \quad \text{(Why?)}$$

Then, after an application of Theorem 3.11 to rewrite $E(\bar{Y}_{.j}^2)$ and $E(\bar{Y}_{..}^2)$,

$$E(Q_1) = \sum_{j=1}^{k} n_j\{\text{Var}\,(\bar{Y}_{.j}) + [E(\bar{Y}_{.j})]^2\} - n\{\text{Var}\,(\bar{Y}_{..}) + [E(\bar{Y}_{..})]^2\}.$$

Of course,

$$\text{Var}\,(\bar{Y}_{.j}) = \frac{\sigma^2}{n_j}$$

and, regardless of whether H_0 or H_1 is true, $\text{Var}\,(\bar{Y}_{..})$ equals σ^2/n:

$$\text{Var}\,(\bar{Y}_{..}) = \text{Var}\,\left(\frac{1}{n}\sum_{j=1}^{k} n_j\bar{Y}_{.j}\right) = \frac{1}{n^2}\sum_{j=1}^{k} n_j^2\,\text{Var}\,(\bar{Y}_{.j}) = \frac{1}{n^2}\sum_{j=1}^{k} n_j\sigma^2 = \frac{\sigma^2}{n}.$$

Similarly, $E(\bar{Y}_{.j}) = \mu_j$ and $E(\bar{Y}_{..}) = \mu$, no matter what the values of the τ_j's might be.

Substituting all these results into the last expression for $E(Q_1)$ gives

$$E(Q_1) = \sum_{j=1}^{k} n_j\left(\frac{\sigma^2}{n_j} + \mu_j^2\right) - n\left(\frac{\sigma^2}{n} + \mu^2\right)$$

$$= (k-1)\sigma^2 + \sum_{j=1}^{k} n_j(\mu_j - \mu)^2$$

$$= (k-1)\sigma^2 + \sum_{j=1}^{k} n_j\tau_j^2.$$

Notice that when H_0 is true, $\sum_{j=1}^{k} n_j\tau_j^2 = 0$ and

$$E\left(\frac{Q_1}{k-1}\right) = \sigma^2. \tag{11.13}$$

When H_1 is true, though, $\sum_{j=1}^{k} n_j\tau_j^2 > 0$ and

$$E\left(\frac{Q_1}{k-1}\right) = \sigma^2 + \frac{1}{k-1}\sum_{j=1}^{k} n_j\tau_j^2, \tag{11.14}$$

a number *greater* than σ^2.

Equations 11.12, 11.13, and 11.14 dictate the nature of the critical region. When H_0 is true the expected values of $Q_1/(k-1)$ and $Q_2/(n-k)$ are both σ^2, meaning that the F ratio should be close to 1. But when H_1 is true,

$$E\left(\frac{Q_1}{k-1}\right) > E\left(\frac{Q_2}{n-k}\right),$$

and the F ratio will tend to be larger than 1. It follows that the critical region should be in the right-hand tail of the $F_{k-1,\,n-k}$ distribution and that H_0 should be rejected at the α level of significance if the observed F ratio is greater than or equal to $F_{1-\alpha,\,k-1,\,n-k}$. (See Figure 11.1.) Case Study 11.1 illustrates the technique.

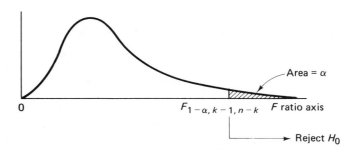

Figure 11.1   *F* distribution with $k - 1$ and $n - k$ degrees of freedom.

CASE STUDY

11.1

A certain fraction of antibiotics injected into the bloodstream are "bound" to serum proteins. This is a phenomenon of considerable pharmacological importance, because as the extent of the binding increases, the systemic uptake of the drug decreases. This, in turn, bears directly on the effectiveness the medication will ultimately have.

In one recent study, determinations were made of the binding percentages characteristic of five widely used antibiotics. Bovine serum was used in each instance, but the results are comparable to what could be expected in human serum (210). (See Table 11.3.)

TABLE 11.3   Serum binding percentages

	Penicillin G	Tetra-cycline	Strepto-mycin	Erythro-mycin	Chloram-phenicol
	29.6	27.3	5.8	21.6	29.2
	24.3	32.6	6.2	17.4	32.8
	28.5	30.8	11.0	18.3	25.0
	32.0	34.8	8.3	19.0	24.2
$t_{.j}$	114.4	125.5	31.3	76.3	111.2
$\bar{y}_{.j}$	28.6	31.4	7.8	19.1	27.8

In terms of the μ and τ_j parameterization, the model equation for these data would be written,

$$Y_{ij} = \mu + \tau_j + \epsilon_{i(j)}, \qquad j = 1, 2, \ldots, 5; i = 1, 2, 3, 4,$$

with τ_j being the differential effect of the jth antibiotic. We will assume that $\epsilon_{i(j)} \sim N(0, \sigma^2)$. A preliminary hypothesis that might be of interest is whether or not the true binding percentages ($\mu_j, j = 1, 2, 3, 4, 5$) for the five antibiotics are all equal. This requires testing

$$H_0: \tau_1 = \tau_2 = \cdots = \tau_5 = 0$$

versus

$$H_1: \text{ not all the } \tau_j\text{'s} = 0.$$

For these data, $k = 5$ and $n = 20$, so the F test calls for H_0 to be rejected at the α level of significance if

$$\frac{(20 - 5) \sum_{j=1}^{5} 4(\bar{y}_{.j} - \bar{y}_{..})^2}{(5 - 1) \sum_{j=1}^{5} \sum_{i=1}^{4} (y_{ij} - \bar{y}_{.j})^2} \geq F_{1-\alpha, 4, 15}.$$

If α is chosen to be 0.05, $F_{0.95, 4, 15} = 3.06$ (see Figure 11.2).

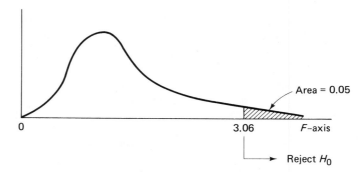

Figure 11.2 F distribution with 4 and 15 degrees of freedom.

From Table 11.3,

$$\bar{y}_{..} = \frac{1}{20} \sum_{j=1}^{5} t_{.j} = \frac{1}{20} (458.7) = 22.94,$$

making the treatment sum of squares 371.19:

$$\sum_{j=1}^{5} (\bar{y}_{.j} - \bar{y}_{..})^2 = \sum_{j=1}^{5} (\bar{y}_{.j} - 22.94)^2 = 371.19.$$

Likewise,

$$\sum_{j=1}^{5} \sum_{i=1}^{4} (y_{ij} - \bar{y}_{.j})^2 = 135.83.$$

Substituting these two sums into the formula for the test statistic gives

$$\frac{15(4)(371.19)}{4(135.83)} = 40.99.$$

> Because 40.99 exceeds the 0.05 critical value (3.06), we reject the null hypothesis that the true binding percentages are all equal, a conclusion hardly unexpected in light of the substantial between-level differences readily apparent in Table 11.3.

QUESTION 11.2.4   In 1965 a silver shortage in the United States prompted Congress to authorize the minting of silverless dimes and quarters. They also recommended that the silver content of half-dollars be reduced from 90% to 40%. Historically, fluctuations in the amount of rare metals found in coins are not uncommon. The data of Table 11.4

TABLE 11.4   Silver content of coins
minted during reign of Manuel I

Early Coinage	Late Coinage
5.9%	5.3%
6.8	5.6
6.4	5.5
7.0	5.1
6.6	6.2
7.7	5.8
7.2	5.8
6.9	
6.2	

concern a possible shift in silver content of a Byzantine coin minted on two separate occasions during the reign of Manuel I (1143–1180). The 16 coins examined (nine belonging to the earlier minting) were part of a large hoard discovered not too long ago in Cyprus. The chemical analysis consisted of first dissolving chips from a particular specimen in a 50% nitric acid solution and then titrating that solution with sodium chloride until all the silver chloride had precipitated out. By weighing the precipitate, the percentage cf silver in the coin could be determined (74). Do an analysis of variance on the data given in Table 11.4. Let 0.05 be the level of significance.

The analysis of variance was introduced in Section 11.1 as a k-sample *extension* of the two-sample t test. The two procedures overlap, though, when k is equal to 2, which raises an obvious question: which procedure is better for testing H_0: $\mu_X = \mu_Y$? The answer, as Example 11.2 shows, is "neither": despite their different test statistics, the two procedures are equivalent—if one rejects H_0, so will the other. (This should have been expected, of course, since both the F test and the t test are GLRT's.)

EXAMPLE 11.2.   Let $X_1, X_2, \ldots, X_n \sim N(\mu_X, \sigma^2)$ and $Y_1, Y_2, \ldots, Y_m \sim N(\mu_Y, \sigma^2)$ be two independent random samples. If these data were subjected to the analysis of variance, the test statistic

$$\frac{\dfrac{Q_1}{k-1}}{\dfrac{Q_2}{n-k}} = \frac{\dfrac{Q_1}{1}}{\dfrac{Q_2}{n+m-2}}$$

would have the F distribution with 1 and $n + m - 2$ degrees of freedom. Now, consider the corresponding t test. By Theorem 8.2, $H_0: \mu_X = \mu_Y$ would be rejected in favor of $H_1: \mu_X \neq \mu_Y$ if either $t \leq -t_{\alpha/2, n+m-2}$ or $t \geq +t_{\alpha/2, n+m-2}$, or, equivalently, if $t^2 \geq t^2_{\alpha/2, n+m-2}$. But

$$T^2 = \left(\frac{\bar{X} - \bar{Y}}{S_p\sqrt{\dfrac{1}{n} + \dfrac{1}{m}}} \right)^2 = \frac{\left[\dfrac{\bar{X} - \bar{Y}}{\sigma\sqrt{\dfrac{1}{n} + \dfrac{1}{m}}} \right]^2}{\dfrac{(n + m - 2)S_p^2}{\sigma^2(n + m - 2)}}$$

$$\sim \frac{[N(0, 1)]^2}{\dfrac{\chi^2_{n+m-2}}{n+m-2}} \sim \frac{\dfrac{\chi^2_1}{1}}{\dfrac{\chi^2_{n+m-2}}{n+m-2}} \sim F_{1, n+m-2}$$

(recall Theorem 7.11). Therefore, the t test will reject H_0 if and only if the F test does.

QUESTION 11.2.5   Verify the conclusion of Example 11.2 by doing a t test on the data of Question 11.2.4. Show that the observed F ratio is the square of the observed t ratio and that the F critical value is the square of the t critical value.

To parallel the treatment given the t test in Chapters 7 and 8, we conclude this section with a few remarks concerning the *distribution* of the F ratio when H_1 is true (recall Appendices 7.1 on page 302 and 8.2 on page 347). Knowing our test statistic's distribution under the alternative hypothesis is important: it makes possible, for example, the construction of a power function. But first we require two preliminary definitions.

DEFINITION 11.1   Let $Y_i \sim N(\mu_i, 1)$ for $i = 1, 2, \ldots, t$. Then

$$Y = \sum_{i=1}^{t} Y_i^2$$

is said to have the *noncentral χ^2 distribution* with t degrees of freedom and noncentrality parameter γ, where

$$\gamma = \frac{1}{2} \sum_{i=1}^{t} \mu_i^2.$$

DEFINITION 11.2   Let V_1 be a noncentral χ^2 random variable with r_1 degrees of freedom and noncentrality parameter γ and let V_2 be an independent (central) χ^2 variable with r_2 degrees of freedom. Then the ratio

$$\frac{V_1/r_1}{V_2/r_2}$$

is said to have a *noncentral F distribution* with r_1 and r_2 degrees of freedom and noncentrality parameter γ.

Substituting $\mu_j - \tau_j$ for μ in part II's expression for Q_1/σ^2 gives

$$\frac{Q_1}{\sigma^2} = \sum_{j=1}^{k} \left(\frac{\bar{Y}_{.j} - \mu}{\frac{\sigma}{\sqrt{n_j}}} \right)^2 - \left(\frac{\bar{Y}_{..} - \mu}{\frac{\sigma}{\sqrt{n}}} \right)^2$$

$$= \sum_{j=1}^{k} \left(\frac{\bar{Y}_{.j} - \mu_j + \tau_j}{\frac{\sigma}{\sqrt{n_j}}} \right)^2 - \left(\frac{\bar{Y}_{..} - \mu}{\frac{\sigma}{\sqrt{n}}} \right)^2.$$

Observe that when H_1 is true, and $E(\bar{Y}_{.j}) = \mu_j$,

$$\frac{\bar{Y}_{.j} - \mu_j + \tau_j}{\frac{\sigma}{\sqrt{n_j}}} \sim N\left(\frac{\tau_j \sqrt{n_j}}{\sigma}, 1 \right).$$

Unaffected by which hypothesis is true, the distribution of the second component of Q_1/σ^2 remains unchanged:

$$\frac{\bar{Y}_{..} - \mu}{\frac{\sigma}{\sqrt{n}}} \sim N(0, 1).$$

By Definition 11.1, then, Q_1/σ^2 has a noncentral χ^2 distribution with $k - 1$ degrees of freedom and noncentrality parameter γ, where

$$\gamma = \frac{1}{2\sigma^2} \sum_{j=1}^{k} n_j \tau_j^2.$$

Finally, since Q_2/σ^2 is always χ_{n-k}^2, it follows immediately from Definition 11.2 that, when H_1 is true,

$$F = \frac{\dfrac{Q_1}{k-1}}{\dfrac{Q_2}{n-k}}$$

has a noncentral F distribution with $k - 1$ and $n - k$ degrees of freedom and noncentrality parameter γ.

Comment. As H_1 gets further away from H_0, as measured by γ, the noncentral F will shift more and more to the right of the central F. Accordingly, the power of the F test will increase. That is,

$$P(F \geq F_{1-\alpha, k-1, n-k}) \longrightarrow 1 \qquad \text{as } \gamma \longrightarrow \infty.$$

The pdf for the noncentral F is not very tractable, but its integral has been evaluated by numerical approximation, and this has allowed the power function for the F test to be tabulated [see, for instance, (106)]. ∎

11.3 COMPUTING FORMULAS FOR THE *F* TEST

In practice, there is a much easier way to calculate the F ratio than indicated in Section 11.2. The expressions for SS_{total} and $SS_{\text{treatments}}$ given in Theorem 11.1 are in forms particularly easy to evaluate with an electronic calculator.

THEOREM 11.1 Let $c = T_{..}^2/n$. Then

$$SS_{total} = \sum_{j=1}^{k} \sum_{i=1}^{n_j} Y_{ij}^2 - c$$

and

$$SS_{treatments} = \sum_{j=1}^{k} \frac{T_{.j}^2}{n_j} - c.$$

PROOF   By definition,

$$SS_{total} = \sum_{j=1}^{k} \sum_{i=1}^{n_j} (Y_{ij} - \bar{Y}_{..})^2 = \sum_{j=1}^{k} \sum_{i=1}^{n_j} (Y_{ij}^2 - 2Y_{ij}\bar{Y}_{..} + \bar{Y}_{..}^2)$$

$$= \sum_{j=1}^{k} \sum_{i=1}^{n_j} Y_{ij}^2 - 2\bar{Y}_{..} \sum_{j=1}^{k} \sum_{i=1}^{n_j} Y_{ij} + n\bar{Y}_{..}^2.$$

But

$$\sum_{j=1}^{k} \sum_{i=1}^{n_j} Y_{ij} = n\bar{Y}_{..},$$

and the first part of the theorem follows:

$$SS_{total} = \sum_{j=1}^{k} \sum_{i=1}^{n_j} Y_{ij}^2 - n\bar{Y}_{..}^2 = \sum_{j=1}^{k} \sum_{i=1}^{n_j} Y_{ij}^2 - c.$$

Similarly,

$$SS_{treatments} = \sum_{j=1}^{k} \sum_{i=1}^{n_j} (\bar{Y}_{.j} - \bar{Y}_{..})^2 = \sum_{j=1}^{k} n_j(\bar{Y}_{.j} - \bar{Y}_{..})^2$$

$$= \sum_{j=1}^{k} n_j(\bar{Y}_{.j}^2 - 2\bar{Y}_{.j}\bar{Y}_{..} + \bar{Y}_{..}^2)$$

$$= \sum_{j=1}^{k} \frac{T_{.j}^2}{n_j} - 2\bar{Y}_{..} \sum_{j=1}^{k} n_j\bar{Y}_{.j} + n\bar{Y}_{..}^2,$$

and, since $\sum_{j=1}^{k} n_j\bar{Y}_{.j} = n\bar{Y}_{..}$,

$$SS_{treatments} = \sum_{j=1}^{k} \frac{T_{.j}^2}{n_j} - n\bar{Y}_{..}^2 = \sum_{j=1}^{k} \frac{T_{.j}^2}{n_j} - c.$$

The error sum of squares can be gotten either directly or by subtraction. The latter approach follows immediately from Equation 11.10:

$$SS_{error} = SS_{total} - SS_{treatments}.$$

It will be left as an exercise to prove that Equation 11.15 gives the error sum of squares directly:

$$SS_{error} = \sum_{j=1}^{k} \left(\sum_{i=1}^{n_j} Y_{ij}^2 - \frac{T_{.j}^2}{n_j} \right). \tag{11.15}$$

Calculations for an F test are typically presented in an *analysis-of-variance* (or *ANOVA*) *table* (see Table 11.5). The first column of an ANOVA table lists

TABLE 11.5   ANOVA table for a completely randomized, one-factor design

Source	df	SS	MS	EMS	F
Treatments	$k - 1$	$\sum_{j=1}^{k} \frac{T_{.j}^2}{n_j} - c$	$\frac{SS_{tr}}{k-1}$	$\sigma^2 + \frac{1}{k-1}\sum_{j=1}^{k} n_j \tau_j^2$	$\frac{MS_{tr}}{MS_{er}}$
Error	$n - k$	$SS_{tot} - SS_{tr}$	$\frac{SS_{er}}{n-k}$	σ^2	
Total	$n - 1$	$\sum_{j=1}^{k}\sum_{i=1}^{n_j} Y_{ij}^2 - c$			

each of the *sources of variation*, as singled out by the model equation. For the completely randomized, one-factor design there are three: treatments, error, and total. Since each source of variation can be used to provide an estimate of σ^2, each has an associated number of degrees of freedom. These are listed in the second column. The third column, denoted *SS*, records the sum of squares corresponding to each source of variation. Next is the mean-square (*MS*) column: entries in this fourth column are each source of variation's sum of squares divided by its degrees of freedom. No mean square is calculated for "total." Still next is the *EMS*, or expected-mean-square, column. For each source of variation (other than "total"), $EMS = E(SS/df)$. Finally, the sixth column displays whatever F ratios are appropriate for the hypotheses being tested. Of course, for the completely randomized, one-factor design there is only one:

$$\frac{SS_{\text{treatments}}/(k-1)}{SS_{\text{error}}/(n-k)} = \frac{MS_{\text{treatments}}}{MS_{\text{error}}}.$$

Often a variety of symbols appear adjacent to the final F ratios to signify their statistical significance. By convention, a single asterisk means that the corresponding H_0 can be rejected at the $\alpha = 0.05$ level of significance. Two asterisks indicate that H_0 can be rejected at $\alpha = 0.01$. A blank, or an *NS* ("not significant"), means that H_0 is accepted at $\alpha = 0.05$.

Comment. The EMS column is not always included in ANOVA tables, particularly when the experimental designs are relatively straightforward, as is the case here and in Chapter 12. In more complicated situations, though, the EMS column is absolutely essential in determining how the analysis is to proceed. ∎

EXAMPLE 11.3. Consider again the antibiotic serum binding data of Table 11.3. For the 20 observations listed,

$$\sum_{j=1}^{5}\sum_{i=1}^{4} y_{ij} = 458.7$$

and

$$\sum_{j=1}^{5}\sum_{i=1}^{4} y_{ij}^2 = 12{,}136.93.$$

This makes the correction factor, c, 10,520.28 and the total sum of squares 1616.65:

$$c = \frac{(458.7)^2}{20} = 10,520.28$$

$$SS_{total} = 12,136.93 - 10,520.28 = 1616.65.$$

Squaring the treatment totals, dividing each by n_j, and subtracting c gives the treatment sum of squares:

$$SS_{treatments} = \frac{(114.4)^2}{4} + \frac{(125.5)^2}{4} + \cdots + \frac{(111.2)^2}{4} - 10,520.28$$

$$= 1480.83.$$

Table 11.6 is the completed ANOVA table.

TABLE 11.6   ANOVA computations

Source	df	SS	MS	EMS	F	
Antibiotics	4	1480.83	370.21	$\sigma^2 + \sum_{j=1}^{5} \tau_j^2$	40.9	**
Error	15	135.82	9.05	σ^2		
Total	19	1616.65				

Note that the entry appearing in the last column is the same value that we got for the observed F ratio in Section 11.2. The two asterisks signify that the test statistic exceeds $F_{0.99, 4, 15} = 3.06$ and that $H_0: \tau_1 = \tau_2 = \cdots = \tau_5 = 0$ should be rejected at the $\alpha = 0.01$ level of significance.

QUESTION 11.3.1   In the fairly recent past there have been two epidemics of acute mercury poisoning in Japanese fishing villages. Since then, mercury pollution has become a hotly debated ecological issue all over the world. Much of the mercury released into the environment originates as a by-product of coal combustion and other industrial processes. It does not become dangerous, though, until it falls into large bodies of water where microorganisms change it into methylmercury, an organic form particularly toxic to humans. Edible fish then become the intermediaries. They ingest and absorb the methylmercury, making themselves a source of contamination. To get some idea of the seriousness of this problem, a research team recently recorded the mercury uptake (in ppm) found in 12 walleyed pike caught in Lake Erie. Three different ages (levels) were singled out (131). (See Table 11.7.) Do the analysis of variance to see whether

TABLE 11.7   Mercury levels (ppm) in walleyed pike

Young of the Year	Yearlings	Two Years and Older
0.60	0.75	1.03
0.64	0.92	0.67
0.62	0.93	0.78
0.44	0.75	0.98

the average mercury uptake levels for these three age groups are significantly different. Can H_0 be rejected at the 0.05 level? at the 0.01 level?

11.4 TESTING SUBHYPOTHESES: ORTHOGONAL CONTRASTS

The F test as described in Sections 11.2 and 11.3 is the usual beginning in an analysis-of-variance problem, but it is seldom the ending. Few researchers, after working months—or even years—to design and carry out an experiment, will be satisfied with a statistician whose conclusion is simply "Yes, the means are all equal," or "No, they are not." In most situations there are other, more specific, inferences an experimenter would like to make. Consider, for example, the serum binding data of Case Study 11.1. Although a total of five antibiotics were tested, it might be that a researcher is especially interested in one particular illness for which only the first two drugs are recommended. It would make sense under those circumstances to single out penicillin G and tetracycline and compare *their* binding percentages apart from all the others: that is, test the hypothesis H_0': $\mu_1 = \mu_2$. Clearly, the ability to do this—break down a broad null hypothesis into smaller, more relevant *subhypotheses*—would add to the analysis of variance a much-needed measure of flexibility.

In general, there are two ways to test subhypotheses, the choice depending, strangely enough, on *when* the H_0' is first specified. If, on the basis of physical considerations, economic factors, past experience, and so on, a particular subhypothesis can be formulated *before any data are taken*, then H_0' can best be tested using an *orthogonal contrast*. On the other hand, if the researcher wishes to do the experiment first and let the results suggest a suitable subhypothesis, the appropriate analysis is any of the so-called *multiple-comparison* techniques. In this section we discuss the construction of orthogonal contrasts; the multiple-comparison problem will be taken up in Section 11.5.

> **DEFINITION 11.3** Let $\mu_1, \mu_2, \ldots, \mu_k$ denote the true means of the k factor levels being sampled. A linear combination, C, of the μ_j's is said to be a *contrast* if the sum of its coefficients is zero. That is, C is a contrast if
>
> $$C = \sum_{j=1}^{k} c_j \mu_j \quad \text{and} \quad \sum_{j=1}^{k} c_j = 0.$$

Contrasts are important because of their relationship to hypothesis tests. Consider the situation just described—testing H_0': $\mu_1 = \mu_2$, or, in a form clearly equivalent, H_0': $\mu_1 - \mu_2 = 0$. Note that in this second format, H_0' is a statement about a contrast—specifically, the contrast C, where

$$C = \mu_1 - \mu_2 = (1)\mu_1 - (1)\mu_2 + (0)\mu_3 + (0)\mu_4 + (0)\mu_5.$$

As another example, suppose there was a valid pharmacological reason for comparing the average level of serum binding for the first two antibiotics to the average level for the last three. Written as a subhypothesis, the statement of no difference would be

$$H_0': \frac{\mu_1 + \mu_2}{2} = \frac{\mu_3 + \mu_4 + \mu_5}{3}.$$

As a contrast, it becomes

$$C = \tfrac{1}{2}\mu_1 + \tfrac{1}{2}\mu_2 - \tfrac{1}{3}\mu_3 - \tfrac{1}{3}\mu_4 - \tfrac{1}{3}\mu_5.$$

In both these cases, the numerical value of the contrast will be 0 if H_0' is true. This suggests that the choice between H_0' and H_1' can be accomplished by first estimating C and then determining, via a significance test, whether that estimate is too far from 0. The details are presented in the next several paragraphs.

We begin by considering some of the mathematical properties of contrasts and their estimates. Since $\bar{Y}_{.j}$ is always an unbiased estimator for μ_j, it seems reasonable to estimate C, a linear combination of population means, with $\hat{C}$, a linear combination of *sample* means:

$$\hat{C} = \sum_{j=1}^{k} c_j \bar{Y}_{.j}.$$

(The coefficients appearing in $\hat{C}$, of course, are the same as those that defined C.) It follows that

$$E(\hat{C}) = \sum_{j=1}^{k} c_j E(\bar{Y}_{.j}) = C$$

and

$$\mathrm{Var}\,(\hat{C}) = \sum_{j=1}^{k} c_j^2 \,\mathrm{Var}\,(\bar{Y}_{.j}) = \sigma^2 \sum_{j=1}^{k} \frac{c_j^2}{n_j}.$$

Comment. Replacing the unknown error variance, σ^2, by its estimate from the ANOVA table—$\hat{\sigma}^2 = MS_{\text{error}}$—gives a formula for the estimated variance of the estimated contrast:

$$S_{\hat{C}}^2 = \widehat{\mathrm{Var}\,(\hat{C})} = MS_{\text{error}} \sum_{j=1}^{k} \frac{c_j^2}{n_j}. \qquad \blacksquare$$

The sampling behavior of $\hat{C}$ is easily derived. By Theorem 7.4, the normality of the Y_{ij}'s insures that $\hat{C}$ is also normal, and by the usual Z transformation, the ratio

$$\frac{\hat{C} - E(\hat{C})}{\sqrt{\mathrm{Var}\,(\hat{C})}} = \frac{\hat{C} - C}{\sqrt{\mathrm{Var}\,(\hat{C})}}$$

is a *standard* normal. Therefore,

$$\left[\frac{\hat{C} - C}{\sqrt{\mathrm{Var}\,(\hat{C})}} \right]^2 \sim \chi_1^2. \qquad (11.16)$$

Of course, if $H_0: \mu_1 = \mu_2 = \cdots = \mu_k$ is true, C is 0, and Equation 11.16 reduces to

$$\frac{\hat{C}^2}{\sigma^2 \sum\limits_{j=1}^{k} \frac{c_j^2}{n_j}} \sim \chi_1^2. \tag{11.17}$$

One final property of contrasts plays a role in the significance testing of subhypotheses. Two contrasts,

$$C_1 = \sum_{j=1}^{k} c_{1j}\mu_j \quad \text{and} \quad C_2 = \sum_{j=1}^{k} c_{2j}\mu_j$$

are said to be *orthogonal* if

$$\sum_{j=1}^{k} \frac{c_{1j}c_{2j}}{n_j} = 0.$$

Similarly, a set of q contrasts, $\{C_i\}_{i=1}^{q}$, are said to be *mutually orthogonal* if

$$\sum_{j=1}^{k} \frac{c_{sj}c_{tj}}{n_j} = 0, \quad \text{for all } s \neq t.$$

(The same definitions apply, or course, to *estimated* contrasts.)

Definition 11.4 and Theorems 11.2 and 11.3, both stated here without proof, summarize the relationship between contrasts and the analysis of variance. In short, the treatment sum of squares can be partitioned into $k - 1$ "contrast" sum of squares, provided the contrasts are mutually orthogonal. These $k - 1$ sums of squares can then be used to form $k - 1$ F ratios—and to test $k - 1$ subhypotheses.

DEFINITION 11.4 Let $C_i = \sum\limits_{j=1}^{k} c_{ij}\mu_j$ be any contrast. The sum of squares associated with C_i is given by

$$SS_{C_i} = \frac{\hat{C}_i^2}{\sum\limits_{j=1}^{k} \frac{c_{ij}^2}{n_j}},$$

where $\hat{C}_i = \sum\limits_{j=1}^{k} c_{ij}\bar{Y}_{.j}$.

THEOREM 11.2 Let $\{C_i = \sum\limits_{j=1}^{k} c_{ij}\mu_j\}_{i=1}^{k-1}$ be a set of $k - 1$ mutually orthogonal contrasts. Let $\{\hat{C}_i = \sum\limits_{j=1}^{k} c_{ij}\bar{Y}_{.j}\}_{i=1}^{k-1}$ be their estimators. Then

$$SS_{\text{treatments}} = \sum_{j=1}^{k} \sum_{i=1}^{n_j} (\bar{Y}_{.j} - \bar{Y}_{..})^2$$

$$= SS_{C_1} + SS_{C_2} + \cdots + SS_{C_{k-1}}.$$

THEOREM 11.3 Let C be a contrast having the same coefficients as the subhypothesis $H_0': c_1\mu_1 + c_2\mu_2 + \cdots + c_k\mu_k = 0$, where $\sum_{i=1}^{k} c_i = 0$. As before, let

$$SS_{\text{error}} = \sum_{j=1}^{k} \sum_{i=1}^{n_j} (Y_{ij} - \bar{Y}_{.j})^2 \quad \text{and} \quad n = \sum_{j=}^{k} n_j.$$

Then, if H_0' is true,

$$F = \frac{\dfrac{SS_C}{1}}{\dfrac{SS_{\text{error}}}{n-k}} \sim F_{1, n-k}.$$

Also, if H_0' is not true, the F ratio will have a noncentral F distribution with $\gamma > 0$, implying that H_0' should be rejected at level α if $F \geq F_{1-\alpha, 1, n-k}$.

Comment. Theorem 11.2 is not meant to imply that only mutually orthogonal contrasts can, or should, be tested. It is simply a statement of a partitioning relationship that exists between $SS_{\text{treatments}}$ and the sum of squares for mutually orthogonal C_i's. In any given experiment, the contrasts that should be singled out are those the experimenter has some prior reason to test.   ∎

CASE STUDY

11.2

As a rule, infants are not able to walk by themselves until they are almost 14 months old. A recent study, however, investigated the possibility of reducing that time through the use of special "walking" exercises (208). A total of 23 infants were included in the experiment—all were one-week-old white males. They were randomly divided into four groups, and for seven weeks each group followed a different training program. Group A received special walking and placing exercises for 12 minutes each day. Group B also had daily 12-minute exercise periods but were not given the special walking and placing exercises. Groups C and D received no special instruction. The progress of Groups A, B, and C was checked every week; the progress of Group D was checked only once, at the end of the study.

After seven weeks the formal training ended and the parents were told they could continue with whatever procedure they desired. Listed in Table 11.8 are the ages (in months) at which each of the 23 children first walked alone. Table 11.9 shows the analysis-of-variance com-

TABLE 11.8 Age when infants first walked alone (months)

	Group A	Group B	Group C	Group D
	9.00	11.00	11.50	13.25
	9.50	10.00	12.00	11.50
	9.75	10.00	9.00	12.00
	10.00	11.75	11.50	13.50
	13.00	10.50	13.25	11.50
	9.50	15.00	13.00	
$t_{.j}$	60.75	68.25	70.25	61.75
$\bar{y}_{.j}$	10.12	11.38	11.71	12.35

TABLE 11.9 ANOVA computations

Source	df	SS	MS	EMS	F	
Exercises	3	14.77	4.92	$\sigma^2 + \frac{1}{3} \sum_{j=1}^{4} n_j \tau_j^2$	2.14	NS
Error	19	43.70	2.30	σ^2		
Total	22	58.47				

putations and the 2.14 *F* ratio. Based on 3 and 19 degrees of freedom, the $\alpha = 0.05$ critical value is 3.13, so $H_0: \mu_A = \mu_B = \mu_C = \mu_D$ is accepted.

Comment. At this point the analysis would usually end; with the overall H_0 being accepted, there is no pressing need to look at contrasts. We will continue with the subhypothesis procedures, however, to illustrate the application of Theorem 11.3.

Recall that groups A and B spent equal amounts of time exercising but followed different regimens. Consequently, a test of $H'_0: \mu_A = \mu_B$ versus $H'_1: \mu_A \neq \mu_B$ would be an obvious way to assess the effectiveness of the special walking and placing exercises. The associated contrast, of course, would be $C_1 = \mu_A - \mu_B$. Similarly, a test of $H''_0: \mu_C = \mu_D (C_2 = \mu_C - \mu_D)$ would provide an evaluation of the psychological effect of periodic progress checks.

From Definition 11.4 and the data in Table 11.8,

$$SS_{C_1} = \frac{\left[1\left(\dfrac{60.75}{6}\right) - 1\left(\dfrac{68.25}{6}\right)\right]^2}{\dfrac{1^2}{6} + \dfrac{(-1)^2}{6}} = 4.68$$

and

$$SS_{C_2} = \frac{\left[1\left(\frac{70.25}{6}\right) - 1\left(\frac{61.75}{5}\right) \right]^2}{\frac{1^2}{6} + \frac{(-1)^2}{5}} = 1.12.$$

TABLE 11.10   Subhypothesis computations

Subhypothesis	Contrast	SS	F	
H_0': $\mu_A = \mu_B$	$C_1 = \mu_A - \mu_B$	4.68	2.03	NS
H_0'': $\mu_C = \mu_D$	$C_2 = \mu_C - \mu_D$	1.12	0.49	NS

Dividing these sums of squares by the error mean square (2.30) gives F ratios of $4.68/2.30 = 2.03$ and $1.12/2.30 = 0.49$, neither of which is significant at the $\alpha = 0.05$ level ($F_{0.95, 1, 19} = 4.38$). (See Table 11.10.) ∎

QUESTION 11.4.1   Verify that $C_3 = \frac{11}{12}\mu_A + \frac{11}{12}\mu_B - \mu_C - \frac{5}{6}\mu_D$ is mutually orthogonal to the C_1 and C_2 of Case Study 11.2. Find SS_{C_3} and illustrate the statement of Theorem 11.2.

QUESTION 11.4.2   For many years sodium nitrite has been used as a curing agent for bacon, and until recently it was thought to be perfectly harmless. But now it appears that during frying, sodium nitrite induces the formation of nitrosopyrrolidine (NPy), a substance suspected of being a carcinogen. In one study focusing on this problem, measurements were made of the amount of NPy (in ppb) recovered after the frying of three slices of four commercially available brands of bacon (161). Do the analysis of variance for the data in Table 11.11 and partition the treatment sum of squares into

TABLE 11.11   NPy recovered from bacon (ppb)

	Brand		
A	B	C	D
20	75	15	25
40	25	30	30
18	21	21	31

a complete set of three mutually orthogonal contrasts. Let the first contrast test H_0': $\mu_A = \mu_B$ and the second, H_0'': $(\mu_A + \mu_B)/2 = (\mu_C + \mu_D)/2$. Do all tests at the 0.05 level of significance.

11.5 MULTIPLE COMPARISONS: TUKEY'S METHOD

Imagine a consumer researcher working for, say, Procter and Gamble, who faces the task of comparing six new laundry detergents in a whiteness test. We have seen that the routine way to begin such a comparison would be with the analysis of variance, testing H_0: $\mu_1 = \mu_2 = \cdots = \mu_6$. What comes next, though, depends on how much was known, or expected, about the detergents *before* the data were collected.[1] Specifically, there may be bits of peripheral information about some or all of the six products to suggest the formation of one or more contrasts. If there *are*, then the methodology of Section 11.4 applies. For example, detergents 1 and 4 may be much cheaper than all the others, in which case it would make sense to test the subhypothesis $H'_0: \mu_1 = \mu_4$. But suppose there is *no* prior information or expectation about any of the six—and, therefore, no reason to specify in advance any particular subhypotheses. How, then, should we proceed?

Perhaps the most tempting course of action would be to wait until the results were in and then begin testing the extremes. If, for instance, detergent 1 rated out to be the best of the six, it might seem reasonable, as a way of "confirming" its superiority, to set up a contrast for testing

$$H'_0: \mu_1 = \frac{\mu_2 + \mu_3 + \mu_4 + \mu_5 + \mu_6}{5}$$

or, perhaps, a set of *five* contrasts to test $H_0^1: \mu_1 = \mu_2$, $H_0^2: \mu_1 = \mu_3$, . . . , $H_0^5: \mu_1 = \mu_6$. Using the contrast approach here, though, raises some disturbing questions. The problem is that by singling out as the groups to compare those whose $\bar{y}_{.j}$'s were markedly different in the first place, we are prejudicing the outcome. This is obvious if we look at the probability of committing a Type I error for (1) two groups chosen at random and (2) the two extreme groups. Let i and j denote two arbitrary factor levels chosen in advance. Let c^* denote the smallest value of $\hat{C} = \bar{Y}_{.i} - \bar{Y}_{.j}$ for which $H'_0: \mu_i = \mu_j$ can be rejected at level α. Then, by definition,

$$P(\bar{Y}_{.i} - \bar{Y}_{.j} \geq c^* | H_0: \mu_1 = \mu_2 = \cdots = \mu_k) = \alpha.$$

But, clearly,

$$P(\bar{Y}_{\max} - \bar{Y}_{\min} \geq c^* | H_0) > P(\bar{Y}_{.i} - \bar{Y}_{.j} \geq c^* | H_0)$$

—an inequality that should serve as fair warning against proceeding in this fashion.

An alternative approach, perhaps, would be to use the contrast technique on *all* comparisons of a certain kind—say, all possible $\binom{6}{2} = 15$ pairwise hypothesis tests: $H_0^1: \mu_1 = \mu_2$, $H_0^2: \mu_1 = \mu_3$, . . . , $H_0^{15}: \mu_5 = \mu_6$. While this adroitly sidesteps the selection problem of the first method, it brings up some disturbing questions of its own about the meaning of α. We know that for any one given test, α is the probability of incorrectly rejecting a true null hypothesis. But suppose that not one but m tests are done, each at level α. The probability of committing *at least one* Type I error among these m is much larger than the nominal α associated with each one. If the tests are independent,

P (at least one Type I error is made in m tests)

$$= 1 - P \text{ (no Type I errors are made in } m \text{ tests)}$$
$$= 1 - (1 - \alpha)^m.$$

[1] A paraphrase of Senator Howard Baker's famous Watergate query would be appropriate here: "How much did the statistician know, and when did he know it?"

Let α' denote the "overall" Type I error for an experiment—that is, the probability of rejecting at least one true H_0. Table 11.12 shows, for the independent case,

TABLE 11.12   Values of α'

m	$\alpha = 0.10$	$\alpha = 0.05$	$\alpha = 0.01$
1	0.10	0.05	0.01
2	0.19	0.10	0.02
5	0.41	0.23	0.05
10	0.65	0.40	0.10
15	0.80	0.54	0.14
20	0.88	0.64	0.18

the rapid deterioration of α' as a function of m. Thus, if an experimenter did 15 independent tests, each at the $\alpha = 0.05$ level, his chances of committing at least one Type I error would not be 0.05, or anything even close to 0.05, but, rather, 0.54.

Whether or not the behavior of α' is worth worrying about (not all statisticians think it is), it does suggest an interesting question: is there any way to test a large, maybe even unspecified, number of subhypotheses and still keep α' fixed (and small)? Or, what is equivalent, is there any valid way to formulate and test hypotheses *after* seeing the data? The answer is "Yes"; in fact, there are quite a few ways. The *multiple-comparison problem*, as this has come to be known, has received a good deal of attention from mathematical statisticians in recent years. In this section, we develop the simplest formulation of one of the earliest multiple-comparison procedures, a method due to Tukey.

Definition 11.5 gives a background result. The derivation of Tukey's method hinges on a distribution we have not seen yet, the *studentized range*.

DEFINITION 11.5   Let $Y_1, Y_2, \ldots, Y_k$ be k independent $N(\mu, \sigma^2)$ random variables, and let R be their *range*:

$$R = \max_i Y_i - \min_i Y_i.$$

Let S^2 be an estimator of σ^2 having v degrees of freedom and let S^2 and the Y_i's be independent. (In the completely randomized, one-factor design, S^2 will be MS_{error} and v will be $n - k$.) The *studentized range*, $Q_{k,v}$, is taken to be the ratio

$$Q_{k,v} = \frac{R}{S}.$$

Table A.5 in the Appendix gives values of $Q_{\alpha,k,v}$, the $[100(1 - \alpha)]$th percentile of $Q_{k,v}$, for $\alpha = 0.05$ and 0.01 and for various k and v.

Theorem 11.4 outlines Tukey's procedure. Notice the restrictions: n_j must be the same for all j, and only pairwise tests are included. (Although more difficult to prove, there are other, more general, multiple-comparison procedures that apply to *any* contrast and do not require equal sample sizes [see, for instance, (153) or (110)].

> **THEOREM 11.4** Let $\bar{y}_{.j}, j = 1, 2, \ldots, k$, be the k sample means in a completely randomized, one-factor design. Let $n_j = r$ be the (constant) sample size and let $\mu_j, j = 1, 2, \ldots, k$, be the true means. The probability is $1 - \alpha$ that all $\binom{k}{2}$ pairwise contrasts, $C_{ij} = \mu_i - \mu_j$, will simultaneously satisfy the inequalities
>
> $$\bar{y}_{.i} - \bar{y}_{.j} - D\sqrt{MS_{\text{error}}} < \mu_i - \mu_j$$
> $$< \bar{y}_{.i} - \bar{y}_{.j} + D\sqrt{MS_{\text{error}}},$$
>
> where $D = Q_{\alpha, k, rk-k}/\sqrt{r}$. If, for a given i and j, zero is not contained in the above inequality, $H_0': \mu_i = \mu_j$ can be rejected in favor of $H_1': \mu_i \neq \mu_j$ at the α level of significance.

PROOF   Let $W_t = \bar{Y}_{.t} - \mu_t$; then $W_t \sim N(0, \sigma^2/r)$. From the definition of the studentized range,

$$\frac{\max_t W_t - \min_t W_t}{\sqrt{\dfrac{MS_{\text{error}}}{r}}} \sim Q_{k, rk-k},$$

which implies that

$$P\left(\frac{\max_t W_t - \min_t W_t}{\sqrt{\dfrac{MS_{\text{error}}}{r}}} < Q_{\alpha, k, rk-k}\right) = 1 - \alpha$$

or, equivalently,

$$P(\max_t W_t - \min_t W_t < D\sqrt{MS_{\text{error}}}) = 1 - \alpha, \tag{11.18}$$

where $D = Q_{\alpha, k, rk-k}/\sqrt{r}$. But if Equation 11.18 is true, it must also be true that

$$P(|W_i - W_j| < D\sqrt{MS_{\text{error}}}) = 1 - \alpha, \qquad \text{for } all \ i \text{ and } j. \tag{11.19}$$

Rewriting Equation 11.19 gives the statement of the theorem:

$$P(-D\sqrt{MS_{\text{error}}} < W_i - W_j < D\sqrt{MS_{\text{error}}}) = 1 - \alpha, \qquad \text{for } all \ i \text{ and } j,$$

or, after simplifying,

$$P(\bar{Y}_{.i} - \bar{Y}_{.j} - D\sqrt{MS_{\text{error}}} < \mu_i - \mu_j < \bar{Y}_{.i} - \bar{Y}_{.j} + D\sqrt{MS_{\text{error}}})$$
$$= 1 - \alpha, \qquad \text{for } all \ i \text{ and } j.$$

EXAMPLE 11.4. Table 11.13 lists the average serum binding percentages for the $k = 5$ antibiotics of Case Study 11.1. Each average was based on $n_j = r = 4$ replicates.

TABLE 11.13 Average serum binding percentages

	Penicillin G	Tetracycline	Streptomycin	Erythromycin	Chloramphenicol
$\bar{y}_{.j}$	28.6	31.4	7.8	19.1	27.8

Suppose we want to make all $\binom{5}{2} = 10$ pairwise comparisons while maintaining an overall Type I error, α', of 0.05. Recall that when the analysis of variance was done on these data, it was found that $MS_{error} = 9.05$ (see Table 11.6). Also, from Table A.5 in the Appendix, $Q_{0.05,5,15} = 4.37$. This makes $D = 4.37/\sqrt{4} = 2.185$ and $D\sqrt{MS_{error}} = 6.58$.

For each of the ten pairwise subhypothesis tests, $H'_0: \mu_i = \mu_j$ versus $H'_1: \mu_i \neq \mu_j$, Table 11.14 lists the corresponding estimated contrast, $\bar{y}_{.i} - \bar{y}_{.j}$, and its 95% "Tukey"

TABLE 11.14 Analysis of pairwise subhypothesis tests

Pairwise Difference	$\bar{y}_{.i} - \bar{y}_{.j}$	Tukey Interval	Conclusion
$\mu_1 - \mu_2$	-2.8	$(\ -9.38,\quad 3.78)$	NS
$\mu_1 - \mu_3$	20.8	$(\ 14.22,\quad 27.38)$	Reject
$\mu_1 - \mu_4$	9.5	$(\ 2.92,\quad 16.08)$	Reject
$\mu_1 - \mu_5$	0.8	$(\ -5.78,\quad 7.38)$	NS
$\mu_2 - \mu_3$	23.6	$(\ 17.02,\quad 30.18)$	Reject
$\mu_2 - \mu_4$	12.3	$(\ 5.72,\quad 18.88)$	Reject
$\mu_2 - \mu_5$	3.6	$(\ -2.98,\quad 10.18)$	NS
$\mu_3 - \mu_4$	-11.3	$(-17.88,\quad -4.72)$	Reject
$\mu_3 - \mu_5$	-20.0	$(-26.58,\quad -13.42)$	Reject
$\mu_4 - \mu_5$	-8.7	$(-15.28,\quad -2.12)$	Reject

confidence interval. As the last column indicates, seven of the subhypotheses are rejected (those whose Tukey intervals do not contain 0) and three are accepted.

QUESTION 11.5.1   Intravenous infusion fluids produced by three different pharmaceutical companies (Cutter, Abbott, and McGaw) were tested for their concentrations of particulate contaminants. Six samples were inspected from each company. The figures listed in Table 11.15 are, for each sample, the number of particles per liter greater than 5

TABLE 11.15 Number of contaminant particles in infusion fluids made by three firms

Cutter	Abbott	McGaw
255	105	577
264	288	515
342	98	214
331	275	413
234	221	401
217	240	260

microns in diameter (182). Do the analysis of variance to test $H_0: \mu_C = \mu_A = \mu_M$ and then test each of the three pairwise subhypotheses by constructing 95% Tukey confidence intervals.

11.6 DATA TRANSFORMATIONS

The three assumptions required by the analysis of variance have already been mentioned: the $\epsilon_{i(j)}$'s must be independent, normally distributed (for each j), and have the same variance. In practice, the three are not equally difficult to satisfy, nor do their violations have equal consequences for the validity of the F test. Independence is certainly a critical property for the $\epsilon_{i(j)}$'s to have, but randomizing the order in which the observations are taken tends to eliminate systematic bias and thereby to satisfy the assumption, at least to a good approximation. Normality is a much more difficult property to induce or even to verify (recall Section 9.4); fortunately, departures from normality, unless extreme, do not seriously compromise the probabilistic integrity of the F test (more will be said about this in Chapter 13).

If the final assumption is violated, though, and all the Y_{ij}'s do *not* have the same variance, the effect on certain inference procedures—for example, the construction of confidence intervals for individual means—can be more unsettling. However, it is possible in some situations to "stabilize" the level-to-level variances by a suitable *data transformation*. In particular, if the variance of the observations in the jth level is a known function of the mean for that same level, we can appeal to Theorem 11.5.

THEOREM 11.5 Suppose $Y_{ij} \sim f_Y(y_{ij}; \mu_j)$, $i = 1, 2, \ldots, n_j; j = 1, 2, \ldots,$ k, where Var $(Y_{ij}) = g(\mu_j)$. Then the transformed variables $Z_{ij} = A(Y_{ij})$ will have a nearly constant variance for all j if

$$A(Y_{ij}) = c_1 \int \frac{1}{\sqrt{g(y_{ij})}} \, dy_{ij} + c_2,$$

where c_1 and c_2 are arbitrary constants.

PROOF   We wish to find a transformation, A, which when applied to the Y_{ij}'s will generate a new set of variables having a constant variance. That is, we seek $A(Y_{ij}) = Z_{ij}$, where Var $(Z_{ij}) = c_1^2$, a constant. By Taylor's theorem,

$$Z_{ij} \doteq A(\mu_j) + (Y_{ij} - \mu_j)A'(\mu_j).$$

Of course, $E(Z_{ij}) = A(\mu_j)$, since $E(Y_{ij} - \mu_j) = 0$. Also,

$$\text{Var}\,(Z_{ij}) = E[Z_{ij} - E(Z_{ij})]^2$$
$$= E[(Y_{ij} - \mu_j)A'(\mu_j)]^2$$
$$= [A'(\mu_j)]^2 \, \text{Var}\,(Y_{ij}) = [A'(\mu_j)]^2 g(\mu_j).$$

Solving for $A'(\mu_j)$ gives

$$A'(\mu_j) = \frac{\sqrt{\mathrm{Var}\,(Z_{ij})}}{\sqrt{g(\mu_j)}} = \frac{c_1}{\sqrt{g(\mu_j)}}.$$

For Y_{ij} in the neighborhood of μ_j, it follows that

$$A(Y_{ij}) = c_1 \int \frac{1}{\sqrt{g(y_{ij})}}\,dy_{ij} + c_2,$$

and the theorem is proved.

The next two examples show the origin of two of the more commonly used data transformations, the square root and the arc sine. The former is appropriate if the given measurements are Poisson, the latter if the Y_{ij}'s are binomial.

EXAMPLE 11.5. Suppose the response variable for the jth level in a completely randomized, one-factor design is Poisson with parameter μ_j,

$$Y_{ij} \sim f_Y(y_{ij}; \mu_j) = \frac{e^{-\mu_j}\mu_j^{y_{ij}}}{y_{ij}!}.$$

For this particular pdf, the variance is *equal* to the mean (recall Theorem 4.2):

$$\mathrm{Var}\,(Y_{ij}) = E(Y_{ij}) = \mu_j = g(\mu_j).$$

Therefore, from Theorem 11.5,

$$A(Y_{ij}) = c_1 \int \frac{1}{\sqrt{y_{ij}}}\,dy_{ij} + c_2 = 2c_1\sqrt{Y_{ij}} + c_2$$

or, letting $c_1 = \frac{1}{2}$ and $c_2 = 0$ to make the transformation as simple as possible,

$$A(Y_{ij}) = Z_{ij} = \sqrt{Y_{ij}}. \tag{11.20}$$

Equation 11.20 implies that if the data are known in advance to be Poisson, each of the observations should be replaced by its square root before we proceed with the analysis of variance.

EXAMPLE 11.6. Suppose Y_{ij} is binomial with parameters n and p_j,

$$Y_{ij} \sim f_Y(y_{ij}; np_j) = \binom{n}{y_{ij}}p_j^{y_{ij}}(1 - p_j)^{n-y_{ij}}.$$

Then, since $E(Y_{ij}) = np_j = \mu_j$,

$$\mathrm{Var}\,(Y_{ij}) = np_j(1 - p_j) = \mu_j\left(1 - \frac{\mu_j}{n}\right) = g(\mu_j).$$

In this case, the stabilizing transformation involves the inverse sine:

$$A(Y_{ij}) = c_1 \int \frac{1}{\sqrt{y_{ij}(1 - y_{ij}/n)}}\,dy_{ij} + c_2$$

$$= c_1 2\sqrt{n}\,\arcsin\left(\frac{y_{ij}}{n}\right)^{1/2} + c_2$$

or, what is equivalent,

$$A(Y_{ij}) = \arcsin\left(\frac{y_{ij}}{n}\right)^{1/2}.$$

QUESTION 11.6.1   An experimenter wants to do an analysis of variance on a set of data involving five treatment groups, each with three replicates. She has computed $\bar{y}_{.j}$ and s_j for each group and gotten the results listed in Table 11.16. What should the experimenter do

TABLE 11.16   Summary of data for analysis of variance

Treatment Group

	1	2	3	4	5
$\bar{y}_{.j}$	9.0	4.0	16.0	9.0	1.0
s_j	3.0	2.0	4.0	3.0	1.0

before computing the various sums of squares necessary to carry out the *F* test? Be as quantitative as possible.

REVIEW EXERCISES FOR CHAPTER 11

1. There is a possibility that a teacher's opinion of the academic abilities of her students may be translated into actions and attitudes that help produce those expectations (that is, the opinion may be a self-fulfilling prophecy). Not too long ago an experiment was set up (144) to investigate that hypothesis. The data given here, while fictitious, are based on the findings of that study. Children in the first grade at a certain school were all given a standard IQ test, but the children's teachers were told it was a special test for predicting whether a child would show sudden spurts of intellectual growth in the near future. The experimenters then divided the children into three groups *at random* but informed the teachers that, according to the test, the children in group I probably would not demonstrate any pronounced intellectual growth for the next year, those in group II would develop at a moderate rate, while those in group III could be expected to advance considerably during the coming year. A year later the children were again given a standard IQ test. The differences in the two IQ scores (second test — first test) for each child are shown below.

IQ increases

Group I	Group II	Group III
3	10	20
2	4	10
6	11	19
10	14	15
11	5	9
5	3	18

Follow the method described in Case Study 11.1 and do an analysis of variance on these data. Use the 0.05 level of significance. Do the data support the contention that teachers' attitudes can affect IQ?

2. Carry out the details to show that

$$\frac{1}{\sigma^2}\left\{\sum_{j=1}^{k} n_j[(\bar{Y}_{\cdot j} - \mu)^2 + (\bar{Y}_{\cdot\cdot} - \mu)^2 - 2(\bar{Y}_{\cdot j} - \mu)(\bar{Y}_{\cdot\cdot} - \mu)]\right\}$$

$$= \sum_{j=1}^{k}\left(\frac{\bar{Y}_{\cdot j} - \mu}{\sigma/\sqrt{n_j}}\right)^2 - \left(\frac{\bar{Y}_{\cdot\cdot} - \mu}{\sigma/\sqrt{N}}\right)^2$$

(see part II of the derivation of the *F* test in Section 11.2).

3. Of the many pesticides that leave residues capable of killing wildlife, dieldrin is the second most prevalent. The following data were collected as part of an experiment to characterize the nature of dieldrin poisoning (173). A total of 33 Japanese quail were randomly divided into three groups of size eleven: the first group was fed a diet that contained 250 ppm of dieldrin; the second group, 50 ppm; and the third, 10 ppm. The data recorded were the concentrations of dieldrin found in the brains of the 17 quail that died.

Brain residues (ppm of dieldrin)

Dosage (ppm of dieldrin)

250	50	10
7.01	6.23	19.52
12.69	12.55	32.03
9.99	22.56	21.91
23.00	19.48	21.61
11.03	31.29	
19.33	24.90	
32.94		

(a) Graph the data (use the format of Case Study 8.3).

(b) Let μ_1, μ_2, and μ_3 denote the true average brain residues associated with these three diets under these particular conditions. Test

$$H_0: \mu_1 = \mu_2 = \mu_3$$

versus

$$H_1: \text{not all the } \mu_j\text{'s are equal}$$

using the procedure illustrated in Case Study 11.1. Let $\alpha = 0.05$.

(c) Other data related to this experiment revealed a strong relationship between dosage level and life span: birds fed higher concentrations of dieldrin tended to die sooner. What does this fact, together with the conclusion of part (b), suggest about the mechanism of dieldrin poisoning?

4. Suppose an experimenter has taken three independent measurements on each of five treatment levels and intends to use the analysis of variance to test

$$H_0: \tau_1 = \tau_2 = \tau_3 = \tau_4 = \tau_5 (= 0)$$

versus

$$H_1: \text{not all the } \tau_j\text{'s are } 0,$$

where the τ_j's are the differential effects of the treatment levels. Two of the possible alternatives in H_1 are

$$H_1^*: \tau_1 = -1, \tau_2 = 2, \tau_3 = 0, \tau_4 = 1, \tau_5 = -2$$

and

$$H_1^{**}: \tau_1 = -3, \tau_2 = 2, \tau_3 = 1, \tau_4 = 0, \tau_5 = 0.$$

Against which alternative will the F test have greater power? Explain. Is H_1^{***}: $\tau_1 = 2, \tau_2 = 1, \tau_3 = 1, \tau_4 = -3, \tau_5 = 0$ an "admissible" member of H_1?

5. Do an analysis of variance on the Mark Twain–Quintus Curtius Snodgrass data of Case Study 8.1. Verify that the observed F ratio is the square of the observed t ratio (3.86).

6. If the random variable Y has a noncentral χ^2 distribution with r degrees of freedom and noncentrality parameter γ (see Definition 11.1), then its moment-generating function is given by

$$M_Y(t) = \frac{1}{(1 - 2t)^{r/2}} e^{2t\gamma/(1-2t)}, \qquad t < \tfrac{1}{2}; 0 < \gamma$$

[for a derivation, see (78)]. Find $E(Y)$ and Var (Y).

7. Suppose $Y_1, Y_2, \ldots, Y_n$ are independent noncentral χ^2 random variables having $r_1, r_2, \ldots, r_n$ degrees of freedom, respectively, and with noncentrality parameters, $\gamma_1, \gamma_2, \ldots, \gamma_n$. Find the distribution of $W = Y_1 + Y_2 + \cdots + Y_n$.

8. The data given in Case Study 8.2 were only part of a larger investigation focusing on the chronology of the Poverty Point culture as revealed by the thermoluminescence of baked clay. The "dates" for four different archeological sites (including Terral Lewis and Jaketown) are listed below.

Estimated baked clay dates (years B.C.)

Terral Lewis	Poverty Point	Teoc Creek	Jaketown
1492	120	1280	1346
1169	841	964	942
883	1303	967	908
988	975	641	858

Do an analysis of variance on these data. Show the ANOVA table. Define the parameters being tested. Let $\alpha = 0.05$.

9. Each of five varieties of corn is planted in three plots in a large field. The respective yields, in bushels per acre, are indicated below.

Var. 1	Var. 2	Var. 3	Var. 4	Var. 5
46.2	49.2	60.3	48.9	52.5
51.9	58.6	58.7	51.4	54.0
48.7	57.4	60.4	44.6	49.3

Test whether the differences among the average yields are statistically significant. Show the ANOVA table. Let 0.05 be the level of significance.

10. A car manufacturer is trying to decide which of three catalytic converters to install on its new models. All three meet the government emission standards but they may not all give the same mileage. Ten new cars, all alike, are selected for the test. Converter A is installed on four of the cars, while converters B and C are each put on three. The cars are then driven over a specially engineered track that simulates city driving. The mileage estimates are listed below.

Converter A	Converter B	Converter C
21.6	22.8	23.9
23.2	25.6	24.6
20.5	24.7	23.8
21.7		

Test whether all three converters are equally economical. Let $\alpha = 0.05$.

11. Derive the computing formula for the error sum of squares given in Equation 11.15:

$$SS_{\text{error}} = \sum_{j=1}^{k} \left(\sum_{i=1}^{n_j} Y_{ij}^2 - \frac{T_{.j}^2}{n_j} \right).$$

Then use the computing formula to double-check the value gotten for SS_{error} in Review Exercise 10.

12. Suppose k independent random samples, each of size r, are taken from an $N(\mu_j, \sigma^2)$ pdf, $j = 1, 2, \ldots, k$, where the μ_j's and σ^2 are all unknown. We wish to set up the generalized-likelihood-ratio test for

$$H_0 : \mu_1 = \mu_2 = \cdots = \mu_k \ (= \mu)$$

versus

$$H_1 : \text{not all the } \mu_j\text{'s are equal.}$$

(a) Write out $L(\omega)$ and $L(\Omega)$.
(b) Show that the MLE's for μ and σ^2 under ω are

$$\hat{\mu} = \frac{1}{rk} \sum_{j=1}^{k} \sum_{i=1}^{r} y_{ij} = \bar{y}_{..}$$

and

$$\hat{\sigma}^2 = \frac{1}{rk} \sum_{j=1}^{k} \sum_{i=1}^{r} (y_{ij} - \bar{y}_{..})^2.$$

(c) Show that the MLE's for μ_j and σ^2 under Ω are

$$\hat{\mu}_j = \frac{1}{r} \sum_{i=1}^{r} y_{ij} = \bar{y}_{.j}, \qquad j = 1, 2, \ldots, k,$$

and

$$\hat{\sigma}^2 = \frac{1}{rk} \sum_{j=1}^{k} \sum_{i=1}^{r} (y_{ij} - \bar{y}_{.j})^2.$$

(d) Show that the generalized likelihood ratio reduces to

$$\lambda = \frac{L(\hat{\omega})}{L(\hat{\Omega})} = \left[\frac{\sum_{j=1}^{k} \sum_{i=1}^{r} (y_{ij} - \bar{y}_{.j})^2}{\sum_{j=1}^{k} \sum_{i=1}^{r} (y_{ij} - \bar{y}_{..})^2} \right]^{rk/2}.$$

13. Refer to Review Exercise 12. Express $\lambda^{2/rk}$ as a function of the treatment sum of squares and the error sum of squares. Show that rejecting $H_0: \mu_1 = \mu_2 = \cdots = \mu_k$ whenever λ is too small (recall Definition 6.2) is equivalent to rejecting H_0 when the observed F ratio,

$$F = \frac{Q_1/(k-1)}{Q_2/k(r-1)},$$

is too large.

14. The cathode warm-up time (in seconds) was determined for three different types of X-ray tubes using 15 observations on each type. The results are listed below.

Warm-up times (sec)

Tube Type

A		B		C	
19	27	20	24	16	14
23	31	20	25	26	18
26	25	32	29	15	19
18	22	27	31	18	21
20	23	40	24	19	17
20	27	24	25	17	19
18	29	22	32	19	18
35		18		18	

Do an analysis of variance on these data and test the hypothesis that the three tube types require the same average warm-up time. Include a pair of orthogonal contrasts in your ANOVA table. Define one of the contrasts so it tests $H'_0: \mu_A = \mu_C$. What does the other contrast test? Check to see that the sums of squares associated with your two contrasts verify the statement of Theorem 11.2.

15. Refer to the IQ data of Review Exercise 1. Suppose the experimenter had reason to believe, before collecting any data, that students in groups II and III would show, on the average, the same increase in IQ. Set up and test the appropriate contrast. Let $\alpha = 0.05$.

16. Test the hypothesis that the average of the true yields for the first three varieties of corn described in Review Exercise 9 is the same as the average for the last two.

17. One of the assumptions underlying the analysis of variance is that the errors in the responses are normally distributed (recall Equation 11.2). Investigate the reasonableness of this assumption in the case of the cathode warm-up data of Review Exercise 14. First, use Equation 11.7 to find the estimates, $\hat{\mu}$ and $\hat{\tau}_j$, $j = 1, 2, 3$. Then calculate the set of *residuals*, $y_{ij} - \hat{\mu} - \hat{\tau}_j$, $i = 1, 2, \ldots, 15$; $j = 1, 2, 3$. Make a histogram of the residuals. Does the shape of the histogram lend credibility to the normality assumption? What sort of formal analysis would be appropriate here?

18. Use Tukey's method to make all the pairwise comparisons for the IQ data of Review Exercise 1.

19. Construct 95% Tukey confidence intervals for all $\binom{5}{2} = 10$ pairwise differences, $\mu_i - \mu_j$, for the data of Review Exercise 9. Summarize the results by plotting the five sample averages on a horizontal axis and drawing straight lines under varieties whose average yields are not significantly different.

20. A commercial film processor is experimenting with two kinds of fully automatic color developers. Six sheets of exposed film are put through each developer. The number of flaws on each negative visible with the naked eye is then counted.

Number of Visible Flaws

Developer A	Developer B
1	8
4	6
5	4
6	9
3	11
7	10

Assume the number of flaws on a given negative is a Poisson random variable. Make an appropriate data transformation and do the indicated analysis of variance.

21. Three air-to-surface missile launchers are tested for their accuracy. The same gun crew fires four rounds with each launcher, each round consisting of 20 missiles. A "hit" is scored if the missile lands within 10 yards of the target. The table below gives the number of hits registered in each round.

Number of Hits Per Round

Launcher A	Launcher B	Launcher C
13	15	9
11	16	11
10	18	10
14	17	8

Compare the accuracy of these three launchers by using the analysis of variance, after making a suitable data transformation. Let $\alpha = 0.05$.

CHAPTER TWELVE

Randomized Block Designs

Pictured is a 5-by-5 Latin square laid out at Bettgelert Forest in 1929 for studying the effect of exposure on Sitka spruce, Norway spruce, Japanese larch, European larch, Pinus contorta, and beech. (A Latin square is a special type of experimental design that extends to two dimensions the important notion of "blocking" that is introduced in this chapter. Latin squares are particularly useful in agricultural research.)

12.1 INTRODUCTION

In any experiment, reducing the magnitude of the experimental error is a highly desirable objective: the smaller σ^2 is, the better will be our chances of rejecting a false null hypothesis. Basically, there are two ways to reduce experimental error. The nonstatistical approach is simply to refine the experimental technique—use better equipment, minimize subjective error, and so on. The statistical method, which can often produce results much more dramatic, is by collecting the data in "blocks," in what is referred to as a *randomized block design*. Such designs are the subject of this chapter.

Historically, it was Fisher who first advanced the notion of blocking. He saw it initially as a statistical defense against the obfuscating effects of soil heterogeneity in agricultural experiments. Suppose, for example, a researcher wishes to compare the yields of four different varieties of corn. Figure 12.1(a) shows the simplest

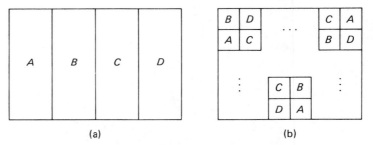

(a)   (b)

Figure 12.1 Two different experimental designs.

experimental layout: variety A is planted in the leftmost portion of the field, variety B is planted next to A, and so on. Even to a city slicker the statistical hazards involved in using this design should be obvious. Suppose, for example, there was a soil *gradient* in the field, with the best soil being in the westernmost part (where variety A was planted). Then if variety A achieved the highest yield, we would not know whether to attribute its success to its inherent quality or to its location (or to some combination of both).

A more sensible *modus operandi* is pictured in Figure 12.1(b). There the field is divided up into a number of smaller "blocks," each block being still further parceled into four "plots." All four varieties are planted in each block, one to a plot, with the plot assignments being chosen at random. Notice that the geographical contiguity of the four plots within a given block insures that the environmental conditions from plot to plot will be relatively uniform and will not lead to any biasing of the observed yields. What the analysis of variance will then do is "pool" from block to block the *within*-block information concerning the treatment differences while bypassing the *between*-block differences—that is, the heterogeneity in the experimental environment. As a result, the treatment comparisons can be made with greater precision. Analytically, where the total sum of squares was partitioned into *two* components in a completely randomized, one-factor design, it will be split into *three* separate sums in a randomized block design: one for treatments, another for blocks, and a third for experimental error.

It did not take long for scientists to realize that the benefits of blocking could be extended well beyond the confines of agricultural experimentation. In medical research, blocks are often made up of subjects of the same age, sex, and overall physical condition. A common practice in animal studies is to form blocks out of littermates. Industrial experiments often require that "time" be a blocking criterion: measurements taken by personnel on the day shift might be considered one block and those taken by the night shift a second block. In some sense the ultimate form of blocking, although one not always physically possible, is to apply the entire set of treatment levels to each subject, thus making each subject his own block.

The next section begins with a development of the analysis of variance for the general randomized block design, where t factor levels are administered within each of b blocks. As before, the hypotheses to be tested are $H_0: \mu_1 = \mu_2 \cdots = \mu_t$ versus H_1: not all the μ_j's are equal. Section 12.2 concludes with a set of case studies illustrating the blocking concept in a variety of different settings.

12.2 THE *F* TEST FOR A RANDOMIZED BLOCK DESIGN

As indicated in Section 12.1, the typical data structure for a randomized block design is a matrix with b rows and t columns, the former representing the b blocks, the latter the t levels of the treatment. We will let $\gamma_{ij}, i = 1, 2, \ldots, b; j = 1, 2, \ldots, t$, denote the true average response associated with the application of level j to block i. Similarly, ω_i and μ_j will denote the true average (separate) responses for block i and level j, respectively. The model equation for Y_{if}, the observation recorded for the ith block and jth level, is a straightforward extension of Equation 11.5:

$$Y_{ij} = \mu + \beta_i + \tau_j + \epsilon_{ij}, \qquad i = 1, \ldots, b; j = 1, \ldots, t, \qquad (12.1)$$

where

$$\mu = \frac{1}{bt} \sum_{i=1}^{b} \sum_{j=1}^{t} \gamma_{ij} = \text{overall mean},$$

$$\beta_i = \omega_i - \mu = \text{differential effect of the } i\text{th block} \\ \text{(relative to } \mu\text{)},$$

$$\tau_j = \mu_j - \mu = \text{differential effect of the } j\text{th treatment} \\ \text{(relative to } \mu\text{)},$$

$$\epsilon_{ij} = \text{experimental error}.$$

The constraints imposed on the parameters of Equation 12.1 will vary somewhat from problem to problem. The block effect and treatment effect, for example, might be either *fixed* or *random*, the choice depending on how the levels for the b blocks and t treatments were selected. If the levels for a factor are in any way preselected or unique, then that factor is said to be *fixed*; otherwise, if the levels included in the experiment are just an arbitrary sample of all possible levels, the factor is said to be *random*. In a typical randomized block design, the treatment effect will be fixed and the block effect random. The τ_j's, then, will be a set of constants for which

$$\sum_{j=1}^{t} \tau_j = 0$$

(recall the analogous constraint for the completely randomized, one-factor design, as stated in Question 11.2.2). The β_i's, on the other hand, will be random variables—specifically, $\beta_i \sim N(0, \sigma_B^2)$, $i = 1, 2, \ldots, b$, where σ_B^2 is unknown. The assumption about the ϵ_{ij}'s will be the same as it was in the completely randomized, one-factor design:

$$\epsilon_{ij} \sim N(0, \sigma^2), \qquad i = 1, \ldots, b; j = 1, \ldots, t.$$

(More will be said about the distinction between fixed and random effects in the case studies.)

Example 12.1 parallels Example 11.1 in showing what the components of Equation 12.1 mean numerically.

EXAMPLE 12.1.   Suppose the (unknown) γ_{ij}'s for a randomized block design with three treatment levels and two blocks are the entries appearing in Table 12.1. (It is assumed here that both "treatments" and "blocks" are fixed).

TABLE 12.1   A set of γ_{ij}'s

		Treatment Level		
		1	2	3
Block	1	14	10	6
	2	15	15	6

Then

$$\mu = \frac{1}{2(3)}(14 + 10 + \cdots + 6) = \frac{66}{6} = 11.0,$$

$$\mu_1 = \frac{29}{2} = 14.5; \quad \mu_2 = \frac{25}{2} = 12.5; \quad \mu_3 = \frac{12}{2} = 6.0,$$

$$\omega_1 = \frac{30}{3} = 10.0; \qquad \omega_2 = \frac{36}{3} = 12.0.$$

It follows that τ_1, the differential effect for treatment level 1, is $14.5 - 11.0 = 3.5$. Similarly, $\tau_2 = 12.5 - 11.0 = 1.5$, and $\tau_3 = 6.0 - 11.0 = -5.0$. The differential effects of the blocks are -1.0 and $+1.0$, respectively: $\beta_1 = 10.0 - 11.0 = -1.0$ and $\beta_2 = 12.0 - 11.0 = +1.0$. Notice that

$$\sum_{j=1}^{3} \tau_j = 0 = \sum_{i=1}^{2} \beta_i. \tag{12.2}$$

QUESTION 12.2.1   Differentiate the least-squares function

$$L = \sum_{i=1}^{b} \sum_{j=1}^{t} (y_{ij} - \mu - \beta_i - \tau_j)^2$$

with respect to all $bt + 1$ parameters and use the two constraints indicated in Equation 12.2 to show that $\hat{\mu} = \bar{y}_{..}$, $\hat{\beta}_i = \bar{y}_{i.} - \bar{y}_{..}$, and $\hat{\tau}_j = \bar{y}_{.j} - \bar{y}_{..}$.

Following the procedure established for the completely randomized, one-factor design, the first step in deriving the F test for $H_0: \tau_1 = \tau_2 = \cdots = \tau_t = 0$ is to partition the total sum of squares, $\sum_{i=1}^{b} \sum_{j=1}^{t} (Y_{ij} - \bar{Y}_{..})^2$, into components related to the terms in the model equation. The appropriate identity to start with (see Equation 11.8) is

$$(Y_{ij} - \bar{Y}_{..}) = (\bar{Y}_{i.} - \bar{Y}_{..}) + (\bar{Y}_{.j} - \bar{Y}_{..}) + (Y_{ij} - \bar{Y}_{i.} - \bar{Y}_{.j} + \bar{Y}_{..}). \quad (12.3)$$

Squaring both sides of Equation 12.3 and summing over i and j gives

$$\sum_{i=1}^{b} \sum_{j=1}^{t} (Y_{ij} - \bar{Y}_{..})^2 = \sum_{i=1}^{b} \sum_{j=1}^{t} (\bar{Y}_{i.} - \bar{Y}_{..})^2 + \sum_{i=1}^{b} \sum_{j=1}^{t} (\bar{Y}_{.j} - \bar{Y}_{..})^2$$

$$+ \sum_{i=1}^{b} \sum_{j=1}^{t} (Y_{ij} - \bar{Y}_{i.} - \bar{Y}_{.j} + \bar{Y}_{..})^2 \quad (12.4)$$

or

$$Q = Q_1 + Q_2 + Q_3.$$

(It will be left as an exercise to verify that the three cross-product terms are identically 0.)

Three of the four terms in Equation 12.4 have obvious interpretations: Q measures the total variability in the data, Q_1 measures the block effect, and Q_2 the treatment effect. To understand Q_3, notice that

$$Y_{ij} - \bar{Y}_{i.} - \bar{Y}_{.j} + \bar{Y}_{..} = Y_{ij} - (\bar{Y}_{i.} - \bar{Y}_{..}) - (\bar{Y}_{.j} - \bar{Y}_{..}) - \bar{Y}_{..}$$

> = observed response − estimated differential block effect − estimated differential treatment effect − estimated overall effect

> = experimental error.

Therefore, the decomposition achieved in Equation 12.4 can be written

$$SS_{\text{total}} = SS_{\text{blocks}} + SS_{\text{treatments}} + SS_{\text{error}}.$$

Theorem 12.1 gives the distributions for Q_2 and Q_3 under H_0 and their expected values under both H_0 and H_1. The proof closely parallels what was done in parts II and III of Section 11.2 and will not be given. (A similar result holds for Q_1 but will not be needed.)

THEOREM 12.1

(a) When $H_0: \tau_1 = \tau_2 = \cdots = \tau_t = 0$ is true,

$$\frac{Q_2}{\sigma^2} \sim \chi^2_{t-1}$$

and $E[Q_2/(t-1)] = \sigma^2$. When H_1: not all the τ_j's $= 0$ is true, Q_2/σ^2 has a noncentral χ^2 distribution and

$$E\left(\frac{Q_2}{t-1}\right) = \sigma^2 + \frac{b}{t-1}\sum_{j=1}^{t}\tau_j^2.$$

(b) Regardless of what is true about the β_i's or the τ_j's,

$$\frac{Q_3}{\sigma^2} \sim \chi^2_{(b-1)(t-1)}$$

and $E\{Q_3/[(b-1)(t-1)]\} = \sigma^2$.

It follows from Theorem 12.1, Theorem 7.11, and Definition 11.2 that

$$\frac{\dfrac{Q_2}{(t-1)\sigma^2}}{\dfrac{Q_3}{(b-1)(t-1)\sigma^2}} = \frac{MS_{\text{treatments}}}{MS_{\text{error}}} \sim \frac{\dfrac{\chi^2_{t-1}}{t-1}}{\dfrac{\chi^2_{(b-1)(t-1)}}{(b-1)(t-1)}}$$

will have a (central) F distribution with $t-1$ and $(b-1)(t-1)$ degrees of freedom when $H_0: \tau_1 = \tau_2 = \cdots = \tau_t = 0$ is true, and a noncentral F distribution when H_1: not all the τ_j's $= 0$ is true. Therefore, the null hypothesis that the means for the t treatment levels are all equal should be rejected at the α level of significance if

$$\frac{MS_{\text{treatments}}}{MS_{\text{error}}} \geq F_{1-\alpha, t-1, (b-1)(t-1)}. \tag{12.5}$$

While Equation 12.5 is sufficient to carry out the analysis of variance, the F ratio can be gotten much easier with the help of computing formulas for $SS_{\text{treatments}}$ and SS_{error}. Theorem 12.2 is the analog of Theorem 11.1. Its proof is a straightforward exercise in summation algebra and will not be given.

THEOREM 12.2   Let $c = T_{..}^2/bt$. Then

$$SS_{\text{total}} = \sum_{i=1}^{b}\sum_{j=1}^{t} Y_{ij}^2 - c,$$

$$SS_{\text{blocks}} = \sum_{i=1}^{b}\frac{T_{i.}^2}{t} - c,$$

$$SS_{\text{treatments}} = \sum_{j=1}^{t}\frac{T_{.j}^2}{b} - c,$$

and

$$SS_{\text{error}} = SS_{\text{total}} - SS_{\text{blocks}} - SS_{\text{treatments}}.$$

The next several examples illustrate the analysis of variance calculations for a randomized block design. Notice the different blocking criteria being utilized. Case Study 12.1 employs the most direct sort of blocking: each subject, having been measured three times, functions as its own block. In Case Study 12.2 a pretest is the blocking mechanism, while in Case Study 12.3, time, in the form of months, delineates one block from another.

CASE STUDY

12.1

Hypnotic age regression is a procedure whereby an adult subject is instructed under hypnosis to return to an earlier chronological age. Although this process can recapture *behavioral* patterns of the earlier age, it is not known whether it can reinstate *perceptual* patterns. A recent study addressed itself to that point by using the Ponzo illusion pictured in Figure 12.2.

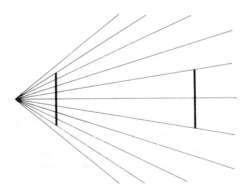

Figure 12.2   The Ponzo illusion.

When the two vertical lines are equal in length, most people will perceive the right-most one as being shorter. Furthermore, studies have shown that the strength of the illusion is a function of age: it becomes more and more pronounced from the childhood years through adolescence; then, in the late teens, it levels off.

The Ponzo illusion was shown to eight college students. These same students were then regressed under hypnosis to age nine and to age five. At each age, their perceptions of the illusion were measured. If hypnosis can, in fact, recapture perceptual patterns, the illusion strengths at these three ages should be significantly different.

The Ponzo illusion was drawn on a set of 21 cards. On each card the leftmost line was 4 inches long; the rightmost line varied from card to card, ranging from $2\frac{1}{2}$ to 5 inches in increments of $\frac{1}{8}$ inch. The cards were shown to each subject in random order. In each instance the subject was instructed to point to the longer line. If the subject perceived the rightmost line to be longer when it was actually $4\frac{6}{8}$ inches but shorter when it was $4\frac{5}{8}$ inches, the strength of the illusion was defined to be $\frac{5}{8} + \frac{1}{2}(\frac{6}{8} - \frac{5}{8}) = 0.69$ inch. The results for the eight subjects are listed in Table 12.2 (132).

TABLE 12.2 Illusion strengths

Subject	Awake	Regressed to Age Nine	Regressed to Age Five	$t_{i.}$
1	0.81	0.69	0.56	2.06
2	0.44	0.31	0.44	1.19
3	0.44	0.44	0.44	1.32
4	0.56	0.44	0.44	1.44
5	0.19	0.19	0.31	0.69
6	0.94	0.44	0.69	2.07
7	0.44	0.44	0.44	1.32
8	0.06	0.19	0.19	0.44
$t_{.j}$	3.88	3.14	3.51	10.53

Here the "subjects" would be considered a random effect (since the eight being used were presumably in no way special, but just an arbitrary sample from the population of all college students) and the "regressed ages" a fixed effect (the three treatment levels—awake, age nine, and age five—were preselected because of their particular relevance to the expected strength of the illusion). The model equation would be

$$Y_{ij} = \mu + \beta_i + \tau_j + \epsilon_{ij}, \qquad i = 1, \ldots, 8; j = 1, 2, 3.$$

From Theorem 12.2,

$$c = \frac{(10.53)^2}{8(3)} = 4.6200.$$

Also, $\displaystyle\sum_{i=1}^{8} \sum_{j=1}^{3} y_{ij}^2 = 5.5889$. Therefore,

$$SS_{\text{total}} = 5.5889 - 4.6200 = 0.9689,$$

$$SS_{\text{treatments}} = \frac{(3.88)^2}{8} + \frac{(3.14)^2}{8} + \frac{(3.51)^2}{8} - 4.6200 = 0.0343,$$

$$SS_{\text{blocks}} = \frac{(2.06)^2}{3} + \cdots + \frac{(0.44)^2}{3} - 4.6200 = 0.7709,$$

and, by subtraction,

$$SS_{\text{error}} = 0.9689 - 0.0343 - 0.7709 = 0.1637.$$

The analysis of variance is detailed in Table 12.3. For any randomized block design, the degrees of freedom for blocks, treatments, error, and total will be $b - 1$, $t - 1$, $(b - 1)(t - 1)$, and $bt - 1$, respectively.

TABLE 12.3 ANOVA computations

Source	df	SS	MS	EMS	F
Subjects	7	0.7709	0.1101		
Ages	2	0.0343	0.0172	$\sigma^2 + \dfrac{8}{3-1} \sum\limits_{j=1}^{3} \tau_j^2$	1.47
Error	14	0.1637	0.0117	σ^2	
Total	23	0.9689			

If we elect to test $H_0 : \mu_1 = \mu_2 = \mu_3$ (or $H_0 : \tau_1 = \tau_2 = \tau_3 = 0$) at the $\alpha = 0.05$ level of significance, our conclusion is to *accept* the null hypothesis (see Figure 12.3).

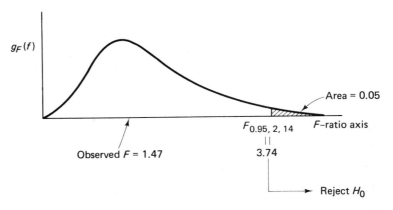

Figure 12.3 *F* distribution with 2 and 14 degrees of freedom.

QUESTION 12.2.2   In recent years a number of research projects in extrasensory perception have examined the possibility that hypnosis may be helpful in bringing out ESP in persons who did not think they had any. The obvious way to test such a hypothesis is with a self-paired design : the ESP ability of a subject when he is awake is compared to his ability when hypnotized. In one study of this sort, 15 college students were each asked to guess the identity of 200 Zener cards (see Case Study 4.5). The same "sender"— that is, the person concentrating on the card—was used for each trial. For 100 of the trials both the student and the sender were awake; for the other 100 both were hypnotized. If chance were the only factor involved, the expected number of correct

TABLE 12.4   Number of correct responses
(out of 100) in ESP experiment

Student	Sender and Student in Waking State	Sender and Student in Hypnotic State
1	18	25
2	19	20
3	16	26
4	21	26
5	16	20
6	20	23
7	20	14
8	14	18
9	11	18
10	22	20
11	19	22
12	29	27
13	16	19
14	27	27
15	15	21
$\bar{y}._{j}$	18.9	21.7

identifications in each set of 100 trials would be 20. The observed average numbers of correct guesses for subjects awake and subjects hypnotized were 18.9 and 21.7, respectively (22). Use the analysis of variance to determine whether that difference is statistically significant at the 0.05 level.

CASE STUDY

12.2

Acrophobia is a fear of heights. It can be treated in a number of different ways. Using contact desensitization, a therapist demonstrates some task that would be difficult for someone with acrophobia to do, such as looking over a ledge or standing on a ladder. Then he guides the subject through the very same maneuver, always keeping in physical contact. Another method of treatment is demonstration participation. Here the therapist tries to *talk* the subject through the task; no physical contact is made. A third technique, live modeling, requires the subject simply to *watch* the task being done—he does not attempt it himself.

These three techniques were compared in a recent study involving 15 volunteers, all of whom had a history of severe acrophobia (141). It was realized at the outset, though, that the affliction was much more incapacitating in some subjects than in others, and that this heterogeneity might compromise the therapy comparison. Accordingly, the experiment began with each subject being given the Height Avoidance Test (HAT), a series of 44 tasks related to ladder climbing. A subject received a "point" for each task successfully completed. On the basis of their final scores the 15 volunteers were divided into five blocks (A, B, C, D, and E), each of size three. The subjects in Block A had the *lowest* scores (that is, the most severe acrophobia), those in block B the second lowest scores, and so on.

Each of the three therapies was then assigned at random to one of the three subjects in each block. When the counseling sessions were over, the subjects retook the HAT. Table 12.5 lists the *changes* in their scores (score after therapy − score before therapy). Our objective will be to test the equality of the therapies at the $\alpha = 0.01$ level of significance.

TABLE 12.5 HAT score changes

Block	Contact Desensitization	Therapy Demonstration Participation	Live Modeling	$t_{i.}$
A	8	2	−2	8
B	11	1	0	12
C	9	12	6	27
D	16	11	2	29
E	24	19	11	54
$t_{.j}$	68	45	17	130

Since $c = (130)^2/15 = 1126.7$ and $\sum_{i=1}^{5} \sum_{j=1}^{3} y_{ij}^2 = 1894$, it follows that

$$SS_{\text{total}} = 1894 = 1126.7 = 767.3,$$

$$SS_{\text{blocks}} = \frac{(8)^2}{3} + \cdots + \frac{(54)^2}{3} - 1126.7 = 438.0,$$

$$SS_{\text{treatments}} = \frac{(68)^2}{5} + \frac{(45)^2}{5} + \frac{(17)^2}{5} - 1126.7 = 260.9,$$

giving an error sum of squares of 68.4:

$$SS_{error} = 767.3 - 438.0 - 260.9 = 68.4.$$

The analysis of variance is summarized in Table 12.6. As the two asterisks signify, $H_0 : \tau_1 = \tau_2 = \tau_3 = 0$ can be rejected at the 0.01 level: $F_{0.99, 2, 8} = 8.65$.

TABLE 12.6 ANOVA computations

Source	df	SS	MS	EMS	F	
Blocks	4	438.0	109.5			
Therapies	2	260.9	130.4	$\sigma^2 + \dfrac{5}{3-1} \sum\limits_{j=1}^{3} \tau_j^2$	15.2	**
Error	8	68.4	8.6	σ^2		
Total	14	767.3				

The Tukey multiple-comparison technique of Section 11.5 can also be applied to a randomized block design. In the notation of Theorem 11.4, the values here for k and r would be 3 and 5, respectively. The definition of D is slightly different, however, since the associated studentized range is no longer $Q_{k, rk-k}$ but rather $Q_{k, (b-1)(t-1)}$, a change reflecting the difference in the number of degrees of freedom available for estimating σ^2. Table 12.7 shows the three pairwise therapy differences and their 95% Tukey intervals. The value used for D was 1.81:

$$D = \frac{Q_{0.05, 3, 8}}{\sqrt{5}} = \frac{4.04}{2.24} = 1.81.$$

TABLE 12.7 Therapy differences and 95% Tukey intervals

Pairwise Difference	$\bar{y}_{.i} - \bar{y}_{.j}$	Tukey Interval	Conclusion
$\mu_1 - \mu_2$	4.6	(−0.7, 9.9)	NS
$\mu_1 - \mu_3$	10.2	(4.9, 15.5)	Reject
$\mu_2 - \mu_3$	5.6	(0.3, 10.9)	Reject

Thus we would conclude, *after the fact*, at the 0.05 level that the therapeutic values of contact desensitization and demonstration participation are not significantly different but that both contact desensitization and demonstration participation are significantly better than live modeling.

QUESTION 12.2.3   A comparison was made of the efficiency of four different unit-dose injection systems. A group of pharmacists and nurses were the "blocks." For each system, the /
were to remove the unit from its outer package, assemble it, and simulate an injection.
In addition to the standard system of using a disposable syringe and needle to draw
the medication from a vial, the other systems tested were Vari-Ject (CIBA Pharmaceutical), Unimatic (Squibb), and Tubex (Wyeth). Table 12.8 lists the average
times (in seconds) needed to implement each of the systems (146). State the model

TABLE 12.8   Average times (sec) for implementing injection systems

Subject	Standard	Vari-Ject	Unimatic	Tubex	$t_{i.}$
1	35.6	17.3	24.4	25.0	102.3
2	31.3	16.4	22.4	26.0	96.1
3	36.2	18.1	22.8	25.3	102.4
4	31.1	17.8	21.0	24.0	94.4
5	39.4	18.8	23.3	24.2	105.7
6	34.7	17.0	21.8	26.2	99.7
7	34.1	14.5	23.0	24.0	95.6
8	36.5	17.9	24.1	20.9	99.4
9	32.2	14.6	23.5	23.5	93.8
10	40.7	16.4	31.3	36.9	125.3
$t_{.j}$	351.8	168.8	238.1	256.0	1014.7

equation for these data and indicate what assumptions are being made about its
components. Do the analysis of variance to test $H_0 : \tau_1 = \cdots = \tau_4 = 0$ versus H_1:
not all the τ_j's $= 0$ at the $\alpha = 0.05$ level of significance. Also, use Tukey's method to
test all six pairwise differences. Note:

$$\sum_{i=1}^{10} \sum_{j=1}^{4} y_{ij}^2 = 27{,}792.39.$$

CASE STUDY

12.3   Case Study 1.3 described an investigation designed to "measure" the
so-called Transylvania effect—the possible influence of the full moon
on human behavior. Table 12.9 reproduces those data (admission
rates to the emergency room of a mental health clinic) and gives the
appropriate row and column totals.

Table 12.10 summarizes the ANOVA calculations. For 2 and 22
degrees of freedom the 0.05 critical value is 3.44. Thus, our conclusion
at the 0.05 level would be to accept $H_0 : \mu_1 = \mu_2 = \mu_3$: a lunar
"effect" has not been demonstrated.

TABLE 12.9   Mental health clinic emergency room admission rates

Admission Rates (patients/day)

Month	(1) Before Full Moon	(2) During Full Moon	(3) After Full Moon	$t_{i.}$
Aug	6.4	5.0	5.8	17.2
Sep	7.1	13.0	9.2	29.3
Oct	6.5	14.0	7.9	28.4
Nov	8.6	12.0	7.7	28.3
Dec	8.1	6.0	11.0	25.1
Jan	10.4	9.0	12.9	32.3
Feb	11.5	13.0	13.5	38.0
Mar	13.8	16.0	13.1	42.9
Apr	15.4	25.0	15.8	56.2
May	15.7	13.0	13.3	42.0
Jun	11.7	14.0	12.8	38.5
Jul	15.8	20.0	14.5	50.3
$t_{.j}$	131.0	160.0	137.5	

TABLE 12.10   ANOVA computations

Source	df	SS	MS	EMS	F
Months	11	451.08	41.01		
Lunar cycles	2	38.59	19.30	$\sigma^2 + \dfrac{12}{3-1}\sum\limits_{j=1}^{3} \tau_j^2$	3.22
Error	22	132.08	6.00	σ^2	
Total	35	621.75			

Testing the general H_0 is not the only appropriate way to analyze these data, though. An a priori subhypothesis is clearly suggested by the circumstances of the problem—specifically, it would make sense to test whether the admission rate during the full moon is different than the *average* of the rates for "before" and "after." What needs to be considered, then, is $H_0' : \mu_2 = (\mu_1 + \mu_3)/2$.

Following the procedure outlined in Section 11.4, the contrast associated with H_0' is

$$C = -\tfrac{1}{2}\mu_1 + \mu_2 - \tfrac{1}{2}\mu_3$$

while its estimate is

$$\hat{C} = -\tfrac{1}{2}(10.92) + 1(13.3) - \tfrac{1}{2}(11.46)$$
$$= 2.11.$$

From Definition 11.4, the sum of squares associated with C is 35.62:

$$SS_C = \frac{(2.11)^2}{\dfrac{\tfrac{1}{4}}{12} + \dfrac{1}{12} + \dfrac{\tfrac{1}{4}}{12}} = 35.62.$$

Dividing SS_C by the mean square for error gives an F ratio with 1 and 22 degrees of freedom:

$$\frac{35.62/1}{132.08/22} = 5.93.$$

But, for the 0.05 level, $F_{0.95,1,22} = 4.30$. Thus, contrary to our acceptance of H_0, we would *reject* H_0' and conclude that the admission rate *is* significantly higher during a full moon than it is for the rest of the month.

Comment. It is always more than a little disconcerting when two valid statistical techniques applied to the same data lead to opposite conclusions. That such apparent contradictions occur, though, should not be unexpected. Different methods of analysis simply utilize the data in different ways. Disagreements from time to time are inevitable. ∎

QUESTION 12.2.4   Heart rates were monitored (10) for six tree shrews (*Tupaia glis*) during three different stages of sleep: LSWS (light slow-wave sleep), DSWS (deep slow-wave sleep), and REM (rapid-eye-movement sleep). (See Table 12.11.)

TABLE 12.11   Heart rates (beats/5 seconds)

Tree Shrew	LSWS	DSWS	REM
1	14.1	11.7	15.7
2	26.0	21.1	21.5
3	20.9	19.7	18.3
4	19.0	18.2	17.0
5	26.1	23.2	22.5
6	20.5	20.7	18.9

(a) Do the analysis of variance to test the equality of the heart rates during these three phases of sleep. Let $\alpha = 0.05$.

(b) Because of the marked physiological difference between REM sleep and LSWS and DSWS sleep, it was decided before the data were collected to test the REM rate against the average of the other two. Test the appropriate subhypothesis with a contrast. Use the 0.05 level of significance. Also, find a second contrast orthogonal to the first and verify that the sum of the sum of squares for the two contrasts equals $SS_{\text{treatments}}$.

12.3 THE PAIRED t TEST

An important special case of a randomized block design arises when the number of treatment levels is reduced to $t = 2$. The simplicity of the data structure then allows the F test to be replaced by the more familiar one-sample t test. Specifically, if X_i and Y_i denote the responses in the ith block to treatment level 1 and treatment level 2, respectively, we will define D_i to be the within-block difference—that is, $D_i = X_i - Y_i$. If there are b blocks, the sample mean and the sample variance of the D_i's have formulas analogous to their Chapter 7 counterparts:

$$\bar{D} = \frac{1}{b} \sum_{i=1}^{b} D_i$$

and

$$S_D^2 = \frac{1}{b-1} \sum_{i=1}^{b} (D_i - \bar{D})^2 = \frac{b \sum_{i=1}^{b} D_i^2 - \left(\sum_{i=1}^{b} D_i \right)^2}{b(b-1)}.$$

Let μ_X and μ_Y denote the true means of levels 1 and 2, and let $\mu_D = \mu_X - \mu_Y$. Testing the null hypothesis $H_0: \mu_X = \mu_Y$ is then the same as testing $H_0: \mu_D = 0$. If it can be assumed that the D_i's are an independent random sample from an $N(\mu_D, \sigma^2)$ distribution, we can look to Theorem 7.15 to provide an appropriate critical region.

Comment. Testing $H_0: \mu_D = 0$ is a one-sample problem analytically but a two-sample problem conceptually. However, it is not analogous to the data structure presented in Chapter 8, because here the two samples are *dependent*, X_i and Y_i being measured on members of the same block. To emphasize these distinctions, the application of Theorem 7.15 in this context is referred to as a *paired t test*.

CASE STUDY

12.4

Blood coagulates only after a complex chain of chemical reactions has taken place. The final step is the transformation of fibrinogen, a protein found in the plasma, into fibrin, which forms the clot. The fibrinogen-fibrin reaction is triggered by another protein, thrombin, which itself is formed under the influence of still other proteins, including one called prothrombin.

A person's blood-clotting ability is typically expressed in terms of a "prothrombin time," which is defined to be the interval between the initiation of the prothrombin-thrombin reaction and the formation of the final clot. Factors affecting the length of a person's prothrombin time have profound medical significance. Recently a study was undertaken to determine what effect, if any, *aspirin* has on this particular function. Twelve adult males participated (204). Their prothrombin times were measured *before* and *three hours after* each was given two aspirin tablets (650 mg) (Table 12.12).

TABLE 12.12   Blood-clotting study

Prothrombin Times (sec)

Subject	Before Aspirin (x_i)	After Aspirin (y_i)	$d_i = x_i - y_i$
1	12.3	12.0	0.3
2	12.0	12.3	−0.3
3	12.0	12.5	−0.5
4	13.0	12.0	1.0
5	13.0	13.0	0
6	12.5	12.5	0
7	11.3	10.3	1.0
8	11.8	11.3	0.5
9	11.5	11.5	0
10	11.0	11.5	−0.5
11	11.0	11.0	0
12	11.3	11.5	−0.2
			1.3

If μ_X and μ_Y denote the true average prothrombin times *before* and *after* aspirin, respectively, and if $\mu_D = \mu_X - \mu_Y$, the hypotheses to be tested are

$$H_0: \mu_D = 0$$

versus

$$H_1: \mu_D \neq 0.$$

We will let 0.05 be the level of significance.

From Table 12.12,

$$\sum_{i=1}^{12} d_i = 1.3 \quad \text{and} \quad \sum_{i=1}^{12} d_i^2 = 2.97.$$

Therefore,

$$\bar{d} = \frac{1}{12}(1.3) = 0.108$$

and

$$s_D^2 = \frac{12(2.97) - (1.3)^2}{12(11)} = 0.257.$$

Since $b = 12$, the critical values for the test statistic will be the 2.5th and 97.5th percentiles of the Student t distribution with 11 degrees of freedom: ± 2.20. The appropriate decision rule, then, from Theorem 7.15, would be

$$\text{Reject } H_0: \mu_D = 0 \text{ if } \frac{\bar{d}}{s_D/\sqrt{12}} \text{ is either } \begin{cases} \leq -2.20 \\ \text{or} \\ \geq +2.20. \end{cases}$$

In this case, the t ratio is

$$\frac{\bar{d}}{s_D/\sqrt{12}} = \frac{0.108}{\sqrt{0.257}/\sqrt{12}} = 0.74,$$

and our conclusion is to *accept* H_0—aspirin has no demonstrable effect on prothrombin time.

QUESTION 12.3.1   Tracheobronchial clearance can be measured by having a subject inhale a radioactive aerosol and then metering at some later time the radiation level in his lungs. In one such experiment, seven pairs of monozygotic twins inhaled aerosols of radioactive Teflon particles. One member of each twin pair lived in a rural area; the other was a city-dweller. The objective of the study was to see what effect environment had on tracheobronchial clearance. Table 12.13 lists the percentages of the radioactivity retained in the lungs of each subject one hour after the initial inhalation (19). Define the appropriate μ_D for these data and test $H_0: \mu_D = 0$ versus $H_1: \mu_D \neq 0$ using a paired t test. Let $\alpha = 0.01$.

TABLE 12.13   Radioactivity retained in the lungs

Twin Pair	Rural	Urban
	Retention Percent	
1	10.1	28.1
2	51.8	36.2
3	33.5	40.7
4	32.8	38.8
5	69.0	71.0
6	38.9	47.0
7	54.6	57.0

QUESTION 12.3.2   Show that the paired t test is equivalent to the F test in a randomized block design when the number of treatment levels is two. Hint: Consider the distribution of $T^2 = b\bar{D}^2/s_D^2$ and follow the approach outlined in Example 11.2.

REVIEW EXERCISES FOR CHAPTER 12

1. Triticale is a hybrid cereal grain derived from wheat and rye. Only recently cultivated, its nutritional properties are largely unknown. Not long ago an experiment was set up to compare the protein content in two different varieties of triticale, 6TA–204 and Rosner. Both were grown at four widely separated sites in central California. The table below shows the average protein content (expressed as a dry-weight percentage) measured for the eight crops (148).

Percent protein in two triticale varieties

Location	6TA–204	Rosner
El Centro	16.2	16.6
Five Points	15.6	16.7
Davis	16.9	16.8
Tulelake	15.9	16.7

Test whether the protein yields for the 6TA–204 and Rosner varieties of triticale are significantly different. Use the 0.05 level of significance. Show the ANOVA table.

2. Rat poison is normally made by mixing the active chemical ingredients with ordinary cornmeal. In many urban areas, though, rats can find food that they prefer to cornmeal, so that the poison is left untouched. One solution is to make the cornmeal more palatable by adding food supplements such as peanut butter or meat. This is effective, but the cost is very high and the food supplements tend to spoil rather quickly. The purpose of the experiment described here was to see whether *artificial* food supplements might be similarly enticing. The study took place in Milwaukee and consisted of five two-week surveys. For each survey, approximately 3200 baits were placed around garbage-storage areas. About 800 of the baits were plain cornmeal; a second 800 were cornmeal mixed with artificial butter-vanilla flavoring; a third set of 800 were cornmeal mixed with artificial roast beef flavor; the remaining baits were cornmeal mixed with an artificial bread flavor. The different kinds of baits were all placed fairly close to one another so that a rat would have equal access to all four. After two weeks the test sites were inspected and the number of baits that were gone was recorded. Then a different set of locations in the same general area was selected and the experiment was repeated for another two weeks. Altogether the study was replicated five times and lasted ten weeks (82). The results are tabulated at the top of the next page. Do an analysis of variance on these data. Is there anything about these percentages to suggest that a series of rat-poison surveys should *not* be done as a completely randomized design? Explain.

Percentage of Baits Accepted

Survey Number	Plain	Butter-Vanilla	Roast Beef	Bread
1	13.8	11.7	14.0	12.6
2	12.9	16.7	15.5	13.8
3	25.9	29.8	27.8	25.0
4	18.0	23.1	23.0	16.9
5	15.2	20.2	19.0	13.7

3. Show that the three cross-product terms that arise when Equation 12.3 is squared are all identically zero, thereby verifying Equation 12.4.

4. Accurate laboratory methods for determining the percentage content of solids in heterogeneous foods are important in maintaining product quality as well as in complying with government regulations. In a recent study (40), three techniques were compared for measuring the percent solids in uncreamed cottage cheese. All three methods were used on each of nine different samples. The results are tabulated below.

Percent solids determined

Sample	Method 1	Method 2	Method 3
1	17.766	17.707	17.677
2	19.167	19.222	19.224
3	19.071	19.113	19.092
4	19.636	19.646	19.520
5	28.703	28.784	28.704
6	19.072	19.131	19.024
7	18.382	18.490	18.362
8	18.576	18.675	18.530
9	18.756	18.870	18.637

Do the methods give significantly different results?

5. Derive the computing formulas given in Theorem 12.2 for SS_{total}, SS_{blocks}, and $SS_{treatments}$.

6. In general, does a "large" value for SS_{blocks} (relative to the magnitude of SS_{error}) indicate that an experiment should or should not have been done as a randomized block design? Explain.

7. Generalize from Theorem 12.1 to get an expression for the expected mean square for "subjects" in the Ponzo illusion problem (Case Study 12.1). Also, set up an F ratio to test the subject "effect." Let $\alpha = 0.05$.

8. For a randomized block design with t treatment levels and b (random) blocks, define

$$\sigma_B^2 = \frac{1}{b-1} \sum_{i=1}^{b} \beta_i^2.$$

Estimate σ^2 and σ_B^2 for the data of Review Exercise 4 by setting the mean square for error and the mean square for blocks equal to their respective expected mean squares (see Review Exercise 7) and then solving the two equations.

9. Refer to the rat-poison data of Review Exercise 2. Partition the treatment sum of squares into three orthogonal contrasts. Let one contrast test the hypothesis that the true acceptance percentage for the plain cornmeal is equal to the true acceptance percentage for the cornmeal with artificial roast beef flavoring. Let a second contrast compare the effectiveness of the "butter-vanilla" and "bread" baits. What does the third contrast test? Do all testing at the $\alpha = 0.10$ level of significance.

10. Construct the three 95% Tukey confidence intervals comparing methods 1, 2, and 3 described in Review Exercise 4.

11. Dry ice can be used as a "bait" for deer flies. The following experiment was set up to determine the importance of the bait's location. Insect traps were positioned at four different places (A, B, C, and D) in a wooded area. The four sites were not chosen randomly. Each was selected because it represented a certain type of location. One trap was on the north slope of a hill, another was in a valley, and so on. Each of the traps was baited with dry ice, and after 24 hours the deer flies that had been killed in each one were counted. The experiment was replicated on five different days. The locations of the traps remained the same from day to day. The table below gives the numbers of dead flies counted (2).

Number of dead deer flies

		Locations			
		A	B	C	D
	May 27	11	19	6	35
	June 3	22	23	10	62
Dates	June 4	130	90	26	202
	June 5	189	118	46	104
	June 15	23	9	6	25

(a) State the mathematical model for this design. Define its terms and state its assumptions.

(b) Do an analysis of variance on these data. Show the ANOVA table, including the *EMS* column. State your conclusions.

(c) Include a complete set of orthogonal contrasts to partition the location sum of squares. One of the contrasts should compare the average effect at locations C and D to the average effect at locations A and B; another should compare the effect of location A to the effect of location B. Also, include a contrast in the ANOVA table comparing the average effect of June to the effect of May.

(d) Suppose the experimenter wanted to do a follow-up study and compare the four locations using a completely randomized, one-way design. On the basis of the data given here, would you recommend such a design? Explain.

12. Analyze the triticale data of Review Exercise 1 with a paired t test.

13. A paint manufacturer is experimenting with a new latex derivative that may improve the chalkiness of its house paint. The problem is that it might also affect the tint, thereby requiring the company to rework its blending procedures. As a test, a quality control engineer draws a sample from each of seven batches of Osage Orange. Each sample is then split into two aliquots, with the latex derivative being added to one of the two. Both samples are then examined with a spectroscope. The data are shown below in standardized lumen units (if the tint of the aliquot was exactly right, it would register a reading of 1.00).

Spectroscopy readings

Batch	Osage Orange	Osage Orange + Additive
1	1.10	1.06
2	1.05	1.02
3	1.08	1.17
4	0.98	1.21
5	1.01	1.01
6	0.96	1.23
7	1.02	1.19

Analyze these data using (a) a paired t test and (b) a randomized block design. Show the equivalence of the results (see Question 12.3.2). Let 0.05 be the level of significance.

14. Let $D_1, D_2, \ldots, D_b$ be within-block differences as defined in Section 12.3. Assume that $D_i \sim N(\mu_D, \sigma_D^2)$, for $i = 1, 2, \ldots, b$. Derive a formula for a $100(1 - \alpha)\%$ confidence interval for μ_D. Use your formula, together with the data of Case Study 12.4, to construct a 95% confidence interval for the true average prothrombin time difference ("before aspirin" − "after aspirin").

15. Suppose that b independent replicate measurements are made on each of t treatment levels. The resulting total sum of squares could be written

$$SS_{\text{total}} = (t - 1)MS_{\text{treatments}} + t(b - 1)MS_{\text{error}}$$

$$= SS_{\text{total}} \text{ (completely randomized design).}$$

If the same t treatment levels were compared using b blocks, the corresponding total sum of squares could then be expressed as

$$SS_{\text{total}} = (t - 1)MS_{\text{treatments}} + (b - 1)MS_{\text{blocks}} + (t - 1)(b - 1)MS_{\text{error}}$$

$$= SS_{\text{total}} \text{ (randomized block design).}$$

Assume that "blocks" are a random effect. Set $E[SS_{\text{total}}(\text{completely randomized design})]$ equal to $E[SS_{\text{total}}(\text{randomized block design})]$ and get an expression for the error variance in the completely randomized design as a function of the error variance and the block variance in the randomized block design. To simplify the notation, let

$$\sigma_T^2 = \frac{1}{t - 1} \sum_{j=1}^{t} \tau_j^2 \quad \text{and} \quad \sigma_B^2 = \frac{1}{b - 1} \sum_{i=1}^{b} \beta_i^2.$$

16. Use the information in Table 12.3, together with your answer to Review Exercise 15, to get an estimated error variance had the Ponzo illusion problem been done as a completely randomized design. Suppose the experiment, in some form, were to be repeated. Would you recommend doing it as a completely randomized design or as a randomized block design? Explain.

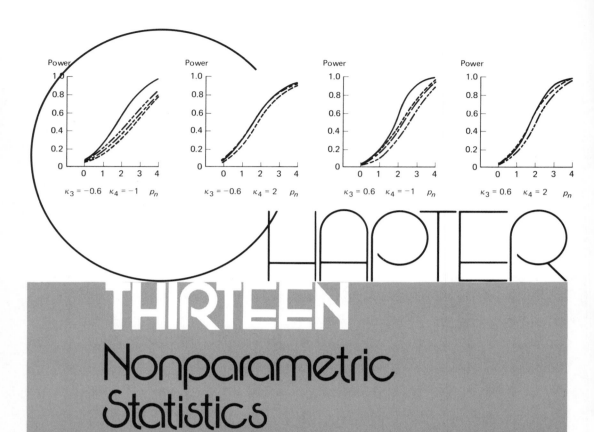

CHAPTER THIRTEEN

Nonparametric Statistics

The figures show a comparison of the power functions of the one-sample t test (solid line) and the sign test (dashed lines) for three different sets of hypotheses, various degrees of nonnormality, a sample size of 10, and a level of significance of 0.05. (The parameter p_n measures the shift from H_0 to H_1; k_3 and k_4 measure the extent of nonnormality in the sampled population.)

13.1 INTRODUCTION

Behind every confidence interval and hypothesis test we have looked at thus far has been a very specific set of assumptions about the origin of the data from which the inference was to be made. For example, in doing a two-sample t test for proportions—$H_0: p_X = p_Y$ versus $H_1: p_X \neq p_Y$—it is assumed that n independent and identically distributed Bernoulli random variables make up the X-sample and m independent and identically distributed Bernoullis the Y-sample. Of course, the most common assumption made in data analysis is that each set of observations is a random sample from a *normal distribution*. This is the condition that was specified in earlier chapters when the inference procedure being discussed involved either the t, the χ^2, or the F distribution.

The need to make such assumptions raises an obvious question: what is affected when what we assume to be true about the data is not? Certainly the test statistic is the same and the computational formulas do not change. The answer, of course, lies with the sampling distribution of the test statistic. In the case of a one-sample t test, $T = (\bar{Y} - \mu_0)/(S/\sqrt{n})$ will no longer behave exactly like a Student t variable with $n - 1$ degrees of freedom if the Y_i's are not $N(\mu_0, \sigma^2)$. As a result, the *actual* probability of committing a Type I error will not necessarily equal the *nominal* probability. That is,

$$\int_{-\infty}^{-t_{\alpha/2, n-1}} f_T(t \mid H_0) \, dt + \int_{t_{\alpha/2, n-1}}^{\infty} f_T(t \mid H_0) \, dt \neq \alpha, \tag{13.1}$$

where $f_T(t \mid H_0)$ is the (unknown) sampling distribution of $T = (\bar{Y} - \mu_0)/(S/\sqrt{n})$ under the assumption that $\mu = \mu_0$ and $f_Y(y)$ is something other than

$$(1/\sqrt{2\pi}\,\sigma)e^{-(1/2)[(y-\mu_0)/\sigma]^2}.$$

Statisticians have sought to overcome the problems implicit in Equation 13.1 in two very different ways. On the one hand, efforts have been made to show that even though T will not have an *exact* Student t distribution when the Y_i's are not normal, $f_T(t \mid H_0)$ will still, in many cases, be similar enough to that distribution so that the actual α is, for all intents and purposes, equal to the nominal α. On the whole, these efforts have been quite successful: unless the Y_i's are markedly nonnormal (and n is very small), the distribution of T will, indeed, have a shape much like that of a Student t, implying that in many cases the assumption of normality is not all that critical after all. This is summarized by saying that the t *test is robust with respect to departures from normality.*

EXAMPLE 13.1.   Questions of robustness tend to be prohibitively difficult to handle in any formal mathematical way. The usual method of attack is through Monte Carlo methods. For example, a study of the robustness of the two-sample t test might begin by generating two independent random samples of size 5 from an *exponential* distribution with λ equal to 1. From the resulting x_i's and y_i's, $\bar{x}$, $\bar{y}$, and s_p are calculated and substituted into the usual two-sample t statistic: $t = (\bar{x} - \bar{y})/(s_p\sqrt{\frac{1}{5} + \frac{1}{5}})$. The resulting value is then compared to what would have been the two critical values.

$\pm t_{\alpha/2, 8}$, and it is recorded whether or not the observed t fell into the rejection region. The same process is then repeated over and over again, each time starting with a different set of random samples. After, say, a thousand such trials, we compute α', the actual proportion of observed t's falling outside the interval $(-t_{\alpha/2, 8}, +t_{\alpha/2, 8})$. If α' is numerically close to α, we will conclude that the two-sample t test is robust against that particular kind of nonnormality.

Table 13.1 summarizes the results of several Monte Carlo studies focusing on the two-sample t test (14). The notation $N(\mu, \sigma^2)(n)$ means that n random observations were drawn from a normal pdf with mean μ and variance σ^2; $U(\mu, \sigma^2)(n)$ represents a random sample of size n drawn from a uniform pdf with mean μ and variance σ^2. In each instance a total of 1000 sets of samples were generated. The nominal level of significance was $\alpha = 0.05$.

As Table 13.1 shows, certain departures from the t-test assumptions have little effect on α—for example, when both random samples are *uniform* variables. Other violations, though, take more of a toll—see, for example, line 2, where five $N(0, 4)$ x_i's and fifteen $N(0, 1)$ y_i's combined to give an α' of 0.160, more than three times the nominal 0.05.

TABLE 13.1   Proportion of t ratios falling in the $\alpha = 0.05$
critical region (1000 trials)

Samples Drawn		Proportion of t Ratios in Critical Region
X	Y	
$N(0, 1)(5)$	$N(0, 4)(15)$	0.064
$N(0, 4)(5)$	$N(0, 1)(15)$	0.160
$U(0, 1)(5)$	$U(0, 1)(5)$	0.051
$U(0, 1)(5)$	$U(0, 4)(5)$	0.071

A second way of dealing with Equation 13.1 is to replace the t test with procedures having test statistics whose distribution under H_0 remains the same, regardless of how the population sampled may change. Inference procedures having this sort of latitude are called *nonparametric*, or more appropriately, *distribution free*.

The number of nonparametric procedures proposed since the early 1940s has been enormous and continues to grow. We do not presume to give any sort of survey of those procedures here; for that, the reader is referred to more specialized texts, such as (59) or (79). Our more limited objective is to introduce some of the basic methodology of nonparametric statistics in the context of problems whose "parametric" solutions we have already seen. In so doing we will state, and in some cases derive, a nonparametric solution to the paired-data problem, the one-sample location problem, and both the experimental design models of Chapters 11 and 12.

13.2 THE SIGN TEST

Among the simplest of all nonparametric procedures is the *sign test*. It finds a number of different applications: if a set of Y_i's come from a *symmetric* (but not necessarily normal) pdf, a sign test can be used for the one-sample location problem, $H_0: \tilde{\mu} = \tilde{\mu}_0$ versus $H_1: \tilde{\mu} \neq \tilde{\mu}_0$, where $\tilde{\mu}$ denotes the median of the population being sampled. The same test is also appropriate in a paired-data situation where the normality of the X_i's and Y_i's is questionable.

Consider the latter situation: $(X_1, Y_1), (X_2, Y_2), \ldots, (X_n, Y_n)$ is a random sample of paired observations. Let $p = P(Y_i > X_i)$. We wish to test whether or not the X's are shifted in location relative to the Y's. In terms of p, the hypotheses reduce to

$$H_0: p = \tfrac{1}{2} \qquad \text{(no shift in location)}$$

versus

$$H_1: p \neq \tfrac{1}{2} \qquad \text{(shift in location)}.$$

Let each (X_i, Y_i) pair be replaced by either a plus sign or a minus sign: a plus sign if $Y_i > X_i$ and a minus sign if $Y_i < X_i$. [It will be assumed that both $f_X(x)$ and $f_Y(y)$ are continuous, so the event of a tie has probability zero.] It follows that if H_0 is true, each (X_i, Y_i) pair constitutes a Bernoulli trial, and Y_+, the number of plus signs, will have a binomial pdf with parameters n and $\tfrac{1}{2}$. The appropriate decision rule, then, is a direct extension of the argument given in Section 6.2: H_0 should be rejected if y_+ is either less than or equal to Y_+^* or greater than or equal to Y_+^{**}, where

$$P(Y_+ \leq Y_+^*) = P(Y_+ \geq Y_+^{**}) = \frac{\alpha}{2}. \qquad (13.2)$$

Comment. For small n, the discreteness of the binomial pdf might make it impossible to find a Y_+^* and a Y_+^{**} satisfying Equation 13.2 exactly. For large n, that becomes less of a problem: we simply appeal to the DeMoivre-Laplace limit theorem and approximate, say, Y_+^{**} by solving

$$z_{1-\alpha/2} = \frac{Y_+^{**} - \tfrac{1}{2} \cdot n}{\sqrt{n/4}}.$$

CASE STUDY

13.1

Children with severe learning problems often have electroencephalograms and behavior patterns similar to those of children with petit-mal, a mild form of epilepsy. This led to speculation that drugs helpful in treating petit-mal might also be useful as "learning facilitators" (170). To test that hypothesis, ten children, ranging in age from 8 to 14 and all having a history of learning and behavioral problems, were recruited to participate in a six-week study. For three of those weeks a child was given a placebo; for the other three weeks, ethosuximide, a widely prescribed anticonvulsant. After each three-week period the children were given several parts of the standard Wechsler IQ test. Because a child might be expected to do better on the test the second time he took it, the order in which the placebo and the ethosuximide were administered was randomized: some children were given the placebo for the first three weeks while others began the study by taking the ethosuximide.

Table 13.2 shows the two verbal IQ scores recorded for each subject. The last entry in the fourth column indicates the value of the test statistic, Y_+.

TABLE 13.2 IQ scores in ethosuximide experiment

Child	IQ After Placebo (x_i)	IQ After Ethosuximide (y_i)	$y_i > x_i$?
1	97	113	+
2	106	113	+
3	106	101	−
4	95	119	+
5	102	111	+
6	111	122	+
7	115	121	+
8	104	106	+
9	90	110	+
10	96	126	+
			$y_+ = 9$

Suppose we wish to test

$$H_0: p = \tfrac{1}{2}$$

versus

$$H_1: p \neq \tfrac{1}{2}$$

at the $\alpha = 0.10$ level of significance, where $p = P(Y_i > X_i)$. The two-sided alternative is used here to allow for the possibility that ethosuximide might actually have an adverse effect on IQ.

For the binomial pdf with parameters $n = 10$ and $p = \tfrac{1}{2}$,

$$P(Y_+ \leq 2) = \sum_{y=0}^{2} \binom{10}{y}\left(\frac{1}{2}\right)^y\left(\frac{1}{2}\right)^{10-y} \doteq 0.05$$

and

$$P(Y_+ \geq 8) = \sum_{y=8}^{10} \binom{10}{y}\left(\frac{1}{2}\right)^y\left(\frac{1}{2}\right)^{10-y} \doteq 0.05.$$

Thus, to maintain a Type I error probability of approximately 0.10, we should reject H_0 if y_+ is either less than or equal to 2 or greater than or equal to 8. In point of fact, the latter was true ($y_+ = 9$), so our conclusion is to *reject* the null hypothesis: it would appear that ethosuximide *does* have an effect on IQ, and a beneficial one at that.

QUESTION 13.2.1   What could be concluded if the "after ethosuximide" IQ's were significantly higher than the "after placebo" IQ's but if all ten children had been given the placebo first?

QUESTION 13.2.2   One reason cited for the mental deterioration so often seen in the very elderly is the reduction in cerebral blood flow that accompanies the aging process. Addressing itself to this notion, a study was recently done in a rest home to see whether cyclandelate, a vasodilator, might be able to stimulate the cerebral circulation and thereby slow down the rate of mental deterioration (6). The drug was given to 11 subjects on a daily basis. To measure its physiological effect, radioactive tracers were used to determine each subject's mean circulation time (MCT) at the start of the experiment and four months later when the study was discontinued. (The MCT is the length of time it takes blood to travel from the carotid artery to the jugular vein.) The results are shown in Table 13.3. Test the appropriate hypothesis with a sign test. Use a one-sided alternative.

TABLE 13.3   Cerebral circulation experiment

	Mean Circulation Time (sec)	
Subject	Before (x_i)	After (y_i)
J.B.	15	13
M.B.	12	8
A.B.	12	12.5
M.B.	14	12
J.L.	13	12
S.M.	13	12.5
M.M.	13	12.5
S.McA.	12	14
A.McL.	12.5	12
F.S.	12	11
P.W.	12.5	10

13.3 THE WILCOXON SIGNED RANK TEST

Although the sign test is a bona fide nonparametric procedure, its extreme simplicity makes it somewhat atypical. The *Wilcoxon signed rank test* introduced in this section is more representative of nonparametric procedures as a whole. Like the sign test, it can be adapted to several different data structures. Here, we will pose it within the framework of a one-sample test for location, where it becomes an alternative to the one-sample t test. The procedure carries over immediately to the paired-data problem and with only minor modifications can also serve as a two-sample test for location and a two-sample test for dispersion (provided the two populations have equal locations). Historically, the Wilcoxon signed rank test was one of the first nonparametric procedures to be developed (it dates back to 1945), and it remains one of the most widely used.

Let $Y_1, Y_2, \ldots, Y_n$ be a random sample of size n from a pdf $f_Y(y)$ that is both continuous and symmetric. Let $\tilde{\mu}$ be the median of $f_Y(y)$. We wish to test

$$H_0: \tilde{\mu} = \tilde{\mu}_0$$

versus

$$H_1: \tilde{\mu} \neq \tilde{\mu}_0,$$

where $\tilde{\mu}_0$ is some prespecified value for $\tilde{\mu}$.

The Wilcoxon statistic is based on the magnitudes, and directions, of the deviations of the Y_i's from $\tilde{\mu}_0$. Let $|Y_1 - \tilde{\mu}_0|, |Y_2 - \tilde{\mu}_0|, \ldots, |Y_n - \tilde{\mu}_0|$ denote the set of absolute deviations of the Y_i's from $\tilde{\mu}_0$. These can be ordered from smallest to largest, and we will define R_i to be the *rank* of $|Y_i - \tilde{\mu}_0|$ in the set $\{|Y_j - \tilde{\mu}_0|\}_{j=1}^{n}$. (The smallest $|Y_j - \tilde{\mu}_0|$ is assigned a rank of 1, the second smallest, a rank of 2, and so on up to n.)

Associated with each R_i will be a sign indicator, Z_i, where

$$Z_i = \begin{cases} 0, & \text{if } Y_i - \tilde{\mu}_0 < 0 \\ 1, & \text{if } Y_i - \tilde{\mu}_0 > 0. \end{cases}$$

We will define the Wilcoxon signed rank statistic, W, to be the linear combination,

$$W = \sum_{i=1}^{n} Z_i R_i.$$

To illustrate this terminology, consider the case where $n = 3$ and $y_1 = 6.0$, $y_2 = 4.9$, and $y_3 = 11.2$. Suppose the problem is to test

$$H_0: \tilde{\mu} = 10.0$$

versus

$$H_1: \tilde{\mu} \neq 10.0.$$

Note that $|y_1 - \tilde{\mu}_0| = 4.0, |y_2 - \tilde{\mu}_0| = 5.1$, and $|y_3 - \tilde{\mu}_0| = 1.2$. Since $1.2 < 4.0 < 5.1$, it follows that $r_1 = 2, r_2 = 3$, and $r_3 = 1$. Also, $z_1 = 0, z_2 = 0$, and $z_3 = 1$. Combining the r_i's and the z_i's we have that

$$\begin{aligned} w &= \sum_{i=1}^{n} z_i r_i \\ &= (0)(2) + (0)(3) + (1)(1) \\ &= 1. \end{aligned}$$

Comment. Notice that W is based on the *ranks* of the deviations from $\tilde{\mu}_0$ and not on the deviations themselves. For this example, the value of W would remain unchanged if y_2 were 4.9, 3.6, or $-10,000$. In each case, r_2 would be 3 and Z_2 would be 0. If the test statistic *did* depend on the magnitude of the deviations, it would have been necessary to specify a particular distribution for $f_Y(y)$, and the resulting procedure would no longer be nonparametric. ∎

It should be clear that W ranges from 0 (all deviations negative) to $\sum_{i=1}^{n} = [n(n+1)]/2$ (all deviations positive). Intuitively, if H_0 were true we would expect the Y_i's to be distributed at random above and below $\tilde{\mu}_0$, meaning that W should tend to take on values close to $[n(n+1)]/4$. The next theorem enables us to determine how close W has to get to either 0 or $[n(n+1)]/2$ before H_0 can be rejected.

> **THEOREM 13.1**   For $\{Y_i\}_{i=1}^{n}$, $\{R_i\}_{i=1}^{n}$, and $\{Z_i\}_{i=1}^{n}$ defined as above, the probability distribution of $W = \sum_{i=1}^{n} Z_i R_i$, when $H_0: \tilde{\mu} = \tilde{\mu}_0$ is true, is given by
>
> $$P(W = w) = f_W(w) = \left(\frac{1}{2^n}\right) \cdot c(w),$$
>
> where $c(w)$ is the coefficient of e^{wt} in the expansion of
>
> $$\prod_{i=1}^{n} (1 + e^{it}).$$

PROOF   The statement and proof of Theorem 13.1 are typical of many nonparametric results. Closed-form expressions for sampling distributions are seldom possible: the combinatorial nature of nonparametric test statistics lends itself more readily to a generating-function format.

To begin, note that if H_0 is true, the distribution of $W = \sum_{i=1}^{n} Z_i R_i$ is equivalent to the distribution of $U = \sum_{i=1}^{n} U_i$, where

$$U_i = \begin{cases} 0 & \text{with probability } \frac{1}{2} \\ i & \text{with probability } \frac{1}{2}. \end{cases}$$

Therefore, W and U have the same moment-generating function. Also, since the Z_i's are independent, so are the U_i's, and from Theorem 3.19,

$$M_U(t) = M_W(t)$$

$$= \prod_{i=1}^{n} M_{U_i}(t)$$

$$= \prod_{i=1}^{n} E(e^{U_i t})$$

$$= \prod_{i=1}^{n} \left(\frac{1}{2} e^{0t} + \frac{1}{2} e^{it}\right)$$

$$= \left(\frac{1}{2^n}\right) \prod_{i=1}^{n} (1 + e^{it}). \tag{13.3}$$

Now, consider the *structure* of $f_W(w)$, the pdf for the Wilcoxon statistic. In forming W, we can prefix R_1 by either a plus sign or a zero; similarly for R_2, $R_3, \ldots,$ and R_n. It follows that since each R_i can take on two different values, the total number of ways to make signed rank sums is 2^n. Under H_0, of course, each of these arrangements is equally likely so, in general terms, the probability distribution of W has the form

$$P(W = w) = f_W(w) = \frac{c(w)}{2^n}, \tag{13.4}$$

where $c(w)$ is the number of ways to assign $+$'s and "zeroes" to the first n integers so that $\sum_{i=1}^{n} Z_i R_i$ has the value w.

The conclusion of Theorem 13.1 follows immediately by comparing the form of $f_W(w)$ to Equation 13.3 and to the general expression for a moment-generating function. By definition,

$$M_W(t) = E(e^{Wt}) = \sum_{w=1}^{n(n+1)/2} e^{wt} f_W(w),$$

but from Equations 13.3 and 13.4 we can write

$$\sum_{w=1}^{n(n+1)/2} e^{wt} f_W(w) = \left(\frac{1}{2^n}\right) \prod_{i=1}^{n} (1 + e^{it}) = \sum_{w=1}^{n(n+1)/2} e^{wt} \cdot \frac{c(w)}{2^n}.$$

We prove the theorem by recognizing that $c(w)$ is the coefficient of e^{wt} in the expansion of $\prod_{i=1}^{n} (1 + e^{it})$.

It may help to clarify these ideas by considering a numerical example. Suppose $n = 4$. Then, by Equation 13.3, the moment-generating function for W becomes

$$M_W(t) = \left(\frac{1 + e^t}{2}\right)\left(\frac{1 + e^{2t}}{2}\right)\left(\frac{1 + e^{3t}}{2}\right)\left(\frac{1 + e^{4t}}{2}\right)$$

$$= \left(\frac{1}{16}\right)\{1 + e^t + e^{2t} + 2e^{3t} + 2e^{4t} + 2e^{5t} + 2e^{6t} + 2e^{7t} + e^{8t} + e^{9t} + e^{10t}\}.$$

Thus, the probability that W equals, say, 2 is $\frac{1}{16}$ (since the coefficient of e^{2t} is 1); the probability that W equals 7 is $\frac{2}{16}$, and so on. The first two columns of Table 13.4 show the complete probability distribution of W, as given by the expansion of $M_W(t)$. The last column enumerates the particular assignments of $+$'s and 0's that generate each value of W.

Cumulative tail area probabilities,

$$P(W \le w_1^*) = \sum_{w=0}^{w_1^*} f_W(w) \quad \text{and} \quad P(W \ge w_2^*) = \sum_{w=w_2^*}^{n(n+1)/2} f_W(w),$$

are listed in Table A.6 of the Appendix for sample sizes ranging from $n = 4$ to $n = 12$. Knowing these probabilities, it is easy to construct decision rules for either one-sided or two-sided alternatives. For example, suppose n is 7 and we wish to test

TABLE 13.4 Probability distribution of W

w	$f_W(w) = P(W = w)$	r_i 1	2	3	4
0	$\frac{1}{16}$	0	0	0	0
1	$\frac{1}{16}$	+	0	0	0
2	$\frac{1}{16}$	0	+	0	0
3	$\frac{2}{16}$	$\{\begin{smallmatrix}+\\0\end{smallmatrix}$	$\begin{smallmatrix}+\\0\end{smallmatrix}$	$\begin{smallmatrix}0\\+\end{smallmatrix}$	$\begin{smallmatrix}0\\0\end{smallmatrix}\}$
4	$\frac{2}{16}$	$\{\begin{smallmatrix}+\\0\end{smallmatrix}$	$\begin{smallmatrix}0\\0\end{smallmatrix}$	$\begin{smallmatrix}+\\0\end{smallmatrix}$	$\begin{smallmatrix}0\\+\end{smallmatrix}\}$
5	$\frac{2}{16}$	$\{\begin{smallmatrix}+\\0\end{smallmatrix}$	$\begin{smallmatrix}0\\+\end{smallmatrix}$	$\begin{smallmatrix}0\\+\end{smallmatrix}$	$\begin{smallmatrix}+\\0\end{smallmatrix}\}$
6	$\frac{2}{16}$	$\{\begin{smallmatrix}+\\0\end{smallmatrix}$	$\begin{smallmatrix}+\\+\end{smallmatrix}$	$\begin{smallmatrix}+\\0\end{smallmatrix}$	$\begin{smallmatrix}0\\+\end{smallmatrix}\}$
7	$\frac{2}{16}$	$\{\begin{smallmatrix}+\\0\end{smallmatrix}$	$\begin{smallmatrix}+\\0\end{smallmatrix}$	$\begin{smallmatrix}0\\+\end{smallmatrix}$	$\begin{smallmatrix}+\\+\end{smallmatrix}\}$
8	$\frac{1}{16}$	+	0	+	+
9	$\frac{1}{16}$	0	+	+	+
10	$\frac{1}{16}$	+	+	+	+
	1				

$$H_0 : \tilde{\mu} = \tilde{\mu}_0$$
versus
$$H_1 : \tilde{\mu} \neq \tilde{\mu}_0$$

at the $\alpha = 0.05$ level of significance. The critical region would be the set of w-values less than or equal to 2 or greater than or equal to 26: $C = \{w : w \leq 2 \text{ or } w \geq 26\}$. This follows immediately from Table A.6, since

$$\sum_{w \in C} f_W(w) = 0.023 + 0.023 \doteq 0.05.$$

CASE STUDY

13.2

Swell sharks (*Cephaloscyllium ventriosum*) are small, reef-dwelling sharks that inhabit the California coastal waters south of Monterey Bay. There is a second population of these fish living nearby in the vicinity of Catalina Island, but it has been hypothesized (65) that the two populations never mix. Separating Santa Catalina from the mainland is a deep basin, which, according to the "separation" hypothesis, is an inpenetrable barrier for these particular fish.

One way to test this theory would be compare the morphology of sharks caught in the two regions. If there were no mixing, we would expect a certain number of differences to have evolved. Table 13.5 lists the total length (*TL*), the height of the first dorsal fin (*HDI*), and the ratio *TL/HDI* for ten male swell sharks caught near Santa Catalina.

TABLE 13.5 Measurements made on ten sharks caught near Santa Catalina

Total Length (mm)	Height of First Dorsal Fin (mm)	*TL/HDI*
906	68	13.32
875	67	13.06
771	55	14.02
700	59	11.86
869	64	13.58
895	65	13.77
662	49	13.51
750	52	14.42
794	55	14.44
787	51	15.43

It has been estimated on the basis of past data that the median *TL/HDI* ratio for male swell sharks caught *off the coast* is 14.60. Is this figure consistent with the data of Table 13.5? In more formal terms, if $\tilde{\mu}$ denotes the true median *TL/HDI* ratio for the Santa Catalina population, can we reject $H_0 : \tilde{\mu} = 14.60$, and thereby lend support to the separation theory?

Table 13.6 gives the values of *TL/HDI* $(= Y_i)$, $Y_i - \tilde{\mu}_0 = Y_i - 14.60$, $|Y_i - 14.60|$, R_i, Z_i, and $R_i Z_i$ for the ten Santa Catalina sharks.

TABLE 13.6 Computations for Wilcoxon signed rank test

TL/HDI $(= y_i)$	$y_i - 14.60$	$|y_i - 14.60|$	r_i	z_i	$r_i z_i$
13.32	−1.28	1.28	8	0	0
13.06	−1.54	1.54	9	0	0
14.02	−0.58	0.58	3	0	0
11.86	−2.74	2.74	10	0	0
13.58	−1.02	1.02	6	0	0
13.77	−0.83	0.83	4.5	0	0
13.51	−1.09	1.09	7	0	0
14.42	−0.18	0.18	2	0	0
14.44	−0.16	0.16	1	0	0
15.43	+0.83	0.83	4.5	1	4.5

Comment. When two or more numbers being ranked are equal, each is assigned the *average* of the ranks they would otherwise have received. In the data of Table 13.6 there are two sharks whose $|y_i - 14.60|$ value is 0.83. Had these two deviations been just slightly different, they would have been given ranks 4 and 5, but since they are equal, each is assigned a rank of 4.5 [$= (4+5)/2$]. ∎

By summing the last column we see that $w = 4.5$. According to Table A.6 in the Appendix the $\alpha = 0.05$ decision rule for testing

$$H_0: \quad \tilde{\mu} = 14.60$$

versus

$$H_1: \quad \tilde{\mu} \neq 14.60$$

requires that H_0 be rejected if w is either less than or equal to 8 or greater than or equal to 47. (Why is the alternative hypothesis two-sided here?) (Note: the *exact* level of significance associated with $C = \{w: w \leq 8 \text{ or } w \geq 47\}$ is $0.024 + 0.024 = 0.048$.) Thus, we should *reject* H_0, since the observed w was less than 8. These particular data, then, would support the separation hypothesis.

QUESTION 13.3.1   The average daily energy expenditures for eight elderly women were estimated on the basis of information received from a battery-powered heart rate monitor that each subject wore. Two overall averages were calculated for each woman, one for the summer months and one for the winter months (150), as shown in Table 13.7. Let

TABLE 13.7   Average daily energy expenditures (kcal)

Subject	Summer (x_i)	Winter (y_i)
1	1458	1424
2	1353	1501
3	2209	1495
4	1804	1739
5	1912	2031
6	1366	934
7	1598	1401
8	1406	1339

μ_D denote the location difference between the summer and winter energy-expenditure populations. Compute $y_i - x_i$, $i = 1, 2, \ldots, 8$, and use the Wilcoxon signed rank procedure to test

$$H_0: \quad \mu_D = 0$$

versus

$$H_1: \quad \mu_D \neq 0.$$

Let $\alpha = 0.15$.

The usefulness of Table A.6 in the Appendix is clearly limited to tests where the sample size is small, less than or equal to 12. To accommodate a larger n, we can define a normalized test statistic, W', where

$$W' = \frac{W - E(W)}{\sqrt{\text{Var}(W)}}$$

and $E(W)$ and Var (W) are the expected value and variance of W *when H_0 is true* (see Theorem 13.2). It can be shown that as n gets large, the distribution of W' converges to the standard normal. Furthermore, for n even as small as 13, $f_{W'}(w')$ and the $N(0, 1)$ are remarkably similar. Therefore, to test

$$H_0: \tilde{\mu} = \tilde{\mu}_0$$

versus

$$H_1: \tilde{\mu} \neq \tilde{\mu}_0$$

at, say, the $\alpha = 0.05$ level of significance, we would reject H_0 if w' was either less than or equal to -1.96 or greater than or equal to $+1.96$.

THEOREM 13.2 When H_0 is true, the mean and standard deviation of the Wilcoxon signed rank statistic, W, are given by

$$E(W) = \frac{n(n+1)}{4}.$$

and

$$\text{Var}(W) = \frac{n(n+1)(2n+1)}{24}.$$

Also, for $n > 12$, the distribution of

$$W' = \frac{W - [n(n+1)]/4}{\sqrt{[n(n+1)(2n+1)]/24}}$$

can be adequately approximated by the standard normal.

PROOF   We will derive $E(W)$ and Var (W); for a proof of the asymptotic normality, see (78). Recall that W has the same distribution as $U = \sum\limits_{i=1}^{n} U_i$, where

$$U_i = \begin{cases} 0 & \text{with probability } \frac{1}{2} \\ i & \text{with probability } \frac{1}{2}. \end{cases}$$

Therefore,

$$E(W) = E\left(\sum_{i=1}^{n} U_i\right) = \sum_{i=1}^{n} E(U_i)$$

$$= \sum_{i=1}^{n} \left(0 \cdot \frac{1}{2} + i \cdot \frac{1}{2}\right) = \sum_{i=1}^{n} \frac{i}{2}$$

$$= \frac{n(n+1)}{4}.$$

Similarly,

$$\text{Var}\,(W) = \text{Var}\,(U) = \sum_{i=1}^{n} \text{Var}\,(U_i),$$

since the U_i's are independent. But

$$\text{Var}\,(U_i) = E(U_i^2) - [E(U_i)]^2$$
$$= \frac{i^2}{2} - \left(\frac{i}{2}\right)^2 = \frac{i^2}{4},$$

making

$$\text{Var}\,(W) = \sum_{i=1}^{n} \frac{i^2}{4} = \left(\frac{1}{4}\right)\left[\frac{n(n+1)(2n+1)}{6}\right]$$
$$= \frac{n(n+1)(2n+1)}{24}.$$

CASE STUDY

13.3

Methadone is a drug widely used in the treatment of heroin addiction; another is cyclazocine. Recently a study was done (138) to evaluate the effectiveness of the latter in reducing a person's psychological dependence on heroin. The subjects were 14 males, all chronic heroin addicts. Each was asked a battery of questions that compared his feelings when he was using heroin to his feelings when he was "clean." The resultant Q-scores ranged from a possible minimum of 11 to a possible maximum of 55, as shown in Table 13.8. (From the way the questions were worded, higher scores represented *less* psychological dependence.)

TABLE 13.8 Q-scores of heroin addicts after
cyclazocine therapy

51	43
53	45
43	27
36	21
55	26
55	22
39	43

The median score for addicts *not* treated with cyclazocine is known from past experience to be 28. Can we conclude on the basis of the data in Table 13.8. that cyclazocine is an effective treatment?

Since high Q-scores represent *less* dependence on heroin (and assuming cyclazocine would not tend to worsen an addict's condition), the alternative hypothesis should be one-sided *to the right*:

$$H_0: \tilde{\mu} = 28$$

versus

$$H_1: \tilde{\mu} > 28.$$

We will set α equal to 0.05.

Table 13.9 shows the computations necessary to find W—in this case, 95.0. With n being larger than 12, $W' = [W - E(W)]/\sqrt{\text{Var}(W)}$

TABLE 13.9   Computations to find W

Q-Score (y_i)	$(y_i - 28)$	$\|y_i - 28\|$	r_i	z_i	$r_i z_i$
51	+23	23	11	1	11
53	+25	25	12	1	12
43	+15	15	8	1	8
36	+8	8	5	1	5
55	+27	27	13.5	1	13.5
55	+27	27	13.5	1	13.5
39	+11	11	6	1	6
43	+15	15	8	1	8
45	+17	17	10	1	10
27	−1	1	1	0	0
21	−7	7	4	0	0
26	−2	2	2	0	0
22	−6	6	3	0	0
43	+15	15	8	1	8
					95.0

has approximately a standard normal distribution, and the 0.05 decision rule becomes

Reject $H_0: \tilde{\mu} = 28$ if $w' \geq 1.64$.

By Theorem 13.2, $E(W) = [14(14 + 1)]/4 = 52.5$ and Var $(W) = [14(14 + 1)(28 + 1)]/24 = 253.75$. Therefore,

$$w' = \frac{95.0 - 52.5}{\sqrt{253.75}} = 2.67,$$

implying that we should *reject* H_0—it would appear that cyclazocine therapy *is* helpful in reducing heroin dependence.

QUESTION 13.3.2   Analyze the Shoshoni rectangle data (Case Study 1.2) using the normal approxima-tion to the signed rank test. Let $\alpha = 0.05$.

13.4 THE KRUSKAL-WALLIS TEST

In the final two sections of this chapter are the nonparametric counterparts for the two analysis-of-variance models introduced in Chapters 11 and 12. Neither of these procedures, the *Kruskal-Wallis test* and the *Friedman test*, will be derived. We will simply state the procedures and illustrate them with examples.

First, we consider the so-called *k-sample problem*. Suppose $k \, (\geq 2)$ independent random samples of sizes $n_1, n_2, \ldots, n_k$ are drawn, representing k populations having the same shape but possibly different locations. Our objective is to test whether the population medians are all the same—that is,

$$H_0 : \ \tilde{\mu}_1 = \tilde{\mu}_2 = \cdots = \tilde{\mu}_k$$

versus

$$H_1 : \text{not all the } \tilde{\mu}_j\text{'s are equal.}$$

The Kruskal-Wallis procedure for testing H_0 is really quite simple, involving considerably fewer computations than the analysis of variance. The first step is to rank the entire set of $n = \sum_{i=1}^{k} n_i$ observations (from smallest to largest). Then the rank sum, $R_{.j}$, is calculated for each sample. Table 13.10 shows the notation we will be using: it follows the same conventions as the dot notation of Chapter 11. The only real difference is the addition of r_{ij}, the symbol for the rank corresponding to y_{ij}.

TABLE 13.10   Notation for Kruskal-Wallis procedure

	Sample			
	1	2	$\cdots$	k
	$y_{11}(r_{11})$	$y_{12}(r_{12})$		$y_{1k}(r_{1k})$
	$y_{21}(r_{21})$			
	$\vdots$	$\vdots$	$\cdots$	$\vdots$
	$y_{n_1 1}(r_{n_1 1})$	$y_{n_2 2}(r_{n_2 2})$		$y_{n_k k}(r_{n_k k})$
Totals	$r_{.1}$	$r_{.2}$		$r_{.k}$

The Kruskal-Wallis statistic, B, is defined as

$$B = \frac{12}{n(n+1)} \sum_{j=1}^{k} \frac{R_{\cdot j}^2}{n_j} - 3(n+1).$$

Notice how B resembles the computing formula for $SS_{\text{treatments}}$ in the analysis of variance. Here $\sum_{j=1}^{k} (R_{\cdot j}^2/n_j)$, and thus B, get larger and larger as the differences between the population medians increase. [Recall that a similar explanation was given for $SS_{\text{treatments}}$ and $\sum_{j=1}^{k} (T_{\cdot j}^2/n_j)$]. Theorem 13.3 gives the distribution of the Kruskal-Wallis statistic and the appropriate critical region for testing H_0 versus H_1.

THEOREM 13.3   If $H_0: \tilde{\mu}_1 = \tilde{\mu}_2 = \cdots = \tilde{\mu}_k$ is true,

$$B = \frac{12}{n(n+1)} \sum_{j=1}^{k} \frac{R_{\cdot j}^2}{n_j} - 3(n+1)$$

has approximately a χ_{k-1}^2 distribution and H_0 should be rejected at the α level of significance if $b \geq \chi_{1-\alpha, k-1}^2$.

CASE STUDY

13.4

On December 1, 1969, a lottery was held in Selective Service head-quarters in Washington, D.C., to determine the draft status of all 19-year-old males. It was the first time such a procedure had been used since World War II. Priorities were established according to a person's birthday. Each of the 366 possible birthdates was written on a slip of paper and put into a small capsule. The capsules were then put into a large bowl, mixed, and drawn out one by one. By agreement, persons whose birthday corresponded to the first capsule drawn would have the highest draft priority; those whose birthday corresponded to the second capsule drawn, the second highest priority, and so on. Table 13.11 shows the order in which the 366 birthdates were drawn (159). The first date was September 14 (001); the last, June 8 (366).

We can think of the observed sequence of draft priorities as ranks from 1 to 366. If the lottery is random, the average of these ranks for each of the months should be approximately equal. If the lottery is *not* random, we would expect to see certain months having a preponderance of high ranks and other months a preponderance of low

TABLE 13.11 1969 draft lottery, highest priority (001) to lowest priority (366)

Date	Jan	Feb	Mar	Apr	May	Jun	Jul	Aug	Sep	Oct	Nov	Dec
1	305	086	108	032	330	249	093	111	225	359	019	129
2	159	144	029	271	298	228	350	045	161	125	034	328
3	251	297	267	083	040	301	115	261	049	244	348	157
4	215	210	275	081	276	020	279	145	232	202	266	165
5	101	214	293	269	364	028	188	054	082	024	310	056
6	224	347	139	253	155	110	327	114	006	087	076	010
7	306	091	122	147	035	085	050	168	008	234	051	012
8	199	181	213	312	321	366	013	048	184	283	097	105
9	194	338	317	219	197	335	277	106	263	342	080	043
10	325	216	323	218	065	206	284	021	071	220	282	041
11	329	150	136	014	037	134	248	324	158	237	046	039
12	221	068	300	346	133	272	015	142	242	072	066	314
13	318	152	259	124	295	069	042	307	175	138	126	163
14	238	004	354	231	178	356	331	198	001	294	127	026
15	017	089	169	273	130	180	322	102	113	171	131	320
16	121	212	166	148	055	274	120	044	207	254	107	096
17	235	189	033	260	112	073	098	154	255	288	143	304
18	140	292	332	090	278	341	190	141	246	005	146	128
19	058	025	200	336	075	104	227	311	177	241	203	240
20	280	302	239	345	183	360	187	344	063	192	185	135
21	186	363	334	062	250	060	027	291	204	243	156	070
22	337	290	265	316	326	247	153	339	160	117	009	053
23	118	057	256	252	319	109	172	116	119	201	182	162
24	059	236	258	002	031	358	023	036	195	196	230	095
25	052	179	343	351	361	137	067	286	149	176	132	084
26	092	365	170	340	357	022	303	245	018	007	309	173
27	355	205	268	074	296	064	289	352	233	264	047	078
28	077	299	223	262	308	222	088	167	257	094	281	123
29	349	285	362	191	226	353	270	061	151	229	099	016
30	164		217	208	103	209	287	333	315	038	174	003
31	211		030		313		193	011		079		100
Totals:	6,236	5,886	7,000	6,110	6,447	5,872	5,628	5,377	4,719	5,656	4,462	3,768

ranks. This suggests that we can test the lottery for randomness by using the Kruskal-Wallis statistic.

Substituting the $r_{.j}$'s shown at the bottom of Table 13.11 into the formula for B gives

$$b = \frac{12}{366(367)} \left[\frac{(6236)^2}{31} + \cdots + \frac{(3768)^2}{31} \right] - 3(367)$$

$$= 25.95.$$

By Theorem 13.3, B has approximately a χ^2 distribution with 11 degrees of freedom (when $H_0: \tilde{\mu}_{\text{Jan}} = \tilde{\mu}_{\text{Feb}} = \cdots = \tilde{\mu}_{\text{Dec}}$ is true). Thus, if we choose to test H_0 at the $\alpha = 0.01$ level of significance, we should reject the null hypothesis if $b \geq 24.7$ (see Table A.3 in the Appendix). But b did exceed that value, so the hypothesis of randomness would be rejected—looking at the data more closely, we can see that there was a tendency for birthdays late in the year to come early in the draft.

An even more resounding rejection of the randomness hypothesis can be gotten by dividing the 12 months into two half-years—the first, January through June; the second, July through December. Then the hypotheses to be tested are

$$H_0: \tilde{\mu}_1 = \tilde{\mu}_2$$

versus

$$H_1: \tilde{\mu}_1 \neq \tilde{\mu}_2.$$

Table 13.12, derived from Table 13.11, summarizes the data appropriately. Substituting these values into the formula for the

TABLE 13.12 Summary of 1969 draft lottery by six-month periods

	Jan–June (1)	July–Dec (2)
$r_{.j}$	37,551	29,610
n_j	182	184

Kruskal-Wallis statistic gives a b value (with 1 degree of freedom) of 16.85:

$$b = \frac{12}{366(367)} \left[\frac{(37,551)^2}{182} + \frac{(29,610)^2}{184} \right] - 3(367)$$

$$= 16.85.$$

The significance of 16.85 can be gauged by recalling the moments of a chi square variable. If $B \sim \chi_1^2$, then $E(B) = 1$ and $\text{Var}(B) = 2$ (see Question 7.5.2). Therefore, the observed b value is more than 11 standard deviations away from its mean:

$$\frac{16.85 - 1}{\sqrt{2}} = 11.2.$$

Analyzed this way, there can be little doubt that the lottery was not random!

QUESTION 13.4.1   What might account for the lack of randomness evident in the 1969 draft lottery?

QUESTION 13.4.2   Analyze the serum binding data of Case Study 11.1 using the Kruskal-Wallis test. Let $\alpha = 0.05$.

13.5 THE FRIEDMAN TEST

The nonparametric analog of the analysis of variance for a randomized block design is *Friedman's test*, a procedure based on within-block *ranks*. Its form is similar to that of the Kruskal-Wallis statistic, and, like its predecessor, it has approximately a χ^2 distribution when H_0 is true. Theorem 13.4 gives a formal statement.

THEOREM 13.4 Suppose $k \, (\geq 2)$ treatments are ranked independently within b blocks. Let $r_{.j}, j = 1, 2, \ldots, k$ be the rank sum of the jth treatment. The null hypothesis that the population medians of the k treatments are all equal is rejected at the α level of significance (approximately) if

$$g = \frac{12}{bk(k+1)} \sum_{j=1}^{k} r_{.j}^2 - 3b(k+1) \geq \chi_{1-\alpha, k-1}^2.$$

Baseball rules allow a batter considerable leeway in how he is permitted to run from home plate to second base. Two of the possibilities are the narrow-angle and the wide-angle paths diagrammed in

Figure 13.1. As a means of comparing the two, time trials were held involving 22 players (201). Each player ran both paths. Recorded for each runner was the time it took to go from a point 35 feet from home plate to a point 15 feet from second base. These are listed in the second and fourth columns of Table 13.13. In the third and fifth

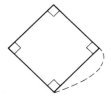

Narrow-angle   Wide-angle

Figure 13.1 Batter's path from home plate to second base.

TABLE 13.13   Times (sec) required to round first base

Player	Narrow-Angle	Rank	Wide-Angle	Rank
1	5.50	1	5.55	2
2	5.70	1	5.75	2
3	5.60	2	5.50	1
4	5.50	2	5.40	1
5	5.85	2	5.70	1
6	5.55	1	5.60	2
7	5.40	2	5.35	1
8	5.50	2	5.35	1
9	5.15	2	5.00	1
10	5.80	2	5.70	1
11	5.20	2	5.10	1
12	5.55	2	5.45	1
13	5.35	1	5.45	2
14	5.00	2	4.95	1
15	5.50	2	5.40	1
16	5.55	2	5.50	1
17	5.55	2	5.35	1
18	5.50	1	5.55	2
19	5.45	2	5.25	1
20	5.60	2	5.40	1
21	5.65	2	5.55	1
22	6.30	2	6.25	1
		39		27

columns are the within-block ranks for each path. A rank of 1 was assigned to the better path, the one enabling the runner to round first base faster.

If $\tilde{\mu}_1$ and $\tilde{\mu}_2$ denote the true median rounding times associated with the narrow-angle and wide-angle paths, respectively, the hypotheses to be tested are

$$H_0: \tilde{\mu}_1 = \tilde{\mu}_2$$

versus

$$H_1: \tilde{\mu}_1 \neq \tilde{\mu}_2.$$

Let $\alpha = 0.05$. By Theorem 13.4, the Friedman statistic (under H_0) will have approximately a χ_1^2 distribution, and the decision rule will be,

Reject H_0 if $g \geq 3.84$.

But

$$g = \frac{12}{22(2)(3)} [(39)^2 + (27)^2] - 3(22)(3)$$

$$= 6.54,$$

implying that the two paths are *not* equivalent, the wide-angle path being significantly better.

QUESTION 13.5.1   Use Friedman's test to analyze the acrophobia data of Case Study 12.2. Let $\alpha = 0.05$.

REVIEW EXERCISES FOR CHAPTER 13

1. Synovial fluid is the clear, viscid secretion that lubricates joints and tendons. For some ailments, its hydrogen-ion concentration (pH) has diagnostic importance. In healthy adults the average pH for synovial fluid is 7.39. Listed below are synovial pH's measured for fluids drawn from the knees of 44 patients with various arthritic conditions (181).

Synovial fluid pH

7.02	7.26	7.31	7.14	7.45	7.32	7.21	7.36	7.36
7.35	7.25	7.24	7.20	7.39	7.40	7.33	7.09	6.60
7.32	7.35	7.34	7.41	7.28	6.99	7.28	7.32	7.29
7.33	7.38	7.32	7.77	7.34	7.10	7.35	6.95	7.31
7.15	7.20	7.34	7.12	7.22	7.30	7.24	7.35	

Is an abnormal synovial pH a symptom of arthritis? Do a sign test on

$$H_0: \tilde{\mu} = 7.39$$

versus

$$H_1: \tilde{\mu} \neq 7.39,$$

where $\tilde{\mu}$ denotes the synovial fluid pH typical of adults with arthritis. Let $\alpha = 0.01$.

2. Draw a histogram of the synovial pH data given in Review Exercise 1. If the shape of the histogram is suitable, do an appropriate *parametric* location test on the data.

3. Let $X_1, X_2, \ldots, X_{22}$ be a random sample of normally distributed random variables with an unknown mean μ and a known variance of 6.0. We wish to test

$$H_0: \mu = 10$$

versus

$$H_1: \mu > 10.$$

Construct a large sample sign test having a Type I error probability of 0.05. What will the power of the test be if $\mu = 11$?

4. Consider again the hypotheses defined in Review Exercise 3. Suppose H_0 and H_1 had been tested with a Z statistic, $(\bar{X} - 10)/(\sqrt{6}/\sqrt{22})$. What would the power of that procedure be if μ was 11?

5. Suppose $n = 7$ paired observations, (X_i, Y_i), are recorded, $i = 1, 2, \ldots, 7$. Let $p = P(Y_i > X_i)$. Write out the entire probability distribution for Y_+, the number of positive differences among the set of $Y_i - X_i$'s, $i = 1, 2, \ldots, 7$, assuming that $p = \frac{1}{2}$. What α levels are possible for testing $H_0: p = \frac{1}{2}$ versus $H_1: p > \frac{1}{2}$?

6. Analyze the Shoshoni rectangle data (Case Study 1.2) with a sign test. Let $\alpha = 0.05$.

7. Classifying the potential danger of a nuclear reactor emergency requires a number of on-site measurements, not the least important of which is wind speed. To compare the performance of two different portable anemometers, a health physicist takes paired observations during 20 randomly selected days in the period from April 1 to July 31. The resulting wind speeds (in mph) are listed below.

Date	Anemometer #1	Anemometer #2	Date	Anemometer #1	Anemometer #2
April 3	12	15	May 19	24	21
April 5	17	19	May 24	12	6
April 6	1	4	May 28	9	12
April 9	6	1	June 9	6	4
April 15	22	19	June 10	2	1
April 26	6	8	June 30	6	12
April 29	2	5	July 19	4	7
May 4	6	9	July 22	11	9
May 7	12	14	July 26	9	7
May 18	2	1	July 28	2	5

Analyze these data with a sign test. Use the large-sample approximation for the distribution of the test statistic. Let $\alpha = 0.05$.

8. Construct a power curve for the hypothesis test on the anemometer data given in Review Exercise 7.

9. In a marketing research test, 28 adult males were asked to shave one side of their face with one brand of razor blade and the other side with a second brand. They were to use the blades for seven days and then decide which was giving the smoother shave. Suppose that 19 of the subjects preferred blade A. Use a sign test to determine whether it can be claimed, at the 0.05 level, that the two blades are significantly different.

10. Suppose a random sample of size 36, $X_1, X_2, \ldots, X_{36}$, is drawn from a uniform pdf defined over the interval $(0, \theta)$, where θ is unknown. Set up a large-sample sign test for deciding whether or not the 25th percentile of the X distribution is equal to 6. Let

$\alpha = 0.05$. With what probability will your procedure commit a Type II error if 7 is the true 25th percentile?

11. To measure the effect on coordination associated with mild intoxication, 13 subjects were each given 15.7 ml of ethyl alcohol per square meter of body surface area and asked to write a certain phrase as many times as they could in the space of one minute (118). The number of correctly written letters was then counted and scaled, with a scale value of 0 representing the score a subject not under the influence of alcohol would be expected to achieve. Negative scores indicate *decreased* writing speeds; positive scores, *increased* writing speeds.

Subject	Score	Subject	Score
1	−6	8	0
2	10	9	−7
3	9	10	5
4	−8	11	−9
5	−6	12	−10
6	−2	13	−2
7	20		

Use the signed rank test to determine whether the level of alcohol provided in this study has any effect on writing speed. Let $\alpha = 0.05$. Omit Subject 8 from your calculations.

12. Two manufacturing processes are available for annealing a certain kind of copper tubing, the primary difference being in the temperature required. The critical response variable is the resulting tensile strength. To compare the methods, 15 pieces of tubing were broken into pairs. One piece from each pair was randomly selected to be annealed at a moderate temperature, the other piece at a high temperature. The resulting tensile strengths (in tons/sq in.) are listed below.

Tensile strengths (tons/sq in.)

Pair	Moderate Temperature	High Temperature
1	16.5	16.9
2	17.6	17.2
3	16.9	17.0
4	15.8	16.1
5	18.4	18.2
6	17.5	17.7
7	17.6	17.9
8	16.1	16.0
9	16.8	17.3
10	15.8	16.1
11	16.8	16.5
12	17.3	17.6
13	18.1	18.4
14	17.9	17.2
15	16.4	16.5

Analyze these data with a Wilcoxon signed rank test. Use a two-sided alternative.

13. Use the Wilcoxon signed rank test on the *Bacillus subtilis* data described in Review Exercise 57 at the end of Chapter 7.

14. Suppose the population being sampled is symmetric and we wish to test $H_0: \tilde{\mu} = \tilde{\mu}_0$. Both the sign test and the signed rank test would be valid. Which procedure, if either, would you expect to have greater power? Why?

15. Use the expansion of

$$\prod_{i=1}^{n} (1 + e^{it})$$

to find the pdf of W when $n = 5$. What α levels are available for testing $H_0: \tilde{\mu} = \tilde{\mu}_0$ versus $H_1: \tilde{\mu} > \tilde{\mu}_0$?

16. The Wilcoxon signed rank statistic,

$$W = \sum_{i=1}^{n} Z_i R_i,$$

can be generalized by replacing the ranks with a set of nonnegative numbers, $V_1 \leq V_2 \leq \cdots \leq V_n$. That is, if $R_i = j$, it gets replaced by V_j. Show that if the V_i's are identically 1, a test based on the generalized Wilcoxon statistic is equivalent to the sign test.

17. Refer to the definition of the generalized Wilcoxon statistic given in Review Exercise 16. Suppose the V_i's, $i = 1, 2, \ldots, n$, are chosen so that

$$\int_0^{V_i} \frac{1}{\sqrt{2\pi}} e^{-z^2/2} \, dz = \frac{i}{n+1}.$$

The V_i's defined in this way are referred to as *normal scores* [see (78)]. Let $n = 5$. Compute the mean and the variance of the generalized Wilcoxon statistic if the V_i's are normal scores.

18. Geologists are always concerned with the reasons why one well is better than another in terms of water quality, productivity, and longevity. One of the possible factors related to a well's success is the rock formation in which it is dug. Listed below are productivity ranks from lowest (1) to highest (80) assigned to 80 water wells dug in central Pennsylvania (167). The five types of rock in which the wells were dug were upper sandy dolomite, Nittany dolomite, Bellefonte dolomite, limestone, and shale.

Productivity ranks for water wells dug in five rock types

Upper Sandy Dolomite	Nittany Dolomite	Bellefonte Dolomite	Limestone	Shale
9	12	1	5	8
18	24	2	13	10
28	29	3	21	14
41	33	4	22	19
43	39	6	27	23
44	46	7	31	25
45	49	11	42	26
50	53	15	54	35

Upper Sandy Dolomite	Nittany Dolomite	Bellefonte Dolomite	Limestone	Shale
58	57	16	56	47
59	67	17	66	55
60	72	20	69	
61	77	30		
62	78	32		
63	79	34		
65	80	36		
68		37		
70		38		
71		40		
73		48		
74		51		
75		52		
76		64		

Analyze these data with a Kruskal-Wallis test. Let $\alpha = 0.01$.

19. Analyze the first-walking-time data of Case Study 11.2 with a Kruskal-Wallis test. Use the 0.05 level of significance.

20. The production of a certain organic chemical requires the addition of ammonium chloride. The manufacturer can conveniently obtain the ammonium chloride in any one of three forms—powdered, moderately ground, and coarse. To see what effect, if any, the quality of the NH_4Cl has, they decide to run the reaction seven times with each form of ammonium chloride. The resulting yields (in pounds) are listed below.

Organic chemical yields (lb)

Powdered NH$_4$Cl	Moderately Ground NH$_4$Cl	Coarse NH$_4$Cl
146	150	141
152	144	138
149	148	142
161	155	146
158	154	139
149	150	145
154	148	137

Compare the yields with a Kruskal-Wallis test. Let $\alpha = 0.05$.

21. Show that the Kruskal-Wallis statistic, B, as defined in Theorem 13.3 can also be written

$$B = \sum_{j=1}^{k} \left(\frac{n - n_j}{n} \right) Z_j^2,$$

where

$$Z_j = \frac{\dfrac{R_{\cdot j}}{n_j} - \dfrac{n+1}{2}}{\sqrt{\dfrac{(n+1)(n-n_j)}{12n_j}}}.$$

[It can also be shown that (1) $E(R_{.j}/n_j) = (n + 1)/2$, (2) Var $(R_{.j}/n_j) = [(n + 1) \cdot (n - n_j)]/12n_j$, and (3) Z_j is approximately an $N(0, 1)$ variable if n_j is sufficiently large. These results, together with the expression showing B to be a function of the sum of squared Z_j's, and the fact that the Z_j's are subject to a constraint

$$\left(\sum_{j=1}^{k} R_{.j} = \frac{n(n + 1)}{2} \right)$$

—meaning the sum would have $k - 1$, rather than k, degrees of freedom—would be the basis for a proof of Theorem 13.3.]

22. When $k = 2$, H_0: $\tilde{\mu}_1 = \tilde{\mu}_2$ is often tested with the *Wilcoxon rank sum* statistic. Suppose the data consist of a random sample of n_1 X's and an independent random sample of n_2 Y's. We can then rank the combined sample and define W_S to be the sum of the ranks associated with the Y's. Then it can be shown that when H_0 is true, the statistic

$$\frac{W_S - [n_2(n_1 + n_2 + 1)/2]}{\sqrt{n_1 n_2(n_1 + n_2 + 1)/12}}$$

has approximately a standard normal pdf [see (59) for details]. Do a Wilcoxon rank sum test on the Mark Twain–Quintus Curtius Snodgrass data of Case Study 8.1.

23. Show that when $k = 2$, the Kruskal-Wallis test is equivalent to the large-sample Wilcoxon rank sum test described in Review Exercise 22. "Verify" your result by doing a Kruskal-Wallis test on the data of Case Study 8.1.

24. The preening habits of male and female fruit flies (*Drosophila melanogaster*) were compared in a controlled environment. The study was carried out in a small chamber that was divided into two parts by a glass partition. Ten flies of one sex were put into one side of the chamber and a single fly of the same sex into the other. (All the

Average preening time
per bout (sec)

Male (x)	Female (y)
2.3	3.7
1.9	5.4
3.3	2.2
2.9	11.7
2.2	2.8
1.3	2.4
2.2	4.0
2.4	2.8
2.1	2.0
1.2	2.8
2.0	2.4
2.7	2.4
2.3	2.9
1.9	10.7
1.2	3.2

fruit flies in the chamber were of the same sex so that courtship rituals would not be a factor.) For a period of three minutes the preening behavior of the single fly was carefully monitored. Eventually the experiment was done 30 times, 15 times with male fruit flies and 15 times with female fruit flies. At the bottom of page 534 are the average lengths of time spent by each of the 30 single flies in a preening "bout" (30). Analyze these data.

25. A civil rights group plans to file suit against a certain city because of its alleged non-compliance with a school busing order. The black-white ratio in this particular community is 20:80, and all the public schools are expected to reflect, as closely as possible, that ratio. The table below lists the actual percentages of black students in two randomly selected sets of schools—the first set is comprised of schools located in predominantly black neighborhoods; the second set, schools located in white neighborhoods.

Percentage of black students

Schools in Black Neighborhoods	Schools in White Neighborhoods
36	21
28	14
41	11
32	30
46	29
39	6
24	18
32	25
45	23

Rank these percentages and use an appropriate procedure to test the civil rights group's claim.

26. The following data come from a field trial set up to assess the effects of different amounts of potash on the breaking strength of cotton fibers (29). The experiment was done in three blocks. The five treatment levels—36, 54, 72, 108, and 144 lbs. of potash per acre—were assigned randomly within each block. The variable recorded was the Pressley strength index.

Pressley strength index for cotton fibers

Treatment (pounds of potash/acre)

		36	54	72	108	144
	1	7.62	8.14	7.76	7.17	7.46
Blocks	2	8.00	8.15	7.73	7.57	7.68
	3	7.93	7.87	7.74	7.80	7.21

Compare the effects of the different levels of potash applications using Friedman's test. Let $\alpha = 0.05$. Test the same hypothesis using an analysis of variance.

27. Compare the rat-poison acceptance percentages listed in Review Exercise 2 in Chapter 12 with a Friedman test. Let $\alpha = 0.05$.

28. Until its recent indictment as a possible carcinogen, cyclamate was a widely used sweetener in soft drinks. The data below show a comparison of three laboratory methods for determining the percentage of sodium cyclamate in commercially produced orange drink. All three procedures were applied to each of 12 samples (152).

Percent sodium cyclamate (w/w)

Method

Sample	Picryl Chloride	Davies	AOAC
1	0.598	0.628	0.632
2	0.614	0.628	0.630
3	0.600	0.600	0.622
4	0.580	0.612	0.584
5	0.596	0.600	0.650
6	0.592	0.628	0.606
7	0.616	0.628	0.644
8	0.614	0.644	0.644
9	0.604	0.644	0.624
10	0.608	0.612	0.619
11	0.602	0.628	0.632
12	0.614	0.644	0.616

Use Friedman's test to determine whether the three methods give significantly different results.

29. Use Friedman's test to analyze the Transylvania-effect data given in Case Study 12.3.

30. Suppose k treatments are to be applied within each of b blocks. Let $\bar{R}_{..}$ denote the average of the bk ranks and let $\bar{R}_{.j} = (1/b)R_{.j}$. Show that the Friedman statistic given in Theorem 13.4 can also be written

$$G = \frac{12b}{k(k+1)} \sum_{j=1}^{k} (\bar{R}_{.j} - \bar{R}_{..})^2.$$

What analysis-of-variance expression does this resemble?

Appendix

TABLE A.1 Cumulative areas under the standard normal distribution

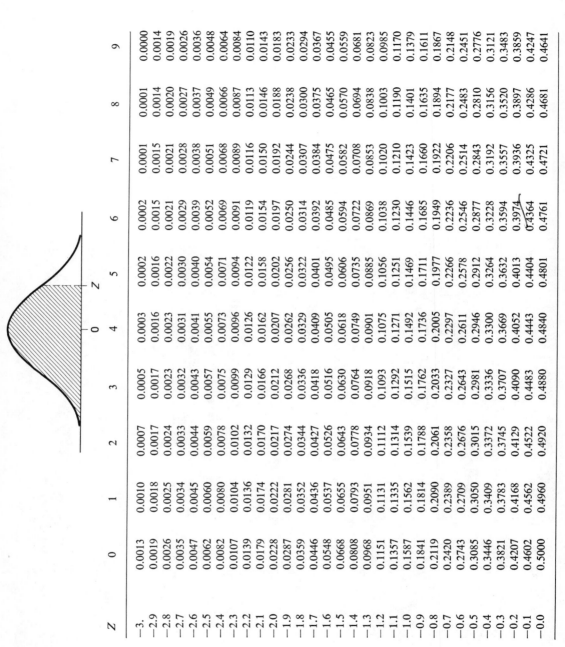

Z	0	1	2	3	4	5	6	7	8	9
−3.	0.0013	0.0010	0.0007	0.0005	0.0003	0.0002	0.0002	0.0001	0.0001	0.0000
−2.9	0.0019	0.0018	0.0017	0.0017	0.0016	0.0016	0.0015	0.0015	0.0014	0.0014
−2.8	0.0026	0.0025	0.0024	0.0023	0.0023	0.0022	0.0021	0.0021	0.0020	0.0019
−2.7	0.0035	0.0034	0.0033	0.0032	0.0031	0.0030	0.0029	0.0028	0.0027	0.0026
−2.6	0.0047	0.0045	0.0044	0.0043	0.0041	0.0040	0.0039	0.0038	0.0037	0.0036
−2.5	0.0062	0.0060	0.0059	0.0057	0.0055	0.0054	0.0052	0.0051	0.0049	0.0048
−2.4	0.0082	0.0080	0.0078	0.0075	0.0073	0.0071	0.0069	0.0068	0.0066	0.0064
−2.3	0.0107	0.0104	0.0102	0.0099	0.0096	0.0094	0.0091	0.0089	0.0087	0.0084
−2.2	0.0139	0.0136	0.0132	0.0129	0.0126	0.0122	0.0119	0.0116	0.0113	0.0110
−2.1	0.0179	0.0174	0.0170	0.0166	0.0162	0.0158	0.0154	0.0150	0.0146	0.0143
−2.0	0.0228	0.0222	0.0217	0.0212	0.0207	0.0202	0.0197	0.0192	0.0188	0.0183
−1.9	0.0287	0.0281	0.0274	0.0268	0.0262	0.0256	0.0250	0.0244	0.0238	0.0233
−1.8	0.0359	0.0352	0.0344	0.0336	0.0329	0.0322	0.0314	0.0307	0.0300	0.0294
−1.7	0.0446	0.0436	0.0427	0.0418	0.0409	0.0401	0.0392	0.0384	0.0375	0.0367
−1.6	0.0548	0.0537	0.0526	0.0516	0.0505	0.0495	0.0485	0.0475	0.0465	0.0455
−1.5	0.0668	0.0655	0.0643	0.0630	0.0618	0.0606	0.0594	0.0582	0.0570	0.0559
−1.4	0.0808	0.0793	0.0778	0.0764	0.0749	0.0735	0.0722	0.0708	0.0694	0.0681
−1.3	0.0968	0.0951	0.0934	0.0918	0.0901	0.0885	0.0869	0.0853	0.0838	0.0823
−1.2	0.1151	0.1131	0.1112	0.1093	0.1075	0.1056	0.1038	0.1020	0.1003	0.0985
−1.1	0.1357	0.1335	0.1314	0.1292	0.1271	0.1251	0.1230	0.1210	0.1190	0.1170
−1.0	0.1587	0.1562	0.1539	0.1515	0.1492	0.1469	0.1446	0.1423	0.1401	0.1379
−0.9	0.1841	0.1814	0.1788	0.1762	0.1736	0.1711	0.1685	0.1660	0.1635	0.1611
−0.8	0.2119	0.2090	0.2061	0.2033	0.2005	0.1977	0.1949	0.1922	0.1894	0.1867
−0.7	0.2420	0.2389	0.2358	0.2327	0.2297	0.2266	0.2236	0.2206	0.2177	0.2148
−0.6	0.2743	0.2709	0.2676	0.2643	0.2611	0.2578	0.2546	0.2514	0.2483	0.2451
−0.5	0.3085	0.3050	0.3015	0.2981	0.2946	0.2912	0.2877	0.2843	0.2810	0.2776
−0.4	0.3446	0.3409	0.3372	0.3336	0.3300	0.3264	0.3228	0.3192	0.3156	0.3121
−0.3	0.3821	0.3783	0.3745	0.3707	0.3669	0.3632	0.3594	0.3557	0.3520	0.3483
−0.2	0.4207	0.4168	0.4129	0.4090	0.4052	0.4013	0.3974	0.3936	0.3897	0.3859
−0.1	0.4602	0.4562	0.4522	0.4483	0.4443	0.4404	0.4364	0.4325	0.4286	0.4247
−0.0	0.5000	0.4960	0.4920	0.4880	0.4840	0.4801	0.4761	0.4721	0.4681	0.4641

TABLE A.1 Cumulative areas under the standard normal distribution (cont.)

Z	0	1	2	3	4	5	6	7	8	9
0.0	0.5000	0.5040	0.5080	0.5120	0.5160	0.5199	0.5239	0.5279	0.5319	0.5359
0.1	0.5398	0.5438	0.5478	0.5517	0.5557	0.5596	0.5636	0.5675	0.5714	0.5753
0.2	0.5793	0.5832	0.5871	0.5910	0.5948	0.5987	0.6026	0.6064	0.6103	0.6141
0.3	0.6179	0.6217	0.6255	0.6293	0.6331	0.6368	0.6406	0.6443	0.6480	0.6517
0.4	0.6554	0.6591	0.6628	0.6664	0.6700	0.6736	0.6772	0.6808	0.6844	0.6879
0.5	0.6915	0.6950	0.6985	0.7019	0.7054	0.7088	0.7123	0.7157	0.7190	0.7224
0.6	0.7257	0.7291	0.7324	0.7357	0.7389	0.7422	0.7454	0.7486	0.7517	0.7549
0.7	0.7580	0.7611	0.7642	0.7673	0.7703	0.7734	0.7764	0.7794	0.7823	0.7852
0.8	0.7881	0.7910	0.7939	0.7967	0.7995	0.8023	0.8051	0.8078	0.8106	0.8133
0.9	0.8159	0.8186	0.8212	0.8238	0.8264	0.8289	0.8315	0.8340	0.8365	0.8389
1.0	0.8413	0.8438	0.8461	0.8485	0.8508	0.8531	0.8554	0.8577	0.8599	0.8621
1.1	0.8643	0.8665	0.8686	0.8708	0.8729	0.8749	0.8770	0.8790	0.8810	0.8830
1.2	0.8849	0.8869	0.8888	0.8907	0.8925	0.8944	0.8962	0.8980	0.8997	0.9015
1.3	0.9032	0.9049	0.9066	0.9082	0.9099	0.9115	0.9131	0.9147	0.9162	0.9177
1.4	0.9192	0.9207	0.9222	0.9236	0.9251	0.9265	0.9278	0.9292	0.9306	0.9319
1.5	0.9332	0.9345	0.9357	0.9370	0.9382	0.9394	0.9406	0.9418	0.9430	0.9441
1.6	0.9452	0.9463	0.9474	0.9484	0.9495	0.9505	0.9515	0.9525	0.9535	0.9545
1.7	0.9554	0.9564	0.9573	0.9582	0.9591	0.9599	0.9608	0.9616	0.9625	0.9633
1.8	0.9641	0.9648	0.9656	0.9664	0.9671	0.9678	0.9686	0.9693	0.9700	0.9706
1.9	0.9713	0.9719	0.9726	0.9732	0.9738	0.9744	0.9750	0.9756	0.9762	0.9767
2.0	0.9772	0.9778	0.9783	0.9788	0.9793	0.9798	0.9803	0.9808	0.9812	0.9817
2.1	0.9821	0.9826	0.9830	0.9834	0.9838	0.9842	0.9846	0.9850	0.9854	0.9857
2.2	0.9861	0.9864	0.9868	0.9871	0.9874	0.9878	0.9881	0.9884	0.9887	0.9890
2.3	0.9893	0.9896	0.9898	0.9901	0.9904	0.9906	0.9909	0.9911	0.9913	0.9916
2.4	0.9918	0.9920	0.9922	0.9925	0.9927	0.9929	0.9931	0.9932	0.9934	0.9936
2.5	0.9938	0.9940	0.9941	0.9943	0.9945	0.9946	0.9948	0.9949	0.9951	0.9952
2.6	0.9953	0.9955	0.9956	0.9957	0.9959	0.9960	0.9961	0.9962	0.9963	0.9964
2.7	0.9965	0.9966	0.9967	0.9968	0.9969	0.9970	0.9971	0.9972	0.9973	0.9974
2.8	0.9974	0.9975	0.9976	0.9977	0.9977	0.9978	0.9979	0.9979	0.9980	0.9981
2.9	0.9981	0.9982	0.9982	0.9983	0.9984	0.9984	0.9985	0.9985	0.9986	0.9986
3.	0.9987	0.9990	0.9993	0.9995	0.9997	0.9998	0.9998	0.9999	0.9999	1.0000

SOURCE: B. W. Lindgren, *Statistical Theory* (New York: Macmillan, 1962), pp. 392–393.

TABLE A.2 Upper percentiles of Student t distributions

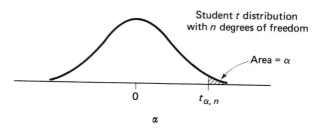

Student t distribution with n degrees of freedom

Area = α

0 $t_{\alpha, n}$

df	0.20	0.15	0.10	0.05	0.025	0.01	0.005
1	1.376	1.963	3.078	6.3138	12.706	31.821	63.657
2	1.061	1.386	1.886	2.9200	4.3027	6.965	9.9248
3	0.978	1.250	1.638	2.3534	3.1825	4.541	5.8409
4	0.941	1.190	1.533	2.1318	2.7764	3.747	4.6041
5	0.920	1.156	1.476	2.0150	2.5706	3.365	4.0321
6	0.906	1.134	1.440	1.9432	2.4469	3.143	3.7074
7	0.896	1.119	1.415	1.8946	2.3646	2.998	3.4995
8	0.889	1.108	1.397	1.8595	2.3060	2.896	3.3554
9	0.883	1.100	1.383	1.8331	2.2622	2.821	3.2498
10	0.879	1.093	1.372	1.8125	2.2281	2.764	3.1693
11	0.876	1.088	1.363	1.7959	2.2010	2.718	3.1058
12	0.873	1.083	1.356	1.7823	2.1788	2.681	3.0545
13	0.870	1.079	1.350	1.7709	2.1604	2.650	3.0123
14	0.868	1.076	1.345	1.7613	2.1448	2.624	2.9768
15	0.866	1.074	1.341	1.7530	2.1315	2.602	2.9467
16	0.865	1.071	1.337	1.7459	2.1199	2.583	2.9208
17	0.863	1.069	1.333	1.7396	2.1098	2.567	2.8982
18	0.862	1.067	1.330	1.7341	2.1009	2.552	2.8784
19	0.861	1.066	1.328	1.7291	2.0930	2.539	2.8609
20	0.860	1.064	1.325	1.7247	2.0860	2.528	2.8453
21	0.859	1.063	1.323	1.7207	2.0796	2.518	2.8314
22	0.858	1.061	1.321	1.7171	2.0739	2.508	2.8188
23	0.858	1.060	1.319	1.7139	2.0687	2.500	2.8073
24	0.857	1.059	1.318	1.7109	2.0639	2.492	2.7969
25	0.856	1.058	1.316	1.7081	2.0595	2.485	2.7874
26	0.856	1.058	1.315	1.7056	2.0555	2.479	2.7787
27	0.855	1.057	1.314	1.7033	2.0518	2.473	2.7707
28	0.855	1.056	1.313	1.7011	2.0484	2.467	2.7633
29	0.854	1.055	1.311	1.6991	2.0452	2.462	2.7564
30	0.854	1.055	1.310	1.6973	2.0423	2.457	2.7500
31	0.8535	1.0541	1.3095	1.6955	2.0395	2.453	2.7441
32	0.8531	1.0536	1.3086	1.6939	2.0370	2.449	2.7385
33	0.8527	1.0531	1.3078	1.6924	2.0345	2.445	2.7333
34	0.8524	1.0526	1.3070	1.6909	2.0323	2.441	2.7284

				α			
df	0.20	0.15	0.10	0.05	0.025	0.01	0.005
35	0.8521	1.0521	1.3062	1.6896	2.0301	2.438	2.7239
36	0.8518	1.0516	1.3055	1.6883	2.0281	2.434	2.7195
37	0.8515	1.0512	1.3049	1.6871	2.0262	2.431	2.7155
38	0.8512	1.0508	1.3042	1.6860	2.0244	2.428	2.7116
39	0.8510	1.0504	1.3037	1.6849	2.0227	2.426	2.7079
40	0.8507	1.0501	1.3031	1.6839	2.0211	2.423	2.7045
41	0.8505	1.0498	1.3026	1.6829	2.0196	2.421	2.7012
42	0.8503	1.0494	1.3020	1.6820	2.0181	2.418	2.6981
43	0.8501	1.0491	1.3016	1.6811	2.0167	2.416	2.6952
44	0.8499	1.0488	1.3011	1.6802	2.0154	2.414	2.6923
45	0.8497	1.0485	1.3007	1.6794	2.0141	2.412	2.6896
46	0.8495	1.0483	1.3002	1.6787	2.0129	2.410	2.6870
47	0.8494	1.0480	1.2998	1.6779	2.0118	2.408	2.6846
48	0.8492	1.0478	1.2994	1.6772	2.0106	2.406	2.6822
49	0.8490	1.0476	1.2991	1.6766	2.0096	2.405	2.6800
50	0.8489	1.0473	1.2987	1.6759	2.0086	2.403	2.6778
51	0.8448	1.0471	1.2984	1.6753	2.0077	2.402	2.6758
52	0.8486	1.0469	1.2981	1.6747	2.0067	2.400	2.6738
53	0.8485	1.0467	1.2978	1.6742	2.0058	2.399	2.6719
54	0.8484	1.0465	1.2975	1.6736	2.0049	2.397	2.6700
55	0.8483	1.0463	1.2972	1.6731	2.0041	2.396	2.6683
56	0.8481	1.0461	1.2969	1.6725	2.0033	2.395	2.6666
57	0.8480	1.0460	1.2967	1.6721	2.0025	2.393	2.6650
58	0.8479	1.0458	1.2964	1.6716	2.0017	2.392	2.6633
59	0.8478	1.0457	1.2962	1.6712	2.0010	2.391	2.6618
60	0.8477	1.0455	1.2959	1.6707	2.0003	2.390	2.6603
61	0.8476	1.0454	1.2957	1.6703	1.9997	2.389	2.6590
62	0.8475	1.0452	1.2954	1.6698	1.9990	2.388	2.6576
63	0.8474	1.0451	1.2952	1.6694	1.9984	2.387	2.6563
64	0.8473	1.0449	1.2950	1.6690	1.9977	2.386	2.6549
65	0.8472	1.0448	1.2948	1.6687	1.9972	2.385	2.6537
66	0.8471	1.0447	1.2945	1.6683	1.9966	2.384	2.6525
67	0.8471	1.0446	1.2944	1.6680	1.9961	2.383	2.6513
68	0.8470	1.0444	1.2942	1.6676	1.9955	2.382	2.6501
69	0.8469	1.0443	1.2940	1.6673	1.9950	2.381	2.6491
70	0.8468	1.0442	1.2938	1.6669	1.9945	2.381	2.6480
71	0.8468	1.0441	1.2936	1.6666	1.9940	2.380	2.6470
72	0.8467	1.0440	1.2934	1.6663	1.9935	2.379	2.6459
73	0.8466	1.0439	1.2933	1.6660	1.9931	2.378	2.6450
74	0.8465	1.0438	1.2931	1.6657	1.9926	2.378	2.6640
75	0.8465	1.0437	1.2930	1.6655	1.9922	2.377	2.6431
76	0.8464	1.0436	1.2928	1.6652	1.9917	2.376	2.6421
77	0.8464	1.0435	1.2927	1.6649	1.9913	2.376	2.6413
78	0.8463	1.0434	1.2925	1.6646	1.9909	2.375	2.6406
79	0.8463	1.0433	1.2924	1.6644	1.9905	2.374	2.6396

α

df	0.20	0.15	0.10	0.05	0.025	0.01	0.005
80	0.8462	1.0432	1.2922	1.6641	1.9901	2.374	2.6388
81	0.8461	1.0431	1.2921	1.6639	1.9897	2.373	2.6380
82	0.8460	1.0430	1.2920	1.6637	1.9893	2.372	2.6372
83	0.8460	1.0430	1.2919	1.6635	1.9890	2.372	2.6365
84	0.8459	1.0429	1.2917	1.6632	1.9886	2.371	2.6357
85	0.8459	1.0428	1.2916	1.6630	1.9883	2.371	2.6350
86	0.8458	1.0427	1.2915	1.6628	1.9880	2.370	2.6343
87	0.8458	1.0427	1.2914	1.6626	1.9877	2.370	2.6336
88	0.8457	1.0426	1.2913	1.6624	1.9873	2.369	2.6329
89	0.8457	1.0426	1.2912	1.6622	1.9870	2.369	2.6323
90	0.8457	1.0425	1.2910	1.6620	1.9867	2.368	2.6316
91	0.8457	1.0424	1.2909	1.6618	1.9864	2.368	2.6310
92	0.8456	1.0423	1.2908	1.6616	1.9861	2.367	2.6303
93	0.8456	1.0423	1.2907	1.6614	1.9859	2.367	2.6298
94	0.8455	1.0422	1.2906	1.6612	1.9856	2.366	2.6292
95	0.8455	1.0422	1.2905	1.6611	1.9853	2.366	2.6286
96	0.8454	1.0421	1.2904	1.6609	1.9850	2.366	2.6280
97	0.8454	1.0421	1.2904	1.6608	1.9848	2.365	2.6275
98	0.8453	1.0420	1.2903	1.6606	1.9845	2.365	2.6270
99	0.8453	1.0419	1.2902	1.6604	1.9843	2.364	2.6265
100	0.8452	1.0418	1.2901	1.6602	1.9840	2.364	2.6260
∞	0.84	1.04	1.28	1.64	1.96	2.33	2.58

SOURCE: *Scientific Tables*, 6th ed. (Basel, Switzerland: J. R. Geigy, 1962), pp. 32–33.

TABLE A.3 Upper and lower percentiles of χ^2 distributions

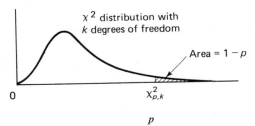

				p				
df	0.010	0.025	0.050	0.10	0.90	0.95	0.975	0.99
1	0.000157	0.000982	0.00393	0.0158	2.706	3.841	5.024	6.635
2	0.0201	0.0506	0.103	0.211	4.605	5.991	7.378	9.210
3	0.115	0.216	0.352	0.584	6.251	7.815	9.348	11.345
4	0.297	0.484	0.711	1.064	7.779	9.488	11.143	13.277
5	0.554	0.831	1.145	1.610	9.236	11.070	12.832	15.086
6	0.872	1.237	1.635	2.204	10.645	12.592	14.449	16.812
7	1.239	1.690	2.167	2.833	12.017	14.067	16.013	18.475
8	1.646	2.180	2.733	3.490	13.362	15.507	17.535	20.090
9	2.088	2.700	3.325	4.168	14.684	16.919	19.023	21.666
10	2.558	3.247	3.940	4.865	15.987	18.307	20.483	23.209
11	3.053	3.816	4.575	5.578	17.275	19.675	21.920	24.725
12	3.571	4.404	5.226	6.304	18.549	21.026	23.336	26.217
13	4.107	5.009	5.892	7.042	19.812	22.362	24.736	27.688
14	4.660	5.629	6.571	7.790	21.064	23.685	26.119	29.141
15	5.229	6.262	7.261	8.547	22.307	24.996	27.488	30.578
16	5.812	6.908	7.962	9.312	23.542	26.296	28.845	32.000
17	6.408	7.564	8.672	10.085	24.769	27.587	30.191	33.409
18	7.015	8.231	9.390	10.865	25.989	28.869	31.526	34.805
19	7.633	8.907	10.117	11.651	27.204	30.144	32.852	36.191
20	8.260	9.591	10.851	12.443	28.412	31.410	34.170	37.566
21	8.897	10.283	11.591	13.240	29.615	32.671	35.479	38.932
22	9.542	10.982	12.338	14.041	30.813	33.924	36.781	40.289
23	10.196	11.688	13.091	14.848	32.007	35.172	38.076	41.638
24	10.856	12.401	13.848	15.659	33.196	36.415	39.364	42.980
25	11.524	13.120	14.611	16.473	34.382	37.652	40.646	44.314
26	12.198	13.844	15.379	17.292	35.563	38.885	41.923	45.642
27	12.879	14.573	16.151	18.114	36.741	40.113	43.194	46.963
28	13.565	15.308	16.928	18.939	37.916	41.337	44.461	48.278
29	14.256	16.047	17.708	19.768	39.087	42.557	45.722	49.588
30	14.953	16.791	18.493	20.599	40.256	43.773	46.979	50.892
31	15.655	17.539	19.281	21.434	41.422	44.985	48.232	52.191
32	16.362	18.291	20.072	22.271	42.585	46.194	49.480	53.486
33	17.073	19.047	20.867	23.110	43.745	47.400	50.725	54.776
34	17.789	19.806	21.664	23.952	44.903	48.602	51.966	56.061

				p				
df	0.010	0.025	0.050	0.10	0.90	0.95	0.975	0.99
35	18.509	20.569	22.465	24.797	46.059	49.802	53.203	57.342
36	19.233	21.336	23.269	25.643	47.212	50.998	54.437	58.619
37	19.960	22.106	24.075	26.492	48.363	52.192	55.668	59.892
38	20.691	22.878	24.884	27.343	49.513	53.384	56.895	61.162
39	21.426	23.654	25.695	28.196	50.660	54.572	58.120	62.428
40	22.164	24.433	26.509	29.051	51.805	55.758	59.342	63.691
41	22.906	25.215	27.326	29.907	52.949	56.942	60.561	64.950
42	23.650	25.999	28.144	30.765	54.090	58.124	61.777	66.206
43	24.398	26.785	28.965	31.625	55.230	59.304	62.990	67.459
44	25.148	27.575	29.787	32.487	56.369	60.481	64.201	68.709
45	25.901	28.366	30.612	33.350	57.505	61.656	65.410	69.957
46	26.657	29.160	31.439	34.215	58.641	62.830	66.617	71.201
47	27.416	29.956	32.268	35.081	59.774	64.001	67.821	72.443
48	28.177	30.755	33.098	35.949	60.907	65.171	69.023	73.683
49	28.941	31.555	33.930	36.818	62.038	66.339	70.222	74.919
50	29.707	32.357	34.764	37.689	63.167	67.505	71.420	76.154

SOURCE: *Scientific Tables*, 6th ed. (Basel, Switzerland: J. R. Geigy, 1962), p. 36.

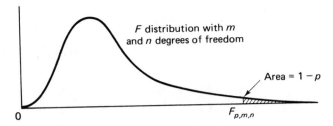

The figure above illustrates the percentiles of F distribution in Table A.4. Table A.4 is used with permission from Wilfrid J. Dixon and Frank J. Massey, Jr., *Introduction to Statistical Analysis*, 2nd. ed. (New York: McGraw-Hill, 1957), pp. 389–404.

TABLE A.4 Percentiles of F distributions

n	p	m 1	2	3	4	5	6	7	8	9	10	11	12	p
1	.0005	$.0^662$	$.0^350$	$.0^238$	$.0^294$	.016	.022	.027	.032	.036	.039	.042	.045	.0005
	.001	$.0^525$	$.0^210$	$.0^260$	.013	.021	.028	.034	.039	.044	.048	.051	.054	.001
	.005	$.0^462$	$.0^251$	.018	.032	.044	.054	.062	.068	.073	.078	.082	.085	.005
	.010	$.0^325$	.010	.029	.047	.062	.073	.082	.089	.095	.100	.104	.107	.010
	.025	$.0^215$	.026	.057	.082	.100	.113	.124	.132	.139	.144	.149	.153	.025
	.05	$.0^262$	.054	.099	.130	.151	.167	.179	.188	.195	.201	.207	.211	.05
	.10	.025	.117	.181	.220	.246	.265	.279	.289	.298	.304	.310	.315	.10
	.25	.172	.389	.494	.553	.591	.617	.637	.650	.661	.670	.680	.684	.25
	.50	1.00	1.50	1.71	1.82	1.89	1.94	1.98	2.00	2.03	2.04	2.05	2.07	.50
	.75	5.83	7.50	8.20	8.58	8.82	8.98	9.10	9.19	9.26	9.32	9.36	9.41	.75
	.90	39.9	49.5	53.6	55.8	57.2	58.2	58.9	59.4	59.9	60.2	60.5	60.7	.90
	.95	161	200	216	225	230	234	237	239	241	242	243	244	.95
	.975	648	800	864	900	922	937	948	957	963	969	973	977	.975
	.99	405^1	500^1	540^1	562^1	576^1	586^1	593^1	598^1	602^1	606^1	608^1	611^1	.99
	.995	162^2	200^2	216^2	225^2	231^2	234^2	237^2	239^2	241^2	242^2	243^2	244^2	.995
	.999	406^3	500^3	540^3	562^3	576^3	586^3	593^3	598^3	602^3	606^3	609^3	611^3	.999
	.9995	162^4	200^4	216^4	225^4	231^4	234^4	237^4	239^4	241^4	242^4	243^4	244^4	.9995
2	.0005	$.0^550$	$.0^350$	$.0^242$	.011	.020	.029	.037	.044	.050	.056	.061	.065	.0005
	.001	$.0^520$	$.0^210$	$.0^268$	.016	.027	.037	.046	.054	.061	.067	.072	.077	.001
	.005	$.0^450$	$.0^250$	.020	.038	.055	.069	.081	.091	.099	.106	.112	.118	.005
	.01	$.0^320$	.010	.032	.056	.075	.092	.105	.116	.125	.132	.139	.144	.01
	.025	$.0^213$	.026	.062	.094	.119	.138	.153	.165	.175	.183	.190	.196	.025
	.05	$.0^250$	.053	.105	.144	.173	.194	.211	.224	.235	.244	.251	.257	.05
	.10	.020	.111	.183	.231	.265	.289	.307	.321	.333	.342	.350	.356	.10
	.25	.133	.333	.439	.500	.540	.568	.588	.604	.616	.626	.633	.641	.25
	.50	.667	1.00	1.13	1.21	1.25	1.28	1.30	1.32	1.33	1.34	1.35	1.36	.50
	.75	2.57	3.00	3.15	3.23	3.28	3.31	3.34	3.35	3.37	3.38	3.39	3.39	.75
	.90	8.53	9.00	9.16	9.24	9.29	9.33	9.35	9.37	9.38	9.39	9.40	9.41	.90
	.95	18.5	19.0	19.2	19.2	19.3	19.3	19.4	19.4	19.4	19.4	19.4	19.4	.95
	.975	38.5	39.0	39.2	39.2	39.3	39.3	39.4	39.4	39.4	39.4	39.4	39.4	.975
	.99	98.5	99.0	99.2	99.2	99.3	99.3	99.4	99.4	99.4	99.4	99.4	99.4	.99
	.995	198	199	199	199	199	199	199	199	199	199	199	199	.995
	.999	998	999	999	999	999	999	999	999	999	999	999	999	.999
	.9995	200^1	200^1	200^1	200^1	200^1	200^1	200^1	200^1	200^1	200^1	200^1	200^1	.9995
3	.0005	$.0^646$	$.0^350$	$.0^244$	.012	.023	.033	.043	.052	.060	.067	.074	.079	.0005
	.001	$.0^519$	$.0^210$	$.0^271$	.018	.030	.042	.053	.063	.072	.079	.086	.093	.001
	.005	$.0^446$	$.0^250$	.021	.041	.060	.077	.092	.104	.115	.124	.132	.138	.005
	.01	$.0^319$	.010	.034	.060	.083	.102	.118	.132	.143	.153	.161	.168	.01
	.025	$.0^212$	.026	.065	.100	.129	.152	.170	.185	.197	.207	.216	.224	.025
	.05	$.0^246$	.052	.108	.152	.185	.210	.230	.246	.259	.270	.279	.287	.05
	.10	.019	.109	.185	.239	.276	.304	.325	.342	.356	.367	.376	.384	.10
	.25	.122	.317	.424	.489	.531	.561	.582	.600	.613	.624	.633	.641	.25
	.50	.585	.881	1.00	1.06	1.10	1.13	1.15	1.16	1.17	1.18	1.19	1.20	.50
	.75	2.02	2.28	2.36	2.39	2.41	2.42	2.43	2.44	2.44	2.44	2.45	2.45	.75
	.90	5.54	5.46	5.39	5.34	5.31	5.28	5.27	5.25	5.24	5.23	5.22	5.22	.90
	.95	10.1	9.55	9.28	9.12	9.01	8.94	8.89	8.85	8.81	8.79	8.76	8.74	.95
	.075	17.4	16.0	15.4	15.1	14.9	14.7	14.6	14.5	14.5	14.4	14.4	14.3	.975
	.99	34.1	30.8	29.5	28.7	28.2	27.9	27.7	27.5	27.3	27.2	27.1	27.1	.99
	.995	55.6	49.8	47.5	46.2	45.4	44.8	44.4	44.1	43.9	43.7	43.5	43.4	.995
	.999	167	149	141	137	135	133	132	131	130	129	129	128	.999
	.9995	266	237	225	218	214	211	209	208	207	206	204	204	.9995

Read $.0^356$ as .00056, 200^1 as 2000, 162^4 as 1620000, etc.

TABLE A.4 Percentiles of the F distributions (cont.)

p \ m	15	20	24	30	40	50	60	100	120	200	500	∞	p	n
.0005	.051	.058	062	.066	.069	.072	.074	.077	.078	.080	.081	.083	.0005	1
.001	.060	.067	.071	.075	.079	.082	.084	.087	.088	.089	.091	.092	.001	
.005	.093	.101	.105	.109	.113	.116	.118	.121	.122	.124	.126	.127	.005	
.01	.115	.124	.128	.132	.137	.139	.141	.145	.146	.148	.150	.151	.01	
.025	.161	.170	.175	.180	.184	.187	.189	.193	.194	.196	.198	.199	.025	
.05	.220	.230	.235	.240	.245	.248	.250	.254	.255	.257	.259	.261	.05	
.10	.325	.336	.342	.347	.353	.356	.358	.362	.364	.366	.368	.370	.10	
.25	.698	.712	.719	.727	.734	.738	.741	.747	.749	.752	.754	.756	.25	
.50	2.09	2.12	2.13	2.15	2.16	2.17	2.17	2.18	2.18	2.19	2.19	2.20	.50	
.75	9.49	9.58	9.63	9.67	9.71	9.74	9.76	9.78	9.80	9.82	9.84	9.85	.75	
.90	61.2	61.7	62.0	62.3	62.5	62.7	62.7	62.8	63.0	63.1	63.2	63.3	.90	
.95	246	248	249	250	251	252	252	253	253	254	254	254	.95	
.975	985	993	997	100^1	101^1	101^1	101^1	101^1	101^1	102^1	102^1	102^1	.975	
.99	616^1	621^1	623^1	626^1	629^1	630^1	631^1	633^1	634^1	635^1	636^1	637^1	.99	
.995	246^2	248^2	249^2	250^2	251^2	252^2	253^2	253^2	254^2	254^2	254^2	255^2	.995	
.999	616^3	621^3	623^3	626^3	629^3	630^3	631^3	633^3	634^3	635^3	636^3	637^3	.999	
.9995	246^4	248^4	249^4	250^4	251^4	252^4	252^4	253^4	253^4	253^4	254^4	254^4	.9995	
.0005	.076	.088	.094	.101	.108	.113	.116	.122	.124	.127	.130	.132	.0005	2
.001	.088	.100	.107	.114	.121	.126	.129	.135	.137	.140	.143	.145	.001	
.005	.130	.143	.150	.157	.165	.169	.173	.179	.181	.184	.187	.189	.005	
.01	.157	.171	.178	.186	.193	.198	.201	.207	.209	.212	.215	.217	.01	
.025	.210	.224	.232	.239	.247	.251	.255	.261	.263	.266	.269	.271	.025	
.05	.272	.286	.294	.302	.309	.314	.317	.324	.326	.329	.332	.334	.05	
.10	.371	.386	.394	.402	.410	.415	.418	.424	.426	.429	.433	.434	.10	
.25	.657	.672	.680	.689	.697	.702	.705	.711	.713	.716	.719	.721	.25	
.50	1.38	1.39	1.40	1.41	1.42	1.42	1.43	1.43	1.43	1.44	1.44	1.44	.50	
.75	3.41	3.43	3.43	3.44	3.45	3.45	3.46	3.47	3.47	3.48	3.48	3.48	.75	
.90	9.42	9.44	9.45	9.46	9.47	9.47	9.47	9.48	9.48	9.49	9.49	9.49	.90	
.95	19.4	19.4	19.5	19.5	19.5	19.5	19.5	19.5	19.5	19.5	1℃ 5	19.5	.95	
.975	39.4	39.4	39.5	39.5	39.5	39.5	39.5	39.5	39.5	39.5	39.5	39.5	.975	
.99	99.4	99.4	99.5	99.5	99.5	99.5	99.5	99.5	99.5	99.5	99.5	99.5	.99	
.995	199	199	199	199	199	199	199	199	199	199	199	200	.995	
.999	999	999	999	999	999	999	999	999	999	999	999	999	.999	
.9995	200^1	200^1	200^1	200^1	200^1	200^1	200^1	200^1	200^1	200^1	200^1	200^1	.9995	
.0005	.093	.109	.117	.127	.136	.143	.147	.156	.158	.162	.166	.169	.0005	3
.001	.107	.123	.132	.142	.152	.158	.162	.171	.173	.177	.181	.184	.001	
.005	.154	.172	.181	.191	.201	.207	.211	.220	.222	.227	.231	.234	.005	
.01	.185	.203	.212	.222	.232	.238	.242	.251	.253	.258	.262	.264	.01	
.025	.241	.259	.269	.279	.289	.295	.299	.308	.310	.314	.318	.321	.025	
.05	.304	.323	.332	.342	.352	.358	.363	.370	.373	.377	.382	.384	.05	
.10	.402	.420	.430	.439	.449	.455	.459	.467	.469	.474	.476	.480	.10	
.25	.658	.675	.684	.693	.702	.708	.711	.719	.721	.724	.728	.730	.25	
.50	1.21	1.23	1.23	1.24	1.25	1.25	1.25	1.26	1.26	1.26	1.27	1.27	.50	
.75	2.46	2.46	2.46	2.47	2.47	2.47	2.47	2.47	2.47	2.47	2.47	2.47	.75	
.90	5.20	5.18	5.18	5.17	5.16	5.15	5.15	5.14	5.14	5.14	5.14	5.13	.90	
.95	8.70	8.66	8.63	8.62	8.59	8.58	8.57	8.55	8.55	8.54	8.53	8.53	.95	
.975	14.3	14.2	14.1	14.1	14.0	14.0	14.0	14.0	13.9	13.9	13.9	13.9	.975	
.99	26.9	26.7	26.6	26.5	26.4	26.4	26.3	26.2	26.2	26.2	26.1	26.1	.99	
.995	43.1	42.8	42.6	42.5	42.3	42.2	42.1	42.0	42.0	41.9	41.9	41.8	.995	
.999	127	126	126	125	125	125	124	124	124	124	124	123	.999	
.9995	203	201	200	199	199	198	198	197	197	197	196	196	.9995	

n	p	1	2	3	4	5	6	7	8	9	10	11	12	p
4	.0005	$.0^644$	$.0^350$	$.0^246$	.013	.024	.036	.047	.057	.066	.075	.082	.089	.0005
	.001	$.0^518$	$.0^210$	$.0^273$	.019	.032	.046	.058	.069	.079	.089	.097	.104	.001
	.005	$.0^444$	$.0^250$	.022	.043	.064	.083	.100	.114	.126	.137	.145	.153	.005
	.01	$.0^318$	.010	.035	.063	.088	.109	.127	.143	.156	.167	.176	.185	.01
	.025	$.0^211$	.026	.066	.104	.135	.161	.181	.198	.212	.224	.234	.243	.025
	05	$.0^244$	.052	.110	.157	.193	.221	.243	.261	.275	.288	.298	.307	.05
	.10	.018	.108	.187	.243	.284	.314	.338	.356	.371	.384	.394	.403	.10
	.25	.117	.309	.418	.484	.528	.560	.583	.601	.615	.627	.637	.645	.25
	.50	.549	.828	.941	1.00	1.04	1.06	1.08	1.09	1.10	1.11	1.12	1.13	.50
	.75	1.81	2.00	2.05	2.06	2.07	2.08	2.08	2.08	2.08	2.08	2.08	2.08	.75
	.90	4.54	4.32	4.19	4.11	4.05	4.01	3.98	3.95	3.94	3.92	3.91	3.90	.90
	.95	7.71	6.94	6.59	6.39	6.26	6.16	6.09	6.04	6.00	5.96	5.94	5.91	.95
	.975	12.2	10.6	9.98	9.60	9.36	9.20	9.07	8.98	8.90	8.84	8.79	8.75	.975
	.99	21.2	18.0	16.7	16.0	15.5	15.2	15.0	14.8	14.7	14.5	14.4	14.4	.99
	.995	31.3	26.3	24.3	23.2	22.5	22.0	21.6	21.4	21.1	21.0	20.8	20.7	.995
	.999	74.1	61.2	56.2	53.4	51.7	50.5	49.7	49.0	48.5	48.0	47.7	47.4	.999
	.9995	106	87.4	80.1	76.1	73.6	71.9	70.6	69.7	68.9	68.3	67.8	67.4	.9995
5	.0005	$.0^643$	$.0^350$	$.0^247$	.014	.025	.038	.050	.061	.070	.081	.089	.096	.0005
	.001	$.0^517$	$.0^210$	$.0^275$	.019	.034	.048	.062	.074	.085	.095	.104	.112	.001
	.005	$.0^443$	$.0^250$	.022	.045	.067	.087	.105	.120	.134	.146	.156	.165	.005
	.01	$.0^317$	.010	.035	.064	.091	.114	.134	.151	.165	.177	.188	.197	.01
	.025	$.0^211$	.025	.067	.107	.140	.167	.189	.208	.223	.236	.248	.257	.025
	.05	$.0^243$	.052	.111	.160	.198	.228	.252	.271	.287	.301	.313	.322	.05
	.10	.017	.108	.188	.247	.290	.322	.347	.367	.383	.397	.408	.418	.10
	.25	.113	.305	.415	.483	.528	.560	.584	.604	.618	.631	.641	.650	.25
	.50	.528	.799	.907	.965	1.00	1.02	1.04	1.05	1.06	1.07	1.08	1.09	.50
	.75	1.69	1.85	1.88	1.89	1.89	1.89	1.89	1.89	1.89	1.89	1.89	1.89	.75
	.90	4.06	3.78	3.62	3.52	3.45	3.40	3.37	3.34	3.32	3.30	3.28	3.27	.90
	.95	6.61	5.79	5.41	5.19	5.05	4.95	4.88	4.82	4.77	4.74	4.71	4.68	.95
	.975	10.0	8.43	7.76	7.39	7.15	6.98	6.85	6.76	6.68	6.62	6.57	6.52	.975
	.99	16.3	13.3	12.1	11.4	11.0	10.7	10.5	10.3	10.2	10.1	9.96	9.89	.99
	.995	22.8	18.3	16.5	15.6	14.9	14.5	14.2	14.0	13.8	13.6	13.5	13.4	.995
	.999	47.2	37.1	33.2	31.1	29.7	28.8	28.2	27.6	27.2	26.9	26.6	26.4	.999
	.9995	63.6	49.8	44.4	41.5	39.7	38.5	37.6	36.9	36.4	35.9	35.6	35.2	.9995
6	.0005	$.0^643$	$.0^350$	$.0^247$	.014	.026	.039	.052	.064	.075	.085	.094	.103	.0005
	.001	$.0^517$	$.0^210$	$.0^275$	.020	.035	.050	.064	.078	.090	.101	.111	.119	.001
	.005	$.0^443$	$.0^250$	.022	.045	.069	.090	.109	.126	.140	.153	.164	.174	.005
	.01	$.0^317$	.010	.036	.066	.094	.118	.139	.157	.172	.186	.197	.207	.01
	.025	$.0^211$	.025	.068	.109	.143	.172	.195	.215	.231	.246	.258	.268	.025
	.05	$.0^243$	.052	.112	.162	.202	.233	.259	.279	.296	.311	.324	.334	.05
	.10	.017	.107	.189	.249	.294	.327	.354	.375	.392	.406	.418	.429	.10
	.25	.111	.302	.413	.481	.524	.561	.586	.606	.622	.635	.645	.654	.25
	.50	.515	.780	.886	.942	.977	1.00	1.02	1.03	1.04	1.05	1.05	1.06	.50
	.75	1.62	1.76	1.78	1.79	1.79	1.78	1.78	1.78	1.77	1.77	1.77	1.77	.75
	.90	3.78	3.46	3.29	3.18	3.11	3.05	3.01	2.98	2.96	2.94	2.92	2.90	.90
	.95	5.99	5.14	4.76	4.53	4.39	4.28	4.21	4.15	4.10	4.06	4.03	4.00	.95
	.975	8.81	7.26	6.60	6.23	5.99	5.82	5.70	5.60	5.52	5.46	5.41	5.37	.975
	.99	13.7	10.9	9.78	9.15	8.75	8.47	8.26	8.10	7.98	7.87	7.79	7.72	.99
	.995	18.6	14.5	12.9	12.0	11.5	11.1	10.8	10.6	10.4	10.2	10.1	10.0	.995
	.999	35.5	27.0	23.7	21.9	20.8	20.0	19.5	19.0	18.7	18.4	18.2	18.0	.999
	.9995	46.1	34.8	30.4	28.1	26.6	25.6	24.9	24.3	23.9	23.5	23.2	23.0	.9995

TABLE A.4 Percentiles of the F distributions (cont.)

p \ m	15	20	24	30	40	50	60	100	120	200	500	∞	p	n
.0005	.105	.125	.135	.147	.159	.166	.172	.183	.186	.191	.196	.200	.0005	**4**
.001	.121	.141	.152	.163	.176	.183	.188	.200	.202	.208	.213	.217	.001	
.005	.172	.193	.204	.216	.229	.237	.242	.253	.255	.260	.266	.269	.005	
.01	.204	.226	.237	.249	.261	.269	.274	.285	.287	.293	.298	.301	.01	
.025	.263	.284	.296	.308	.320	.327	.332	.342	.346	.351	.356	.359	.025	
.05	.327	.349	.360	.372	.384	.391	.396	.407	.409	.413	.418	.422	.05	
.10	.424	.445	.456	.467	.478	.485	.490	.500	.502	.508	.510	.514	.10	
.25	.664	.683	.692	.702	.712	.718	.722	.731	.733	.737	.740	.743	.25	
.50	1.14	1.15	1.16	1.16	1.17	1.18	1.18	1.18	1.18	1.19	1.19	1.19	.50	
.75	2.08	2.08	2.08	2.08	2.08	2.08	2.08	2.08	2.08	2.08	2.08	2.08	.75	
.90	3.87	3.84	3.83	3.83	3.82	3.80	3.80	3.79	3.78	3.78	3.77	3.76	.90	
.95	5.86	5.80	5.77	5.75	5.72	5.70	5.69	5.66	5.66	5.65	5.64	5.63	.95	
.975	8.66	8.56	8.51	8.46	8.41	8.38	8.36	8.32	8.31	8.29	8.27	8.26	.975	
.99	14.2	14.0	13.9	13.8	13.7	13.7	13.7	13.6	13.6	13.5	13.5	13.5	.99	
.995	20.4	20.2	20.0	19.9	19.8	19.7	19.6	19.5	19.5	19.4	19.4	19.3	.995	
.999	46.8	46.1	45.8	45.4	45.1	44.9	44.7	44.5	44.4	44.3	44.1	44.0	.999	
.9995	66.5	65.5	65.1	64.6	64.1	63.8	63.6	63.2	63.1	62.9	62.7	62.6	.9995	
.0005	.115	.137	.150	.163	.177	.186	.192	.205	.209	.216	.222	.226	.0005	**5**
.001	.132	.155	.167	.181	.195	.204	.210	.223	.227	.233	.239	.244	.001	
.005	.186	.210	.223	.237	.251	.260	.266	.279	.282	.288	.294	.299	.005	
.01	.219	.244	.257	.270	.285	.293	.299	.312	.315	.322	.328	.331	.01	
.025	.280	.304	.317	.330	.344	.353	.359	.370	.374	.380	.386	.390	.025	
.05	.345	.369	.382	.395	.408	.417	.422	.432	.437	.442	.448	.452	.05	
.10	.440	.463	.476	.488	.501	.508	.514	.524	.527	.532	.538	.541	.10	
.25	.669	.690	.700	.711	.722	.728	.732	.741	.743	.748	.752	.755	.25	
.50	1.10	1.11	1.12	1.12	1.12	1.13	1.13	1.14	1.14	1.14	1.15	1.15	.50	
.75	1.89	1.88	1.88	1.88	1.88	1.88	1.87	1.87	1.87	1.87	1.87	1.87	.75	
.90	3.24	3.21	3.19	3.17	3.16	3.15	3.14	3.13	3.12	3.12	3.11	3.10	.90	
.95	4.62	4.56	4.53	4.50	4.46	4.44	4.43	4.41	4.40	4.39	4.37	4.36	.95	
.975	6.43	6.33	6.28	6.23	6.18	6.14	6.12	6.08	6.07	6.05	6.03	6.02	.975	
.99	9.72	9.55	9.47	9.38	9.29	9.24	9.20	9.13	9.11	9.08	9.04	9.02	.99	
.995	13.1	12.9	12.8	12.7	12.5	12.5	12.4	12.3	12.3	12.2	12.2	12.1	.995	
.999	25.9	25.4	25.1	24.9	24.6	24.4	24.3	24.1	24.1	23.9	23.8	23.8	.999	
.9995	34.6	33.9	33.5	33.1	32.7	32.5	32.3	32.1	32.0	31.8	31.7	31.6	.9995	
.0005	.123	.148	.162	.177	.193	.203	.210	.225	.229	.236	.244	.249	.0005	**6**
.001	.141	.166	.180	.195	.211	.222	.229	.243	.247	.255	.262	.267	.001	
.005	.197	.224	.238	.253	.269	.279	.286	.301	.304	.312	.318	.324	.005	
.01	.232	.258	.273	.288	.304	.313	.321	.334	.338	.346	.352	.357	.01	
.025	.293	.320	.334	.349	.364	.375	.381	.394	.398	.405	.412	.415	.025	
.05	.358	.385	.399	.413	.428	.437	.444	.457	.460	.467	.472	.476	.05	
.10	.453	.478	.491	.505	.519	.526	.533	.546	.548	.556	.559	.564	.10	
.25	.675	.696	.707	.718	.729	.736	.741	.751	.753	.758	.762	.765	.25	
.50	1.07	1.08	1.09	1.10	1.10	1.11	1.11	1.11	1.12	1.12	1.12	1.12	.50	
.75	1.76	1.76	1.75	1.75	1.75	1.75	1.74	1.74	1.74	1.74	1.74	1.74	.75	
.90	2.87	2.84	2.82	2.80	2.78	2.77	2.76	2.75	2.74	2.73	2.73	2.72	.90	
.95	3.94	3.87	3.84	3.81	3.77	3.75	3.74	3.71	3.70	3.69	3.68	3.67	.95	
.975	5.27	5.17	5.12	5.07	5.01	4.98	4.96	4.92	4.90	4.88	4.86	4.85	.975	
.99	7.56	7.40	7.31	7.23	7.14	7.09	7.06	6.99	6.97	6.93	6.90	6.88	.99	
.995	9.81	9.59	9.47	9.36	9.24	9.17	9.12	9.03	9.00	8.95	8.91	8.88	.995	
.999	17.6	17.1	16.9	16.7	16.4	16.3	16.2	16.0	16.0	15.9	15.8	15.7	.999	
.9995	22.4	21.9	21.7	21.4	21.1	20.9	20.7	20.5	20.4	20.3	20.2	20.1	.9995	

n	p	1	2	3	4	5	6	7	8	9	10	11	12	p
7	.0005	$.0^642$	$.0^350$	$.0^248$	.014	.027	.040	.053	.066	.078	.088	.099	.108	.0005
	.001	$.0^517$	$.0^210$	$.0^276$	.020	.035	.051	.067	.081	.093	.105	.115	.125	.001
	.005	$.0^442$	$.0^250$	.023	.046	.070	.093	.113	.130	.145	.159	.171	.181	.005
	.01	$.0^317$	.010	.036	.067	.096	.121	.143	.162	.178	.192	.205	.216	.01
	.025	$.0^210$	.025	.068	.110	.146	.176	.200	.221	.238	.253	.266	.277	.025
	.05	$.0^242$	.052	.113	.164	.205	.238	.264	.286	.304	.319	.332	.343	.05
	.10	.017	.107	.190	.251	.297	.332	.359	.381	.399	.414	.427	.438	.10
	.25	.110	.300	.412	.481	.528	.562	.588	.608	.624	.637	.649	.658	.25
	.50	.506	.767	.871	.926	.960	.983	1.00	1.01	1.02	1.03	1.04	1.04	.50
	.75	1.57	1.70	1.72	1.72	1.71	1.71	1.71	1.70	1.69	1.69	1.69	1.68	.75
	.90	3.59	3.26	3.07	2.96	2.88	2.83	2.78	2.75	2.72	2.70	2.68	2.67	.90
	.95	5.59	4.74	4.35	4.12	3.97	3.87	3.79	3.73	3.68	3.64	3.60	3.57	.95
	.975	8.07	6.54	5.89	5.52	5.29	5.12	4.99	4.90	4.82	4.76	4.71	4.67	.975
	.99	12.2	9.55	8.45	7.85	7.46	7.19	6.99	6.84	6.72	6.62	6.54	6.47	.99
	.995	16.2	12.4	10.9	10.0	9.52	9.16	8.89	8.68	8.51	8.38	8.27	8.18	.995
	.999	29.2	21.7	18.8	17.2	16.2	15.5	15.0	14.6	14.3	14.1	13.9	13.7	.999
	.9995	37.0	27.2	23.5	21.4	20.2	19.3	18.7	18.2	17.8	17.5	17.2	17.0	.9995
8	.0005	$.0^342$	$.0^350$	$.0^248$	.014	.027	.041	.055	.068	.081	.092	.102	.112	.0005
	.001	$.0^517$	$.0^210$	$.0^276$	.020	.036	.053	.068	.083	.096	.109	.120	.130	.001
	.005	$.0^442$	$.0^250$	.027	.047	.072	.095	.115	.133	.149	.164	.176	.187	.005
	.01	$.0^317$	.010	.036	.068	.097	.123	.146	.166	.183	.198	.211	.222	.01
	.025	$.0^210$	.025	.069	.111	.148	.179	.204	.226	.244	.259	.273	.285	.025
	.05	$.0^242$	.052	.113	.166	.208	.241	.268	.291	.310	.326	.339	.351	.05
	.10	.017	.107	.190	.253	.299	.335	.363	.386	.405	.421	.435	.445	.10
	.25	.109	.298	.411	.481	.529	.563	.589	.610	.627	.640	.654	.661	.25
	.50	.499	.757	.860	.915	.948	.971	.988	1.00	1.01	1.02	1.02	1.03	.50
	.75	1.54	1.66	1.67	1.66	1.66	1.65	1.64	1.64	1.64	1.63	1.63	1.62	.75
	.90	3.46	3.11	2.92	2.81	2.73	2.67	2.62	2.59	2.56	2.54	2.52	2.50	.90
	.95	5.32	4.46	4.07	3.84	3.69	3.58	3.50	3.44	3.39	3.35	3.31	3.28	.95
	.975	7.57	6.06	5.42	5.05	4.82	4.65	4.53	4.43	4.36	4.30	4.24	4.20	.975
	.99	11.3	8.65	7.59	7.01	6.63	6.37	6.18	6.03	5.91	5.81	5.73	5.67	.99
	.995	14.7	11.0	9.60	8.81	8.30	7.95	7.69	7.50	7.34	7.21	7.10	7.01	.995
	.999	25.4	18.5	15.8	14.4	13.5	12.9	12.4	12.0	11.8	11.5	11.4	11.2	.999
	.9995	31.6	22.8	19.4	17.6	16.4	15.7	15.1	14.6	14.3	14.0	13.8	13.6	.9995
9	.0005	$.0^641$	$.0^350$	$.0^248$	.015	.027	.042	.056	.070	.083	.094	.105	.115	.0005
	.001	$.0^517$	$.0^210$	$.0^277$	.021	.037	.054	.070	.085	.099	.112	.123	.134	.001
	.005	$.0^442$	$.0^250$	.023	.047	.073	.096	.117	.136	.153	.168	.181	.192	.005
	.01	$.0^317$	.010	.037	.068	.098	.125	.149	.169	.187	.202	.216	.228	.01
	.025	$.0^210$	.025	.069	.112	.150	.181	.207	.230	.248	.265	.279	.291	.025
	.05	$.0^240$	.052	.113	.167	.210	.244	.272	.296	.315	.331	.345	.358	.05
	.10	.017	.107	.191	.254	.302	.338	.367	.390	.410	.426	.441	.452	.10
	.25	.108	.297	.410	.480	.529	.564	.591	.612	.629	.643	.654	.664	.25
	.50	.494	.749	.852	.906	.939	.962	.978	.990	1.00	1.01	1.01	1.02	.50
	.75	1.51	1.62	1.63	1.63	1.62	1.61	1.60	1.60	1.59	1.59	1.58	1.58	.75
	.90	3.36	3.01	2.81	2.69	2.61	2.55	2.51	2.47	2.44	2.42	2.40	2.38	.90
	.95	5.12	4.26	3.86	3.63	3.48	3.37	3.29	3.23	3.18	3.14	3.10	3.07	.95
	.975	7.21	5.71	5.08	4.72	4.48	4.32	4.20	4.10	4.03	3.96	3.91	3.87	.975
	.99	10.6	8.02	6.99	6.42	6.06	5.80	5.61	5.47	5.35	5.26	5.18	5.11	.99
	.995	13.6	10.1	8.72	7.96	7.47	7.13	6.88	6.69	6.54	6.42	6.31	6.23	.995
	.999	22.9	16.4	13.9	12.6	11.7	11.1	10.7	10.4	10.1	9.89	9.71	9.57	.999
	.9995	28.0	19.9	16.8	15.1	14.1	13.3	12.8	12.4	12.1	11.8	11.6	11.4	.9995

p	15	20	24	30	40	50	60	100	120	200	500	∞	p	n
.0005	.130	.157	.172	.188	.206	.217	.225	.242	.246	.255	.263	.268	.0005	7
.001	.148	.176	.191	.208	.225	.237	.245	.261	.266	.274	.282	.288	.001	
.005	.206	.235	.251	.267	.285	.296	.304	.319	.324	.332	.340	.345	.005	
.01	.241	.270	.286	.303	.320	.331	.339	.355	.358	.366	.373	.379	.01	
.025	.304	.333	.348	.364	.381	.392	.399	.413	.418	.426	.433	.437	.025	
.05	.369	.398	.413	.428	.445	.455	.461	.476	.479	.485	.493	.498	.05	
.10	.463	.491	.504	.519	.534	.543	.550	.562	.566	.571	.578	.582	.10	
.25	.679	.702	.713	.725	.737	.745	.749	.760	.762	.767	.772	.775	.25	
.50	1.05	1.07	1.07	1.08	1.08	1.09	1.09	1.10	1.10	1.10	1.10	1.10	.50	
.75	1.68	1.67	1.67	1.66	1.66	1.66	1.65	1.65	1.65	1.65	1.65	1.65	.75	
.90	2.63	2.59	2.58	2.56	2.54	2.52	2.51	2.50	2.49	2.48	2.48	2.47	.90	
.95	3.51	3.44	3.41	3.38	3.34	3.32	3.30	3.27	3.27	3.25	3.24	3.23	.95	
.975	4.57	4.47	4.42	4.36	4.31	4.28	4.25	4.21	4.20	4.18	4.16	4.14	.975	
.99	6.31	6.16	6.07	5.99	5.91	5.86	5.82	5.75	5.74	5.70	5.67	5.65	.99	
.995	7.97	7.75	7.65	7.53	7.42	7.35	7.31	7.22	7.19	7.15	7.10	7.08	.995	
.999	13.3	12.9	12.7	12.5	12.3	12.2	12.1	11.9	11.9	11.8	11.7	11.7	.999	
.9995	16.5	16.0	15.7	15.5	15.2	15.1	15.0	14.7	14.7	14.6	14.5	14.4	.9995	
.0005	.136	.164	.181	.198	.218	.230	.239	.257	.262	.271	.281	.287	.0005	8
.001	.155	.184	.200	.218	.238	.250	.259	.277	.282	.292	.300	.306	.001	
.005	.214	.244	.261	.279	.299	.311	.319	.337	.341	.351	.358	.364	.005	
.01	.250	.281	.297	.315	.334	.346	.354	.372	.376	.385	.392	.398	.01	
.025	.313	.343	.360	.377	.395	.407	.415	.431	.435	.442	.450	.456	.025	
.05	.379	.409	.425	.441	.459	.469	.477	.493	.496	.505	.510	.516	.05	
.10	.472	.500	.515	.531	.547	.556	.563	.578	.581	.588	.595	.599	.10	
.25	.684	.707	.718	.730	.743	.751	.756	.767	.769	.775	.780	.783	.25	
.50	1.04	1.05	1.06	1.07	1.07	1.07	1.08	1.08	1.08	1.09	1.09	1.09	.50	
.75	1.62	1.61	1.60	1.60	1.59	1.59	1.59	1.58	1.58	1.58	1.58	1.58	.75	
.90	2.46	2.42	2.40	2.38	2.36	2.35	2.34	2.32	2.32	2.31	2.30	2.29	.90	
.95	3.22	3.15	3.12	3.08	3.04	3.02	3.01	2.97	2.97	2.95	2.94	2.93	.95	
.975	4.10	4.00	3.95	3.89	3.84	3.81	3.78	3.74	3.73	3.70	3.68	3.67	.975	
.99	5.52	5.36	5.28	5.20	5.12	5.07	5.03	4.96	4.95	4.91	4.88	4.86	.99	
.995	6.81	6.61	6.50	6.40	6.29	6.22	6.18	6.09	6.06	6.02	5.98	5.95	.995	
.999	10.8	10.5	10.3	10.1	9.92	9.80	9.73	9.57	9.54	9.46	9.39	9.34	.999	
.9995	13.1	12.7	12.5	12.2	12.0	11.8	11.8	11.6	11.5	11.4	11.4	11.3	.9995	
.0005	.141	.171	.188	.207	.228	.242	.251	.270	.276	.287	.297	.303	.0005	9
.001	.160	.191	.208	.228	.249	.262	.271	.291	.296	.307	.316	.323	.001	
.005	.220	.253	.271	.290	.310	.324	.332	.351	.356	.366	.376	.382	.005	
.01	.257	.289	.307	.326	.346	.358	.368	.386	.391	.400	.410	.415	.01	
.025	.320	.352	.370	.388	.408	.420	.428	.446	.450	.459	.467	.473	.025	
.05	.386	.418	.435	.452	.471	.483	.490	.508	.510	.518	.526	.532	.05	
.10	.479	.509	.525	.541	.558	.568	.575	.588	.594	.602	.610	.613	.10	
.25	.687	.711	.723	.736	.749	.757	.762	.773	.776	.782	.787	.791	.25	
.50	1.03	1.04	1.05	1.05	1.06	1.06	1.07	1.07	1.07	1.08	1.08	1.08	.50	
.75	1.57	1.56	1.56	1.55	1.55	1.55	1.54	1.54	1.53	1.53	1.53	1.53	.75	
.90	2.34	2.30	2.28	2.25	2.23	2.22	2.21	2.19	2.18	2.17	2.17	2.16	.90	
.95	3.01	2.94	2.90	2.86	2.83	2.80	2.79	2.76	2.75	2.73	2.72	2.71	.95	
.975	3.77	3.67	3.61	3.56	3.51	3.47	3.45	3.40	3.39	3.37	3.35	3.33	.975	
.99	4.96	4.81	4.73	4.65	4.57	4.52	4.48	4.42	4.40	4.36	4.33	4.31	.99	
.995	6.03	5.83	5.73	5.62	5.52	5.45	5.41	5.32	5.30	5.26	5.21	5.19	.995	
.999	9.24	8.90	8.72	8.55	8.37	8.26	8.19	8.04	8.00	7.93	7.86	7.81	.999	
.9995	11.0	10.6	10.4	10.2	9.94	9.80	9.71	9.53	9.49	9.40	9.32	9.26	.9995	

n	p \ m	1	2	3	4	5	6	7	8	9	10	11	12	p
10	.0005	$.0^6 41$	$.0^3 50$	$.0^2 49$	.015	.028	.043	.057	.071	.085	.097	.108	.119	.0005
	.001	$.0^5 17$	$.0^2 10$	$.0^2 77$	.021	.037	.054	.071	.087	.101	.114	.126	.137	.001
	.005	$.0^4 41$	$.0^2 50$	.023	.048	.073	.098	.119	.139	.156	.171	.185	.197	.005
	.01	$.0^3 17$	.010	.037	.069	.100	.127	.151	.172	.190	.206	.220	.233	.01
	.025	$.0^2 10$	.025	.069	.113	.151	.183	.210	.233	.252	.269	.283	.296	.025
	.05	$.0^2 41$	.052	.114	.168	.211	.246	.275	.299	.319	.336	.351	.363	.05
	.10	.017	.106	.191	.255	.303	.340	.370	.394	.414	.430	.444	.457	.10
	.25	.107	.296	.409	.480	.529	.565	.592	.613	.631	.645	.657	.667	.25
	.50	.490	.743	.845	.899	.932	.954	.971	.983	.992	1.00	1.01	1.01	.50
	.75	1.49	1.60	1.60	1.59	1.59	1.58	1.57	1.56	1.56	1.55	1.55	1.54	.75
	.90	3.28	2.92	2.73	2.61	2.52	2.46	2.41	2.38	2.35	2.32	2.30	2.28	.90
	.95	4.96	4.10	3.71	3.48	3.33	3.22	3.14	3.07	3.02	2.98	2.94	2.91	.95
	.975	6.94	5.46	4.83	4.47	4.24	4.07	3.95	3.85	3.78	3.72	3.66	3.62	.975
	.99	10.0	7.56	6.55	5.99	5.64	5.39	5.20	5.06	4.94	4.85	4.77	4.71	.99
	.995	12.8	9.43	8.08	7.34	6.87	6.54	6.30	6.12	5.97	5.85	5.75	5.66	.995
	.999	21.0	14.9	12.6	11.3	10.5	9.92	9.52	9.20	8.96	8.75	8.58	8.44	.999
	.9995	25.5	17.9	15.0	13.4	12.4	11.8	11.3	10.9	10.6	10.3	10.1	9.93	.9995
11	.0005	$.0^6 41$	$.0^3 50$	$.0^2 49$	.015	.028	.043	.058	.072	.086	.099	.111	.121	.0005
	.001	$.0^5 16$	$.0^2 10$	$.0^2 78$	.021	.038	.055	.072	.088	.103	.116	.129	.140	.001
	.005	$.0^4 40$	$.0^2 50$	.023	.048	.074	.099	.121	.141	.158	.174	.188	.200	.005
	.01	$.0^3 16$	.010	.037	.069	.100	.128	.153	.175	.193	.210	.224	.237	.01
	.025	$.0^2 10$	.025	.069	.114	.152	.185	.212	.236	.256	.273	.288	.301	.025
	.05	$.0^2 41$	.052	.114	.168	.212	.248	.278	.302	.323	.340	.355	.368	.05
	.10	.017	.106	.192	.256	.305	.342	.373	.397	.417	.435	.448	.461	.10
	.25	.107	.295	.408	.481	.529	.565	.592	.614	.633	.645	.658	.667	.25
	.50	.486	.739	.840	.893	.926	.948	.964	.977	.986	.994	1.00	1.01	.50
	.75	1.47	1.58	1.58	1.57	1.56	1.55	1.54	1.53	1.53	1.52	1.52	1.51	.75
	.90	3.23	2.86	2.66	2.54	2.45	2.39	2.34	2.30	2.27	2.25	2.23	2.21	.90
	.95	4.84	3.98	3.59	3.36	3.20	3.09	3.01	2.95	2.90	2.85	2.82	2.79	.95
	.975	6.72	5.26	4.63	4.28	4.04	3.88	3.76	3.66	3.59	3.53	3.47	3.43	.975
	.99	9.65	7.21	6.22	5.67	5.32	5.07	4.89	4.74	4.63	4.54	4.46	4.40	.99
	.995	12.2	8.91	7.60	6.88	6.42	6.10	5.86	5.68	5.54	5.42	5.32	5.24	.995
	.999	19.7	13.8	11.6	10.3	9.58	9.05	8.66	8.35	8.12	7.92	7.76	7.62	.999
	.9995	23.6	16.4	13.6	12.2	11.2	10.6	10.1	9.76	9.48	9.24	9.04	8.88	.9995
12	.0005	$.0^6 41$	$.0^3 50$	$.0^2 49$	.015	.028	.044	.058	.073	.087	.101	.113	.124	.0005
	.001	$.0^5 16$	$.0^2 10$	$.0^2 78$	.021	.038	.056	.073	.089	.104	.118	.131	.143	.001
	.005	$.0^4 39$	$.0^2 50$	.023	.048	.075	.100	.122	.143	.161	.177	.191	.204	.005
	.01	$.0^3 16$	.010	.037	.070	.101	.130	.155	.176	.196	.212	.227	.241	.01
	.025	$.0^2 10$	.025	.070	.114	.153	.186	.214	.238	.259	.276	.292	.305	.025
	.05	$.0^2 41$	.052	.114	.169	.214	.250	.280	.305	.325	.343	.358	.372	.05
	.10	.016	.106	.192	.257	.306	.344	.375	.400	.420	.438	.452	.466	.10
	.25	.106	.295	.408	.480	.530	.566	.594	.616	.633	.649	.662	.671	.25
	.50	.484	.735	.835	.888	.921	.943	.959	.972	.981	.989	.995	1.00	.50
	.75	1.46	1.56	1.56	1.55	1.54	1.53	1.52	1.51	1.51	1.50	1.50	1.49	.75
	.90	3.18	2.81	2.61	2.48	2.39	2.33	2.28	2.24	2.21	2.19	2.17	2.15	.90
	.95	4.75	3.89	3.49	3.26	3.11	3.00	2.91	2.85	2.80	2.75	2.72	2.69	.95
	.975	6.55	5.10	4.47	4.12	3.89	3.73	3.61	3.51	3.44	3.37	3.32	3.28	.975
	.99	9.33	6.93	5.95	5.41	5.06	4.82	4.64	4.50	4.39	4.30	4.22	4.16	.99
	.995	11.8	8.51	7.23	6.52	6.07	5.76	5.52	5.35	5.20	5.09	4.99	4.91	.995
	.999	18.6	13.0	10.8	9.63	8.89	8.38	8.00	7.71	7.48	7.29	7.14	7.01	.999
	.9995	22.2	15.3	12.7	11.2	10.4	9.74	9.28	8.94	8.66	8.43	8.24	8.08	.9995

p \ m	15	20	24	30	40	50	60	100	120	200	500	∞	p	n
.0005	.145	.177	.195	.215	.238	.251	.262	.282	.288	.299	.311	.319	.0005	10
.001	.164	.197	.216	.236	.258	.272	.282	.303	.309	.321	.331	.338	.001	
.005	.226	.260	.279	.299	.321	.334	.344	.365	.370	.380	.391	.397	.005	
.01	.263	.297	.316	.336	.357	.370	.380	.400	.405	.415	.424	.431	.01	
.025	.327	.360	.379	.398	.419	.431	.441	.459	.464	.474	.483	.488	.025	
.05	.393	.426	.444	.462	.481	.493	.502	.518	.523	.532	.541	.546	.05	
.10	.486	.516	.532	.549	.567	.578	.586	.602	.605	.614	.621	.625	.10	
.25	.691	.714	.727	.740	.754	.762	.767	.779	.782	.788	.793	.797	.25	
.50	1.02	1.03	1.04	1.05	1.05	1.06	1.06	1.06	1.06	1.07	1.07	1.07	.50	
.75	1.53	1.52	1.52	1.51	1.51	1.50	1.50	1.49	1.49	1.49	1.48	1.48	.75	
.90	2.24	2.20	2.18	2.16	2.13	2.12	2.11	2.09	2.08	2.07	2.06	2.06	.90	
.95	2.85	2.77	2.74	2.70	2.66	2.64	2.62	2.59	2.58	2.56	2.55	2.54	.95	
.975	3.52	3.42	3.37	3.31	3.26	3.22	3.20	3.15	3.14	3.12	3.09	3.08	.975	
.99	4.56	4.41	4.33	4.25	4.17	4.12	4.08	4.01	4.00	3.96	3.93	3.91	.99	
.995	5.47	5.27	5.17	5.07	4.97	4.90	4.86	4.77	4.75	4.71	4.67	4.64	.995	
.999	8.13	7.80	7.64	7.47	7.30	7.19	7.12	6.98	6.94	6.87	6.81	6.76	.999	
.9995	9.56	9.16	8.96	8.75	8.54	8.42	8.33	8.16	8.12	8.04	7.96	7.90	.9995	
.0005	.148	.182	.201	.222	.246	.261	.271	.293	.299	.312	.324	.331	.0005	11
.001	.168	.202	.222	.243	.266	.282	.292	.313	.320	.332	.343	.353	.001	
.005	.231	.266	.286	.308	.330	.345	.355	.376	.382	.394	.403	.412	.005	
.01	.268	.304	.324	.344	.366	.380	.391	.412	.417	.427	.439	.444	.01	
.025	.332	.368	.386	.407	.429	.442	.450	.472	.476	.485	.495	.503	.025	
.05	.398	.433	.452	.469	.490	.503	.513	.529	.535	.543	.552	.559	.05	
.10	.490	.524	.541	.559	.578	.588	.595	.614	.617	.625	.633	.637	.10	
.25	.694	.719	.730	.744	.758	.767	.773	.780	.788	.794	.799	.803	.25	
.50	1.02	1.03	1.03	1.04	1.05	1.05	1.05	1.06	1.06	1.06	1.06	1.06	.50	
.75	1.50	1.49	1.49	1.48	1.47	1.47	1.47	1.46	1.46	1.46	1.45	1.45	.75	
.90	2.17	2.12	2.10	2.08	2.05	2.04	2.03	2.00	2.00	1.99	1.98	1.97	.90	
.95	2.72	2.65	2.61	2.57	2.53	2.51	2.49	2.46	2.45	2.43	2.42	2.40	.95	
.975	3.33	3.23	3.17	3.12	3.06	3.03	3.00	2.96	2.94	2.92	2.90	2.88	.975	
.99	4.25	4.10	4.02	3.94	3.86	3.81	3.78	3.71	3.69	3.66	3.62	3.60	.99	
.995	5.05	4.86	4.76	4.65	4.55	4.49	4.45	4.36	4.34	4.29	4.25	4.23	.995	
.999	7.32	7.01	6.85	6.68	6.52	6.41	6.35	6.21	6.17	6.10	6.04	6.00	.999	
.9995	8.52	8.14	7.94	7.75	7.55	7.43	7.35	7.18	7.14	7.06	6.98	6.93	.9995	
.0005	.152	.186	.206	.228	.253	.269	.280	.305	.311	.323	.337	.345	.0005	12
.001	.172	.207	.228	.250	.275	.291	.302	.326	.332	.344	.357	.365	.001	
.005	.235	.272	.292	.315	.339	.355	.365	.388	.393	.405	.417	.424	.005	
.01	.273	.310	.330	.352	.375	.391	.401	.422	.428	.441	.450	.458	.01	
.025	.337	.374	.394	.416	.437	.450	.461	.481	.487	.498	.508	.514	.025	
.05	.404	.439	.458	.478	.499	.513	.522	.541	.545	.556	.565	.571	.05	
.10	.496	.528	.546	.564	.583	.595	.604	.621	.625	.633	.641	.647	.10	
.25	.695	.721	.734	.748	.762	.771	.777	.789	.792	.799	.804	.808	.25	
.50	1.01	1.02	1.03	1.03	1.04	1.04	1.05	1.05	1.05	1.05	1.06	1.06	.50	
.75	1.48	1.47	1.46	1.45	1.45	1.44	1.44	1.43	1.43	1.43	1.42	1.42	.75	
.90	2.11	2.06	2.04	2.01	1.99	1.97	1.96	1.94	1.93	1.92	1.91	1.90	.90	
.95	2.62	2.54	2.51	2.47	2.43	2.40	2.38	2.35	2.34	2.32	2.31	2.30	.95	
.975	3.18	3.07	3.02	2.96	2.91	2.87	2.85	2.80	2.79	2.76	2.74	2.72	.975	
.99	4.01	3.86	3.78	3.70	3.62	3.57	3.54	3.47	3.45	3.41	3.38	3.36	.99	
.995	4.72	4.53	4.43	4.33	4.23	4.17	4.12	4.04	4.01	3.97	3.93	3.90	.995	
.999	6.71	6.40	6.25	6.09	5.93	5.83	5.76	5.63	5.59	5.52	5.46	5.42	.999	
.9995	7.74	7.37	7.18	7.00	6.80	6.68	6.61	6.45	6.41	6.33	6.25	6.20	.9995	

TABLE A.4 Percentiles of the F distributions (cont.)

n	p	1	2	3	4	5	6	7	8	9	10	11	12	p
15	.0005	$.0^641$	$.0^350$	$.0^249$	.015	.029	.045	.061	.076	.091	.105	.117	.129	.0005
	.001	$.0^516$	$.0^210$	$.0^279$	.021	.039	.057	.075	.092	.108	.123	.137	.149	.001
	.005	$.0^439$	$.0^250$	.023	.049	.076	.102	.125	.147	.166	.183	.198	.212	.005
	.01	$.0^316$	.010	.037	.070	.103	.132	.158	.181	.202	.219	.235	.249	.01
	.025	$.0^210$	.025	.070	.116	.156	.190	.219	.244	.265	.284	.300	.315	.025
	.05	$.0^241$	.051	.115	.170	.216	.254	.285	.311	.333	.351	.368	.382	.05
	.10	.016	.106	.192	.258	.309	.348	.380	.406	.427	.446	.461	.475	.10
	.25	.105	.293	.407	.480	.531	.568	.596	.618	.637	.652	.667	.676	.25
	.50	.478	.726	.826	.878	.911	.933	.948	.960	.970	.977	.984	.989	.50
	.75	1.43	1.52	1.52	1.51	1.49	1.48	1.47	1.46	1.46	1.45	1.44	1.44	.75
	.90	3.07	2.70	2.49	2.36	2.27	2.21	2.16	2.12	2.09	2.06	2.04	2.02	.90
	.95	4.54	3.68	3.29	3.06	2.90	2.79	2.71	2.64	2.59	2.54	2.51	2.48	.95
	.975	6.20	4.76	4.15	3.80	3.58	3.41	3.29	3.20	3.12	3.06	3.01	2.96	.975
	.99	8.68	6.36	5.42	4.89	4.56	4.32	4.14	4.00	3.89	3.80	3.73	3.67	.99
	.995	10.8	7.70	6.48	5.80	5.37	5.07	4.85	4.67	4.54	4.42	4.33	4.25	.995
	.999	16.6	11.3	9.34	8.25	7.57	7.09	6.74	6.47	6.26	6.08	5.93	5.81	.999
	.9995	19.5	13.2	10.8	9.48	8.66	8.10	7.68	7.36	7.11	6.91	6.75	6.60	.9995
20	.0005	$.0^640$	$.0^350$	$.0^250$	.015	.029	.046	.063	.079	.094	.109	.123	.136	.0005
	.001	$.0^516$	$.0^210$	$.0^279$	.022	.039	.058	.077	.095	.112	.128	.143	.156	.001
	.005	$.0^439$	$.0^250$	.023	.050	.077	.104	.129	.151	.171	.190	.206	.221	.005
	.01	$.0^316$	.010	.037	.071	.105	.135	.162	.187	.208	.227	.244	.259	.01
	.025	$.0^210$	.025	.071	.117	.158	.193	.224	.250	.273	.292	.310	.325	.025
	.05	$.0^240$	.051	.115	.172	.219	.258	.290	.318	.340	.360	.377	.393	.05
	.10	.016	.106	.193	.260	.312	.353	.385	.412	.435	.454	.472	.485	.10
	.25	.104	.292	.407	.480	.531	.569	.598	.622	.641	.656	.671	.681	.25
	.50	.472	.718	.816	.868	.900	.922	.938	.950	.959	.966	.972	.977	.50
	.75	1.40	1.49	1.48	1.47	1.45	1.44	1.43	1.42	1.41	1.40	1.39	1.39	.75
	.90	2.97	2.59	2.38	2.25	2.16	2.09	2.04	2.00	1.96	1.94	1.91	1.89	.90
	.95	4.35	3.49	3.10	2.87	2.71	2.60	2.51	2.45	2.39	2.35	2.31	2.28	.95
	.975	5.87	4.46	3.86	3.51	3.29	3.13	3.01	2.91	2.84	2.77	2.72	2.68	.975
	.99	8.10	5.85	4.94	4.43	4.10	3.87	3.70	3.56	3.46	3.37	3.29	3.23	.99
	.995	9.94	6.99	5.82	5.17	4.76	4.47	4.26	4.09	3.96	3.85	3.76	3.68	.995
	.999	14.8	9.95	8.10	7.10	6.46	6.02	5.69	5.44	5.24	5.08	4.94	4.82	.999
	.9995	17.2	11.4	9.20	8.02	7.28	6.76	6.38	6.08	5.85	5.66	5.51	5.38	.9995
24	.0005	$.0^640$	$.0^350$	$.0^250$	.015	.030	.046	.064	.080	.096	.112	.126	.139	.0005
	.001	$.0^516$	$.0^210$	$.0^279$	.022	.040	.059	.079	.097	.115	.131	.146	.160	.001
	.005	$.0^440$	$.0^250$	.023	.050	.078	.106	.131	.154	.175	.193	.210	.226	.005
	.01	$.0^316$	.010	.038	.072	.106	.137	.165	.189	.211	.231	.249	.264	.01
	.025	$.0^210$	.025	.071	.117	.159	.195	.227	.253	.277	.297	.315	.331	.025
	.05	$.0^240$	.051	.116	.173	.221	.260	.293	.321	.345	.365	.383	.399	.05
	.10	.016	.106	.193	.261	.313	.355	.388	.416	.439	.459	.476	.491	.10
	.25	.104	.291	.406	.480	.532	.570	.600	.623	.643	.659	.671	.684	.25
	.50	.469	.714	.812	.863	.895	.917	.932	.944	.953	.961	.967	.972	.50
	.75	1.39	1.47	1.46	1.44	1.43	1.41	1.40	1.39	1.38	1.38	1.37	1.36	.75
	.90	2.93	2.54	2.33	2.19	2.10	2.04	1.98	1.94	1.91	1.88	1.85	1.83	.90
	.95	4.26	3.40	3.01	2.78	2.62	2.51	2.42	2.36	2.30	2.25	2.21	2.18	.95
	.975	5.72	4.32	3.72	3.38	3.15	2.99	2.87	2.78	2.70	2.64	2.59	2.54	.975
	.99	7.82	5.61	4.72	4.22	3.90	3.67	3.50	3.36	3.26	3.17	3.09	3.03	.99
	.995	9.55	6.66	5.52	4.89	4.49	4.20	3.99	3.83	3.69	3.59	3.50	3.42	.995
	.999	14.0	9.34	7.55	6.59	5.98	5.55	5.23	4.99	4.80	4.64	4.50	4.39	.999
	.9995	16.2	10.6	8.52	7.39	6.68	6.18	5.82	5.54	5.31	5.13	4.98	4.85	.9995

p \ m	15	20	24	30	40	50	60	100	120	200	500	∞	p	n
.0005	.159	.197	.220	.244	.272	.290	.303	.330	.339	.353	.368	.377	.0005	15
.001	.181	.219	.242	.266	.294	.313	.325	.352	.360	.375	.388	.398	.001	
.005	.246	.286	.308	.333	.360	.377	.389	.415	.422	.435	.448	.457	.005	
.01	.284	.324	.346	.370	.397	.413	.425	.450	.456	.469	.483	.490	.01	
.025	.349	.389	.410	.433	.458	.474	.485	.508	.514	.526	.538	.546	.025	
.05	.416	.454	.474	.496	.519	.535	.545	.565	.571	.581	.592	.600	.05	
.10	.507	.542	.561	.581	.602	.614	.624	.641	.647	.658	.667	.672	.10	
.25	.701	.728	.742	.757	.772	.782	.788	.802	.805	.812	.818	.822	.25	
.50	1.00	1.01	1.02	1.02	1.03	1.03	1.03	1.04	1.04	1.04	1.04	1.05	.50	
.75	1.43	1.41	1.41	1.40	1.39	1.39	1.38	1.38	1.37	1.37	1.36	1.36	.75	
.90	1.97	1.92	1.90	1.87	1.85	1.83	1.82	1.79	1.79	1.77	1.76	1.76	.90	
.95	2.40	2.33	2.39	2.25	2.20	2.18	2.16	2.12	2.11	2.10	2.08	2.07	.95	
.975	2.86	2.76	2.70	2.64	2.59	2.55	2.52	2.47	2.46	2.44	2.41	2.40	.975	
.99	3.52	3.37	3.29	3.21	3.13	3.08	3.05	2.98	2.96	2.92	2.89	2.87	.99	
.995	4.07	3.88	3.79	3.69	3.59	3.52	3.48	3.39	3.37	3.33	3.29	3.26	.995	
.999	5.54	5.25	5.10	4.95	4.80	4.70	4.64	4.51	4.47	4.41	4.35	4.31	.999	
.9995	6.27	5.93	5.75	5.58	5.40	5.29	5.21	5.06	5.02	4.94	4.87	4.83	.9995	
.0005	.169	.211	.235	.263	.295	.316	.331	.364	.375	.391	.408	.422	.0005	20
.001	.191	.233	.258	.286	.318	.339	.354	.386	.395	.413	.429	.441	.001	
.005	.258	.301	.327	.354	.385	.405	.419	.448	.457	.474	.490	.500	.005	
.01	.297	.340	.365	.392	.422	.441	.455	.483	.491	.508	.521	.532	.01	
.025	.363	.406	.430	.456	.484	.503	.514	.541	.548	.562	.575	.585	.025	
.05	.430	.471	.493	.518	.544	.562	.572	.595	.603	.617	.629	.637	.05	
.10	.520	.557	.578	.600	.623	.637	.648	.671	.675	.685	.694	.704	.10	
.25	.708	.736	.751	.767	.784	.794	.801	.816	.820	.827	.835	.840	.25	
.50	.989	1.00	1.01	1.01	1.02	1.02	1.02	1.03	1.03	1.03	1.03	1.03	.50	
.75	1.37	1.36	1.35	1.34	1.33	1.33	1.32	1.31	1.31	1.30	1.30	1.29	.75	
.90	1.84	1.79	1.77	1.74	1.71	1.69	1.68	1.65	1.64	1.63	1.62	1.61	.90	
.95	2.20	2.12	2.08	2.04	1.99	1.97	1.95	1.91	1.90	1.88	1.86	1.84	.95	
.975	2.57	2.46	2.41	2.35	2.29	2.25	2.22	2.17	2.16	2.13	2.10	2.09	.975	
.99	3.09	2.94	2.86	2.78	2.69	2.64	2.61	2.54	2.52	2.48	2.44	2.42	.99	
.995	3.50	3.32	3.22	3.12	3.02	2.96	2.92	2.83	2.81	2.76	2.72	2.69	.995	
.999	4.56	4.29	4.15	4.01	3.86	3.77	3.70	3.58	3.54	3.48	3.42	3.38	.999	
.9995	5.07	4.75	4.58	4.42	4.24	4.15	4.07	3.93	3.90	3.82	3.75	3.70	.9995	24
.0005	.174	.218	.244	.274	.309	.331	.349	.384	.395	.416	.434	.449	.0005	
.001	.196	.241	.268	.298	.332	.354	.371	.405	.417	.437	.455	.469	.001	
.005	.264	.310	.337	.367	.400	.422	.437	.469	.479	.498	.515	.527	.005	
.01	.304	.350	.376	.405	.437	.459	.473	.505	.513	.529	.546	.558	.01	
.025	.370	.415	.441	.468	.498	.518	.531	.562	.568	.585	.599	.610	.025	
.05	.437	.480	.504	.530	.558	.575	.588	.613	.622	.637	.649	.659	.05	
.10	.527	.566	.588	.611	.635	.651	.662	.685	.691	.704	.715	.723	.10	
.25	.712	.741	.757	.773	.791	.802	.809	.825	.829	.837	.844	.850	.25	
.50	.983	.994	1.00	1.01	1.01	1.02	1.02	1.02	1.02	1.03	1.03	1.03	.50	
.75	1.35	1.33	1.32	1.31	1.30	1.29	1.29	1.28	1.28	1.27	1.27	1.26	.75	
.90	1.78	1.73	1.70	1.67	1.64	1.62	1.61	1.58	1.57	1.56	1.54	1.53	.90	
.95	2.11	2.03	1.98	1.94	1.89	1.86	1.84	1.80	1.79	1.77	1.75	1.73	.95	
.975	2.44	2.33	2.27	2.21	2.15	2.11	2.08	2.02	2.01	1.98	1.95	1.94	.975	
.99	2.89	2.74	2.66	2.58	2.49	2.44	2.40	2.33	2.31	2.27	2.24	2.21	.99	
.995	3.25	3.06	2.97	2.87	2.77	2.70	2.66	2.57	2.55	2.50	2.46	2.43	.995	
.999	4.14	3.87	3.74	3.59	3.45	3.35	3.29	3.16	3.14	3.07	3.01	2.97	.999	
.9995	4.55	4.25	4.09	3.93	3.76	3.66	3.59	3.44	3.41	3.33	3.27	3.22	.9995	

TABLE A.4 Percentiles of the F distributions (cont.)

n	p	1	2	3	4	5	6	7	8	9	10	11	12	p
30	.0005	$.0^640$	$.0^350$	$.0^250$	.015	.030	.047	.065	.082	.098	.114	.129	.143	.0005
	.001	$.0^516$	$.0^210$	$.0^280$	.022	.040	.060	.080	.099	.117	.134	.150	.164	.001
	.005	$.0^440$	$.0^250$	.024	.050	.079	.107	.133	.156	.178	.197	.215	.231	.005
	.01	$.0^316$	.010	.038	.072	.107	.138	.167	.192	.215	.235	.254	.270	.01
	.025	$.0^210$	.025	.071	.118	.161	.197	.229	.257	.281	.302	.321	.337	.025
	.05	$.0^240$	.051	.116	.174	.222	.263	.296	.325	.349	.370	.389	.406	.05
	.10	.016	.106	.193	.262	.315	.357	.391	.420	.443	.464	.481	.497	.10
	.25	.103	.290	.406	.480	.532	.571	.601	.625	.645	.661	.676	.688	.25
	.50	.466	.709	.807	.858	.890	.912	.927	.939	.948	.955	.961	.966	.50
	.75	1.38	1.45	1.44	1.42	1.41	1.39	1.38	1.37	1.36	1.35	1.35	1.34	.75
	.90	2.88	2.49	2.28	2.14	2.05	1.98	1.93	1.88	1.85	1.82	1.79	1.77	.90
	.95	4.17	3.32	2.92	2.69	2.53	2.42	2.33	2.27	2.21	2.16	2.13	2.09	.95
	.975	5.57	4.18	3.59	3.25	3.03	2.87	2.75	2.65	2.57	2.51	2.46	2.41	.975
	.99	7.56	5.39	4.51	4.02	3.70	3.47	3.30	3.17	3.07	2.98	2.91	2.84	.99
	.995	9.18	6.35	5.24	4.62	4.23	3.95	3.74	3.58	3.45	3.34	3.25	3.18	.995
	.999	13.3	8.77	7.05	6.12	5.53	5.12	4.82	4.58	4.39	4.24	4.11	4.00	.999
	.9995	15.2	9.90	7.90	6.82	6.14	5.66	5.31	5.04	4.82	4.65	4.51	4.38	.9995
40	.0005	$.0^640$	$.0^350$	$.0^250$	.016	.030	.048	.066	.084	.100	.117	.132	.147	.0005
	.001	$.0^516$	$.0^210$	$.0^280$	.022	.042	.061	.081	.101	.119	.137	.153	.169	.001
	.005	$.0^440$	$.0^250$	.024	.051	.080	.108	.135	.159	.181	.201	.220	.237	.005
	.01	$.0^316$	.010	.038	.073	.108	.140	.169	.195	.219	.240	.259	.276	.01
	.025	$.0^399$	.025	.071	.119	.162	.199	.232	.260	.285	.307	.327	.344	.025
	.05	$.0^240$	.051	.116	.175	.224	.265	.299	.329	.354	.376	.395	.412	.05
	.10	.016	.106	.194	.263	.317	.360	.394	.424	.448	.469	.488	.504	.10
	.25	.103	.290	.405	.480	.533	.572	.603	.627	.647	.664	.680	.691	.25
	.50	.463	.705	.802	.854	.885	.907	.922	.934	.943	.950	.956	.961	.50
	.75	1.36	1.44	1.42	1.40	1.39	1.37	1.36	1.35	1.34	1.33	1.32	1.31	.75
	.90	2.84	2.44	2.23	2.09	2.00	1.93	1.87	1.83	1.79	1.76	1.73	1.71	.90
	.95	4.08	3.23	2.84	2.61	2.45	2.34	2.25	2.18	2.12	2.08	2.04	2.00	.95
	.975	5.42	4.05	3.46	3.13	2.90	2.74	2.62	2.53	2.45	2.39	2.33	2.29	.975
	.99	7.31	5.18	4.31	3.83	3.51	3.29	3.12	2.99	2.89	2.80	2.73	2.66	.99
	.995	8.83	6.07	4.98	4.37	3.99	3.71	3.51	3.35	3.22	3.12	3.03	2.95	.995
	.999	12.6	8.25	6.60	5.70	5.13	4.73	4.44	4.21	4.02	3.87	3.75	3.64	.999
	.9995	14.4	9.25	7.33	6.30	5.64	5.19	4.85	4.59	4.38	4.21	4.07	3.95	.9995
60	.0005	$.0^640$	$.0^350$	$.0^251$	.016	.031	.048	.067	.085	.103	.120	.136	.152	.0005
	.001	$.0^516$	$.0^210$	$.0^280$	.022	.041	.062	.083	.103	.122	.140	.157	.174	.001
	.005	$.0^440$	$.0^250$	.024	.051	.081	.110	.137	.162	.185	.206	.225	.243	.005
	.01	$.0^316$	.010	.038	.073	.109	.142	.172	.199	.223	.245	.265	.283	.01
	.025	$.0^399$	.025	.071	.120	.163	.202	.235	.264	.290	.313	.333	.351	.025
	.05	$.0^240$	.051	.116	.176	.226	.267	.303	.333	.359	.382	.402	.419	.05
	.10	.016	.106	.194	.264	.318	.362	.398	.428	.453	.475	.493	.510	.10
	.25	.102	.289	.405	.480	.534	.573	.604	.629	.650	.667	.680	.695	.25
	.50	.461	.701	.798	.849	.880	.901	.917	.928	.937	.945	.951	.956	.50
	.75	1.35	1.42	1.41	1.38	1.37	1.35	1.33	1.32	1.31	1.30	1.29	1.29	.75
	.90	2.79	2.39	2.18	2.04	1.95	1.87	1.82	1.77	1.74	1.71	1.68	1.66	.90
	.95	4.00	3.15	2.76	2.53	2.37	2.25	2.17	2.10	2.04	1.99	1.95	1.92	.95
	.975	5.29	3.93	3.34	3.01	2.79	2.63	2.51	2.41	2.33	2.27	2.22	2.17	.975
	.99	7.08	4.98	4.13	3.65	3.34	3.12	2.95	2.82	2.72	2.63	2:56	2.50	.99
	.995	8.49	5.80	4.73	4.14	3.76	3.49	3.29	3.13	3.01	2.90	2.82	2.74	.995
	.999	12.0	7.76	6.17	5.31	4.76	4.37	4.09	3.87	3.69	3.54	3.43	3.31	.999
	.9995	13.6	8.65	6.81	5.82	5.20	4.76	4.44	4.18	3.98	3.82	3.69	3.57	.9995

TABLE A.4 Percentiles of the F distributions (cont.)

p \ m	15	20	24	30	40	50	60	100	120	200	500	∞	p	n
.0005	.179	.226	.254	.287	.325	.350	.369	.410	.420	.444	.467	.483	.0005	30
.001	.202	.250	.278	.311	.348	.373	.391	.431	.442	.465	.488	.503	.001	
.005	.271	.320	.349	.381	.416	.441	.457	.495	.504	.524	.543	.559	.005	
.01	.311	.360	.388	.419	.454	.476	.493	.529	.538	.559	.575	.590	.01	
.025	.378	.426	.453	.482	.515	.535	.551	.585	.592	.610	.625	.639	.025	
.05	.445	.490	.516	.543	.573	.592	.606	.637	.644	.658	.676	.685	.05	
.10	.534	.575	.598	.623	.649	.667	.678	.704	.710	.725	.735	.746	.10	
.25	.716	.746	.763	.780	.798	.810	.818	.835	.839	.848	.856	.862	.25	
.50	.978	.989	.994	1.00	1.01	1.01	1.01	1.02	1.02	1.02	1.02	1.02	.50	
.75	1.32	1.30	1.29	1.28	1.27	1.26	1.26	1.25	1.24	1.24	1.23	1.23	.75	
.90	1.72	1.67	1.64	1.61	1.57	1.55	1.54	1.51	1.50	1.48	1.47	1.46	.90	
.95	2.01	1.93	1.89	1.84	1.79	1.76	1.74	1.70	1.68	1.66	1.64	1.62	.95	
.975	2.31	2.20	2.14	2.07	2.01	1.97	1.94	1.88	1.87	1.84	1.81	1.79	.975	
.99	2.70	2.55	2.47	2.39	2.30	2.25	2.21	2.13	2.11	2.07	2.03	2.01	.99	
.995	3.01	2.82	2.73	2.63	2.52	2.46	2.42	2.32	2.30	2.25	2.21	2.18	.995	
.999	3.75	3.49	3.36	3.22	3.07	2.98	2.92	2.79	2.76	2.69	2.63	2.59	.999	
.9995	4.10	3.80	3.65	3.48	3.32	3.22	3.15	3.00	2.97	2.89	2.82	2.78	.9995	
.0005	.185	.236	.266	.301	.343	.373	.393	.441	.453	.480	.504	.525	.0005	40
.001	.209	.259	.290	.326	.367	.396	.415	.461	.473	.500	.524	.545	.001	
.005	.279	.331	.362	.396	.436	.463	.481	.524	.534	.559	.581	.599	.005	
.01	.319	.371	.401	.435	.473	.498	.516	.556	.567	.592	.613	.628	.01	
.025	.387	.437	.466	.498	.533	.556	.573	.610	.620	.641	.662	.674	.025	
.05	.454	.502	.529	.558	.591	.613	.627	.658	.669	.685	.704	.717	.05	
.10	.542	.585	.609	.636	.664	.683	.696	.724	.731	.747	.762	.772	.10	
.25	.720	.752	.769	.787	.806	.819	.828	.846	.851	.861	.870	.877	.25	
.50	.972	.983	.989	.994	1.00	1.00	1.01	1.01	1.01	1.01	1.02	1.02	.50	
.75	1.30	1.28	1.26	1.25	1.24	1.23	1.22	1.21	1.21	1.20	1.19	1.19	.75	
.90	1.66	1.61	1.57	1.54	1.51	1.48	1.47	1.43	1.42	1.41	1.39	1.38	.90	
.95	1.92	1.84	1.79	1.74	1.69	1.66	1.64	1.59	1.58	1.55	1.53	1.51	.95	
.975	2.18	2.07	2.01	1.94	1.88	1.83	1.80	1.74	1.72	1.69	1.66	1.64	.975	
.99	2.52	2.37	2.29	2.20	2.11	2.06	2.02	1.94	1.92	1.87	1.83	1.80	.99	
.995	2.78	2.60	2.50	2.40	2.30	2.23	2.18	2.09	2.06	2.01	1.96	1.93	.995	
.999	3.40	3.15	3.01	2.87	2.73	2.64	2.57	2.44	2.41	2.34	2.28	2.23	.999	
.9995	3.68	3.39	3.24	3.08	2.92	2.82	2.74	2.60	2.57	2.49	2.41	2.37	.9995	
.0005	.192	.246	.278	.318	.365	.398	.421	.478	.493	.527	.561	.585	.0005	60
.001	.216	.270	.304	.343	.389	.421	.444	.497	.512	.545	.579	.602	.001	
.005	.287	.343	.376	.414	.458	.488	.510	.559	.572	.602	.633	.652	.005	
.01	.328	.383	.416	.453	.495	.524	.545	.592	.604	.633	.658	.679	.01	
.025	.396	.450	.481	.515	.555	.581	.600	.641	.654	.680	.704	.720	.025	
.05	.463	.514	.543	.575	.611	.633	.652	.690	.700	.719	.746	.759	.05	
.10	.550	.596	.622	.650	.682	.703	.717	.750	.758	.776	.793	.806	.10	
.25	.725	.758	.776	.796	.816	.830	.840	.860	.865	.877	.888	.896	.25	
.50	.967	.978	.983	.989	.994	.998	1.00	1.00	1.01	1.01	1.01	1.01	.50	
.75	1.27	1.25	1.24	1.22	1.21	1.20	1.19	1.17	1.17	1.16	1.15	1.15	.75	
.90	1.60	1.54	1.51	1.48	1.44	1.41	1.40	1.36	1.35	1.33	1.31	1.29	.90	
.95	1.84	1.75	1.70	1.65	1.59	1.56	1.53	1.48	1.47	1.44	1.41	1.39	.95	
.975	2.06	1.94	1.88	1.82	1.74	1.70	1.67	1.60	1.58	1.54	1.51	1.48	.975	
.99	2.35	2.20	2.12	2.03	1.94	1.88	1.84	1.75	1.73	1.68	1.63	1.60	.99	
.995	2.57	2.39	2.29	2.19	2.08	2.01	1.96	1.86	1.83	1.78	1.73	1.69	.995	
.999	3.08	2.83	2.69	2.56	2.41	2.31	2.25	2.11	2.09	2.01	1.93	1.89	.999	
.9995	3.30	3.02	2.87	2.71	2.55	2.45	2.38	2.23	2.19	2.11	2.03	1.98	.9995	

TABLE A.4   Percentiles of the F distributions (cont.)

n	p	1	2	3	4	5	6	7	8	9	10	11	12	p
120	.0005	$.0^6 40$	$.0^3 50$	$.0^2 51$	.016	.031	.049	.067	.087	.105	.123	.140	.156	.0005
	.001	$.0^5 16$	$.0^2 10$	$.0^2 81$	.023	.042	.063	.084	.105	.125	.144	.162	.179	.001
	.005	$.0^4 39$	$.0^2 50$	.024	.051	.081	.111	.139	.165	.189	.211	.230	.249	.005
	.01	$.0^3 16$	.010	.038	.074	.110	.143	.174	.202	.227	.250	.271	.290	.01
	.025	$.0^3 99$	.025	.072	.120	.165	.204	.238	.268	.295	.318	.340	.359	.025
	.05	$.0^2 39$	.051	.117	.177	.227	.270	.306	.337	.364	.388	.408	.427	.05
	.10	.016	.105	.194	.265	.320	.365	.401	.432	.458	.480	.500	.518	.10
	.25	.102	.288	.405	.481	.534	.574	.606	.631	.652	.670	.685	.699	.25
	.50	.458	.697	.793	.844	.875	.896	.912	.923	.932	.939	.945	.950	.50
	.75	1.34	1.40	1.39	1.37	1.35	1.33	1.31	1.30	1.29	1.28	1.27	1.26	.75
	.90	2.75	2.35	2.13	1.99	1.90	1.82	1.77	1.72	1.68	1.65	1.62	1.60	.90
	.95	3.92	3.07	2.68	2.45	2.29	2.18	2.09	2.02	1.96	1.91	1.87	1.83	.95
	.975	5.15	3.80	3.23	2.89	2.67	2.52	2.39	2.30	2.22	2.16	2.10	2.05	.975
	.99	6.85	4.79	3.95	3.48	3.17	2.96	2.79	2.66	2.56	2.47	2.40	2.34	.99
	.995	8.18	5.54	4.50	3.92	3.55	3.28	3.09	2.93	2.81	2.71	2.62	2.54	.995
	.999	11.4	7.32	5.79	4.95	4.42	4.04	3.77	3.55	3.38	3.24	3.12	3.02	.999
	.9995	12.8	8.10	6.34	5.39	4.79	4.37	4.07	3.82	3.63	3.47	3.34	3.22	.9995
∞	.0005	$.0^6 39$	$.0^3 50$	$.0^2 51$	.016	.032	.050	.069	.088	.108	.127	.144	.161	.0005
	.001	$.0^5 16$	$.0^2 10$	$.0^2 81$	.023	.042	.063	.085	.107	.128	.148	.167	.185	.001
	.005	$.0^4 39$	$.0^2 50$	.024	.052	.082	.113	.141	.168	.193	.216	.236	.256	.005
	.01	$.0^3 16$	.010	.038	.074	.111	.145	.177	.206	.232	.256	.278	.298	.01
	.025	$.0^3 98$	.025	.072	.121	.166	.206	.241	.272	.300	.325	.347	.367	.025
	.05	$.0^2 39$	.051	.117	.178	.229	.273	.310	.342	.369	.394	.417	.436	.05
	.10	.016	.105	.195	.266	.322	.367	.405	.436	.463	.487	.508	.525	.10
	.25	.102	.288	.404	.481	.535	.576	.608	.634	.655	.674	.690	.703	.25
	.50	.455	.693	.789	.839	.870	.891	.907	.918	.927	.934	.939	.945	.50
	.75	1.32	1.39	1.37	1.35	1.33	1.31	1.29	1.28	1.27	1.25	1.24	1.24	.75
	.90	2.71	2.30	2.08	1.94	1.85	1.77	1.72	1.67	1.63	1.60	1.57	1.55	.90
	.95	3.84	3.00	2.60	2.37	2.21	2.10	2.01	1.94	1.88	1.83	1.79	1.75	.95
	.975	5.02	3.69	3.12	2.79	2.57	2.41	2.29	2.19	2.11	2.05	1.99	1.94	.975
	.99	6.63	4.61	3.78	3.32	3.02	2.80	2.64	2.51	2.41	2.32	2.25	2.18	.99
	.995	7.88	5.30	4.28	3.72	3.35	3.09	2.90	2.74	2.62	2.52	2.43	2.36	.995
	.999	10.8	6.91	5.42	4.62	4.10	3.74	3.47	3.27	3.10	2.96	2.84	2.74	.999
	.9995	12.1	7.60	5.91	5.00	4.42	4.02	3.72	3.48	3.30	3.14	3.02	2.90	.9995

m	15	20	24	30	40	50	60	100	120	200	500	∞	p	n
.0005	.199	.256	.293	.338	.390	.429	.458	.524	.543	.578	.614	.676	.0005	120
.001	.223	.282	.319	.363	.415	.453	.480	.542	.568	.595	.631	.691	.001	
.005	.297	.356	.393	.434	.484	.520	.545	.605	.623	.661	.702	.733	.005	
.01	.338	.397	.433	.474	.522	.556	.579	.636	.652	.688	.725	.755	.01	
.025	.406	.464	.498	.536	.580	.611	.633	.684	.698	.729	.762	.789	.025	
.05	.473	.527	.559	.594	.634	.661	.682	.727	.740	.767	.785	.819	.05	
.10	.560	.609	.636	.667	.702	.726	.742	.781	.791	.815	.838	.855	.10	
.25	.730	.765	.784	.805	.828	.843	.853	.877	.884	.897	.911	.923	.25	
.50	.961	.972	.978	.983	.989	.992	.994	1.00	1.00	1.00	1.01	1.01	.50	
.75	1.24	1.22	1.21	1.19	1.18	1.17	1.16	1.14	1.13	1.12	1.11	1.10	.75	
.90	1.55	1.48	1.45	1.41	1.37	1.34	1.32	1.27	1.26	1.24	1.21	1.19	.90	
.95	1.75	1.66	1.61	1.55	1.50	1.46	1.43	1.37	1.35	1.32	1.28	1.25	.95	
.975	1.95	1.82	1.76	1.69	1.61	1.56	1.53	1.45	1.43	1.39	1.34	1.31	.975	
.99	2.19	2.03	1.95	1.86	1.76	1.70	1.66	1.56	1.53	1.48	1.42	1.38	.99	
.995	2.37	2.19	2.09	1.98	1.87	1.80	1.75	1.64	1.61	1.54	1.48	1.43	.995	
.999	2.78	2.53	2.40	2.26	2.11	2.02	1.95	1.82	1.76	1.70	1.62	1.54	.999	
.9995	2.96	2.67	2.53	2.38	2.21	2.11	2.01	1.88	1.84	1.75	1.67	1.60	.9995	
.0005	.207	.270	.311	.360	.422	.469	.505	.599	.624	.704	.804	1.00	.0005	∞
.001	.232	.296	.338	.386	.448	.493	.527	.617	.649	.719	.819	1.00	.001	
.005	.307	.372	.412	.460	.518	.559	.592	.671	.699	.762	.843	1.00	.005	
.01	.349	.413	.452	.499	.554	.595	.625	.699	.724	.782	.858	1.00	.01	
.025	.418	.480	.517	.560	.611	.645	.675	.741	.763	.813	.878	1.00	.025	
.05	.484	.543	.577	.617	.663	.694	.720	.781	.797	.840	.896	1.00	.05	
.10	.570	.622	.652	.687	.726	.752	.774	.826	.838	.877	.919	1.00	.10	
.25	.736	.773	.793	.816	.842	.860	.872	.901	.910	.932	.957	1.00	.25	
.50	.956	.967	.972	.978	.983	.987	.989	.993	.994	.997	.999	1.00	.50	
.75	1.22	1.19	1.18	1.16	1.14	1.13	1.12	1.09	1.08	1.07	1.04	1.00	.75	
.90	1.49	1.42	1.38	1.34	1.30	1.26	1.24	1.18	1.17	1.13	1.08	1.00	.90	
.95	1.67	1.57	1.52	1.46	1.39	1.35	1.32	1.24	1.22	1.17	1.11	1.00	.95	
.975	1.83	1.71	1.64	1.57	1.48	1.43	1.39	1.30	1.27	1.21	1.13	1.00	.975	
.99	2.04	1.88	1.79	1.70	1.59	1.52	1.47	1.36	1.32	1.25	1.15	1.00	.99	
.995	2.19	2.00	1.90	1.79	1.67	1.59	1.53	1.40	1.36	1.28	1.17	1.00	.995	
.999	2.51	2.27	2.13	1.99	1.84	1.73	1.66	1.49	1.45	1.34	1.21	1.00	.999	
.9995	2.65	2.37	2.22	2.07	1.91	1.79	1.71	1.53	1.48	1.36	1.22	1.00	.9995	

TABLE A.5 Upper percentiles of studentized range distributions

Studentized range distribution with k and v degrees of freedom

Area = α

$Q_{\alpha,k,v}$

v	k / $1-\alpha$	2	3	4	5	6	7	8	9	10	11	12	13	14	15	16
1	0.95	18.0	27.0	32.8	37.1	40.4	43.1	45.4	47.4	49.1	50.6	52.0	53.2	54.3	55.4	56.3
	0.99	90.0	135	164	186	202	216	227	237	246	253	260	266	272	277	282
2	0.95	6.09	8.3	9.8	10.9	11.7	12.4	13.0	13.5	14.0	14.4	14.7	15.1	15.4	15.7	15.9
	0.99	14.0	19.0	22.3	24.7	26.6	28.2	29.5	30.7	31.7	32.6	33.4	34.1	34.8	35.4	36.0
3	0.95	4.50	5.91	6.82	7.50	8.04	8.48	8.85	9.18	9.46	9.72	9.95	10.2	10.4	10.5	10.7
	0.99	8.26	10.6	12.2	13.3	14.2	15.0	15.6	16.2	16.7	17.1	17.5	17.9	18.2	18.5	18.8
4	0.95	3.93	5.04	5.76	6.29	6.71	7.05	7.35	7.60	7.83	8.03	8.21	8.37	8.52	8.66	8.79
	0.99	6.51	8.12	9.17	9.96	10.6	11.1	11.5	11.9	12.3	12.6	12.8	13.1	13.3	13.5	13.7
5	0.95	3.64	4.60	5.22	5.67	6.03	6.33	6.58	6.80	6.99	7.17	7.32	7.47	7.60	7.72	7.83
	0.99	5.70	6.97	7.80	8.42	8.91	9.32	9.67	9.97	10.2	10.5	10.7	10.9	11.1	11.2	11.4
6	0.95	3.46	4.34	4.90	5.31	5.63	5.89	6.12	6.32	6.49	6.65	6.79	6.92	7.03	7.14	7.24
	0.99	5.24	6.33	7.03	7.56	7.97	8.32	8.61	8.87	9.10	9.30	9.49	9.65	9.81	9.95	10.1
7	0.95	3.34	4.16	4.68	5.06	5.36	5.61	5.82	6.00	6.16	6.30	6.43	6.55	6.66	6.76	6.85
	0.99	4.95	5.92	6.54	7.01	7.37	7.68	7.94	8.17	8.37	8.55	8.71	8.86	9.00	9.12	9.24
8	0.95	3.26	4.04	4.53	4.89	5.17	5.40	5.60	5.77	5.92	6.05	6.18	6.29	6.39	6.48	6.57
	0.99	4.74	5.63	6.20	6.63	6.96	7.24	7.47	7.68	7.87	8.03	8.18	8.31	8.44	8.55	8.66
9	0.95	3.20	3.95	4.42	4.76	5.02	5.24	5.43	5.60	5.74	5.87	5.98	6.09	6.19	6.28	6.36
	0.99	4.60	5.43	5.96	6.35	6.66	6.91	7.13	7.32	7.49	7.65	7.78	7.91	8.03	8.13	8.23
10	0.95	3.15	3.88	4.33	4.65	4.91	5.12	5.30	5.46	5.60	5.72	5.83	5.93	6.03	6.11	6.20
	0.99	4.48	5.27	5.77	6.14	6.43	6.67	6.87	7.05	7.21	7.36	7.48	7.60	7.71	7.81	7.91
11	0.95	3.11	3.82	4.26	4.57	4.82	5.03	5.20	5.35	5.49	5.61	5.71	5.81	5.90	5.99	6.06
	0.99	4.39	5.14	5.62	5.97	6.25	6.48	6.67	6.84	6.99	7.13	7.25	7.36	7.46	7.56	7.65
12	0.95	3.08	3.77	4.20	4.51	4.75	4.95	5.12	5.27	5.40	5.51	5.62	5.71	5.80	5.88	5.95
	0.99	4.32	5.04	5.50	5.84	6.10	6.32	6.51	6.67	6.81	6.94	7.06	7.17	7.26	7.36	7.44
13	0.95	3.06	3.73	4.15	4.45	4.69	4.88	5.05	5.19	5.32	5.43	5.53	5.63	5.71	5.79	5.86
	0.99	4.26	4.96	5.40	5.73	5.98	6.19	6.37	6.53	6.67	6.79	6.90	7.01	7.10	7.19	7.27
14	0.95	3.03	3.70	4.11	4.41	4.64	4.83	4.99	5.13	5.25	5.36	5.46	5.55	5.64	5.72	5.79
	0.99	4.21	4.89	5.32	5.63	5.88	6.08	6.26	6.41	6.54	6.66	6.77	6.87	6.96	7.05	7.12

TABLE A.5 Upper percentiles of studentized range distributions (cont.)

v	$1-\alpha$	k 2	3	4	5	6	7	8	9	10	11	12	13	14	15	16
15	0.95	3.01	3.67	4.08	4.37	4.60	4.78	4.94	5.08	5.20	5.31	5.40	5.49	5.58	5.65	5.72
	0.99	4.17	4.83	5.25	5.56	5.80	5.99	6.16	6.31	6.44	6.55	6.66	6.76	6.84	6.93	7.00
16	0.95	3.00	3.65	4.05	4.33	4.56	4.74	4.90	5.03	5.15	5.26	5.35	5.44	5.52	5.59	5.66
	0.99	4.13	4.78	5.19	5.49	5.72	5.92	6.08	6.22	6.35	6.46	6.56	6.66	6.74	6.82	6.90
17	0.95	2.98	3.63	4.02	4.30	4.52	4.71	4.86	4.99	5.11	5.21	5.31	5.39	5.47	5.55	5.61
	0.99	4.10	4.74	5.14	5.43	5.66	5.85	6.01	6.15	6.27	6.38	6.48	6.57	6.66	6.73	6.80
18	0.95	2.97	3.61	4.00	4.28	4.49	4.67	4.82	4.96	5.07	5.17	5.27	5.35	5.43	5.50	5.57
	0.99	4.07	4.70	5.09	5.38	5.60	5.79	5.94	6.08	6.20	6.31	6.41	6.50	6.58	6.65	6.72
19	0.95	2.96	3.59	3.98	4.25	4.47	4.65	4.79	4.92	5.04	5.14	5.23	5.32	5.39	5.46	5.53
	0.99	4.05	4.67	5.05	5.33	5.55	5.73	5.89	6.02	6.14	6.25	6.34	6.43	6.51	6.58	6.65
20	0.95	2.95	3.58	3.96	4.23	4.45	4.62	4.77	4.90	5.01	5.11	5.20	5.28	5.36	5.43	5.49
	0.99	4.02	4.64	5.02	5.29	5.51	5.69	5.84	5.97	6.09	6.19	6.29	6.37	6.45	6.52	6.59
24	0.95	2.92	3.53	3.90	4.17	4.37	4.54	4.68	4.81	4.92	5.01	5.10	5.18	5.25	5.32	5.38
	0.99	3.96	4.54	4.91	5.17	5.37	5.54	5.69	5.81	5.92	6.02	6.11	6.19	6.26	6.33	6.39
30	0.95	2.89	3.49	3.84	4.10	4.30	4.46	4.60	4.72	4.83	4.92	5.00	5.08	5.15	5.21	5.27
	0.99	3.89	4.45	4.80	5.05	5.24	5.40	5.54	5.65	5.76	5.85	5.93	6.01	6.08	6.14	6.20
40	0.95	2.86	3.44	3.79	4.04	4.23	4.39	4.52	4.63	4.74	4.82	4.91	4.98	5.05	5.11	5.16
	0.99	3.82	4.37	4.70	4.93	5.11	5.27	5.39	5.50	5.60	5.69	5.77	5.84	5.90	5.96	6.02
60	0.95	2.83	3.40	3.74	3.98	4.16	4.31	4.44	4.55	4.65	4.73	4.81	4.88	4.94	5.00	5.06
	0.99	3.76	4.28	4.60	4.82	4.99	5.13	5.25	5.36	5.45	5.53	5.60	5.67	5.73	5.79	5.84
120	0.95	2.80	3.36	3.69	3.92	4.10	4.24	4.36	4.48	4.56	4.64	4.72	4.78	4.84	4.90	4.95
	0.99	3.70	4.20	4.50	4.71	4.87	5.01	5.12	5.21	5.30	5.38	5.44	5.51	5.56	5.61	5.66
∞	0.95	2.77	3.31	3.63	3.86	4.03	4.17	4.29	4.39	4.47	4.55	4.62	4.68	4.74	4.80	4.85
	0.99	3.64	4.12	4.40	4.60	4.76	4.88	4.99	5.08	5.16	5.23	5.29	5.35	5.40	5.45	5.49

SOURCE: Olive Jean Dunn and Virginia A. Clark, *Applied Statistics: Analysis of Variance and Regression* (New York: Wiley, 1974), pp. 371–372.

TABLE A.6   Upper and lower percentiles of the Wilcoxon signed rank statistic, W

	w_1^*	w_2^*	$P(W \leq w_1^*) = P(W \geq w_2^*)$
$n = 4$	0	10	0.062
	1	9	0.125
$n = 5$	0	15	0.031
	1	14	0.062
	2	13	0.094
	3	12	0.156
$n = 6$	0	21	0.016
	1	20	0.031
	2	19	0.047
	3	18	0.078
	4	17	0.109
	5	16	0.156
$n = 7$	0	28	0.008
	1	27	0.016
	2	26	0.023
	3	25	0.039
	4	24	0.055
	5	23	0.078
	6	22	0.109
	7	21	0.148
$n = 8$	0	36	0.004
	1	35	0.008
	2	34	0.012
	3	33	0.020
	4	32	0.027
	5	31	0.039
	6	30	0.055
	7	29	0.074
	8	28	0.098
	9	27	0.125
$n = 9$	1	44	0.004
	2	43	0.006
	3	42	0.010
	4	41	0.014
	5	40	0.020
	6	39	0.027
	7	38	0.037
	8	37	0.049
	9	36	0.064
	10	35	0.082
	11	34	0.102
	12	33	0.125

SOURCE: Wilfrid J. Dixon and Frank J. Massey, Jr., *Introduction to Statistical Analysis*, 2nd. ed. (New York: McGraw-Hill, 1957), pp. 443–444.

TABLE A.6 Upper and lower pecentiles of the Wilcoxon signed rank statistic, W (cont.)

	w_1^*	w_2^*	$P(W \le w_1^*) = P(W \ge w_2^*)$
$n = 10$	3	52	0.005
	4	51	0.007
	5	50	0.010
	6	49	0.014
	7	48	0.019
	8	47	0.024
	9	46	0.032
	10	45	0.042
	11	44	0.053
	12	43	0.065
	13	42	0.080
	14	41	0.097
	15	40	0.116
	16	39	0.138
$n = 11$	5	61	0.005
	6	60	0.007
	7	59	0.009
	8	58	0.012
	9	57	0.016
	10	56	0.021
	11	55	0.027
	12	54	0.034
	13	53	0.042
	14	52	0.051
	15	51	0.062
	16	50	0.074
	17	49	0.087
	18	48	0.103
	19	47	0.120
	20	46	0.139
$n = 12$	7	71	0.005
	8	70	0.006
	9	69	0.008
	10	68	0.010
	11	67	0.013
	12	66	0.017
	13	65	0.021
	14	64	0.026
	15	63	0.032
	16	62	0.039
	17	61	0.046
	18	60	0.055
	19	59	0.065
	20	58	0.076
	21	57	0.088
	22	56	0.102
	23	55	0.117
	24	54	0.133

Answers to Selected Questions and Review Exercises

CHAPTER 2

Questions

Section 2.2

1. In general, $\{9, \text{no } 9 \text{ or } 7, \text{no } 9 \text{ or } 7, \ldots, 9\}$ **4.** $P(A \cup B) = P(A) + P(B) - P(A \cap B)$ **5.** 27
6. $A \cap B = \{(x, y): 2 < x < 3, 2 < y < 3\}$

Section 2.4

1. (a) $\frac{10}{216}$; (b) $\frac{212}{216}$ **2.** Cipher: $z\,a\,b\,c\ldots$; Text: $e\,f\,g\,h\ldots$ **3.** $\frac{12}{21}$

Section 2.5

1. Hint: Show that $\int_{-60}^{0} [(60 + x)/3600]\,dx + \int_{0}^{60} [(60 - x)/3600]\,dx = 1$ **2.** 0.0625   **4.** 188

Section 2.6

1. No   **2.** $1/n$ **3.** $\frac{2}{3}$ **4.** 0.7005   **5.** (a) 0.275; (b) $\frac{9}{55}$

Section 2.8

3. $\frac{9}{17}$ **4.** 0.491   **5.** 21

Section 2.9

1. 7^8 **2.** $2(n - r - 1)/n(n - 1)$ **3.** (a) $7!/7^7$; (b) $1/7^6$; (c) $6/7^5$
4. (a) $8!$; (b) $2(4!)^2$; (c) $8!/4!4!$; (d) $8!/4!$ **6.** (a) $40\big/\binom{52}{5}$; (b) $40(4^4 - 1)\big/\binom{52}{5}$ **7.** $13 \cdot 48 \big/ \binom{52}{5}$
10. Hint: For one of the ways, consider the binomial expansion of $(1 + 1)^n$.

Section 2.10

3. $\frac{2}{7}$

Review Exercises

1. $S = \{$sss, ssf, sfs, sff, fss, fsf, ffs, fff$\}$, $A = \{$sfs, fss$\}$, $B = \{$fff$\}$
2. If "abc" denotes the assignment of collar "a" to the male, collar "b" to the female, and collar "c" to the cub, then $S = \{$abc, acb, bac, bca, cab, cba$\}$.
4. $\{(17850, 17851), (17850, 17852), \ldots, (17854, 17855)\}$. Altogether the sample space has 15 members.
5. To maximize his survival probability the prisoner should put a single white chip in one of the urns and the other 19 chips in the other urn.
6. (a) $\bigcup_{i=1}^{k} E_i = E_1$; (b) $\bigcap_{i=1}^{k} E_i = E_k$ **7.** (a) True; (b) True; (c) False; (d) True; (e) True
9. $S = \{$ab, ac, ad, ae, af, bc, bd, be, bf, cd, ce, cf, de, df, ef$\}$ **10.** $A = (A_{11} \cap A_{21}) \cup (A_{12} \cap A_{22})$
12. $A \cap B = \{2\}$, $A \cup B = \{-4, -3, 2\}$ **13.** $A = \{(b, c) \mid b^2 < c\}$ **16.** .7   **17.** .7   **19.** 15%   **20.** $\frac{6}{20}$ **21.** $\frac{6}{36}$
22. $c = \frac{3}{4}$ **23.** (a) $\frac{3}{8000}$; (b) .578   **25.** $\frac{120}{729}$ **26.** $P(A) = 2.5e^{-1.5} \doteq 0.56$ **27.** $\frac{3}{5}$
29. $P(A) = (1 - e^{-4})/(1 - e^{-9}) \doteq 0.982$ **31.** No. Probability of success = .71   **32.** 0.98   **34.** $\frac{34}{70}$ **36.** 0.43
37. 0.14   **38.** $\frac{28}{38}$ **39.** 0.69   **40.** (a) 0.23
41. No. Hint: Write out the "new" sample space and assign probabilities to its members. Then apply Definition 2.4.
43. Yes. $P(A \cap B) = \frac{5}{33} = (\frac{5}{11})(\frac{11}{33}) = P(A)P(B)$ **45.** Dependent. $P(A \mid B) = 0 \neq P(A)$ **46.** 0.543
49. $P(A) = \frac{1}{2}$, $P(B) = \frac{1}{3}$, $P(C) = \frac{1}{6}$; $P(A \cap B) = \frac{1}{6}$, $P(A \cap C) = \frac{1}{12}$, $P(B \cap C) = \frac{1}{18}$; $P(A \cap B \cap C) = \frac{1}{36}$

50. $P(A \cap B \cap C) = 0 \neq P(A)P(B)P(C) = \frac{1}{8}$ **51.** (a) 0.88; (b) 0.952
53. $A \otimes B$ is a solid cylinder with a height of 1 and a base whose radius is 1. **55.** 0.368
57. Hint: Let $P_i = P$ (client eventually goes bankrupt if stock is currently selling for \$$(5 + i)$ per share). Then
$P_1 = \frac{1}{2} + \frac{1}{2}P_2$ and $P_2 = P_1^2$. (Why?) Solve for P_1.
58. 12 **60.** $P(A \text{ wins}) = \frac{8}{14}$ **63.** (a) $(\frac{5}{6})^{k-1}(\frac{1}{6})$ **65.** $\frac{150}{671}$ **66.** $2^{20} = 1,048,576$
67. (a) $\binom{44}{3}\binom{6}{3}/\binom{50}{6}$; (b) $\binom{44}{2}\binom{6}{4}/\binom{50}{6}$; (c) $\binom{44}{1}\binom{6}{5}/\binom{50}{6}$; (d) $\binom{44}{0}\binom{6}{6}/\binom{50}{6}$ **69.** 95,550
71. (a) 4!3!3!; (b) 3!4!3!3!; (c) 10! **73.** 9 **74.** $(49 \cdot 48! \cdot 4!)/52! \doteq 0.00018$ **75.** (a) $\binom{9}{2}4$; (b) $9 \cdot 8 \cdot 4$
76. 0.37 **77.** (a) $1/\binom{4}{2}\binom{5}{2}\binom{4}{2}$; (b) No **79.** $2^7 - 1$ **80.** 22,848
81. (a) $\binom{54}{7}/\binom{98}{7}$; (b) $\binom{54}{6}\binom{44}{1}/\binom{98}{7}$; (c) $\binom{54}{5}\binom{44}{2}/\binom{98}{7}$ **83.** 288 **85.** 0.0014 **86.** 0.40 **89.** $\binom{32}{13}/\binom{52}{13}$
90. 0.015 **91.** $\frac{54}{256}$ **93.** 15 **94.** $4^4\binom{32}{4}/\binom{48}{12}$ **95.** 2^8 **97.** $\binom{12}{2}\binom{5}{3}4^3/\binom{52}{5}$ **99.** $\binom{8}{2}/\binom{10}{4}$
100. $\binom{25}{2}\binom{20}{2}/\binom{45}{4}$ **103.** 0.997 **105.** $(0.2401)^2(0.7599)^{44} \doteq 0.0000003$ **108.** 0.61 **109.** (a) 2; (b) 0.033
111. (a) 0.172; (b) 0.122 **113.** (a) 0.575; (b) 0.593

CHAPTER 3

Questions

Section 3.2

2. (a) $F_X(x) = \begin{cases} 0, & x < -60 \\ \frac{1}{3600}(60x + \frac{1}{2}x^2) + \frac{1}{2}, & -60 \le x < 0 \\ \frac{1}{3600}(60x - \frac{1}{2}x^2) + \frac{1}{2}, & 0 \le x < 60 \\ 1, & 60 \le x \end{cases}$

(b) $\frac{5}{9}$

3. $F_X(t) = \begin{cases} 0, & t < a \\ (t-a)/(b-a), & a \le t \le b \\ 1, & t > b \end{cases}$

Section 3.3

1. (a)

			y	
		0	1	2
x	0	$\frac{16}{36}$	$\frac{8}{36}$	$\frac{1}{36}$
	1	$\frac{8}{36}$	$\frac{2}{36}$	0
	2	$\frac{1}{36}$	0	0

(b) $f_Z(0) = \frac{16}{36}$, $f_Z(1) = \frac{16}{36}$, $f_Z(2) = \frac{4}{36}$
2. (a) $c = 1$; (b) $\frac{1}{8}$
3. $F_Z(z) = \begin{cases} z^2/2, & 0 < z \le 1 \\ 2z - z^2/2 - 1, & 1 < z \le 2 \end{cases}$
4. (a) $f_X(x) = 1, 0 < x < 1$; (b) $f_Y(y) = -\ln y, 0 < y < 1$

Section 3.4

1. $f_{Y_1,\ldots,Y_n}(y_1, \ldots, y_n) = \lambda^{-n}e^{-(1/\lambda)\sum_{i=1}^{n} y_i}, y_i > 0$

Section 3.5

1. Hint: $P(\frac{1}{2} < X < \frac{3}{4}) = P(2 < Y < \frac{5}{2})$ 2. $f_Y(y) = \frac{1}{3}, -3 < y < 0$ 4. $f_Y(y) = (1/y^2)e^{-1/y}, y > 0$
5. $f_Y(y) = \frac{3}{16}\pi^{-3/2}\sqrt{y}, 0 < y < 4\pi$ 7. $e^{-(r+s)}(r + s)^z/z!, z = 0, 1, 2, \ldots$ 8. $\frac{1}{2}z^2e^{-z}, z > 0$

Section 3.6

2. $ne^{-ny}, y > 0$

Section 3.7

2. (a) $\frac{1}{4}(3 - x), 0 < x < 2$; (b) $(\frac{1}{2})[(6 - x - y)/(3 - x)], 2 < y < 4$

Section 3.8

1. $\frac{5}{3}$ 2. c 4. $f_X(x)$ must be symmetric   5. 1   6. $\sum_{i=1}^{n} a_i = 1$ 8. $E(Z) = \frac{2}{36}$

Section 3.9

1. $\frac{12}{25}$ 2. $\frac{1}{12}$

Section 3.10

2. $np + n^2p^2 - np^2$ 4. No

Review Exercises

1. $\binom{3}{x}(\frac{1}{4})^x(\frac{3}{4})^{3-x}$ 3. (a) 0.135; (b) 0.118; (c) 0.517   4. (a) 2.57; (b) 48.73
5. (a) 0.165; (b) $\binom{x-1}{2}(0.45)^3(0.55)^{x-3}$ 7. (a) $(\frac{1}{4})^{x-1}(\frac{3}{4})$; (b) $(\frac{1}{4})^4$
8. $1 - F_X(7) \doteq 0.001$. A small value for $1 - F_X(7)$ would tend to support the claim that the new treatment is better.
9. $\binom{6}{x}(\frac{1}{3})^x(\frac{2}{3})^{6-x}$ 10. (a) $f_X(x) = \binom{4}{x}\binom{7}{4-x}/\binom{11}{4}, x = 0, 1, 2, 3, 4$ 15. 3   17. (a) 0.648; (b) 0.005
19. $F_X(x) = \begin{cases} \frac{1}{2} + x + \frac{1}{2}x^2, & -1 < x < 0 \\ \frac{1}{2} + x - \frac{1}{2}x^2, & 0 \le x < 1 \end{cases}$ 21. $F_X(x) = \begin{cases} \frac{1}{2}x - \frac{1}{4}, & \frac{1}{2} \le x < 1 \\ x^2/4, & 1 \le x \le 2 \end{cases}$
22. $F_X(1) = \frac{1}{27}$; P(exactly 2 are released in less than a year) $= 0.004$
23. (a) $\ln 2$; (b) 0.223; (c) 0.223; (d) $f_X(x) = 1/x, 1 < x < e$ 25. (a) $1 - 4(x^2/2 + x/2 + \frac{1}{4})e^{-2x}$; (c) 0.938
26. $f_{X,Y}(x, y) = \begin{cases} 1/(4\pi), & x^2 + y^2 \le 4 \\ 0, & \text{otherwise} \end{cases}$
27. $\left[\binom{4}{x}\binom{6}{y}\binom{5}{3 - x - y}\right]/\binom{15}{3}$ 28. (a) $f_Y(y) = \frac{1}{36}\{5^{3-y}/y!(3 - y)!\}$; (c) $f_Y(y) = ye^{-y}$
29. $\left[\binom{4}{x}\binom{4}{y}\binom{44}{4 - x - y}\right]/\binom{52}{4}$ 30. $c = \frac{1}{10}$; $f_X(1) = \frac{1}{10}, f_X(2) = \frac{6}{10}, f_X(3) = \frac{3}{10}$
35. $1/(\pi\sqrt{A^2 - t^2}), -A < t < A$ 37. $\sqrt{y}/18, 0 < y < 9$
40. Hint: Let $X = -\lambda \ln Y$, where Y is a uniform random variable defined over $(0, 1)$. Find $F_X(x)$ and differentiate.
41. $(\sqrt{2}a/m^{3/2})(\sqrt{t}\,e^{-(2b/m)t}), t > 0$ 43. 0.347   47. $f_Y(y) = 2yf_X(y^2)$ 49. $\frac{5}{16}$
53. (b) $f_Y(y) = 4y - 4y^3, 0 < y < 1$ and $f_X(x) = 4x^3, 0 < x < 1$; (c) $f_{Y|0.8}(y) = 3.125y, 0 < y < 0.8$; (d) $\frac{9}{15}$
55. (a) $\frac{6}{5}(2x^2 + x), 0 < x < 1$; (b) $\frac{27}{224}$; (c) 0.598   59. (a) 0.400; (b) 0; (c) $e^{-y} \cdot e^x$ 61. $\frac{1}{3}(2y + 2)$
62. $\frac{91}{35}$ 63. 1   64. 2.53   66. (a) $\frac{6}{15}$; (b) $\frac{4}{3}$ 67. $\frac{9}{4}$ 68. (a) $\frac{9}{5}$; (b) $\frac{1}{5}$ 69. 2   71. 2
73. (a) $1/\lambda$; (b) $\frac{3}{2}$; (c) Mean does not exist; (d) 1; (e) 1   75. 0.90   77. $\frac{701}{64}$ 78. $\min_a g(a) = \text{Var}(X)$
79. (a) 7; (b) $\frac{161}{36}$; (c) $\frac{1}{3}$ 81. $E(X_2') = 0.61$ 83. (a) $4np_1 + 6mp_2$; (b) $16np_1q_1 + 36mp_2q_2$
85. (a) $\frac{1}{20}$; (b) $\frac{2}{3}$; (c) $\frac{3}{80}$; (d) $\frac{1}{6}$ 87. 250   91. λ^{-2}; (b) $\frac{3}{4}$; (d) $\frac{1}{2}$ 95. $[(e^t - 1)/t]^2$ 96. $pe^{(7/2)t}/(1 - qe^{t/2})$
97. Hint: Examine the form of the product, $M_X(t) \cdot M_Y(t)$, and look at the coefficients of t and $t^2/2$.
98. $npq(q - p)$

CHAPTER 4

Questions

Section 4.2

1. 0.099   **2.** λ **4.** Yes. The Poisson assumption seems reasonable here.

Section 4.3

1. (1) 0.1004; (2) 0.1210
2. As the sample size increases, the sample proportion of successes (X/n) provides a better and better approximation for the true proportion–indeed, the former *converges* to the latter.
3. 0.0465. Hypnosis seems to improve guessing ability.   **4.** 0.0013
5. Hint: First find the moment generating function for a standard normal. Then write $X = \mu + \sigma Z$ and apply Theorem 3.19.
6. 216   **7.** Probability of molar of length ≥ 9 mm $\doteq 0.04$. It is unlikely the skull belongs to genus *Papio*.

Section 4.4

1. Hint: Write $E(e^{tN}) = \sum_{n=1}^{\infty} pe^{tn}(1-p)^{n-1} = (p/q)\sum_{n=1}^{\infty}(qe^t)^n$.

Section 4.5

1. 0.647

Section 4.6

1. Hint: Use Property 3 of Theorem 4.9.
3. $(1 - t/\lambda)^{-r}$ **4.** $M_{X_1+X_2}(t) = M_{X_1}(t) \cdot M_{X_2}(t) = (1 - t/\lambda)^{-r} \cdot (1 - t/\lambda)^{-s} = (1 - t/\lambda)^{-r-s}$

Review Exercises

1. 6.9×10^{-12} **3.** (a) 0.195; (b) 0.430

5.

Hits	Expected Number
0	227.3
1	211.4
2	98.3
3	30.5
4	7.1
5+	1.3

6. Hint: By inspection, write down the pdf's of X and Y. Then enumerate what the event $X = 2Y$ includes and use the independence property of X and Y to get a final probability.
7. 0.472   **8.** Z is Poisson with $\lambda = 10$ **9.** (a) 0.865; (b) binomial, $n = 20,000$, $p = 1/10,000$ **11.** 0.127
13. Hint: Consider the product of the moment generating functions for X and Y.
14. Hint: Let X denote the number of occurrences in the interval $(a, a + y)$. Note that (1) $P(X = 0) = e^{-\lambda y}$ and (2) $X = 0$ if and only if $Y > y$.
15. 0.036   **17.** Mean $= 1/\lambda$; median $= (\ln 2)/\lambda$ **19.** 0.602   **21.** 0.908
23. (a) 0.0158; (b) 0.1446; (c) 0.3629; (d) 0.3629; (e) 0.1140; (f) 1.0000   **24.** $\sqrt{\pi}/4$
25. 0.1802 (Using the continuity correction)   **26.** (a) 0.1357; (b) 0.0579; (c) 0.3094
27. Statistical significance $= P(Z \geq 13.3) \doteq 0$. Conclusion: People postpone dying until after their birthday.

29. $\mu \pm \sigma$ **30.** Hint: Use the result given in Question 4.3.5 together with the statement of Theorem 3.19.
31. No. Probability mileage exceeding 25,000 is 0.8413.   **33.** 0.894   **35.** (a) 0.067; (b) 0.159
36. (a) 0.0062; (b) 0.3372; (c) 0.5934   **37.** 8.71; (b) Variability
39. (a) Normal distribution: A and $F - 6.7\%$, B and $D - 24.2\%$, and $C - 38.3\%$; (b) Uniform distribution: B, C, and $D - 28.9\%$, A and $F - 6.7\%$
41. (a) 583.8; (b) Hint: Use the formula for the standard deviation of a binomial pdf.
42. $E(Y^{2k+1}) = 0$ for $k = 0, 1, 2, \ldots$ **44.** 130   **45.** 24.8   **46.** $(144.5 - 124)/\sigma = 1.28$ **48.** 0.064   **50.** 2
53. $\frac{125}{1296}$ **54.** 14.7   **57.** $X = Y + r$, so by Theorem 3.19 $M_X(t) = e^{rt}M_Y(t)$ **59.** $f_Z(z) = 2\lambda^{-3}z^2e^{-z/\lambda}$
61. Choose a random sample $y_1, \ldots, y_r$ from a uniform distribution over [0, 1]. Take $z = -\lambda \ln(y_1 y_2 \cdots y_r)$.
62. Take $\lambda = 1000$ in Review Exercise 59.
63. Hint: Write $F_{Z^2}(t) = P(Z^2 \leq t) = P(-\sqrt{t} \leq Z \leq \sqrt{t}) = 2 \cdot P(0 \leq Z \leq \sqrt{t})$ and differentiate.
64. (a) $F_T(t) = 1 - e^{-H_T(t)}$; (b) $f_T(t) = h_T(t) \cdot e^{-H_T(t)}$
65. If $f_X(x) = (1/\lambda)e^{-x/\lambda}$, then $h_X(x) = 1/\lambda$. Thus, the hazard rate is constant and no "wear out" is predicted.
66. $h_X(x) = (\beta/\alpha) \cdot x^{\beta-1}$ **67.** (c) $f_X(x) = (1/9!5^{10})x^9e^{-x/5}$

CHAPTER 5

Questions

Section 5.3

1. 0.41

Section 5.4

1. $(n + 1)Y_{\min}$ **3.** $\sum_{i=1}^{n} a_i = 1$ **5.** $W = Y/n$ is preferable because of its smaller variance.

Section 5.6

1. Cramer–Rao lower bound $= \lambda^2/n$ **2.** Cramer–Rao lower bound $= \text{Var}(\bar{Y}) = \lambda/n$

Section 5.7

2. Yes

Section 5.8

1. y/n **3.** $\frac{5}{24}$ **5.** $Y_{\min}$ **6.** MLE $= -1 - (n/\ln \prod_{i=1}^{n} y_i)$

Section 5.9

1. Intervals based on $W_1 = (n + 1) \cdot Y_{\min}$ would be wider.

Section 5.10

2. 71

Review Exercises

1. Estimated occurrence rate $= \bar{y} = 1.33$ cases/month. Yes, $\bar{Y}$ is unbiased for the Poisson parameter.
2. $a + b$ must equal 1   **4.** Let $W = 3X^2$ **5.** Yes. $\lim_{n \to \infty} E(W_n) = \lim_{n \to \infty} [n/(n + 1)]\theta = \theta$ **7.** 66
9. $E(Y_i) = \{n!/(i - 1)!(n - i)!\} \cdot \{\Gamma(i + 1)\Gamma(n - i + 1)/\Gamma(n + 2)\} \cdot \theta$

14. Relative efficiency of W_1 to $W_2 = (\lambda/n)/\lambda = 1/n$ **16.** Approximately 27

20. Hint: Write $(W_n - \theta)^2$ as $[(W_n - E(W_n)) - (\theta - E(W_n))]^2$ **22.** Yes **24.** $\prod\limits_{i=1}^{n} X_i$ **26.** $\sum\limits_{i=1}^{n} X_i$

29. $(1/2n) \sum\limits_{i=1}^{n} x_i$ **30.** 1.69 **36.** 2.17 **37.** Method of moments estimate $= 2(\bar{x}) = 100$; $MLE = x_{\max} = 92$.

39. Yes, if the confidence interval is not symmetric around the point estimate.

40. 0.67σ **42.** $(0.45, 0.47)$ **45.** $(0.251, 0.379)$ **46.** 1024 **49.** $(0.008, 0.91)$

CHAPTER 6

Questions

Section 6.2

1. Let $p = $ true probability of a mosquito preferring red light to white light. Reject $H_0: p = \frac{1}{2}$ at the 0.01 level. The observed Z ratio is 10.5; the corresponding critical values are ± 2.58. We are assuming that each mosquito makes its choice independently of the movement of the others.

3. From $1 - e^{-W^*/1000} = 0.05$, $W^* = 50$, approximately.

Section 6.3

1. No

2. When $p = 0.201$, $1 - \beta = 0.15$; for $p = 0.202$, $1 - \beta = 0.34$; for $p = 0.204$, $1 - \beta = 0.79$; for $p = 0.206$, $1 - \beta = 0.98$. At the H_0 value ($p = 0.200$), $\beta = 1 - \alpha = 0.05$.

3. We would like the power function to be as "steep" as possible.

Section 6.4

1. $\max\limits_{\omega} L(p) = p_0^k (1 - p_0)^{\sum\limits_{i=1}^{k} n_i - k}$; $\max\limits_{\Omega} L(p) = \left(k \Big/ \sum\limits_{i=1}^{k} n_i \right)^k \left[1 - \left(k \Big/ \sum\limits_{i=1}^{k} n_i \right) \right]^{\sum\limits_{i=1}^{k} n_i - k}$

Review Exercises

1. Let $p = $ true proportion of people dying in the 3-month period preceding their birthday. Test $H_0: p = \frac{1}{4}$ versus $H_1: p < \frac{1}{4}$. Reject H_0 if $(y - np_0)/\sqrt{np_0(1 - p_0)} \leq -2.33$. Conclusion: Reject H_0. The value of the test statistic is -10.8.

3. Yes **4.** $\alpha = 0.10$; $\beta = 0.72$ when $\theta = 2.5$

5. Test $H_0: p = 0.40$ versus $H_1: p > 0.40$. Critical value is 2.33. Test statistic equals 0.91. Conclusion: Accept H_0.

7. $\alpha = 0.5$; $\beta = 0.33$ when the urn is 60% white; $\beta = 0.18$ when the urn is 70% white.

8. Hint: Use the binomial distribution to give the probability of drawing y white chips.

10. $\alpha = \int_0^{0.3} ze^{-z}\, dz = 0.037$ **11.** $\alpha = \sum\limits_{y=7}^{12} \binom{12}{y} (1/3)^y (2/3)^{12-y} = 0.07$

13. P value $= P(Z \geq 1.76) = 1 - 0.9608 = 0.0392$

14. Because the Poisson is discrete, it will not be possible, in general, to find a critical region whose associated probability under H_0 is exactly α. (*See* Review Exercise 9.)

15. (a) 0.0808; (b) 0.8413

17. As n increases, the "overlap" between the H_0 and the H_1 distributions for the test statistic decreases.

19. $k = 1.98$, that being the value that satisfies the equation $\int_k^2 (5y^4/2^5)\, dy = 0.05$

22. $\alpha = 0.064$; $\beta = 0.107$. Type I errors would be considered more serious than Type II errors.
24. Recall Question 3.3.3.
26. Reject H_0. The observed test statistic (6.41) exceeds the 0.05 critical value (1.64).
29. Reject H_0 if $1/4y_1 y_2 \leq k$. (To find k, see Example 3.12.)

CHAPTER 7

Questions

Section 7.2

2. 14.7   **3.** $\hat{\mu} = \bar{y}$, $\widehat{\sigma^2} = [n/(n-1)]s^2$

Section 7.3

3. (a) 0.29; (b) 0.43; (c) 0.23

Section 7.4

1. $N(0, 1) \doteq \left(\sum\limits_{i=1}^{k} Y_i - k/2 \right) \Big/ \sqrt{k/12}$, where the Y_i's are independent uniform random variables chosen from the interval (0, 1).

Section 7.5

3. 12   **4.** (0.0050, 0.0183)
5. Hint: The appropriate hypotheses to be tested are $H_0: \sigma^2 = 1.1$ versus $H_1: \sigma^2 < 1.1$.

Section 7.7

1. Reject $H_0: \mu = \mu_0$ in favor of $H_1: \mu < \mu_0$ if $(\bar{y} - \mu_0)/(s/\sqrt{n}) \leq -t_{\alpha, n-1}$ **2.** (0.617, 0.705)

Review Exercises

1. (a) $\bar{x} = 0.484$, $s = 0.240$; (b) 0.472
2. The 1-standard deviation interval is (7.3, 20.3) and contains 65% of the observations; the 2-standard deviation interval is (0.8, 26.8) and contains 96% of the observations.
3. (a) Expected proportion = 0.819; (b) Observed proportion = 0.833   **4.** Yes
5. (a) 15%; (b) Normality assumption does not seem tenable. For example, there is too much area in the right tail for normality.
7. $\sum\limits_{i=1}^{n} (X_i - \mu)^2$ is sufficient
8. Hint: Let X and Y denote a cylinder diameter and a piston diameter, respectively. Note that P(piston needs to be reworked) $= P(Y > X) = P(Y - X > 0)$. Standardize $Y - X$.
9. (a) 0.027; (b) $\sigma = 0.224$ **10.** Hint: Use Theorems 3.7, 3.12, 3.13, and standardize.
11. (a) $E(W) = 50$, Var $(W) = 29$; (b) 0.176   **15.** 2663   **17.** $P(Z > 3.16) \doteq 0.0008$
18. $P(Y_i \geq 1) \doteq 0.37$; $P(\bar{Y} \geq 1) \doteq 0.16$ **19.** 0.885   **21.** (a) -1; (b) 0.522
23. $P(Y \geq 7) \doteq P(Y \geq 6.5) \doteq 0.22$ **25.** 0.121   **26.** 0.8030
28. (a) Statistical significance $= P(Y \geq 8) \doteq P(Y \geq 7.5) \doteq P(Z \geq 2.60) = 0.005$
29. (a) $\chi^2_{p,n} = \sqrt{2n} \cdot z_p + n$; (b) 178.4
32. (a) From Theorem 3.19, $M_Y(t) = M_{2n\bar{X}/\theta}(t) = M_{\bar{X}}[(2n/\theta) \cdot t] = [M_{X_i}(2t/\theta)]^n = \{1/[1 - \theta(2t/\theta)]\}^n = [1/(1 - 2t)]^n$

33. Hint: "Invert" the distribution result that $(n-1)S^2/\sigma^2 \sim \chi^2_{n-1}$.
35. (a) $(14.914, 59.528)$; (b) $(3.862, 7.715)$ 37. 0.592
38. Reject H_0 since the test statistic equals 85.567, which exceeds the critical value 19.023.
39. Hint: "Standardize" $(n-1)S^2/\sigma^2_0$.   41. 9   42. $(5.14, 6.78)$
44. Expected value $= n/(n-2)$ for $n > 2$, infinite otherwise.
45. $E(Y) = r/(r+s)$, Var $(Y) = rs/(r+s+1)(r+s)^2$ 47. (a) $(30.82, 40.01)$
49. $n^{r/2}[1 \cdot 3 \cdot \ldots \cdot (r-1)/(n-r)(n-r+2) \cdot \ldots \cdot (n-2)]$
50. (a) $(\bar{y} - 1.96 \cdot \sigma/\sqrt{n}, \bar{y} + 1.96 \cdot \sigma/\sqrt{n})$; (b) $n = 15.37\sigma^2/L^2$ 51. (a) $(4.597, 6.237)$ 54. $\pi/2$
55. $(0.237, 0.731)$
57. Accept H_0 since the t statistic has value -1.723, which is larger than $-t_{18,0.025} = -2.110$.
59. Reject H_0 if $\bar{y} \geq \mu_0 + 1.64\sigma/\sqrt{n}$ 60. (a) 0.023; (b) 0.115
62. Reject H_0 if $(\bar{y} - \mu_0)/(\sigma/\sqrt{n})$ is either (1) ≤ -1.96 or (2) $\geq +1.96$.
63. Reject H_0. The value of the test statistic (-3.27) is less than the 0.05 critical value (-1.64). Thus, the coating appears to be beneficial.   65. 9

CHAPTER 8

Questions

Section 8.2

1. The numerator is a linear combination of independent normals. (*See* Theorem 7.4)
2. Two-sided. There is no prior information that would suggest on which side of μ_Y, μ_X should lie–assuming the two are not equal.
3. The absolute value of the difference between $\bar{x}$ and $\bar{y}$ would have to be greater than or equal to 427.0.

Section 8.3

1. Hint: Use Theorems 7.8 and 7.11.
2. The $F_{9,7}$ critical values are 0.238 and 4.82. Therefore, we accept $H_0: \sigma^2_X = \sigma^2_Y$ because the observed F ratio is $0.0000955/0.0002103 = 0.45$.

Section 8.4

1. The difference between p_W and p_H may have an economic origin. Wallace supporters, on the average, come from a lower socioeconomic class than Humphrey supporters, so it would be more difficult for them to afford the Metro sticker.
2. 0.5222

Section 8.5

1. $(-218.6, 457.6)$ 2. $((71,234.0/50,326.3) \cdot F_{0.005,3,3}, (71,234.0/50,326.3) \cdot F_{0.995,3,3})$ 3. $(-0.020, 0.084)$

Review Exercises

1. Observed t ratio $= 2.45$ (with 29 degrees of freedom). The one-sided 0.05 critical value is 1.6991. Conclusion: The average age at death of short presidents is significantly greater than the average age at death of tall presidents (at the $\alpha = 0.05$ level of significance).
3. Observed t ratio $= -1.28$ (with 10 degrees of freedom). Since $\pm t_{0.025,10} = \pm 2.2281$, accept $H_0: \mu_A = \mu_B$.
5. (Hint: Use Theorem 7.4 to get the mean and variance for $Y - X$ and $\bar{Y} - \bar{X}$.) (a) $P(Y - X < 0) = P(Z < -0.26) = 0.3974$; (b) $P(\bar{Y} - \bar{X} < 0) = P(Z < -0.81) = 0.2090$

6. Observed t ratio $= 0.93$ (with 14 degrees of freedom). Accept the null hypothesis that carpeting has no effect on airborne bacteria levels, since $\pm t_{0.025,\,14} = \pm 2.1448$.

7. The absolute value of $\bar{x} - \bar{y}$ would have to equal or exceed 3.2.

8. Taking a simple average of s_X^2 and s_Y^2 would not be reasonable if the two sample sizes were different. More weight should be given to the estimator for σ^2 that is based on the greater number of observations, since that estimator is more precise–that is, it has the smaller variance.

11. Reject H_0: $\mu_X = \mu_Y$ if $(\bar{x} - \bar{y})/[\sigma\sqrt{(1/n) + (1/m)}]$ is either (1) less than or equal to $-z_{\alpha/2}$ or (2) greater than or equal to $+z_{\alpha/2}$.   **12.** 9   **14.** Observed t ratio $= -1.93$ (with 9 degrees of freedom)

15. Hint: Use Theorem 3.7 and the fact that $E(S_X^2) = E(S_Y^2) = \sigma^2$.

17. Observed t ratio $= 1.24$ (with 13 degrees of freedom)

18. (a) Observed F ratio $= 2.410/3.524 = 0.68$ (with 19 and 19 degrees of freedom); accept H_0; (b) Observed t ratio $= 2.79$ (with 38 degrees of freedom). Since $\pm t_{0.025,\,38} = \pm 2.0244$, reject H_0: $\mu_X = \mu_Y$; (c) 0.87

22. Accept H_0. The observed F ratio, $s_Y^2/s_X^2 = 0.289$, lies between $F_{0.025,\,9,\,9} = 0.248$ and $F_{0.975,\,9,\,9} = 4.03$.

23. Observed Z ratio $= 1.76$. If α were 0.10, we would *reject* H_0 ($\pm z_{0.05} = \pm 1.64$); if α were 0.05, we would *accept H_0* ($\pm z_{0.025} = \pm 1.96$).

24. $-2 \ln \lambda = 10.37$. Reject H_0: $p_X = p_Y$ at the $\alpha = 0.01$ level of significance since $-2 \ln \lambda$ is greater than or equal to $\chi^2_{0.99,\,1}$ ($= 6.635$).

26. The player is right. His observed Z ratio is only -0.25.

28. P value $= P(Z \geq 3.12) + P(Z \leq -3.12) \doteq 0.002$

29. Reject the null hypothesis that the two groups of mice are equally aggressive. The observed Z ratio is -3.56. The $\alpha = 0.05$ critical values are ± 1.96.

30. Hint: "Invert" the distribution result given in Theorem 8.1.

32. $((\bar{x} - \bar{y}) - z_{\alpha/2} \cdot \sqrt{(\sigma_X^2/n) + (\sigma_Y^2/m)}, (\bar{x} - \bar{y}) + z_{\alpha/2} \cdot \sqrt{(\sigma_X^2/n) + (\sigma_Y^2/m)})$

33. $((s_X^2/s_Y^2) \cdot F_{0.025,\,9,\,7}, (s_X^2/s_Y^2) \cdot F_{0.975,\,9,\,7}) = (0.52, 10.61)$

34. Hint: Use Theorems 7.7, 7.8, and 7.11.

CHAPTER 9

Questions

Section 9.2

1. $(7!/2!4!1!)(0.2546)^3(0.4908)^4 \doteq 0.101$

Section 9.3

2. Reject H_0 since the test statistic has value 7.431, which exceeds the critical value $\chi^2_{0.95,\,1} = 3.841$.

3. Accept the model since the test statistic has value 7.712, which is less than the critical value $\chi^2_{0.99,\,3} = 11.345$.

Section 9.4

1. Accept the model since the test statistic has value 0.749, which is less than the critical value $\chi^2_{0.99,\,3} = 11.345$.

2. Reject H_0 since the test statistic has value 14.825, which exceeds the critical value $\chi^2_{0.95,\,6} = 12.592$.

4. Accept the model since the test statistic has value 0.228, which is less than the critical value $\chi^2_{0.95,\,2} = 5.991$.

Section 9.5

1. The criteria are independent since the test statistic has value 2.126, which is less than the critical value $\chi^2_{0.95,\,1} = 3.841$.

2. The factors are dependent, since the test statistic has value 19.572, which exceeds the critical value $\chi^2_{0.99,\,3} = 11.345$.

3. The criteria are dependent, since the test statistic has value 12.698, which exceeds the critical value $\chi^2_{0.95,\,1} = 3.841$.

Review Exercises

1. $1 - (7!/2!2!3!)(1/4)^2(1/3)^2(5/12)^3 \doteq 0.895$ **3.** $(5!/2!2!1!)(0.1587)^2(0.7745)^3(0.0668) \doteq 0.030$
5. (a) $(93!/23!50!20!)(1/4)^{23}(1/2)^{50}(1/4)^{20}$; (b) No. The probability of any specific triple of numbers is quite small.
8. $\hat{p}_i = x_i/n, i = 1, \ldots, k$ **9.** $\frac{20}{9}$

11–13.

Number of White Chips	Expected Number (Hypergeometric)	Expected Number (Binomial)
0	$\frac{100}{3}$	36
1	$\frac{160}{3}$	48
2	$\frac{40}{3}$	16

(a) Accept the hypergeometric model since the test statistic has value 0.969, which is less than the critical value $\chi^2_{0.90,2} = 4.605$; (b) Accept the binomial model since the test statistic has value 3.299, which is less than the critical value $\chi^2_{0.95,2} = 5.991$.
15. (a) The coin is not fair since the chi-square statistic has value 8, which exceeds the critical value $\chi^2_{0.95,1} = 3.841$; (b) The coin is not fair since the (approximately normal) test statistic has value 2.83, which exceeds the critical value, $z_{0.025} = 1.96$; (c) Yes
16. Classes: ≤ 249, 250–259, 260–269, 270–279, ≥ 280.
Expected numbers: 10.6, 13.3, 17.2, 14.9, 14.0.
Reject H_0 since the test statistic has value 12.01, which is greater than the critical value $\chi^2_{0.90,4} = 7.779$.
N.B. An acceptable normal fit results from using $\bar{y} = 260.3$ and $s = 15.3$ as the mean and standard deviation.
17. Reject H_0 since the test statistic has value 8.543, which exceeds the critical value $\chi^2_{0.95,1} = 3.841$.
19. Accept the 9:3:3:1 model since the test statistic has value 1.473, which is less than the critical value $\chi^2_{0.99,3} = 11.345$.
21. Classes: 0–1, 1–2, 2–3.
Expected numbers: 1.85, 12.96, 35.19.
Reject the model since the test statistic has value 23.56, which exceeds the critical value $\chi^2_{0.95,2} = 5.991$.
22. Accept the model since the test statistic has value 1.346, which is less than the critical value $\chi^2_{0.95,3} = 7.815$.
23. Classes: 0, 1, $\ldots$, 9, 10+. Accept the model since the test statistic has value 12.885, which is less than the critical value $\chi^2_{0.95,9} = 16.919$.
24. Accept the model since the test statistic has value 0.612, which is less than the critical value $\chi^2_{0.95,1} = 3.841$.
26. Classes: 0, 1, $\ldots$, 5, 6+. Accept the model since the test statistic has value 1.98, which is less than the critical value $\chi^2_{0.99,4} = 13.277$.
27. (a) Classes: 0, 1, $\ldots$, 5, 6+. $\hat{\lambda} = \frac{683}{270}$ infected plants per quadrat. Reject the Poisson model since the test statistic has value 46.78, which exceeds the critical value $\chi^2_{0.95,5} = 11.071$; (b) The Poisson model is inappropriate because events pertaining to a given quadrat are not independent of those concerning other quadrats.
29. The criteria are dependent. The test statistic has value 50.16, which exceeds the critical value $\chi^2_{0.99,1} = 6.635$.
30. The criteria are independent. The test statistic has value 0.204, which is less than the critical value $\chi^2_{0.95,1} = 3.841$.
31. The criteria are dependent. The test statistic has value 23.79, which exceeds the critical value $\chi^2_{0.99,1} = 6.635$.
33. The criteria are dependent. The test statistic has value 27.24, which exceeds the critical value $\chi^2_{0.99,4} = 13.277$.

CHAPTER 10

Questions

Section 10.2

8. 0.825

Section 10.4

1. (a) $y = 25.232 + 3.291x$; (b) $y(18) = 84.47°F$ **2.** $y = 91.24e^{-0.076x}$

Section 10.5

3. (1.992, 4.590)
4. Accept H_0 since the test statistic has value -1.586, which lies between the critical values, $\pm t_{0.025, 2} = \pm 4.303$.
6. (a) Confidence interval: (110.520, 135.059); (b) Prediction interval: (89.018, 156.561); (c) Committee should be more interested in confidence interval, which concerns industry averages.

Section 10.7

1. Reject H_0 since the statistic has value -4.247, which is less than $-t_{0.025, 7} = -2.365$.

Review Exercises

1. 105/72   **2.** (a) $-8/484$; (b) $-2/15\sqrt{14}$ **4.** (a) 8/450; (b) 0.492   **6.** $1/\sqrt{141}$ **15.** (a) 53/9; (b) 287/81
16. -0.0296 **17.** -0.453 **18.** 0.730   **21.** $E(Y|x) = (n - x)[p_Y/(1 - p_X)]$ **23.** (a) 2/9; (b) 9/4
25. Follows from Theorem 10.5   **27.** 2.5   **29.** $E(Y|x) = 2/3(1 - x)$ **30.** $y = 0.35x + 0.075$
32. $y = 0.955x + 103.513$ **36.** $y = 0.074x^{1.437}$ **37.** Average weight in the year 3000 $\doteq$ 890   **39.** $a - c, b$
42. $y = 468.34e^{-0.039x}$ **43.** $y = 400(\frac{1}{2})^{x/20}$ **44.** (b) $y = 104.141e^{-0.146x}$; (c) 4.757 days
49. Choose $n/2$ points at $x = -5$, $n/2$ at $x = 5$ **51.** (a) (0.158, 0.231); (b) (-11.937, 23.743)
54. Accept H_0 since the t statistic has value 1.631, which is less than the critical value $t_{0.05, 4} = 2.132$.
55. (a) $y = 15.771 + 0.06x$; (b) Reject H_0 since t has value 4.64, which exceeds $t_{0.05, 19} = 1.729$; (c) (0.033, 0.087); (d) (142.65, 189.38)
56. (a) Accept H_0 since the t statistic for a has value 1.653, which is less than $t_{0.025, 8} = 2.306$; (b) Accept H_0 since the t statistic for b has value 1.004, which is less than $t_{0.025, 8} = 2.306$.
57. Reject H_0 since the test statistic has value 4.11, which is greater than $t_{0.025, 18} = 2.101$.
59. (a) (-5.569, 2.569); (b) (6.963, 13.137)
60. (a) $y = 9.406 + 0.081x$; (b) Reject H_0 since the test statistic has value 3.407, which is greater than $t_{0.05, 10} = 1.812$; (c) (0.028, 0.135); (d) (23.51, 27.84); (e) (25.57, 42.03); (f) (-2.358, 21.170)
61. (a) $y = -0.104 + 0.988x$; (b) (13.62, 13.84); (c) (13.29, 14.17)
67. (a) $E(X) = E(Y) = 0$; (b) Var $(X) = 4$, Var $(Y) = 1$; (c) $p = \frac{1}{2}$; (d) $k = 1/2\pi\sqrt{3}$
69. Accept H_0 since the test statistic has value -0.094, which is greater than $t_{0.025, 10} = -2.228$.
71. (0.087, 0.886)

CHAPTER 11

Questions

Section 11.2

1. $\epsilon_{1(1)} = 1.5$. $\sum_{i=1}^{n_j} (Y_{ij} - \bar{Y}_{.j})$ is identically zero. Both sums have an expected value of zero.
4. Observed F ratio $= 22.36$ (with 1 and 14 degrees of freedom). Reject H_0.
5. Observed t ratio $= 4.73$, which is the square root of the observed F ratio.

Section 11.3

1.

Source	df	SS	MS	F
Ages	2	0.2050	0.1025	6.53
Error	9	0.1415	0.0157	
Total	11	0.3465		

Since $F_{0.95, 2, 9} = 4.26$ and $F_{0.99, 2, 9} = 8.02$, we can reject H_0 at the 0.05 level but not at the 0.01 level.

Section 11.4

2.

Source	df	SS	MS	F	
Brands	3	558.92	186.31	0.66	NS
Error	8	2241.33	280.17		
Total	11	2800.25			

Subhypothesis	Contrast	SS	F	
H_0': $\mu_A = \mu_B$	$C_1 = \mu_A - \mu_B$	308.17	1.10	NS
H_0'': $(\mu_A + \mu_B)/2 = (\mu_C + \mu_D)/2$	$C_2 = (\frac{1}{2}\mu_A + \frac{1}{2}\mu_B$			
	$- \frac{1}{2}\mu_C - \frac{1}{2}\mu_D)$	184.08	0.66	NS
H_0''': $\mu_C = \mu_D$	$C_3 = \mu_C - \mu_D$	66.67	0.24	NS

As a partial check of the computations, note that $558.92 = 308.17 + 184.08 + 66.67$. (*See* Theorem 11.2.)

Section 11.5

1.

Source	df	SS	MS	F
Companies	2	113,646.33	56823.16	5.81*
Error	15	146,753.67	9783.58	
Total	17	260,400.00		

$D = 3.67/\sqrt{6}$

Pairwise Difference	95% Tukey Interval	Conclusion
Cutter vs. Abbott	$(-78.9, 217.5)$	NS
Cutter vs. McGaw	$(-271.0, 25.4)$	NS
Abbott vs. McGaw	$(-340.4, -44.0)$	Reject

Section 11.6

1. Each observation should be replaced by its square root.

Review Exercises

1. Observed F ratio $= 7.46$ (with 2 and 15 degrees of freedom). Yes, the data support the contention—H_0 would be rejected since $F_{0.95, 2, 15} = 3.68$.

3. (b) Observed F ratio $= 66.125/70.98 = 0.93$ (with 2 and 14 degrees of freedom); Accept H_0; (c) The effects of dieldrin are cumulative.

4. H_1^{**}, because the τ_j's in H_1^{**} give a larger value for the noncentrality parameter, γ. No, H_1^{***} is inadmissible.

6. $E(Y) = 2\gamma + r$

7. W is a noncentral χ^2 random variable with $r_1 + r_2 + \cdots + r_n$ degrees of freedom and a noncentrality parameter equal to $\gamma_1 + \gamma_2 + \cdots + \gamma_n$. Hint: Use Theorem 3.19 and the expression for $M_Y(t)$ given in Review Exercise 6 to find the moment generating function for W.

9.

Source	df	SS	MS	F
Varieties	4	270.27	67.57	6.39
Error	10	105.68	10.57	
Total	14	375.95		

Reject H_0 ($F_{0.95, 4, 10} = 3.48$)

11. Hint: Begin by expanding the expression $(Y_{ij} - \bar{Y}_{.j})^2$; then combine the resulting cross product term and $\bar{Y}_{.j}^2$ term by summing over i.

14.

Source	df	SS	MS	F
Tubes	2	510.71	255.36	11.56
$\begin{pmatrix} C_1 \\ C_2 \end{pmatrix}$	$\begin{pmatrix} 1 \\ 1 \end{pmatrix}$	$\begin{pmatrix} 264.033 \\ 246.678 \end{pmatrix}$	$\begin{pmatrix} 264.033 \\ 246.678 \end{pmatrix}$	11.95 11.17
Error	42	927.73	22.09	
Total	44	1438.44		

The second contrast is defined by $C_2 = \frac{1}{2}\mu_A - \mu_B + \frac{1}{2}\mu_C$. It tests the subhypothesis $H_0'': (\mu_A + \mu_C)/2 = \mu_B$. At the 0.01 level, the original null hypothesis as well as the two subhypotheses are all rejected.

16. Let $C = \frac{1}{3}\mu_1 + \frac{1}{3}\mu_2 + \frac{1}{3}\mu_3 - \frac{1}{2}\mu_4 - \frac{1}{2}\mu_5$. Then $SS_C = 72.36$ and the observed F ratio is $72.36/10.57 = 6.84$ (with 1 and 10 degrees of freedom).

17. $\hat{\mu} = \bar{y}_{..} = 22.89$; $\hat{\tau}_1 = \bar{y}_{.1} - \bar{y}_{..} = 1.31$; $\hat{\tau}_2 = \bar{y}_{.2} - \bar{y}_{..} = 3.31$; $\hat{\tau}_3 = \bar{y}_{.3} - \bar{y}_{..} = -4.62$. A chi square goodness-of-fit test can be used to test the normality of the residuals (see Example 9.2).

18. Consider 95% Tukey intervals. Then $D = 3.67/\sqrt{6}$ and $MS_{error} = 18.43$.

Comparison	95% Tukey Interval	Conclusion
I vs. II	$(-8.10, 4.76)$	NS
I vs. III	$(-15.43, -2.57)$	Reject
II vs. III	$(-13.76, -0.90)$	Reject

21. Hint: Use the arc sine transformation.

CHAPTER 12

Questions

Section 12.2

2.

Source	df	SS	MS	F
Students	14	400.80	28.63	
States	1	61.63	61.63	7.20*
Error	14	119.87	8.56	
Total	29	582.3		

3.

Source	df	SS	MS	F
Subjects	9	191.71	21.30	
Systems	3	1708.03	569.34	100.95**
Error	27	152.25	5.64	
Total	39	2051.99		

Reject $H_0: \tau_1 = \tau_2 = \tau_3 = \tau_4 = 0$; $D = Q_{0.05,\,4,\,27}/\sqrt{10} = 3.87/\sqrt{10} = 1.22$

Comparison	95% Tukey Interval	Conclusion
Std. vs. V-J	$(15.40, 21.20)$	Reject
Std. vs. Uni.	$(8.47, 14.27)$	Reject
Std. vs. Tubex	$(6.68, 12.48)$	Reject
V-J vs. Uni.	$(-9.83, -4.03)$	Reject
V-J vs. Tubex	$(-11.62, -5.82)$	Reject
Uni. vs. Tubex	$(-4.69, 1.11)$	NS

4. (a) Observed F ratio $= MS_{\text{sleep}}/MS_{\text{error}} = 8.50/2.06 = 4.13$; Reject H_0 since $F_{0.95, 2, 10} = 4.10$; **(b)** Let $C_1 = \frac{1}{2}\mu_{LSWS} + \frac{1}{2}\mu_{DSWS} - 1 \cdot \mu_{REM}$. Then $SS_{C_1} = 4.99$ and the corresponding observed F ratio is $4.99/2.06 = 2.42$ (with 1 and 10 degrees of freedom). Accept H'_0. The contrast orthogonal to C_1 is $C_2 = \mu_{LSWS} - \mu_{DSWS}$. Also, $SS_{C_2} = 12.00$ and $SS_{C_1} + SS_{C_2} = 16.99 = SS_{\text{sleep}}$.

Section 12.3

1. $\mu_D =$ True average urban retention $\% -$ true average rural retention $\%$. Observed t ratio $= 4.01/(10.147/\sqrt{7}) = 1.04$. Accept $H_0: \mu_D = 0$ since $\pm t_{0.005, 6} = \pm 3.7074$.

Review Exercises

2.

Source	df	SS	MS	F
Surveys	4	495.322	123.83	
Flavors	3	56.378	18.79	7.58**
Error	12	29.762	2.48	
Total	19	581.462		

Yes, the overall acceptance percentages vary markedly from survey to survey. If a flavor comparison study were to be set up as a completely randomized design, the error variance would be greatly inflated by what is apparently a very heterogeneous experimental environment.

3. Hint: Do not algebraically expand the cross products term by term. Use the properties of the summation convention and the fact that the sum of a set of observations around their sample mean is identically zero.

4. Yes, the observed F ratio for "methods" is 10.39 (with 2 and 16 degrees of freedom).

6. Values for SS_{blocks} that are very large relative to SS_{error} indicate that the experimental environment is quite heterogeneous. In these cases the randomized block design should be used. Otherwise, the estimated error variance will be so large that the F test's ability to detect differences between treatments would be seriously compromised.

7. $EMS_{\text{subjects}} = \sigma^2 + [3/(8-1)] \sum_{i=1}^{8} \beta_i^2$. Reject $H_0: \beta_1 = \beta_2 = \cdots = \beta_8 = 0$ at the $\alpha = 0.05$ level of significance if $MS_{\text{subjects}}/MS_{\text{error}} \geq F_{0.95, 7, 14}$.

9. $C_1 = \mu_P - \mu_R$; $SS_{C_1} = 18.225$; F ratio $= 7.35$
$C_2 = \mu_{Bv} - \mu_B$; $SS_{C_2} = 38.025$; F ratio $= 15.33$
$C_3 = \mu_P - \mu_{Bv} + \mu_R - \mu_B$; $SS_{C_3} = 0.128$; F ratio $= 0.05$
Reject at the $\alpha = 0.10$ level of significance the null subhypotheses associated with C_1 and C_2. C_3 tests whether the average effect of the Plain and Roast Beef flavors is significantly different from the average effect of the Butter Vanilla and Bread flavors.

12. Observed t ratio $= 2.12$ (with 3 degrees of freedom). Given that $\alpha = 0.05$, accept H_0.

14. $(\bar{d} - t_{\alpha/2, b-1}(s_D/\sqrt{b-1}), \bar{d} + t_{\alpha/2, b-1}(s_D/\sqrt{b-1}))$. For the data of Case Study 12.4, the 95% confidence interval for μ_D is $(-0.21, 0.43)$.

16. 0.0416. The experiment should be done as a randomized block design. The "projected" error variance had the experiment been done as a completely randomized design is 3.6 times as large as the estimated error variance obtained with the randomized block design.

CHAPTER 13

Questions

Section 13.2

1. Nothing. The higher IQ's could have resulted solely from the students' becoming more familiar with the test. Randomizing the testing order is absolutely essential in an experiment of this nature.

2. Let $p = P(Y_i > X_i)$. For $\alpha = 0.05$, we should reject $H_0: p = \frac{1}{2}$ in favor of $H_1: p < \frac{1}{2}$ if $y_+ \leq 2$, since $P(Y_+ \leq 2) = 0.033$ but $P(Y_+ \leq 3) > 0.05$. Here, $y_+ = 2$, so reject H_0.

Section 13.3

1. $W = 9$. Accept H_0.   **2.** $W' = 1.72$. Accept H_0.

Section 13.4

2. $B = 15.3$ (with 4 degrees of freedom). Reject H_0.

Section 13.5

1. $G = 8.4$ (with 2 degrees of freedom). Reject H_0.

Review Exercises

1. Let Y_+ denote the number of Y_i's for which $Y_i - 7.39$ is positive. Omit the observation that *equals* 7.39. Then $y_+ = 4$, based on $n = 43$ observations, giving an observed Z ratio of -5.34. Reject H_0.

2. Hint: The appropriate parametric procedure is the one-sample t test.

3. Let Y_+ denote the number of X_i's exceeding 10. Reject H_0 if $y_+ \geq 15$. When $\mu = 11$, the power of the test is given by $\sum\limits_{k=15}^{22} \binom{22}{k}(0.6591)^k(0.3409)^{22-k} = 0.51$ where $0.6561 = P(X_i > 10 \,|\, \mu = 11) = P(Z > -0.41)$.

4. Power $= P(\bar{X} \geq 10.86 \,|\, \mu = 11) = 0.6064$

6. Let Y_+ denote the number of observations for which $Y_i - 0.618$ is positive. For $n = 20$, we should reject H_0 if either $y_+ \leq 5$ or $y_+ \geq 15$ (this will give an α of 0.042). But $y_+ = 11$, so accept H_0.

9. The approximate, large sample observed Z ratio is 1.89. With ± 1.96 being the 0.05 critical values, we should accept H_0—the blades have not been shown to be significantly different.

10. Hint: Let $p = P(X_i > 6)$. Define Y_+ to be the number of X_i's greater than 6. Test $H_0: p = 0.75$ vs. $H_1: p \neq 0.75$.

12. Observed Z ratio $= 0.99$. Accept H_0.

14. We would expect the signed rank test to have the greater power because it makes use of more of the information in the data.

18. $B = 26.14$ (with 4 degrees of freedom). Reject H_0 and conclude that the productivity of a water well *is* significantly affected by the type of rock in which it is dug.

20. $B = 12.48$ (with 2 degrees of freedom). Reject H_0.

22. Reject H_0 at the $\alpha = 0.01$ level of significance. $W_S = 61$ and the observed Z ratio is -3.02. The 0.01 critical values are ± 2.58.

24. Let $\tilde{\mu}_M$ and $\tilde{\mu}_F$ denote the true median preening times per bout for male and female fruit flies, respectively. Test $H_0: \tilde{\mu}_M = \tilde{\mu}_F$ vs. $H_1: \tilde{\mu}_M \neq \tilde{\mu}_F$. Let $\alpha = 0.05$. Using the Kruskal–Wallis procedure, we get an observed B equal to 10.20. The 0.05 critical value is $\chi^2_{0.95, 1} = 3.841$. Reject H_0.

26. $G = 8.8$ (with 4 degrees of freedom). Accept H_0. With the analysis of variance, the observed F ratio for "treatments" is $0.1831/0.0437 = 4.19$ (with 4 and 8 degrees of freedom). At the 0.05 level, we would reject H_0.

29. $G = 1.50$ (with 2 degrees of freedom).

Bibliography

1. ALEXANDER, S. A., "Price Movements in Speculative Markets: Trends or Random Walks," in *The Random Character of Stock Market Prices*, PAUL H. COOTNER, ed. (The M.I.T. Press, Cambridge, 1964).

2. ANDERSON, J. R., and HOY, J. B., "Relationship between Host Attack Rates and CO_2-Baited Insect Flight Trap Catches of Certain *Symphoromyia* Species," *Journal of Medical Entomology*, vol. 9, 1972, pp. 373–393.

3. APOSTOL, TOM M., *Mathematical Analysis: A Modern Approach to Advanced Calculus* (Addison-Wesley, Reading, 1957).

4. ASH, ROBERT B., *Real Analysis and Probability* (Academic Press, New York, 1972).

5. Automobile and Rubber Industries Tire Committee of the S.A.E. War Engineering Board, Report on Interim Tires and Treads, vol. 1, 1942.

6. BALL, J. A. C., and TAYLOR, A. R., "The Effect of Cyclandelate on Mental Function and Cerebral Blood Flow in Elderly Patients," in *Research on the Cerebral Circulation*, JOHN STIRLING MEYER, HELMUT LECHNER, and OTTO EICHHORN, eds. (Thomas, Springfield, Ill., 1969), pp. 347–363.

7. BARNICOT, N. A., and BROTHWELL, D. R., "The Evaluation of Metrical Data in the Comparison of Ancient and Modern Bones," in *Medical Biology and Etruscan Origins*, G. E. W. WOLSTENHOLME and CECILIA M. O'CONNOR, eds. (Little, Brown, and Co., Boston, 1959), p. 136. Used with permission.

8. BARTLE, ROBERT G., *The Elements of Real Analysis*, 2d ed. (Wiley, New York, 1976).

9. BELLANY, IAN, "Strategic Arms Competition and the Logistic Curve," *Survival*, vol. 16, 1974, pp. 228–230.

10. BERGER, R. J., and WALKER, J. M., "A Polygraphic Study of Sleep in the Tree Shrew (*Tupaia glis*)," *Brain, Behavior and Evolution*, vol. 5, 1972, p. 62. Used with permission.

11. *Biometrika Tables for Statisticians*, vol. 1, E. S. PEARSON and H. O. HARTLEY, eds. (Cambridge University Press, London, 1954), p. 135.

12. BLACKMAN, SHELDON and CATALINA, DON, "The Moon and the Emergency Room," *Perceptual and Motor Skills*, vol. 37, 1973, pp. 624–626. Reprinted with permission of publisher and author.

13. BLUM, JULIUS R., and ROSENBLATT, JUDAH I., *Probability and Statistics* (W. B. Saunders, Philadelphia, 1972), p. 144.

14. BONEAU, C. ALAN, "The Effects of Violations of Assumptions Underlying the t Test," *Psychological Bulletin*, vol. 57, 1960, pp. 49–64.

15. BORTKIEWICZ, L., *Das Gesetz der Kleinen Zahlen* (Teubner, Leipzig, 1898).

16. BOYD, EDITH, "The Specific Gravity of the Human Body," *Human Biology*, vol. 5, 1933, pp. 651–652.

17. BRINEGAR, CLAUDE S., "Mark Twain and the Quintus Curtius Snodgrass Letters: A Statistical Test of Authorship," *Journal of the American Statistical Association*, vol. 58, 1963, pp. 85–96. Used with permission.

18. BULLARD, ROGER W., and SHUMAKE, STEPHEN A., "Food Temperature Preference Response of *Desmodus rotundus*," *Journal of Mammalogy*, vol. 54, 1973, pp. 299–302.

19. CAMNER, PER, and PHILIPSON, KLAS, "Urban Factor and Tracheobronchial Clearance," *Archives of Environmental Health*, vol. 27, 1973, p. 82. Used with permission.

20. CAMPBELL, CATHY, and JOINER, BRIAN L., "How to Get the Answer without Being Sure You've Asked the Question," *American Statistician*, vol. 27, 1973, pp. 229–231.

21. CARVER, W. A., "A Genetic Study of Certain Chlorophyll Deficiencies in Maize," *Genetics*, vol. 12, 1927, pp. 415–440.

22. CASLER, LAWRENCE, "The Effects of Hypnosis on GESP," *Journal of Parapsychology*, vol. 28, 1964, pp. 126–134. Used with permission.

23. CHATTERGEE, S., "Estimating Wildlife Populations," in *Statistics by Example— Finding Models*, FREDERICK MOSTELLER et al., eds. (Addison-Wesley, Reading, Mass., 1973), pp. 25–33.

24. CHOW, G. C., "Statistical Demand Functions and their Use in Forecasting," in *The Demand for Durable Goods*, A. C. HARBERGER, ed. (University of Chicago Press, Chicago, 1960), pp. 149–178.

25. CLARKE, R. D., "An Application of the Poisson Distribution," *Journal of the Institute of Actuaries*, vol. 22, 1946, p. 48.

26. CLASON, CLYDE B., *Exploring the Distant Stars* (G. P. Putnam's Sons, New York, 1958), p. 337.

27. CLEMENS, S. L., *Life on the Mississippi* (Harper & Row, New York, 1917), p. 156.

28. COCHRAN, W. G., "Approximate Significance Levels of the Behrens-Fisher Test," *Biometrics*, vol. 20, 1964, pp. 191–195.

29. COCHRAN, W. G., and COX, GERTRUDE M., *Experimental Designs*, 2d ed. (Wiley, New York, 1957), p. 108.

30. CONNOLLY, KEVIN, "The Social Facilitation of Preening Behaviour in Drosophila Melanogaster," *Animal Behavior*, vol. 16, 1968, pp. 385–391. Used with permission.

31. CRAMER, HARALD, *Mathematical Methods of Statistics* (Princeton University Press, Princeton, N.J., 1946).

32. Daily Stock Price Record, NYSE, January 1978–March 1978 (Standard and Poor, New York, 1978).

33. DAS, S. C., "Fitting Truncated Type III Curves to Rainfall Data," *Australian Journal of Physics*, vol. 8, 1955, pp. 298–304.

34. DAVID, F. N., *Games, Gods, and Gambling* (Hafner, New York, 1962).

35. DAVIS, D. J., "An Analysis of Some Failure Data," *Journal of the American Statistical Association*, vol. 47, 1952, pp. 113–150. Used with permission.

36. DEWEY, G., *Relative Frequency of English Spellings* (Teachers College Press, Columbia University, New York, 1970).

37. DIAZ, JOSE LUIS, and HUTTUNEN, MATTI O., "Persistent Increase in Brain Serotonin Turnover after Chronic Administration of LSD in the Rat," *Science*, vol. 174, 1971, pp. 62–63.

38. DIXON, WILFRID J., and MASSEY, FRANK J., JR., *Introduction to Statistical Analysis*, 2d ed. (McGraw-Hill, New York, 1957). Used with permission.

39. DUBOIS, CORA, ed., *Lowie's Selected Papers in Anthropology* (University of California Press, 1960), pp. 137–142. Used with permission.

40. EMMONS, D. B., LARMOND, ELIZABETH, and BECKETT, D. C., "Determination of Total Solids in Heterogeneous Heat-Sensitive Foods," *Journal of the Association of Official Analytical Chemists*, vol. 54, 1971, pp. 1403–1405. Used with permission.

41. FADELEY, ROBERT CUNNINGHAM, "Oregon Malignancy Pattern Physiographically Related to Hanford, Washington, Radioisotope Storage," *Journal of Environmental Health*, vol. 27, 1965, pp. 883–897. Used with permission.

42. FAIRLEY, WILLIAM B., "Evaluating the 'Small' Probability of a Catastrophic Accident from the Marine Transportation of Liquefied Natural Gas," in *Statistics and Public Policy*, WILLIAM B. FAIRLEY and FREDERICK MOSTELLER, eds. (Addison-Wesley, Reading, Mass., 1977), pp. 331–353.

43. FAIRLEY, WILLIAM B., and MOSTELLER, FREDERICK, "A Conversation about Collins," in *Statistics and Public Policy*, WILLIAM B. FAIRLEY and FREDERICK MOSTELLER, eds. (Addison-Wesley, Reading, Mass., 1977), pp. 369–379.

44. FEARS, THOMAS, SCOTTS, JOSEPH, and SCHNEIDERMAN, MARVIN A., "Skin Cancer, Melanoma, and Sunlight," *American Journal of Public Health*, vol. 66, 1976, pp. 461–464. Used with permission.

45. FELLER, WILLIAM, *An Introduction to Probability Theory and Its Applications*, vol. 1, 2d ed. (Wiley, New York, 1957).

46. FINKBEINER, DANIEL T., *Introduction to Matrices and Linear Transformations* (W. H. Freeman, San Francisco, 1960).

47. FISHBEIN, MORRIS, *Birth Defects* (Lippincott, Philadelphia, 1962), p. 177.

48. FISHER, R. A., "On the 'Probable Error' of a Coefficient of Correlation Deduced from a Small Sample," *Metron*, vol. 1, 1921, pp. 3–32.

49. FISHER, R. A., "The Negative Binomial Distribution," *Annals of Eugenics*, vol. 11, 1941, pp. 182–187.

50. FRASER, D. A. S., *Probability and Statistics: Theory and Applications* (Duxbury, North Scituate, Mass., 1976), pp. 62–63.

51. FREUND, JOHN E., *Mathematical Statistics*, 2d ed. (Prentice-Hall, Englewood Cliffs, N.J., 1971), p. 226.

52. FRY, THORNTON C., *Probability and Its Engineering Uses*, 2d ed. (Van Nostrand-Reinhold, New York, 1965), pp. 206–209.

53. FURUHATA, TANEMOTO, and YAMAMOTO, KATSUICHI, *Forensic Odontology* (Thomas, Springfield, Ill., 1967), p. 84.

54. GABRIEL, K. R., and NEUMANN, J., "On a Distribution of Weather Cycles by Length," *Quarterly Journal of the Royal Meteorological Society*, vol. 83, 1957, pp. 375–380.

55. GALTON, FRANCIS, *Memories of My Life* (Methuen, London, 1908), p. 302.

56. GALTON, FRANCIS, *Natural Inheritance* (Macmillan, London, 1908).

57. GENDREAU, PAUL, et al., "Changes in EEG Alpha Frequency and Evoked Response Latency During Solitary Confinement," *Journal of Abnormal Psychology*, vol. 79, 1972, pp. 54–59.

58. GERBER, ROBERT C., et al., "Kinetics of Aurothiomalate in Serum and Synovial Fluid," *Arthritis and Rheumatism*, vol. 15, 1972, p. 626.

59. GIBBONS, JEAN DICKINSON, *Nonparametric Statistical Inference* (McGraw-Hill, New York, 1971).

60. GOLDMAN, MALCOLM, *Introduction to Probability and Statistics* (Harcourt, Brace & World, New York, 1970), pp. 399–403.

61. GOODMAN, LEO A., "Serial Number Analysis," *Journal of the American Statistical Association*, vol. 47, 1952, pp. 622–634.

62. GRAYBILL, FRANKLIN A., *An Introduction to Linear Statistical Models*, vol. 1 (McGraw-Hill, New York, 1961).

63. GRIFFIN, DONALD R., WEBSTER, FREDERIC A., and MICHAEL, CHARLES R., "The Echolocation of Flying Insects by Bats," *Animal Behavior*, vol. 8, 1960, p. 148. Used with permission.

64. GROSS, NEAL, MASON, WARD S., and MCEACHERN, ALEXANDER W., *Explorations in Role Analysis* (Wiley, New York, 1958), p. 297. Used with permission.

65. GROVER, CHARLES A., "Population Differences in the Swell Shark *Cephaloscyllium ventriosum*," *California Fish and Game*, vol. 58, 1972, pp. 191–197.

66. HACKING, IAN, *The Emergence of Probability* (Cambridge University Press, London, 1974).

67. HAGERMAN, R. L., and SENBET, L. W., "A Test of Accounting Bias and Marketing Structure," *Journal of Business*, vol. 49, 1976, pp. 509–514. Used with permission of The University of Chicago Press.

68. HAGGARD, WILLIAM H., BILTON, THADDEUS H., and CRUTCHER, HAROLD L., "Maximum Rainfall from Tropical Cyclone Systems which Cross the Appalachians," *Journal of Applied Meteorology*, vol. 12, 1973, pp. 50–61. Used with permission.

69. HANSEL, C. E. M., *ESP: A Scientific Evaluation* (Scribner, New York, 1966), pp. 86–89.

70. HARE, EDWARD, PRICE, JOHN, and SLATER, ELIOT, "Mental Disorder and Season of Birth: A National Sample Compared with the General Population," *British Journal of Psychiatry*, vol. 124, 1974, pp. 81–86.

71. HASTINGS, N. A. J., and PEACOCK J. B., *Statistical Distributions* (Butterworth, London, 1975).

72. HAZEL, W. M., and EGLOF, W. K., "Determination of Calcium in Magnesite and Fused Magnesia," *Industrial and Engineering Chemistry (Analytical Edition)*, vol. 18, 1946, pp. 759–760. Copyright by the American Chemical Society. Used with permission.

73. HEATH, CLARK W. and HASTERLIK, ROBERT J., "Leukemia among Children in a Suburban Community," *The American Journal of Medicine*, vol. 34, pp. 796–812.

74. HENDY, M. F., and CHARLES, J. A., "The Production Techniques, Silver Content and Circulation History of the Twelfth-Century Byzantine Trachy," *Archaeometry*, vol. 12, 1970, pp. 13–21. Used with permission.

75. HENNEKENS, C., et al., "Coffee Drinking and Death Due to Coronary Heart Disease," *New England Journal of Medicine*, vol. 294, 1976, pp. 633–636.

76. HERDAN, G., *Language as Choice and Chance* (P. Noordhoff N.V., Groningen, 1956), p. 122.

77. HERSEN, MICHEL, "Personality Characteristics of Nightmare Sufferers," *Journal of Nervous and Mental Diseases*, vol. 153, 1971, pp. 29–31. © 1971 The Williams & Wilkins Co., Baltimore. Used with permission.

78. HOGG, ROBERT V., and CRAIG, ALLEN T., *Introduction to Mathematical Statistics*, 3d ed. (Macmillan, New York, 1970).

79. HOLLANDER, MYLES, and WOLFE, DOUGLAS A., *Nonparametric Statistical Methods* (Wiley, New York, 1973).

80. HORVATH, FRANK S., and REID, JOHN E., "The Reliability of Polygraph Examiner Diagnosis of Truth and Deception," *Journal of Criminal Law, Criminology, and Police Science*, vol. 62, 1971, pp. 276–281. Used with permission.

81. HUDGENS, GERALD A., DENENBERG, VICTOR H., and ZARROW, M. X., "Mice Reared with Rats: Effects of Preweaning and Postweaning Social Interactions upon Adult Behaviour," *Behaviour*, vol. 30, 1968, pp. 259–274. Used with permission.

82. HULBERT, ROGER H., and KRUMBIEGEL, EDWARD R., "Synthetic Flavors Improve Acceptance of Anticoagulant-Type Rodenticides," *Journal of Environmental Health*, vol. 34, 1972, pp. 407–411. Used with permission.

83. HUXTABLE, J., AITKEN, M. J., and WEBER, J. C., "Thermoluminescent Dating of Baked Clay Balls of the Poverty Point Culture," *Archaeometry*, vol. 14, 1972, pp. 269–275. Used with permission.

84. HYNEK, JOSEPH ALLEN, *The UFO Experience: A Scientific Inquiry* (Regnery, Chicago, 1972).

85. IMAI, YOSHITUKA, "Linkage Groups of the Japanese Morning Glory," *Genetics*, vol. 14, 1929, pp. 223–255.

86. JACOBSON, EUGENE, and KOSSOFF, JEROME, "Self-percept and Consumer Attitudes Toward Small Cars," in *Consumer Behavior in Theory and in Action*, STEUART HENDERSON BRITT, ed. (Wiley, New York, 1970), pp. 126–129. Reprinted with permission of John Wiley & Sons, Inc., © 1970.

87. JAMES, ANDREW, and MONCADA, ROBERT, "Many Set Color TV Lounges Show Highest Radiation," *Journal of Environmental Health*, vol. 31, 1969, pp. 359–360. Used with permission.

88. JOHNSON, NORMAN L., and KOTZ, SAMUEL, *Continuous Univariate Distributions*, vol. 1 (Houghton Mifflin, Boston, 1970).

89. JOHNSON, NORMAN L., and KOTZ, SAMUEL, *Continuous Univariate Distributions*, vol. 2 (Houghton Mifflin, Boston, 1970).

90. JOHNSON, NORMAN L., and KOTZ, SAMUEL, *Discrete Distributions* (Houghton Mifflin, Boston, 1969).

91. KENDALL, M. G., "Natural Law in the Social Sciences," *Journal of the Royal Statistical Society*, Series A, vol. 124, 1961, pp. 1–15.

92. KNEAFSEY, JAMES T., *Transportation Economic Analysis* (Heath, Lexington, Mass., 1975), p. 230.

93. KOLMOGOROV, A., *Foundations of the Theory of Probability*, 2d ed. (Chelsea, New York, 1956).

94. KRONOVETER, KENNETH J., and SOMERVILLE, GORDON W., "Airplane Cockpit Noise Levels and Pilot Hearing Sensitivity," *Archives of Environmental Health*, vol. 20, 1970, p. 498. Used with permission.

95. KULLDORFF, GUNNAR, "Estimation of One or Two Parameters of the Exponential Distribution on the Basis of Suitably Chosen Order Statistics," *The Annals of Mathematical Statistics*, vol. 34, 1963, pp. 1419–1431.

96. LARSEN, DIANE K., personal communication.

97. LATHEM, EDWARD CONNERY, ed., *The Poetry of Robert Frost* (Holt, Rinehart and Winston, New York, 1969), p. 362.

98. LAVALLE, IRVING H., *An Introduction to Probability, Decision, and Inference* (Holt, Rinehart and Winston, New York, 1970).

99. LAWRENCE, JOSEPH J., and MAXWELL, MILTON A., "Drinking and Socioeconomic Status," in *Society, Culture, and Drinking Patterns*, DAVID J. PITTMAN and CHARLES R. SNYDER, eds. (Wiley, New York, 1962), p. 143. Used with permission.

100. LEMON, ROBERT E., and CHATFIELD, CHRISTOPHER, "Organization of Song in Cardinals," *Animal Behaviour*, vol. 19, 1971, pp. 1–17. Used with permission.

101. LI, FREDERICK P., "Suicide Among Chemists," *Archives of Environmental Health*, vol. 19, 1969, p. 519.

102. LOTTENBACH, K., "Vasomotor Tone and Vascular Response to Local Cold in Primary Raynaud's Disease," *Angiology*, vol. 22, 1971, pp. 4–8. Used with permission.

103. MAGUIRE, B. A., PEARSON, E. S., and WYNN, A. H. A., "The Time Intervals between Industrial Accidents," *Biometrika*, vol. 39, 1952, pp. 168–180.

104. MAISTROV, L. E., *Probability Theory—A Historical Sketch* (Academic Press, New York, 1974).

105. MALINA, ROBERT M., "Comparison of the Increase in Body Size between 1899 and 1970 in a Specially Selected Group with that in the General Population," *Physical Anthropology*, vol. 37, 1972, pp. 135–141.

106. MANN, H. B., *Analysis and Design of Experiments* (Dover, New York, 1949).

107. McCONNELL, THOMAS R., "Suggestibility in Children as a Function of Chronological Age," *Journal of Abnormal and Social Psychology*, vol. 67, 1963, pp. 286–289.

108. McIntyre, Donald B., "Precision and Resolution in Geochronometry," in *The Fabric of Geology*, Claude C. Albritton, Jr., ed. (Freeman, Cooper, and Co., Stanford, Calif., 1963), pp. 112–134.

109. "Medical News," *Journal of the American Medical Association*, vol. 219, 1972, p. 981.

110. Miller, Rupert G., Jr., *Simultaneous Statistical Inference* (McGraw-Hill, New York, 1966).

111. Miller, Russell R., "Drug Surveillance Utilizing Epidemiologic Methods," *American Journal of Hospital Pharmacy*, vol. 30, 1973, pp. 584–592. Used with permission.

112. Minkoff, Eli C., "A Fossil Baboon from Angola, with a Note on *Australopithecus*," *Journal of Paleontology*, vol. 46, 1972, pp. 836–844. Used with permission.

113. Morgan, Peter J., "A Photogrammetric Survey of Hoseason Glacier, Kemp Coast, Antarctica," *Journal of Glaciology*, vol. 12, 1973, pp. 113–120. Used with permission.

114. Moriarity, Shane, personal communication.

115. Mosteller, Frederick, *Fifty Challenging Problems in Probability with Solutions* (Addison-Wesley, Reading, Mass., 1965), pp. 35–36.

116. Mulcahy, Risteard, McGilvray, J. W., and Hickey, Noel, "Cigarette Smoking Related to Geographic Variations in Coronary Heart Disease Mortality and to Expectation of Life in the Two Sexes," *American Journal of Public Health*, vol. 60, 1970, p. 1516. Used with permission.

117. Munford, A. G., "A Note on the Uniformity Assumption in the Birthday Problem," *American Statistician*, vol. 31, 1977, p. 119.

118. Nash, Harvey, *Alcohol and Caffeine* (Thomas, Springfield, Ill., 1962), p. 96. Courtesy of Charles C Thomas, Publisher, Springfield, Ill.

119. *Newsweek*, March 6, 1978, p. 78.

120. Nye, Francis Iven, *Family Relationships and Delinquent Behavior* (Wiley, New York, 1958), p. 37.

121. Olvin, J. F., "Moonlight and Nervous Disorders," *American Journal of Psychiatry*, vol. 99, 1943, pp. 578–584.

122. Ore, Oystein, "Pascal and the Invention of Probability Theory," *American Mathematical Monthly*, vol. 67, 1960, p. 412.

123. Orringer, Eugene, et al., "Splenectomy in Chronic Thrombocytopenic Purpura," *Journal of Chronic Diseases*, vol. 23, 1970, pp. 117–122. Reprinted with permission from Pergamon Press Ltd.

124. Parzen, Emanuel, *Modern Probability Theory and Its Applications* (Wiley, New York, 1960).

125. Passingham, R. E., "Anatomical Differences between the Neocortex of Man and Other Primates," *Brain, Behavior, and Evolution*, vol. 7, 1973, pp. 337–359. Used with permission.

126. Pearson, Karl, "Notes on the History of Correlation," *Biometrika*, vol. 13, 1920–1921.

127. PEARSON, E. S., and KENDALL, M. G., *Studies in the History of Statistics and Probability* (Griffin, London, 1970).

128. PHILLIPS, DAVID P., "Deathday and Birthday: An Unexpected Connection," in *Statistics: A Guide to the Unknown*, JUDITH M. TANUR, et al., eds. (Holden-Day, San Francisco, 1972), pp. 52–65.

129. PHILLIPS, LAWRENCE D., *Bayesian Statistics for Social Scientists* (Thomas Nelson & Sons, Ltd., London, 1973).

130. PIERCE, GEORGE W., *The Songs of Insects* (Harvard University Press, Cambridge, 1949), pp. 12–21. Reprinted with permission.

131. PILLAY, K. K. S., et al., "Mercury Pollution of Lake Erie Ecosphere," *Environmental Research*, vol. 5, 1972, pp. 172–181.

132. PORTER, JOHN W., et al., "Effect of Hypnotic Age Regression on the Magnitude of the Ponzo Illusion," *Journal of Abnormal Psychology*, vol. 79, 1972, pp. 189–194. Copyright 1972 by the American Psychological Association. Reprinted with permission.

133. PREMACK, DAVID, "Language in Chimpanzee?" *Science*, vol. 172, 1971, pp. 808–822. Copyright 1971 by the American Association for the Advancement of Science. Used with permission.

134. QUETELET, L. A. J., *Lettres sur la Theorie des Probabilites, appliquee aux Sciences Morales et Politiques* (M. Hayez, Imprimeur de L'Academie Royal des Sciences, des Lettres et des Beaux-Arts de Belgique, Bruxelles, 1846), p. 400.

135. RABINOVITCH, NACHUM L., *Probability and Statistical Inference in Ancient and Medieval Jewish Literature* (University of Toronto Press, Toronto, 1973).

136. RAGSDALE, A. C. and BRODY, S., *Journal of Dairy Science*, vol. 5, 1922, p. 214.

137. RAHMAN, N. A., *Practical Exercises in Probability and Statistics* (Hafner, New York, 1972).

138. RESNICK, RICHARD B., FINK, MAX, and FREEDMAN, ALFRED M., "A Cyclazocine Typology in Opiate Dependence," *American Journal of Psychiatry*, vol. 126, 1970, pp. 1256–1260.

139. RICH, CLYDE L., "Is Random Digit Dialing Really Necessary?" *Journal of Marketing Research*, vol. 14, 1977, pp. 300–305.

140. RICHARDSON, LEWIS F., "The Distribution of Wars in Time," *Journal of the Royal Statistical Society*, vol. 107, 1944, pp. 242–250.

141. RITTER, BRUNHILDE, "The Use of Contact Desensitization, Demonstration-plus-participation and Demonstration-alone in the Treatment of Acrophobia," *Behaviour Research and Therapy*, vol. 7, 1969, pp. 157–164. Reprinted with permission from Pergamon Press, Ltd.

142. ROBERTS, CHARLOTTE A., "Retraining of Inactive Medical Technologists—Whose Responsibility?" *American Journal of Medical Technology*, vol. 42, 1976, pp. 115–123.

143. ROCHAT, ROGER W., "Cervical Cancer Screening: The Effect of Infrequently Occurring Disease on the Accuracy of Diagnosis," (presented to the Society for Epidemiological Research, Toronto, Canada, 1976).

144. ROSENTHAL, ROBERT and JACOBSON, LENORE F., "Teacher Expectations for the Disadvantaged," *Scientific American*, vol. 218, 1968, pp. 19–23.

145. ROTH, LEWIS F., "Juvenile Susceptibility of Ponderosa Pine to Dwarf Mistletoe," *Phytopathology*, vol. 64, 1974, pp. 689–692. Used with permission.

146. ROULETTE, AMOS, "An Assessment of Unit Dose Injectable Systems," *American Journal of Hospital Pharmacy*, vol. 29, 1972, p. 61. Used with permission.

147. ROWLEY, WAYNE A., "Laboratory Flight Ability of the Mosquito, *Culex Tarsalis Coq.*," *Journal of Medical Entomology*, vol. 7, 1970, pp. 713–716.

148. RUCKMAN, JOSEPH E., ZSCHEILE, FREDERICK P., JR., and QUALSET, CALVIN O., "Protein, Lysine, and Grain Yields of Triticale and Wheat as Influenced by Genotype and Location," *Journal of Agricultural and Food Chemistry*, vol. 21, 1973, pp. 697–700. Copyright by the American Chemical Society. Reprinted with permission.

149. RUTHERFORD, SIR ERNEST, CHADWICK, JAMES, and ELLIS, C. D., *Radiations from Radioactive Substances* (Cambridge University Press, London, 1951), p. 172.

150. SALVOSA, CARMENCITA B., PAYNE, PHILIP R., and WHEELER, ERICA F., "Energy Expenditure of Elderly People Living Alone or in Local Authority Homes," *American Journal of Clinical Nutrition*, vol. 24, 1971, p. 1468. Used with permission.

151. SAMARAS, THOMAS T., "That Song Put Down Short People, But . . . ," *Science Digest*, vol. 84, 1978, pp. 76–79.

152. SATURLEY, B. A., "Colorimetric Determination of Cyclamate in Soft Drinks, Using Picryl Chloride," *Journal of the Association of Official Analytical Chemists*, vol. 55, 1972, pp. 892–894. Used with permission.

153. SCHEFFE, HENRY, *The Analysis of Variance* (Wiley, New York, 1959).

154. SCHEFFE, HENRY, "Practical Solutions of the Behrens-Fisher Problem," *Journal of the American Statistical Association*, vol. 65, 1970, pp. 1501–1508.

155. SCHELL, EMIL D., "Samuel Pepys, Isaac Newton, and Probability," *The American Statistician*, vol. 14, 1960, pp. 27–30.

156. SCHOENEMAN, ROBERT L., DYER, RANDOLPH H., and EARL, ELAINE M., "Analytical Profile of Straight Bourbon Whiskies," *Journal of the Association of Official Analytical Chemists*, vol. 54, 1971, pp. 1247–1261. Used with permission.

157. *Scientific Tables*, 6th ed. (Geigy, Basle, 1962).

158. SEARLE, SHAYLE R., *Linear Models* (Wiley, New York, 1971).

159. Selective Service System, Office of the Director, Washington, D.C.

160. SELTZER, CARL C., et al., "Smoking Habits and Pain Tolerance," *Archives of Environmental Health*, vol. 29, 1974, pp. 170–172. Used with permission.

161. SEN, NRISINHA, et al., "Effect of Sodium Nitrite Concentration on the Formation of Nitrosopyrrolidine and Dimethylnitrosamine in Fried Bacon," *Journal of Agricultural and Food Chemistry*, vol. 22, 1974, pp. 540–541. Copyright by the American Chemical Society. Reprinted with permission.

162. SHAHIDI, SYED A., et al., "Celery Implicated in High Bacteria Count Salads," *Journal of Environmental Health*, vol. 32, 1970, p. 669.

163. SHARPE, ROGER S. and JOHNSGARD, PAUL A., "Inheritance of Behavioral Characters in F_2 Mallard $\times$ Pintail (*Anas Platyrhynchos L.* $\times$ *Anas Acuta L.*) Hybrids," *Behaviour*, vol. 27, 1966, pp. 259–272.

164. SHAW, G. B., *The Doctor's Dilemma, with a Preface on Doctors* (Brentano's, New York, 1911), p. lxiv.

165. SHORE, NEIL S., GREENE, REGINALD, and KAZEMI, HOMAYOUN, "Lung Dysfunction in Workers Exposed to *Bacillus subtilis* Enzyme," *Environmental Research*, vol. 4, 1971, pp. 512–519. Used with permission.

166. SICHEL, HERBERT S., "The Estimation of Parameters of a Negative Binomial Distribution with Special Reference to Psychological Data," *Psychometrika*, vol. 16, 1951, pp. 107–127. Used with permission.

167. SIDDIQUI, S. H., and PARIZEK, R. R., "Application of Nonparametric Statistical Tests in Hydrogeology," *Ground Water*, vol. 10, 1972, pp. 26–31.

168. SINKOV, ABRAHAM, *Elementary Cryptanalysis: A Mathematical Approach* (Random House, New York, 1968).

169. SKELLAM, J. G., "Random Dispersal in Theoretical Populations," *Biometrika*, vol. 38, 1951, pp. 196–218. Used with permission.

170. SMITH, W. LYNN, "Facilitating Verbal-Symbolic Functions in Children with Learning Problems and 14–6 Positive Spike EEG Patterns with Ethosuximide (Zarontin)," in *Drugs and Cerebral Function*, WALLACE SMITH, ed. (Thomas, Springfield, Ill., 1970), p. 125. Courtesy of Charles C Thomas, Publisher, Springfield, Ill.

171. SRB, ADRIAN M., OWEN, RAY D., and EDGAR, ROBERT S., *General Genetics*, 2d ed. (W. H. Freeman, San Francisco, 1965), pp. 12–16.

172. *State Regulations for Protection Against Radiation* (Tennessee Department of Public Health, Division of Radiological Health, Nashville, 1978), 1200–2–6–.05 (3)(C).

173. STICKEL, W. H., STICKEL, L. F., and SPANN, J. W., "Tissue Residues of Dieldrin in Relation to Mortality in Birds and Mammals," in *Chemical Fallout*, MORTON W. MILLER and GEORGE G. BERG, eds. (Thomas, Springfield, Ill., 1969), pp. 178–179. Courtesy of Charles C Thomas, Publisher, Springfield, Ill.

174. SUKHATME, P. V., "On Fisher and Behren's Test of Significance for the Difference in Means of Two Normal Samples," *Sankhya*, vol. 4, 1938, pp. 39–48.

175. SUTTON, D. H., "Gestation Period," *Medical Journal of Australia*, vol. 1, 1945, pp. 611–613.

176. SZALONTAI, S., and TIMAFFY, M., "Involutional Thrombopathy," in *Age with a Future*, P. FROM HANSEN, ed. (F. A. Davis, Philadelphia, 1964), p. 345.

177. *Tables of the Incomplete Γ-Function*, KARL PEARSON, ed. (Cambridge University Press, London, 1922).

178. *Tennessean* (Nashville), Jan. 20, 1973.

179. *Tennessean* (Nashville), Aug. 30, 1973.

180. TODHUNTER, ISAAC, *A History of the Mathematical Theory of Probability* (Cambridge University Press, Chelsea, 1962).

181. TREUHAFT, PAUL S., and McCARTY, DANIEL J., "Synovial Fluid pH, Lactate, Oxygen and Carbon Dioxide Partial Pressure in Various Joint Diseases," *Arthritis and Rheumatism*, vol. 14, 1971, pp. 476–477.

182. TURCO, SALVATORE, and DAVIS, NEIL, "Particulate Matter in Intravenous Infusion Fluids—Phase 3," *American Journal of Hospital Pharmacy*, vol. 30, 1973, p. 612. Used with permission.

183. ULBRICH, J., *Die Bisamratte* (Heinrich, Dresden, 1930).

184. VAN TWYVER, H., and ALLISON, T., "Sleep in the Armadillo *Dasypus novemcinctus* at Moderate and Low Ambient Temperatures," *Brain, Behavior, and Evolution*, vol. 9, 1974, pp. 107–120. Used with permission.

185. VIDINS, EVA I., FOX, JO ANN E., and BECK, IVAN T., "Transmural Potential Difference (PD) in the Body of the Esophagus in Patients with Esophagitis, Barrett's Epithelium and Carcinoma of the Esophagus," *American Journal of Digestive Diseases*, vol. 16, 1971, pp. 991–999.

186. VINCENT, PAULINE, "Factors Influencing Patient Noncompliance: A Theoretical Approach," *Nursing Research*, vol. 20, 1971, p. 514.

187. VOGT, EVAN Z., and HYMAN, RAY, *Water Witching U.S.A.* (University of Chicago Press, Chicago, 1959), p. 55. Used with permission.

188. VON MISES, RICHARD, *Probability, Statistics and Truth*, 2d ed. (Macmillan, New York, 1957).

189. WALKER, HELEN M., *Studies in the History of Statistical Method* (Williams and Wilkins, Baltimore, 1929).

190. WALTER, WILLIAM G., and STOBER, ANGIE, "Microbial Air Sampling in a Carpeted Hospital," *Journal of Environmental Health*, vol. 30, 1968, p. 405. Used with permission.

191. WEISS, WILLIAM, "Cigarette Smoke Gas Phase and *Paramecium* Survival," *Archives of Environmental Health*, vol. 17, 1968, p. 63.

192. WERNER, MARTHA, STABENAU, JAMES R., and POLLIN, WILLIAM, "Thematic Apperception Test Method for the Differentiation of Families of Schizophrenics, Delinquents, and 'Normals'," *Journal of Abnormal Psychology*, vol. 75, 1970, pp. 139–145.

193. WESTERGAARD, HARALD, *Contributions to the History of Statistics* (Agathen Press, New York, 1968).

194. WHITWORTH, WILLIAM ALLEN, *Choice and Chance* (Hafner, New York, 1965).

195. WILKS, SAMUEL S., *Mathematical Statistics* (Wiley, New York, 1962), pp. 408–411.

196. WILTON, D. P., and FAY, R. W., "Response of Adult *Anopheles Stephensi* to Light of Various Wavelengths," *Journal of Medical Entomology*, vol. 9, 1972, pp. 301–304. Used with permission.

197. WINSLOW, CHARLES, *The Conquest of Epidemic Disease* (Princeton University Press, Princeton, N.J., 1943), p. 303.

198. WIORKOWSKI, JOHN J., "A Curious Aspect of Knockout Tournaments of Size 2^m," *The American Statistician*, vol. 26, 1972, pp. 28–30.

199. WOLF, STEWART, ed., *The Artery and the Process of Arteriosclerosis: Measurement and Modification* (Proceedings of an Interdisciplinary Conference on Fundamental Data on Reactions of Vascular Tissue in Man, April 19–25, 1970, Lindau, West Germany) (Plenum Press, New York, 1972), p. 116.

200. WOOD, ROBERT M., "Giant Discoveries of Future Science," *Virginia Journal of Science*, vol. 21, 1970, pp. 169–177. Used with permission.

201. WOODWARD, W. F., "A Comparison of Base Running Methods in Baseball," M. Sc. Thesis, Florida State University, 1970. Used with permission.

202. WRIGHTSMAN, LAWRENCE S., "Wallace Supporters and Adherence to 'Law and Order'," in *Human Social Behavior*, ROBERT A. BARON and ROBERT M. LIEBERT, eds. (Dorsey Press, Homewood, 1971), pp. 217–225.

203. WYLER, ALLEN R., MINORU, MASUDA, and HOLMES, THOMAS H., "Magnitude of Life Events and Seriousness of Illness," *Psychosomatic Medicine*, vol. 33, 1971, pp. 115–122. Used with permission.

204. YOCHEM, DONALD, and ROACH, DARRELL, "Aspirin: Effect on Thrombus Formation Time and Prothrombin Time of Human Subjects," *Angiology*, vol. 22, 1971, p. 72. Used with permission.

205. YULE, G., "Why Do We Sometimes Get Nonsense-Correlations between Time Series?—A Study in Sampling and the Nature of Time Series," *Journal of the Royal Statistical Society*, vol. 89, 1926, pp. 1–69.

206. YULE, G. U., and KENDALL, M. G., *An Introduction to the Theory of Statistics*, 14th ed. (Charles Griffin, London, 1965).

207. ZARET, THOMAS M., "Predators, Invisible Prey, and the Nature of Polymorphism in the *Cladocera* (Class *Crustacea*)," *Limnology and Oceanography*, vol. 17, 1972, pp. 171–184. Used with permission.

208. ZELAZO, PHILIP R., ZELAZO, NANCY ANN, and KOLB, SARAH, " 'Walking' in the Newborn," *Science*, vol. 176, 1972, pp. 314–315. Copyright 1972 by the American Association for the Advancement of Science. Used with permission.

209. ZELINSKY, DANIEL, *A First Course in Linear Algebra*, 2d ed. (Academic Press, New York, 1973).

210. ZIV, G., and SULMAN, F. G., "Binding of Antibiotics to Bovine and Ovine Serum," *Antimicrobial Agents and Chemotherapy*, vol. 2, 1972, pp. 206–213. Used with permission.

Index